Teaching Mathematical Modelling: Connecting to Research and Practice

International Perspectives on the Teaching and Learning of Mathematical Modelling

For further volumes:
http://www.springer.com/series/10093

Gloria Ann Stillman • Gabriele Kaiser
Werner Blum • Jill P. Brown
Editors

Teaching Mathematical Modelling: Connecting to Research and Practice

Springer

Editors
Gloria Ann Stillman
Education VIC | Faculty of Education
Australian Catholic University
Ballarat, VIC, Australia

Gabriele Kaiser
Faculty of Education
University of Hamburg
Hamburg, Germany

Werner Blum
Department of Mathematics,
University of Kassel
Kassel, Germany

Jill P. Brown
Education VIC | Faculty of Education
Australian Catholic University
Melbourne, VIC, Australia

ISSN 2211-4920
ISSN 2211-4939 (electronic)
ISBN 978-94-007-6539-9
ISBN 978-94-007-6540-5 (eBook)
DOI 10.1007/978-94-007-6540-5
Springer Dordrecht Heidelberg New York London

Library of Congress Control Number: 2013943697

Printed on acid-free paper

Springer is part of Springer Science+Business Media (www.springer.com)

Series Preface

Applications and modelling and their learning and teaching in school and university have become a prominent topic in the last decades in view of the growing worldwide relevance of the usage of mathematics in science, technology and everyday life. However, although there is consensus that modelling should play an important role in mathematics education, the situation in school and university is not satisfactory. Given the worldwide impending shortage of students who are interested in mathematics and science, it is essential to discuss possible changes of mathematics education in school and tertiary education towards the inclusion of real world examples and the competencies to use mathematics to solve real world problems.

This innovative book series established by Springer "*International Perspectives on the Teaching and Learning of Mathematical Modelling*", aims at promoting academic discussion on the teaching and learning of mathematical modelling at various educational levels all over the world. The series will publish books from various theoretical perspectives around the world dealing with Teaching and Learning of Mathematical Modelling in Schooling and at Tertiary level. This series will also enable the *International Community of Teachers of Mathematical Modelling and Applications* (ICTMA), an ICMI affiliated Study Group to publish books coming out of its biennial conference series. ICTMA is a unique worldwide group where not only mathematics educators dealing with education at school level are included but also applied mathematicians interested in teaching and learning modelling at tertiary level are represented as well. Two of these books published by Springer have already appeared.

The planned books will display the worldwide state-of-the-art in this field, most recent educational research results and new theoretical developments and will be of interest for a wide audience. Themes dealt with in the books will be teaching and learning of mathematical modelling in schooling and at tertiary level including the usage of technology in modelling, psychological aspects of modelling and its teaching, modelling competencies, curricular aspects, modelling examples and courses, teacher education and teacher education courses. The book series aims to support the discussion on mathematical modelling and its teaching internationally and will promote the teaching and learning of mathematical modelling all over the world in schools and universities.

The series is supported by an editorial board of internationally well-known scholars, who bring in their long experience in the field as well as their expertise to this series. The members of the editorial board are: Maria Salett Biembengut (Brazil), Werner Blum (Germany), Helen Doerr (USA), Peter Galbraith (Australia), Toshikazu Ikeda (Japan), Mogens Niss (Denmark), and Jinxing Xie (China).

We hope this book series will inspire readers in the present and the future to promote the teaching and learning of mathematical modelling all over the world.

Gloria Ann Stillman
Ballarat, Australia

Gabriele Kaiser
Hamburg, Germany

Contents

Part III Pedagogical Issues for Teaching and Learning

Contributors

Peter Adams The University of Queensland, St Lucia, QLD, Australia

Verónica Vargas-Alejo Universidad de Quintana Roo, Chetumal, México

Jonas Bergman Ärlebäck Department of Mathematics, Linköping University, Linköping, Sweden

Jonei Cerqueira Barbosa Federal University of Bahia, Salvador, BA, Brazil

Michael Besser University of Kassel, Kassel, Germany

Maria Salett Biembengut Pontifícia Universidade Católica do Rio Grande do Sul – PUCRS, Porto Alegre, Brazil

Morten Blomhøj IMFUFA, Department of Science, Systems, and Models, Roskilde University, Roskilde, Denmark

Werner Blum University of Kassel, Kassel, Germany

Jill P. Brown Australian Catholic University, Melbourne, Australia

Raymond Brown Griffith University, Gold Coast, Australia

Nils Buchholtz University of Hamburg, Hamburg, Germany

Ang Keng Cheng National Institute of Education, Nanyang Technological University, Singapore, Singapore

César Cristóbal-Escalante Universidad de Quintana Roo, Chetumal, México

Dirk De Bock Katholieke Universiteit Leuven, Leuven, Belgium

Hogeschool-Universiteit Brussel, Brussels, Belgium

Andréia Maria Pereira de Oliveira State University of Feira de Santana, Feira de Santana, BA, Brazil

Shelley Dole The University of Queensland, St Lucia, QLD, Australia

Ann Downton Australian Catholic University, Melbourne, Australia

Lyn D. English Queensland University of Technology, Brisbane, QLD, Australia

Chan Chun Ming Eric National Institute of Education, Nanyang Technological University, Singapore, Singapore

Rhonda Faragher Australian Catholic University, Canberra, Australia

Loe Feijs Eindhoven University of Technology, Eindhoven, The Netherlands

Rita Borromeo Ferri University of Kassel, Kassel, Germany

Peter Frejd Department of Mathematics, Linköping University, Linköping, Sweden

Juntao Fu Department of Mathematical Sciences, Tsinghua University, Beijing, China

Peter Galbraith The University of Queensland, St Lucia, QLD, Australia

Vince Geiger Australian Catholic University, Brisbane, QLD, Australia

Merrilyn Goos The University of Queensland, St Lucia, QLD, Australia

Gilbert Greefrath University of Münster, Münster, Germany

Susanne Grünewald University of Hamburg, Hamburg, Germany

Lee Ngan Hoe National Institute of Education, Nanyang Technological University, Singapore, Singapore

Jun Hu Eindhoven University of Technology, Eindhoven, The Netherlands

Toshikazu Ikeda Yokohama National University, Yokohama, Japan

Michael Jennings The University of Queensland, St Lucia, QLD, Australia

Gabriele Kaiser Faculty of Education, University of Hamburg, Hamburg, Germany

Tetsushi Kawasaki Gifu University, Gifu-Shi Yanagido 1–1, Japan,

Tinne Hoff Kjeldsen IMFUFA, Department of Science, Systems, and Models, Roskilde University, Roskilde, Denmark

Malte Klimczak Deutsches Institut für Internationale Pädagogische Forschung, Frankfurt am Main, Germany

André Krug University of Kassel, Kassel, Germany

Sebastian Kuntze Ludwigsburg University of Education, Ludwigsburg, Germany

Janeen Lamb Australian Catholic University, Brisbane, QLD, Australia

Geert Langereis Eindhoven University of Technology, Eindhoven, The Netherlands

Richard Lesh Indiana University, Bloomington, IN, USA

Issic K. C. Leung The Hong Kong Institute of Education, Hong Kong, Hong Kong

Matthias Ludwig Goethe-University Frankfurt, Frankfurt, Germany

Akio Matsuzaki Saitama University, Saitama, Japan

Sarah Mesrogli University of Hamburg, Hamburg, Germany

Lars Mouwitz National Centre of Mathematics Education, Gothenburg University, Gothenburg, Sweden

Kit Ee Dawn Ng National Institute of Education, Nanyang Technological University, Singapore, Singapore

Yoshiki Nisawa Kyoto Prefectural Rakuhoku Senior High School, Kyoto, Japan

Daniel Clark Orey Centro de Educação Aberta e à Distância (CEAD), Universidade Federal de Ouro Preto (UFOP), Ouro Preto, MG, Brazil

Trevor Redmond AB Paterson College, Gold Coast, Australia

Xenia-Rosemarie Reit Goethe-University Frankfurt, Frankfurt, Germany

Michael Rieß University of Münster, Münster, Germany

Milton Rosa Centro de Educação Aberta e à Distância (CEAD), Universidade Federal de Ouro Preto (UFOP), Ouro Preto, MG, Brazil

Akihiko Saeki Naruto University of Education, Naruto, Japan

Mark Schofield Edge Hill University, Lancashire, UK

Stanislaw Schukajlow University of Paderborn, Paderborn, Germany

Joanne Sheehy AB Paterson College, Gold Coast, Australia

Hans-Stefan Siller University of Koblenz, Koblenz, Germany

Tan Liang Soon National Institute of Education, Nanyang Technological University, Singapore, Singapore

Peter Stender University of Hamburg, Hamburg, Germany

Gloria Ann Stillman Australian Catholic University, Ballarat, Australia

Wim Van Dooren Katholieke Universiteit Leuven, Leuven, Belgium

Hamsa Venkat School of Education, University of the Witwatersrand, Johannesburg, South Africa

Lieven Verschaffel Centre for Instructional Psychology and Technology, Katholieke Universiteit Leuven, Leuven, Belgium

Jana Visnovska The University of Queensland, Brisbane, QLD, Australia

Christiane Vogl WIKU Salzburg, Salzburg, Austria

Pauline Vos University of Agder, Kristiansand, Norway

Wanty Widjaja Deakin University, Burwood, Australia

Mark Winter University of the Witwatersrand, Johannesburg, South Africa

Jinxing Xie Department of Mathematical Sciences, Tsinghua University, Beijing, China

Akira Yanagimoto Kyoto University of Education, Kyoto, Japan

Noboru Yoshimura Tennoji J.H.S. attached to Osaka Kyoiku University, Osaka, Japan

Chapter 1
Mathematical Modelling: Connecting to Teaching and Research Practices – The Impact of Globalisation

Gloria Ann Stillman, Gabriele Kaiser, Werner Blum, and Jill P. Brown

Abstract ICTMA is a community of researchers from different countries working together to establish a learning culture of mathematical modelling. A clear aim is globalization of knowledge and understandings about research and theory about the teaching of mathematical modelling and applications that the community values. A characteristic of this global community in mathematical modelling and applications education is that it operates from a large variety of different theories and research paradigms when it comes to both research practices and teaching practices. The purpose of this chapter is to show that diversity but at the same time the inherent connections that create a fertile research field.

G.A. Stillman (✉)
Education VIC | Faculty of Education, Australian Catholic University, Ballarat, Australia
e-mail: gloria.stillman@acu.edu.au

G. Kaiser
Faculty of Education, University of Hamburg, Hamburg, Germany
e-mail: gabriele.kaiser@uni-hamburg.de

W. Blum
Department of Mathematics, University of Kassel, Kassel, Germany
e-mail: blum@mathematik.uni-kassel.de

J.P. Brown
Education VIC | Faculty of Education, Australian Catholic University, Melbourne, VIC, Australia
e-mail: jill.brown@acu.edu.au

G.A. Stillman et al. (eds.), *Teaching Mathematical Modelling: Connecting to Research and Practice*, International Perspectives on the Teaching and Learning of Mathematical Modelling, DOI 10.1007/978-94-007-6540-5_1,

1 Introduction

Different members of the ICTMA community have different levels of experience in research, in general, and in research into the teaching of mathematical modelling, in particular, as well as different levels of access to resources both for teaching and researching. However, participation in the ICTMA community is seen as a professional learning experience for all whether experienced or not. Through research conferences it is hoped to foster international collaboration and connection and knowledge networks but another clear aim is globalisation of knowledge and understandings about the research and teaching of mathematical modelling in various levels of schooling, in vocational education contexts and university settings. The ICTMA community is quite fluid with biennial conference location having impact both locally and globally in stimulating and disseminating research, both as a lead up to the conference and post conference. At the same time many in the ICTMA community, or others with similar research and scholarship interests, who are unable to attend continue to theorise or research into the teaching of modelling and applications. It is important that any published contributions stimulated by the conference be connected to this on-going body of research and scholarship as well as previous outputs from the community (e.g., Kaiser et al. 2011; Lesh et al. 2010).

2 Innovative Practices in Research and Teaching

A characteristic of this global community in mathematical modelling and applications education is that its members operate from a large variety of different theories and research paradigms in both research practices and teaching practices. This is, of course, not unique to ICTMA (see e.g., Prediger et al. 2008) but a small community such as this can lose some of its sense of connection if other theoretical ideas or research paradigms are not considered thoughtfully with respect to what is valued, even if eventually these are not embraced. "An important role of theory in research is to provide new ways of conceptualizing practical questions" by transforming the dilemmas and problems of practice into more tractable forms for resolution (Rodríguez et al. 2008, p. 287). Different theoretical lenses and their implications for research practices or teaching practices are examined in this book.

To begin, Galbraith takes the reader on such a journey in the chapter resulting from his plenary lecture at ICTMA 15. The history of ICTMA is outlined highlighting several issues derived from this community. The first set of issues is concerned with the domain. Traditionally, modelling has two concurrent purposes: to solve a particular problem at hand and, over time, to develop modelling skills that individuals can apply to problems in their world. Furthermore, "authenticity" has loomed large as a domain issue in modelling discussions and will be discussed again in several chapters in this book. It is proposed that authenticity be viewed in terms of four dimensions: content authenticity, process authenticity, situation authenticity, and product authenticity. The second set of issues is concerned with the community.

By employing the Vygotskian idea of Zone of Proximal Development (ZPD), Galbraith describes how, within the ICTMA community of practice, both practitioners and researchers, experienced and inexperienced, are represented. By using the diverse expertise of its members, a group ZPD could potentially be generated that inspires visions that none of us alone could produce. The third set of issues is concerned with practice. Research themes in the domain of practice are described, especially focussing on the issues in formulating a mathematical model and the role of metacognitive activity.

Ärlebäck and Frejd explore Sfard's notion (2008) of *commognition* which merges and combines principles from theories of communication and cognition. Thinking is discussed within this perspective as a special type of interpersonal communication and learning as a change in discourses. Here, a discourse is characterised by the meaning and use of language, including not only consensus on its interpretation, but also established conventions for communicating and interacting among members of the discourse. In addition, there are agreed formalised ways of establishing truth within the discourse. When examining mathematical modelling activities in a teaching and learning environment, several kinds of discourses come into play: *everyday discourses* pervaded by personal experiences; *literate discourses* in which communication is characterised by the use of the special discourses (depending on the task) where mathematical models are used, created, developed and modified; and *classroom discourses* encapsulating school norms and rules. When Ärlebäck and Frejd applied this approach operationalized as a research practice to transcripts of the discourse of two students working on a modelling task, not only did they demonstrate that this approach can identify the types of discourse processes that take place when students engage in mathematical modelling but also they felt their analysis highlighted how cognitive and social aspects are simultaneously manifested in modelling activities. In the example used, the discourse is mediated by the social context and the task itself. Even though the task is strongly pre-structured, it is still open enough to provoke discourses on the relevance of different models.

De Oliveira and Barbosa examine *tensions in discourses* when Brazilian teachers attempted to engage learners in mathematical modelling tasks, adding to their previous work (de Oliveria and Barbosa 2010). They analyse teachers' discourses based on Bernstein's theoretical framework. *Discourse* in this instance is defined as an oral or written text produced by an individual in a specific social context. The teachers introduced real problems into their pedagogic practices but the students had difficulties solving these, because they did not understand how to use mathematical content in the solution and they were not able to produce a *legitimate text* (Bernstein 2000) for the development of the modelling task. Teachers then questioned how they should resolve these student difficulties as well as how they should teach (previous and new) mathematical content so such difficulties do not arise when using this content for modelling. These questions represent discontinuities in relation to the present discourses and the new discourse of mathematical modelling in the pedagogic practice. These discontinuities refer to *the tension of students' mathematical performance*. To deal with this tension teachers had taught the previous and new mathematical content during the development of the modelling task in their lessons; but largely it is a timing dilemma as to when teachers should work on mathematical content related to the theme of the modelling task. Others (e.g., Steen and Turner 2007) have pointed this out previously but this dilemma is still not unequivocally resolved by teaching or research.

The chapter by Borromeo Ferri and Lesh is epistemological in nature and has no direct link to practice. It is relevant to the theory of learning and teaching of modelling and as such could spark innovative practices in teaching with a consequential flow-on effect to possible future research agendas. The question they ponder and thus ask the reader to consider is: Should interpretation systems be considered to be models if they only function implicitly? As has been pointed out previously, "in any application of mathematics a mathematical *model* is involved, explicitly or implicitly" (Niss et al. 2007, p. 4). The authors note that implicit or intuitive models occur in situations such as where "sense-making systems must function much more rapidly than is possible using formal, analytic, or completely conscious thinking", or where the information needing to be considered "exceeds the processing power of available models". These "implicit models" rarely are used in textbooks.

Rosa and Orey argue that the application of ethnomathematical techniques and tools of modelling allows us to examine systems taken from reality and offers insight into forms of mathematics done in a holistic way. The pedagogical approach that respects a diversity of cultural forms of mathematics is best represented through ethnomodelling, which is a process of translation and elaboration of problems and the questions taken from reality. Ethnomodels as cultural artefacts are pedagogical tools used to facilitate the understanding and comprehension of systems taken from the reality of cultural groups (Rosa and Orey 2009). Ethnomodels are thus considered to be external representations consistent with mathematical knowledge that are socially constructed and shared by members of specific cultural groups. The intention of Rosa and Orey is to broaden the discussion of possibilities for the inclusion of ethnomathematics and associated ethnomodelling perspectives that respect the social diversity of distinct cultural groups with guarantees for the development of understanding different ways of doing mathematics through dialogue and respect.

Relating a current problem to a similar problem solved previously has long been associated with approaches to problem solving. Saeki and Matsuzaki provide an analytical instantiation of this principle with respect to modelling problems. When attempting to solve a modelling task there might be cases where changing from the initial modelling task where progress has stalled, to a similar modelling task based on prior experiences where some traction is considered possible by the modeler, is advisable. There is switching between the two modelling cycles, and progress in the first modelling cycle based on results or limitations obtained from the second modelling cycle becomes deliberate. Saeki and Matsuzaki use the term "dual modelling cycle" in a way that each problem has its own individual modelling cycle. Another view is that the outworking of a generic modelling cycle has characteristics that differ according to the specifics of a given problem. From the perspective of Saeki and Matsuzaki the purpose in using the dual modelling cycle is its utility in facilitating teaching when a first modelling task is able to be related to a similar modelling task in order to solve the initial task through feedback from the solution of the second.

The three chapters by Buchholtz, Geiger and Rosa and Orey arose from a symposium at ICTMA 15 about the role of theory in research about mathematical modelling. As Buchholtz points out, a complex phenomenon such as the learning and teaching of mathematical modelling cannot be fully described by just a single theory particularly given the diversity of cultural backgrounds, research and educational systems where the teaching and researching of mathematical modelling in education is taking a foothold. Theory acts as a guide for research practices but also is shaped by the findings of that research in a symbiotic process. Geiger refers to a more general theoretical description formulated by Strässer generalising the notion of the didactical triangle and applies it to describe and explain activities in the process of mathematical modelling in a holistic way. He illustrates the applicability of Strässer's tetrahedral description to mathematical modelling and application activity by analysing an episode from his research data. Rosa and Orey see ethnomodelling as an alternative methodological approach to analyse mathematical modelling and mathematical ideas in different cultural contexts and describe the dilemma between an "emic" and an "etic" perspective in this field of research, which occurs when taking ethnomodelling as a research lens. They argue that traditional mathematical modelling does not fully take into account the implications of the cultural aspect of human social systems and suggest ethnomodelling as a pedagogical approach which integrates an "emic" perspective into a mathematical modelling curriculum.

3 Research into, or Evaluation of, Teaching Practice

Research into or evaluation of teaching practice is an on-going interest in the ICTMA community (Niss 2001). At times this takes as its focus evaluation of teaching or learning sequences focussing on modelling activities inserted into more traditional mathematics curricula. In these cases there usually is a design aspect to the teaching materials or the facilitation of activities or both. In other instances it is the teacher's behaviour (their practice) and the students' behaviour in response to this that is described and investigated.

Ang describes a teaching experiment carried out in a Year 9 Singaporean class with a complex modelling problem where a teacher with no experience in teaching mathematical modelling and limited professional development on modelling made an attempt to conduct a modelling activity. The task was carefully designed and generally adhered to the principles of design suggested by Galbraith (2006) although the teacher had no prior knowledge of these but had simply planned the tasks and worksheets "based on [his] own understanding and pedagogical sense". The teacher lacked certain key skills in facilitating modelling activities. The lack of a strong framework probably resulted in the need for the teacher to "over-facilitate" and provide constant intervention in order to help students progress with the task. This experiment also showed that in general, although they may face difficulties in grasping and applying certain mathematical concepts, Singaporean students are not opposed to the idea of engaging in mathematical modelling tasks.

Blomhøj and Kjeldsen provide insights from their involvement in the teaching of modelling projects at university over many years and an evaluation of the facilitation of such a project to illustrate their conceptualisation of this. Their experience from the course supports the notion that the modelling context provides a window to students' understanding and their images of the mathematical concepts they work with as well as to their understanding of a mathematical model and modelling. Dialogue is used deliberately to open such windows. For specific modelling tasks conceptual difficulties seem to be rather stable across cohorts of students. To notice, and to design and teach to overcome such particular learning difficulties is a way of enhancing the students' mathematical learning through modelling activities. The modelling context provides opportunities to challenge and strengthen the students' concept images concerning important mathematical concepts. Modelling projects are designed deliberately to accentuate potential cognitive conflicts in the students' concept images. The project and the related dialogue with teachers encourage students to reflect on the mathematical meaning of a mathematical object as well as on its interpretation and validity as a model. The authors note that such projects can be a didactical vehicle both for developing modelling competency and for enhancing students' conceptual learning of mathematics.

Chan's chapter focuses on an investigation of the engagement of Year 6 students in mathematical modelling in a Singapore school by exploring the mathematical reasoning of a group of students attempting to design an ideal tourism route. Engaging in mathematical modelling activity provides affordances for student-task-teacher interaction and in particular for the modelling task to drive learning. Chan notes that in such an environment, the students' cognitive engagement can be high as they attempt to identify goals and variables, interpret problem situations, interrogate data, make inquiries, monitor their solutions and improve their conceptualisations. This form of instruction differs starkly from learning mathematics in a traditional manner which Chan claims is limited in scope in revealing what students are capable of. Engaging students in mathematical modelling paves the way for teachers to redesign ways where students can solve problems in real-world settings and bring their mathematical reasoning to the fore.

Fu and Xie investigated the effect of a one-semester course aiming to introduce some basic modelling concepts and examples to first year students at a highly academic Chinese university. For 15 weekly lectures the teacher introduced a modelling situation related to the students' daily lives, including basic models arising from engineering, public administration, operations management, marketing and economics. Out of class, students were asked to form groups to independently pose, discuss and build up mathematical models to solve real world problems of interest to them. At the end of the course, each group submitted a report presenting the group's work. Two tests originally designed by Haines et al. (2001) were used to check whether there were any differences between the mathematical modelling skills on entry to the university and after the course. Statistical analyses showed no significant difference between scores achieved on the two tests. Fu and Xie attributed this in part to the already very high scores in the pre-test having a ceiling effect which makes it difficult to improve further.

Geiger, Goos and Dole describe the practice of one teacher who took advantage of a major, potentially disruptive, building development within her school to design a sequence of lessons in which students were challenged to adapt to the changes that were associated with the construction in order to demonstrate their numeracy skills. The teacher was participating in a project which used Goos' (2007) model of numeracy as a basis for planning for teaching and for reflecting upon the effectiveness of practice. The teacher showed initiative taking a professional risk in building an element of her mathematics teaching program around a disruptive event. Her actions paralleled her attempts to promote flexible and adaptive thinking in her students in relation to the use of mathematics in context. Thus, this position enabled her to design an activity that wove the elements of the numeracy model into students' learning in a seamless fashion.

Grünewald reports the evaluation of the effects of a short-term modelling project on the development of modelling competencies by Year 9 students in Germany. To evaluate the modelling project she developed her own modelling test based on test proposals designed by Haines and Crouch (2001) and Houston and Neill (2003), which focussed on the development of the sub-competencies of mathematical modelling of students. In addition, five students were interviewed at the beginning and the end of the project to gain deeper insight into their development of modelling competencies. Results showed that it is possible to foster at least students' sub-competencies of mathematical modelling within such a short-term modelling project.

Matsuzaki and Saeki describe and evaluate the effect of experimental classes for undergraduate university students who were interested in mathematics education or intended to become a mathematics teacher based on evidence of a dual modelling cycle in their approach to tasks. The purpose of the chapter is to provide evidence for their contention that using the dual modelling cycle is useful in facilitating teaching when a first modelling task is able to be related to a similar modelling task in order to solve the initial task through feedback from the solution of the second. The authors verified some stages of a dual modelling cycle empirically. They conclude that the dual modelling cycle framework they proposed appears to have potential for drawing modellers' attention to the possibility of solving the task at hand by first solving a related problem as Polya (1988) suggested.

The posing of tasks that have several solutions, fostering development of several solution methods, and the presentation and discussion of these solution methods in the classroom are characteristic features of Japanese teaching method. How crucial these features are for good performance in mathematics and whether these teaching methods can also be used effectively in other countries are still open questions. As an attempt to address this, Schukajlow and Krug present first results from an explorative video study of pairs of Year 9 German students solving four modelling problems that demand two outcomes. Analysis of videos of one task showed that students had only minor problems while solving the task. If they could find one solution, the second outcome was developed also. Although students validated their results, they rarely compared different outcomes independently. Schukajlow and Krug see the comparison of solution methods as an important part of teaching tasks with multiple solutions that can improve student performance and cognitive flexibility. They recommend that prompting the comparison of different outcomes and solution methods by teachers is crucial for the improvement of modelling competency.

Stillman, Brown and Galbraith evaluate their role as mentors in an extracurricular modelling event where Year 10 and 11 students chose their own situation to model and worked on in small groups assisted by mentors for 2 days. They investigate whether the student experience met their intended purposes as mentors, namely the whole event be considered inherently valuable as a learning experience about modelling and application of mathematics to real situations. Motivation for deciding on the choice of real world problematic situation to model was mainly goal oriented with many indicating the task was considered inherently useful and group members were interested in learning from their modelling experience. The majority of groups reported the outcome of the problem(s) they had formulated in their chosen situation or contextual details they had discovered during their investigation of this situation as their most interesting learning outcome from being involved in the event. Even though the vast majority reported having not participated in similar school activities, all groups were able to choose a situation to model and engage in attempts to model it over the 2 days to an extent that enabled them to report increased understanding of the complexity of the situation and to describe and critique their attempts at mathematical analysis of self initiated questions about the situation. Stillman, Brown and Galbraith conclude the whole event was considered by most participants as inherently valuable as intended.

Yanagimoto and Yoshimura evaluate the use of two different sets of teaching materials with Japanese junior high school students. They demonstrate that the materials can be successfully used with students at this level in a teacher directed whole class approach as an introduction to modelling. They suggest that more open teaching practices such as using group learning would be necessary after the introduction period. The mathematical modelling teaching materials which dealt with societal and environmental issues aroused students' interest and curiosity, and encouraged students to open up to societal issues more, boosting their awareness as members of society. An evaluation showed that with both sets of materials student perception of the utility of mathematics in the real world increased.

4 Pedagogical Issues for Teaching and Learning

Pedagogical issues are the focus of both on-going research in areas where mathematical modelling and applications have been present in curriculum documents for some time (e.g., Australia, Denmark, Germany) and also when a new education system (e.g., Singapore) starts advocating the use of applications and modelling in teaching, learning and/or assessment contexts. It is not surprising that when the first efforts at implementation arise in new educational systems that pedagogical issues for the teaching of modelling soon become the focus. "The pedagogy of applications and modelling intersects with the general pedagogy of mathematics instruction ... but simultaneously involves a range of practices that are not part of the traditional mathematics classroom" (Niss et al. 2007, p. 21). In order to fulfil the goals of teaching applications and mathematical modelling, according to Henn (2007, p. 322) "an adequate Modelling Pedagogy is necessary". This entails "adequate problems, adequate teaching methods and instructional modes, adequate tools and adequate modes of assessment".

4.1 Appropriate Problems for an Adequate Modelling Pedagogy

Brown reports on an empirical study, involving three Year 6 classes, of the extent of student engagement with real world tasks. The focus is on whether applications and mathematical modelling are viewed as being outside of mathematics, as an add-on, or as an integral part of mathematics, requiring students and teachers to operate in a "culture of mathematising as a practice" (Bauersfeld 1993 cited in Yackel and Cobb 1996, p. 459). Two tasks were designed that required students to reflect on their mathematics and make their thinking explicit in keeping with Bauersfeld (1992). Although students clearly engaged with a first task involving the supply of wooden letters for a toy store, for many it was perceived as a (problematic) division problem without an appropriate exact answer. Few engaged with the context or gave any indication that the task required Bauersfeld's "flexible interpretation" (p. 467) with the majority not making any use of data although it should have been apparent that data were easily accessible. A second task about the stock levels of brass numerals in a hardware store resulted in interpretation beyond a "ritualized reading" (Bauersfeld 1992). Students engaged with the context and for several this was sustained. In this task students were more likely to see their role as sense making and independent mathematising. Across the three classes, students were not experienced in interpreting mathematical problem situations nor believing this was a normal part of school mathematics. This lack of experience of shared mathematising and negotiation of understanding and meanings contributed to lack of student belief that they had personal experience and knowledge to bring to the solution of the task. Brown concludes that using tasks requiring students to reflect on their mathematics and make their thinking explicit can contribute to Year 6 students perceiving themselves as playing an important role in interpreting real world problem situations and relating mathematics to the real world.

In 2001, Niss, in noting missing and wanted research in applications and modelling education commented, that "only relatively little attention has been paid to tasks which are complex, extensive, authentic or realistic, and time consuming" (p. 81). Authentic tasks have been an on-going theme in more recent work by Kaiser and colleagues. Kaiser and Stender explore student interest in, and opinion about, the realism of several authentic modelling problems from evaluations of several modelling weeks. Such problems (e.g., the optimal design of location of bus stops along a bus route) are authentic in the eyes of educational researchers and mathematicians but Kaiser and Stender investigated whether these were seen in the same light by average students not especially interested in mathematics as would be encountered in most classrooms not streamed for student ability. Thus, even though the authenticity of real-world problems often creates difficulties (Busse 2011; Stillman and Galbraith 2003), teacher intervention should focus only on supporting students in the case of lacking mathematical techniques or when they are in a dead-end situation. Students who struggle with comprehension or lack of reference to the real world in their mathematics classroom lessons can experience the utility of mathematics in solving such tasks when using mathematics to answer questions that arise from reality.

Buchholtz and Mesrogli also investigate the issue of experiencing solving of complex authentic modelling problems such as chlorination of a swimming pool through the use of 'modelling weeks' for upper secondary school students. They were interested in whether the modelling problems tackled in these events were feasible for students at this level of schooling, what attitudes the students had towards mathematics and whether the event influenced student mathematical interest. Buchholtz and Mesrogli conclude that, although students' mathematical knowledge might not be very high and their modelling competencies differ widely, experiencing helplessness and insecurity is a central aspect and necessary when dealing with mathematical modelling problems.

Jennings and Adams, and Leung provide examples of mathematical modelling of real world cases as exemplars of suitable tasks for teaching modelling. Jennings and Adams discuss mathematical modelling of the pharmacokinetics of alcohol after consumption, and in particular, Blood Alcohol Content levels. Leung addresses the challenge of teaching mathematical modelling skills to students, in non-mathematics majors, from business schools, who will potentially be employed to tackle business problems raised in market competition. In the discussion of methods to estimate customer lifetime value (CLV), the focus is on the dynamical relationship among variables rather than simply setting up a formula from which the subject can be readily solved. Such a dynamical system approach exhibits the logistic nature of the CLV model. The teaching implication of learning this logistic property is that the technique is applicable in other market scenarios.

Winter argues that by using contextual tasks, we are able to gain insight into the nature of initial modelling competencies across the modelling cycle with beginning modellers such as pre-service teachers who have not experienced modelling instruction. There is a strong push globally towards introducing modelling to young children to ensure development of modelling specific competencies from the early years (see English 2003). South Africa's educational policy explicitly emphasises development of learners' skills and abilities to model real life contexts which contain mathematical features. Within mathematical literacy, modelling focuses on the use of basic mathematics from the early years to lower secondary implying that knowledge of elementary mathematics is sufficient for solving real life situations. The elementary mathematics base and use of simple word problems provide links between young children modelling and mathematical literacy modelling.

In South Africa, the advocacy to have contexts for mathematical literacy drawn from personal, educational/occupational, or public situations, coupled with its citizenship perspective provides a rationale for the use of elementary mathematics. The demand for new ways of structuring classroom teaching in mathematical literacy (Antonius et al. 2007) so that tasks with real life contexts are starting points, provided a strong rationale for introducing a university mathematics content course which aimed at developing the pre-service teachers' modelling skills in order that they develop an adequate pedagogy for teaching the modelling expected in mathematical literacy.

One approach to teaching using mathematical modelling and applications is to take a socio-critical perspective (Kaiser and Sriraman 2006) where the goal is the development of a critical understanding of the world through modelling physical and social phenomena. The role of modern textbook resources in providing sources of problems and tasks for promoting this development and how teachers could use these resources to mediate such a learning goal in relation to the applications of mathematics is an under researched area. In their chapter Stillman, Brown, Faragher, Geiger and Galbraith interrogate curriculum documents and textbook tasks to assess the potential role of textbooks in developing this understanding in students through mathematical modelling. Although the analysis is carried out in the context of Australian state curricula, there are implications for educators and researchers in other parts of the world where similar goals are expressed in curriculum documents. Clearly, regular textbooks will have only limited impact in broadening teaching horizons for this purpose if teachers have little knowledge of socio-critical perspectives in a mathematics context and/or little inclination to take up the challenge of incorporating these into modelling tasks.

4.2 Adequate Teaching Methods and Instructional Modes

Kaiser and Stender also address the issue of the adequacy of teaching methods when using authentic modelling tasks in an authentic manner, namely as self-directed modelling activities. As the main goal is to support students in their self-directed modelling activities, the question arises: How can teachers support the students without destroying the independency of these modelling activities? They develop the approach of providing scaffolding as a comprehensive, long-term support combined with interventions as direct and immediate adaptive actions by the teachers form the theoretical framework for their work. In contrast to the work of Leiß (2007) who found that strategic interventions were included infrequently in the intervention-repertoire of the teachers he observed, Kaiser and Stender observed the usage of general strategic interventions. They suggest the request to the students given by the teacher to present the state of their work to the teacher as they enter the group, as prerequisite for an adequate scaffold by the teacher, enabling diagnosis of the state of student work, and as a central part of an effective feedback.

Ikeda notes that from a broad perspective, there are two categories of pedagogical aims of modelling which must be captured in teaching methods that underpin an adequate modelling pedagogy. Firstly modelling itself is treated as an objective and secondly modelling is treated as a means to mathematical knowledge construction. For the first aim the teacher needs to set an appropriate situation so that students realise the necessity of solving a real-world problem, assist students' abstraction processes and show students the necessity of controlling for various assumptions. For the second aim, three teaching principles are suggested: (1) expanding and clarifying real-world situations satisfying a ready-made model, (2) expanding and integrating mathematical knowledge by setting up a concrete situation so that students can consider it, and (3) refining and clarifying the developed mathematical methods by treating instances of the same contexts repeatedly.

Kuntze, Siller and Vogl address the issue of the adequacy of teacher's professional knowledge of modelling as a big idea in mathematics and its relevance in the classroom as an underpinning foundation for developing an adequate modelling pedagogy. They focus on Austrian teachers' self-perceptions of their pedagogical content knowledge (PCK) related to modelling. These self-perceptions were explored directly from the perspectives of views about PCK relevant for modelling-specific teacher-student interactions and indirectly from views about teacher satisfaction with university studies concerning ways of fostering modelling abilities. 'Direct reference' scales focused on self-perceptions of diagnostic knowledge related to the modelling process and on providing modelling-specific help to students. 'Indirect reference' items focused on self-perceptions about modelling-specific PCK learned at university and about PCK relevant for technology use in the modelling process. The teachers' self-perceptions of their PCK related to modelling suggest that there is a need for professional development not only as far as PCK related to modelling is concerned, but also related to the aspect of a pedagogical modelling-specific self-efficacy of teachers. This self-efficacy may be supported by positive experiences of the teachers with modelling tasks in the classroom. In particular, professional development support for teachers related to the use of technology in the modelling process might be helpful. Kuntze, Siller and Vogl conclude that both pre-service teachers' and in-service teachers' professional knowledge concerning modelling should be developed further. Also, there is a need for further research into the structure of professional teacher knowledge concerning modelling to provide an empirical base for the conception of sustainable professional development activities about adequate modelling pedagogy for teachers.

To develop an adequate modelling pedagogy, especially with respect to diagnostic competencies, teachers also need access to knowledge of expected demonstration of modelling competency by students at different grade levels of schooling. Ludwig and Reit attempt to investigate this through their analysis of student solution approaches for a modelling task through grades 6–11. Students in younger grades had more difficulties in solving this modelling task reinforcing the notion that modelling competencies appear to be enhanced by experience and practice but the increasing sophistication of mathematical knowledge with grade level might also be an issue. For all grades there was an obstacle proceeding from the explicit numeric result to its generalisation and this could be commensurate with similar difficulties in purely mathematical contexts. In contrast to the widespread stereotype of mathematically science oriented boys and linguistically gifted girls there was no statistical evidence of differences in the performance of boys and girls in elaborating on a modelling task.

Ng addresses the vexed issue of teacher readiness for mathematical modelling. Both in-service and pre-service teachers with no experience implementing modelling tasks with their classes or having engaged in previous modelling tasks as modellers were the subject of her study. The findings from this study suggest that the challenges in teacher education for fostering a positive modelling climate in systems where modelling is new in the curriculum (e.g., Singapore) are manyfold. Work is needed to help change the mind-set of in-service teachers as to the use of open-ended tasks situated in real world contexts. They may have to first learn to accept and then later scaffold tasks with a non-exhaustive list of solutions which can use various mathematical representations. Although Kuntze (2011) found in-service teachers were more receptive in his German study to intensive modelling activities, this was not confirmed for Singapore teachers as there was qualitative evidence from both samples indicative of blockages to such tasks due to beliefs about mathematics when the teachers were engaged in modelling themselves. Pre-service teachers may also need guidance in anticipating the nature of mathematics applied during modelling tasks by the students. This may be less sophisticated, less organised, and less focused than what the pre-service teachers themselves can produce. In contrast, for in-service teachers the opposite may be true with expectations needing to be raised in terms of the quality of mathematics applied. Work is also needed with these teachers on the range of expected mathematical outcomes for a given task, including the use of more sophisticated mathematical thinking. Discussion of how to scaffold students towards moving to using more sophisticated mathematics is also needed if adequate modelling pedagogy is to be developed in these circumstances.

Redmond, Brown and Sheehy investigate how the discourse of the mathematics classroom impacts on the practices that students engage when modelling mathematics. The principles of collective argumentation (Brown and Renshaw 2000) are used by teachers and students in the educational context studied to guide engagement in the discourse of their mathematics classrooms. The aim of this discourse is to enable students to analyse mathematical contexts, to synthesise strategies to mathematise these tasks, and to communicate solutions and conclusions to others. Redmond, Brown and Sheehy argue that when students engage with the discourse of their mathematics classroom in a manner that promotes the communication of ideas, they employ mathematical modelling practices that reflect the cyclical approaches to modelling employed by mathematicians. Further, these modelling processes and ways of operating should be evidenced in the reviewing and editing that students make public when authoring drafts of assessment reports of their modelling. Such reviewing and editing processes may not only provide insights into student competency in mathematical modelling, but also insights into the way students represent their engagement with the discourse of their mathematics classroom.

According to Tan and Ang, teachers' knowledge for mathematical modelling instruction can determine, to a large extent, how they perceive and respond to curriculum innovation efforts to inject mathematical modelling activities into the secondary mathematics classroom in Singapore. To think about knowledge of mathematical modelling instruction requires going beyond knowing the mathematical modelling process. It may require understanding the complex interplay among aspects of other forms of teacher knowledge in the mathematical modelling teaching and learning environment. Pre-service teachers have expressed concerns about their lack of understanding in teaching mathematical modelling which is only now being introduced into the curriculum which they must teach. Opportunities in pre-service courses to explore mathematical modelling learning experiences were welcomed as none had prior experience with the type of modelling tasks implemented in the study. The pre-service teachers' initial mathematical modelling learning experience was limited to applying particular mathematics topics with rather narrow and known techniques. Hence they were not used to integrating various concepts and techniques in their first mathematical modelling task and their modelling process was often isolated from the real world situation. The tutor's introspections during the modelling discussions provided the basis for forming case stories that could have enabled the pre-service teachers to reflect and pick up nuances about important elements of modelling activities in the various stages of the modelling process. The coupling of such case stories with the learning scaffold structured in the mathematical modelling learning tasks was paramount to the development and transfer of mathematical modelling competencies to other problem contexts. Tan and Ang argue for the necessity of novice teachers in teaching mathematical modelling performing modelling tasks themselves. This is so that the experience and knowledge gained can help them explicate aspects and nuances of the modelling process with respect to novel modelling tasks. Explicating such aspects and nuances of the modelling process is important later when these are needed to be translated into features of modelling tasks for teaching and learning purposes. Thus an adequate modelling pedagogy is brought into being by adequate personal modelling experience.

The translation of one's understanding of a problem situation into a mathematical model constitutes a key step in the process of mathematical modelling according to Van Dooren, De Bock, and Verschaffel. The authors show that university students in their study were very proficient in relating descriptions of realistic situations to models in cases when the situation was described as linear. When the situation was "almost" linear, there was a strong tendency to connect the situation also to linear models (and for inverse linear situations, to some extent, to affine models with a negative slope). These results parallel those of other studies (Van Dooren et al. 2004) showing the "default" role of the linear model. Results also indicated that the representational mode had a strong impact on students' modelling accuracy and on the tendency to inappropriately connect non-linear situations to linear models. A particular representation may highlight aspects of non-linearity that are easily noticed by students and therefore facilitate correct reasoning, but be misleading when representing a situation with another model. An implication for modelling pedagogy is the need for drawing sufficient attention to representations, to matching representations with each other and to linking them to realistic situations, and for explicitly discussing differences between linear and different types of "almost" linear models.

Chapters by Lee, Chan, Ng and Geiger, as a result of the second symposium at ICTMA 15, address the perplexing issue of systemic change in pedagogy so as to adequately incorporate modelling pedagogy in a context where current practices are quite different from these, namely the Singaporean context where modelling and applications are a recent addition to the curriculum. Lee's chapter captured the readiness of participants in a relevant professional development event in designing modelling tasks and the value they attached to problem posing as part of the total student mathematical learning experience. Chan explored the issue of adequacy of short teacher professional development to facilitate teachers effectively carrying out modelling activities in classrooms that have been dominated by teacher-centred pedagogy. Ng presents a preliminary teacher facilitation structure developed for initial professional development in an outreach programme for mathematical modelling in Singapore primary and secondary schools. Findings on teacher readiness in facilitating students' mathematical thinking, reasoning, and communication during mathematical modelling are discussed. Feedback from the teachers with respect to their focuses and concerns about facilitating modelling tasks centred around the need for a balance of scaffolding during open-ended tasks where the teacher has to be aware of when to step in and when to draw away to encourage thinking and promote competencies. Geiger points out that widespread reform in teaching and learning practice within mathematics education is always challenging especially when the proposed reforms are of a very different nature to existing modes of practice. Geiger concludes that supporting teachers to understand, and then to adopt, these new practices will be at the heart of attempts to progress this reform.

4.3 Adequate Tools for an Adequate Modelling Pedagogy

There is no denying that technological tools are "a powerful tool to aid in modelling" (Henn 2007, p. 324) but they are not necessarily so as other factors come into account. The CASI-Project examines the long-term use of digital tools in mathematics teaching with German Year 9 and 10 secondary students. According to Greefrath and Reiß the use of digital tools influences each part of the modelling cycle so the technology is useful in relating to the real world and mathematical world of modelling. The different representations of functional relationships and the change between the different representations and real situation are said to be influenced by digital tool use. Previously, Weigand and Bichler (2010) observed improved school achievements in translation between graph and algebraic expression when using digital tools. However, in the CASI-Project there was little difference between the performance of students translating from a graph to a real situation or vice versa with or without digital tools. Students using digital tools were, however, able to employ multiple solution methods such as availing themselves of computer drawing facilities included in dynamic drawing software.

Lamb and Visnovska explore the adequacy of students' explorations of mathematical models using a computer applet tool. In this instance the students were secondary school teachers in a professional learning course. The researchers were attempting to gauge the teachers' preparedness to facilitate productive mathematical discussions in their classrooms that compare different models. In the particular project about a speed trap on which the teachers were working, they had just commenced learning how to engage in activities where they genuinely analyse data. The teachers considered a range of computer tool options, and used these in different ways that could support monitoring their own students' work during task exploration if using the same project with their students. However, most teachers did not attempt to compare different solutions or explain how they connect with specific statistical ideas. Some teachers clearly expected that if there is a *real* trend in data, it should be equally evident in all models. Teachers' discussion of different models and their difficulty in reconciling conflicting ideas reflected their lack of prior experience with engaging in mathematical discussions grounded in comparison of such models. Developing adequate modelling pedagogy will require supporting teachers to work flexibly with key statistical ideas as well as with the tools through which student learning can be built no matter how adequate the tools might be for modelling in the hands of an expert user.

4.4 Adequate Modes of Assessment for an Adequate Modelling Pedagogy

Wiliam (2007) in the *Second Handbook of Research on Mathematics Teaching and Learning* points out the pivotal role classroom assessment plays in keeping learning on track, in particular formative assessment in the form of feedback. The interdisciplinary research project Co²CA (*Conditions and Consequences of Classroom Assessment*) aims at investigating the impact of different kinds of feedback in competency-oriented mathematics teaching on student performance. The project implements written and oral feedback into teaching through the following principles: feedback is given individually to students in short intervals; it refers to students' solution processes; and students' strengths and difficulties and self-improvement strategies are pointed out. Project teachers were trained to orally intervene in students' working only minimally in order to allow student independence as much as possible. They were informed also about different ways of intervening and supporting, namely through metacognitive, problem content, affective and classroom organisational interventions. Besser, Blum and Klimczak report on an experimental study involving 39 Year 9 classes in this project. The classes were assigned randomly to a control group where no special feedback was given to students, or one of two experimental groups, the first where students received written feedback three times within a 13 lesson sequence and the second where additional oral feedback supported the written form. The research investigated whether these special kinds of formative assessment can help teachers improve students' learning processes when dealing with technical and modelling tasks and whether an implementation in every-day teaching can foster students' performances. Preliminary results showed there were no significant differences between the control group and either experimental group in the post-test but the control group had performed significantly better in the pre-test than the second experimental group. Since analyses of covariance showed no influences of the experimental condition either, the quality of the implementation of the treatment is being investigated further to explain these effects. A significant challenge is to control for both the overall quality of teaching (by analysing many hours of video-taped lessons) and for the quality of written and oral teacher feedback (by developing adequate coding schemes for both forms of feedback).

Great diversity exists internationally in how, and even if, the teaching and learning of applications and modelling are implemented. In parallel, there is similar diversity in summative assessment practices, often politically and culturally influenced. Greefrath and Reiß show that there were some differences (sometimes quite large ones) in their German CASI-project using reality based examination tasks between groups of Year 9 students taught with and without digital tools, but their results were inconclusive regarding advantages of one group over the other. Questions addressing the full modelling cycle are often too complex to be treated in an examination but it is possible to design questions that test certain aspects of reality-based problems such as filtering important information and structuring the situation. The use of digital tools made more complex examination questions accessible for lower achieving students at lower secondary level. These test items were still challenging, but diverse approaches were able to be verified.

Vos describes a study of modelling characteristics in recent mathematics examinations in the Netherlands where modelling is integrated into the mathematics curriculum. The test items showed a wide variety of situations, demonstrating mathematics is everywhere, a valid aspect of modelling. On the other hand to increase test reliability, the complexity of tasks was reduced and students' creativity was limited.

Tasks that are valid with respect to mathematical modelling simulate the work of professional modelers requiring students to undertake activities as described in the modelling cycle. These would be expected to take an extended period of time and be worked on by collaborative teams. In contrast, when a problem situation is offered together with a ready-made mathematical model, structuring, simplifying and mathematising are omitted. An advantage of this mechanistic mathematising format is that all students start at the same point ensuring that subsequent modelling activities have the same mathematical demand because of similar complexities of the models. The presence of a ready-made model raises task reliability but students cannot demonstrate all competencies. The ready-made mathematical model is merely a starting point for 'working mathematically'. In other tasks ready-made models need to be reproduced from the situation. In this format, modelling activities are asked in reverse order, what Vos calls *reproductive mathematising*. The advantage in examinations is that a range of competencies from the beginning of the modelling cycle are covered raising test validity with respect to modelling. Also, all students can continue with the correct model in ensuing tasks, even if unable to reproduce the model. This reduces answer variety, raising test reliability. However, students often short circuit reaching the ready-made model, by reasoning backwards using surface features of the model.

Thus despite the tasks on the examination papers being situated in real life contexts, students were merely required to perform mathematical activities loosely related to the situation described. The modelling cycle was broken into separate steps, reducing the answer range. Modelling activities such as structuring, simplifying and mathematising were avoided. These three reductions led to tasks that start from a situation together with a mathematical model (table, graph, diagram). The models were offered ready-made in the task and students were merely asked for calculations leading to a numerical answer. The Dutch examination papers thus may meet reliability criteria, but with respect to modelling they are limited in validity. Vos recommends that project-based or portfolio assessment which have a high validity in relation to the creative aspects of modelling not be neglected.

5 Applicability at Different Levels of Schooling, Vocational Education, and in Tertiary Education

Much energy has been invested by the ICTMA community and others in arguing the case for, or demonstrating through documented implementation examples, that applications and modelling deserve a reasonable place in mathematics education whether that be at different levels of schooling, in vocational education or tertiary education. "The vertical interconnectedness from primary to tertiary level is indispensable: One cannot start early enough with simple modelling examples" (Henn 2007, p. 323). There is still a space for such advocacy whether it be research based, theoretical or empirically based argument or demonstration of best practice.

English argues that there is a need to build a stronger foundation in the mathematical sciences, one that will equip students for the challenges of the twenty-first century. Citing authoritative sources she lists core competencies that are key elements of productive and innovative workplace practices presumably that would ensure such a foundation. To achieve this aim, she recommends an increased focus on interdisciplinary problem solving that engages students in complex modelling with challenging, life-based scenarios. Unlike many others (e.g., Heilio 2011), however, she sees such modelling having its applicability at much lower levels of schooling where the focus is on future-oriented learning experiences, that is, ones that engage students in the kinds of mathematical and scientific thinking needed for challenges beyond the classroom. The cases she selects to provide evidence-based support for her argument are from two design-based studies, a data modelling one in Year 1 and engineering-based modelling experiences in Year 7.

Downton also sees the necessity of laying an early foundation for modelling advancing the case for the applicability of problem posing, a critical component of mathematical modelling, in the primary years of schooling. Problem posing provides opportunities for young learners to make links between the real world and mathematics. Stillman (in press) concurs, "finding and posing problems are essential ingredients in any education program in schools promoting mathematical modelling". One of the distinguishing features of mathematical modelling is that "modelers find and pose their own problems to solve" (Stillman in press). Downton conducted teaching experiments in Years 1–2 and 3–4 involving problem posing. She provides documented evidence that students as young as 6 years of age are capable of generating questions that can be investigated using mathematical modelling. She argues that the use of problem posing and real world artefacts in the early years of schooling could provide the foundation for later modelling and applications experiences.

In a context where the mathematical modelling has been newly introduced into the high school curriculum, Kawasaki and Nisawa propose the linking of mathematical modelling with other new mathematical content to enable authentic modelling to be conducted. Previously, Kawasaki and Morija (2011) had advocated using modelling experiences to develop Japanese senior high school students' awareness of the interactions between mathematics and science. In secondary education in Japan as in many other countries, mathematics treats functions of only one variable. However, if natural science and social phenomena are formulated as mathematical models, expressions using functions of one variable have limitations. Functions of two variables are necessary to express real case scenarios through mathematics. Kawasaki and Nisawa suggest this content should be introduced into the traditional curriculum. They support their argument by documenting use of a task with Year 12 students involving minimising the packaging of a box, where a two variable function is required. They found that the use of 3D models not only allowed students to deepen their function knowledge but also to gain an appreciation for the complexity of real world problems. Kawasaki and Nisawa argue that in a context where mathematical modelling is new to students its applicability in the classroom cannot be assumed, rather it must be shown to be authentic and the mathematical content needed for this should be added to the curriculum.

If we are to move forward in the field in schooling, then real world examples and modelling need to be valued in mathematics teacher education by the inclusion of examples and full modelling courses in the programs (Kaiser and Maaß 2007). This sentiment is shared by Widjaja who considers what are suitable experiences within the Indonesian teacher education context. She demonstrates the authenticity of the common carpark design task, in this case to redesign a motorbike parking lot. This resonated with students engaged with the task whose most common mode of transport was a motorbike. Living in places where the daily practice sees most road users riding motorbikes is common in many, but not all, places around the world. This task would not have the same authenticity to pre-service teachers in Australia, for example. Wadjaja's message is clear—mathematical modelling is applicable to teacher education—only by experiencing authentic modelling tasks themselves will students develop pedagogical content knowledge about modelling for their future practice. She provides documented evidence of her students developing an awareness of the authenticity of modelling and establishment in the local context that modelling is a critical part of mathematics and therefore applicable in both teacher education and the school curriculum.

Also in the field of teacher education, the document analysis by Biembengut presents a mapping of Brazilian mathematics teacher education courses which have as part of their syllabus a specific modelling subject or any subject that deals with modelling. The guiding research question is: How has mathematical modelling been conceived in mathematics teacher education courses in Brazil? The possibility that mathematical modelling becomes regular classroom practice in Brazilian schools lies in the interest these future teachers have in contributing to a greater development in Basic Education. Biembengut clearly sees this interest being developed (or not) during their teacher education experiences or in professional development programs once they are teaching. As the continental dimensions of Brazil make it difficult to provide professional learning to continually serve all mathematics teachers, documenting the contribution made by professors responsible for these subjects in pre-service courses in improving mathematics education in Brazil is vital.

Three conceptions of modelling were evident in the subjects. Firstly, modelling is a *method of teaching and research*. The goal is to learn mathematics content from the subject and simultaneously learn to do research. The teacher proposes modelling examples to teach or indicates the mathematical content found in these examples (*realistic* or *applied modelling*), and subsequently, teaches the modelling process by inducing students to raise questions and data about the subject or topic, to formulate a hypothesis and then, formulate a mathematical model and solve the issues raised departing from the model and finally, to evaluate the model (*epistemological*). The second conception sees modelling as an alternative way of teaching mathematics, where the objective is the student learns mathematics. Modelling is proposed as it motivates students to learn mathematics from subjects or topics in their context. Beginning with student chosen topics, professors raise issues (*educational modelling*) and solve them while indicating the presence of mathematical content (*contextual*). These topics are authentic and integrated with the development of mathematical theories. This conception was the most prevalent in the documents analysed. In the third conception, modelling becomes the learning environment, and focuses on social issues. Modelling in education is able to show mathematics as a tool for decision making on environmental issues. This conception (*socio-critical modelling*) is complementary in nature to ethno-mathematics. The students seek to deal with issues involving situations related to society, developing a critical positioning regarding the context. This was the least prevalent conception in the documents.

The integration of mathematical modelling in the syllabuses of mathematics teacher education courses reflects support at official levels of education, in almost all Brazilian states, for the applicability of modelling in school education because of the possibility of providing modern youth with better knowledge and skills. Despite insufficient time to develop mathematical modelling fully and its usual placement at the end of teacher education courses, the merit and importance of mathematical modelling teaching in these courses cannot be underestimated if the applicability of modelling in schooling contexts is to be universally realised in Brazil.

Cristóbal and Vargas note that if students do not analyse non-mathematical situations during learning experiences, which demand the use of mathematical knowledge, then they do not think about their knowledge and they do not perform any adaptation that allows them to use knowledge in other situations. Mathematical modelling courses in higher mathematics education such as at university level, where students have to analyse situations that are different to the problems of text-books have been shown by Perrenet and Adan (2010), amongst others, as suitable vehicles for students to develop this awareness of applicability of their mathematics to other situations. In these situations the students need, in addition to mathematical knowledge and problem solving skills, to be able to translate the situation to mathematically tractable problems, to have a broad view of mathematics, to cultivate communication skills, and to use their common sense and intuition. In these situations the students have to use their skills and knowledge, and they have to learn or discover new techniques or concepts. Cristóbal and Vargas explore the applicability of such principles to students studying linear algebra in university mathematics.

Mouwitz raises issues that might cause us to pause and reconsider how we view mathematical modelling being applicable in workplace settings. He argues there is a tacit rationality, with a broader more qualitative mathematical essence, that has another origin and another character and function than traditional classroom mathematics. This kind of rationality is bound to personal acting in complex settings, and to our bodily interactions with the outside world. It is easily identified as a specific quality in the work of experienced craftsmen, tradesmen, professionals and sportsmen, but the author argues that it is also to be found everywhere in ordinary working life if we but have the eyes to see. It is also a necessary interpretation tool when constructing and concretising abstract and general scientific models. Such a tool is seen as a bridging process that needs dialogue and mutual respect for informal knowledge outside the classroom and institutional knowledge. Model thinking and analogical thinking represent two thinking styles that can come into conflict with each other. In many cases model thinking wins in such conflicts, and this can lead to a loss of praxis knowledge in workplace settings. This applies particularly where there is a generational change and the importance of hidden praxis knowledge becomes evident. A new group of practitioners, even though highly educated and with new tools such as computers, may not be able to replace the many years of praxis knowledge accumulated over time (see Göranzon 1993, for examples). The importance of mathematical modelling and its effectiveness are pervasive in workplace settings but as a cautionary tale it is important to point out its boundaries and its connections to practice, both *before* and *after* the model design. Further, Mouwitz cautions that if this is not recognised and highlighted, the "knowledge society" could instead initiate a massive societal de-professionalisation.

Moving beyond both school education and in fact beyond mathematics education per se, Schofield theorises about the applicability of mathematical modelling in university settings. His particular focus is on the use of modelling eliciting activities in university classrooms where students in the future will need to "use mathematics 'in the wild' … [for example, as] responders to natural disasters or unfolding terrorist scenarios" (Hamilton and Hoyles 2006, p. 4). He argues that to be successful in the sense that using these tasks is seen as of essential applicability to these students in their courses, consideration is needed regarding inducting such students into the ways of modelling, in particular the use of collaboration to solve complex tasks, an approach to mathematics which may be alien to some students.

Also with a focus beyond mathematics, Lagereis, Hu and Feijs argue that competency based learning in a project driven industrial design education environment results in a different view of mathematical modelling than that held in conventional tertiary education such as classical design engineering education. They argue that modelling tends to be hidden, however, redesigning tasks can make the usefulness and importance of modelling visible and allow students to be more successful modellers and designers and appreciate the applicability of mathematical modelling in an industrial design environment where the focus is on the design of intelligent systems, products and related services. The authors have noted that previously for industrial design students there is a virtual barrier in the flow of their work when going from science to the natural need for mathematics. It appears that investing in a deeper layer of abstraction is not seen as worth the effort by students. In the new approaches suggested by the authors opportunities for experiencing the competency *descriptive and mathematical modelling* can be provided without students being limited by a lack of technological skills.

6 Conclusion

A comprehensive appraisal of research challenges in the field of applications and modelling is contained in Niss (2001) and Blum et al. (2007). Recent special issues of international journals (e.g., Biehler and Leiß 2010; Kaiser et al. 2006; Stillman et al. 2010) and ICTMA volumes (Kaiser et al. 2011; Lesh et al. 2010) document the work that has been undertaken to address many of these in the ensuing years. In this current volume several studies have been shown to be underway that address several of these challenges as well. The ICTMA community operates from a large variety of different theories and research paradigms when it comes to both research practices and teaching practices but paramount to this global community of researchers is working together to establish a learning culture of mathematical modelling. The diversity of our global community both illuminates and connects our different perspectives on teaching practice and research practices as shown in this volume.

References

Antonius, S., Haines, C., Jensen, T. H., & Niss, M. (with Burkhardt, H.). (2007). In W. Blum, P. L. Galbraith, H. W. Henn, & M. Niss (Eds.), *Modelling and applications in mathematics education: The 14th ICMI study* (pp. 295–308). New York: Springer.

Bauersfeld, H. (1992). Classroom cultures from a social constructivist's perspective. *Educational Studies in Mathematics, 23*, 467–481.

Bernstein, B. (2000). *Pedagogy, symbolic control and identify: Theory, research, critique*. Lanham: Rowman & Littlefield.

Biehler, R., & Leiss, D. (Eds.). (2010). Empirical research on mathematical modelling [Special Issue]. *Journal für Mathematik- Didaktik, 31*(1), 5–8.

Blum, W., Galbraith, P. L., Henn, H.-W., & Niss, M. (Eds.). (2007). *Modelling and applications in mathematics education: The 14th ICMI study*. New York: Springer.

Brown, R. A. J., & Renshaw, P. D. (2000). Collective argumentation: A sociocultural approach to reframing classroom teaching and learning. In H. Cowie & G. van der Aalsvoort (Eds.), *Social interaction in learning and instruction: The meaning of discourse for the construction of knowledge* (pp. 52–66). Amsterdam: Pergamon Press.

Busse, A. (2011). Upper secondary students' handling of real-world contexts. In G. Kaiser, W. Blum, R. Borromco Ferri, & G. Stillman (Eds.), *Trends in the teaching and learning of mathematical modelling* (pp. 37–46). New York: Springer.

de Oliveria, A. M. P., & Barbosa, J. C. (2010). Mathematical modelling and the teachers' tensions. In R. Lesh, P. L. Galbraith, C. R. Haines, & A. Hurford (Eds.), *Modeling students' mathematical modelling competencies* (pp. 511–517). New York: Springer.

English, L. D. (2003). Mathematical modelling with young learners. In S. J. Lamon, W. A. Parker, & S. K. Houston (Eds.), *Mathematical modelling: A way of life* (pp. 3–18). Chichester: Horwood.

Galbraith, P. (2006). Real world problems: Developing principles of design. In P. Grootenboer, R. Zevenbergen, & M. Chinnappan (Eds.), *Identities, cultures and learning spaces* (Vol. 1, pp. 228–236). Adelaide: MERGA.

Goos, M. (2007). *Developing numeracy in the learing areas (middle years)*. Paper presented at the South Australian Literacy and Numeracy Expo, Adelaide.

Göranzon, B. (1993). *The practical intellect: Computers and skills*. Paris: OECD.

Haines, C., & Crouch, R. (2001). Recognizing constructs within mathematical modelling. *Teaching Mathematics and Its Applications, 20*(3), 129–138.

Haines, C., Crouch, R., & Davis, J. (2001). Understanding students' modelling skills. In J. F. Matos, W. Blum, S. K. Houston, & S. P. Carreira (Eds.), *Modelling and mathematics education – ICTMA 9: Application in science and technology* (pp. 366–380). Chichester: Horwood.

Hamilton, E. & Hoyles, C. (2006). *Enhancing and assessing complex reasoning through Models and Modeling: UK-US collaborative research.* To NSF and to the Economics and Social Research Council of England, in collaboration with the University of London. Funded Supplement to 0433373, September 2006. http://tinyur1.com/HamiltonHSDSupplement-pdf. Accessed 24 Dec 2012.

Heilio, M. (2011). Modelling and the educational challenge in industrial mathematics. In G. Kaiser, W. Blum, R. Borromeo Ferri, & G. Stillman (Eds.), *Trends in teaching and learning mathematical modelling: ICTMA 14* (pp. 479–488). New York: Springer.

Henn, H.-W. (2007). Modelling pedagogy – Overview. In W. Blum, P. L. Galbraith, H.-W. Henn, & M. Niss (Eds.), *Modelling and applications in mathematics education: The 14th ICMI study* (pp. 321–324). New York: Springer.

Houston, K., & Neill, N. (2003). Assessing modelling skills. In S. J. Lamon, W. A. Parker, & S. K. Houston (Eds.), *Mathematical modelling: A way of life ICTMA11* (pp. 155–164). Chichester: Horwood.

Kaiser, G., & Maaß, K. (2007). Modelling in lower secondary mathematics classroom—Problems and opportunities. In W. Blum, P. L. Galbraith, H.-W. Henn, & M. Niss (Eds.), *Modelling and applications in mathematics education: The 14th ICMI study* (pp. 99–108). New York: Springer.

Kaiser, G., & Sriraman, B. (2006). A Global survey of international perspectives on modelling in mathematics education. *Zentralblatt für Didaktik der Mathematik (ZDM), 38*(3), 302–310.

Kaiser, G., Blomhøj, M., & Sriraman, B. (Eds.). (2006). Mathematical modelling and applications: Empirical and theoretical perspectives [Special Issue]. *Zentralblatt für Didaktik der Mathematik (ZDM), 38*(2), 82–85.

Kaiser, G., Blum, W., Borromeo Ferri, R., & Stillman, G. (Eds.). (2011). *Trends in teaching and learning mathematical modelling: ICTMA 14* (pp. 479–488). New York: Springer.

Kawasaki, T., & Morija, S. (2011). Using modelling experiences to develop Japanese senior high school students' awareness of the interactions between mathematics and science. In G. Kaiser, W. Blum, R. Borromeo Ferri, & G. Stillman (Eds.), *Trends in the teaching and learning of mathematical modelling* (pp. 603–615). New York: Springer.

Kuntze, S. (2011). In-service and prospective teachers' views about modelling tasks in the mathematics classroom: Results of a quantitative empirical study. In G. Kaiser, W. Blum, R. Borromeo Ferri, & G. Stillman (Eds.), *Trends in the teaching and learning of mathematical modelling* (pp. 279–288). New York: Springer.

Leiß, D. (2007). *Hilf mir es selbst zu tun*. Hildesheim: Franzbecker.

Lesh, R., Galbraith, P. L., Haines, C. R., & Hurford, A. (Eds.). (2010). *Modeling students' mathematical modelling competencies: ICTMA13*. New York: Springer.

Niss, M. (2001). Issues and problems of research on the teaching and learning of applications and modelling. In J. F. Matos, W. Blum, S. K. Houston, & S. P. Carreira (Eds.), *Modelling and mathematics education: ICTMA 9* (pp. 72–88). Chichester: Horwood.

Niss, M., Blum, W., & Galbraith, P. (2007). Introduction. In W. Blum, P. L. Galbraith, H.-W. Henn, & M. Niss (Eds.), *Modelling and applications in mathematics education: The 14th ICMI study* (pp. 3–32). New York: Springer.

Perrenet, J., & Adan, I. (2010). The academic merits of modelling in higher mathematics education. *Mathematics Education Research Journal, 22*(2), 121–140.

Polya, G. (1988). *How to solve it: A new aspect of mathematical method* (1st Princeton Science Library ed.). Princeton: Princeton University Press.

Prediger, S., Arzarello, F., Bosch, M., & Lenfant, A. (2008). Comparing, combining, coordinating-networking strategies for connecting theoretical approaches [Editorial]. *ZDM_The International Journal on Mathematics Education, 40*(2), 163–164.

Rodríguez, E., Bosch, M., & Gascón, J. (2008). A networking method to compare theories: Metacognition in problem solving reformulated within the Anthropological Theory of the Didactic. *ZDM_The International Journal on Mathematics Education, 40*(2), 287–301.

Rosa, M., & Orey, D. C. (2009). Symmetrical freedom quilts: The ethnomathematics of ways of communication, liberation, and arts. *Revista Latinoamericana de Etnomatemática, 2*(2), 52–75.

Sfard, A. (2008). *Thinking as communicating: Human development, the growth of discourses, and mathematizing*. Cambridge: Cambridge University Press.

Steen, L. A., & Turner, R. (with Burkhardt, H.). (2007). Developing mathematical literacy. In W. Blum, P. L. Galbraith, H.-W. Henn, & M. Niss (Eds.), *Modelling and applications in mathematics education: The 14th ICMI study* (pp. 285–294). New York: Springer.

Stillman, G. (in press). Problem finding and problem posing for mathematical modelling. In K.E.D. Ng & N. H. Lee (Eds.), *Mathematical modelling_From theory to practice*. Singapore: World Scientific.

Stillman, G., Brown, J., & Galbraith, P. (Eds.). (2010). Applications and mathematical modelling in mathematics learning and teaching [Special Issue]. *Mathematics Education Research Journal, 22*(2), 1–6.

Stillman, G., & Galbraith, P. (2003). Towards constructing a measure of the complexity of application tasks. In S. J. Lamon, W. A. Parkur, & S. K. Houston (Eds.), *Mathematical modelling: A way of life ICTMAII* (pp. 179–188). Chichester: Horwood.

Van Dooren, W., De Bock, D., Hessels, A., Janssens, D., & Verschaffel, L. (2004). Student's over-reliance on proportionality: Evidence from primary school students solving elementary arithmetic problems. In H.-W. Henn & W. Blum (Eds.), *ICMI study 14: Modelling and applications in mathematics education* (Pre-conference Vol, pp. 291–296). Dortmund: University of Dortmund.

Weigand, H.-G., & Bichler, E. (2010). Symbolic calculators in mathematics lessons: The case of calculus. *International Journal for Technology in Mathematics Education, 17*(1), 3–15.

Wiliam, D. (2007). Keeping learning on track: Classroom assessment and the regulation of learning. In F. K. Lester (Ed.), *Second handbook of research on mathematics teaching and learning* (pp. 1051–1098). Charlotte: National Council of Teachers of Mathematics/IAP.

Yackel, E., & Cobb, P. (1996). Sociomathematical norms, argumentation, and autonomy in mathematics. *Journal for Research in Mathematics Education, 27*(4), 458–477.

Part I
Innovative Practices in Modelling Education Research and Teaching

Chapter 2
From Conference to Community: An ICTMA Journey—*The Ken Houston Inaugural Lecture*

Peter Galbraith

Abstract As a community ICTMA draws on its conference tradition as well as developing new directions in research and practice to enhance its mission of promoting the teaching and learning of mathematical modelling and applications at all levels of education. This chapter reflects on aspects of its mission with respect to the integrity of modelling activity, authenticity of its approach to modelling, characteristics aimed to enhance a supportive and collaborative community, and activity within representative research foci. It concludes by identifying avenues for advocacy aimed at assuring a productive and vibrant future.

1 Introduction

In his plenary address at the First International Conference on the Teaching of Mathematical Modelling in 1983, Henry Pollak pointed out that:

> society provides the time for mathematics to be taught in schools, colleges and universities, not because it is beautiful, which it is, or because it provides great training for the mind, which it does, but because it is so useful. (Pollak 1984, p. xv)

This conviction continues to be reinforced internationally within official documents that set specific educational goals for the learning of mathematics – as in the following:

> Mathematical literacy is defined in PISA as the capacity to identify, understand and engage in mathematics, and to make well-founded judgements about the role that mathematics plays in an individual's current and future private life, occupational life, social life with peers and relatives, and life as a constructive, concerned and reflective citizen. (OECD 2001, p. 22)

P. Galbraith (✉)
School of Education, The University of Queensland, St Lucia, QLD, Australia
e-mail: p.galbraith@uq.edu.au

G.A. Stillman et al. (eds.), *Teaching Mathematical Modelling: Connecting to Research and Practice*, International Perspectives on the Teaching and Learning of Mathematical Modelling, DOI 10.1007/978-94-007-6540-5_2,

> Mathematics aims to ensure that students are confident, creative users and communicators of mathematics, able to investigate, represent and interpret situations in their personal and work lives and as active citizens. (Australian Curriculum Assessment and Reporting Authority 2010, p. 1)

So it is that we find ourselves carrying this torch, at times through conference participation, but mostly as members of small groups of colleagues, or as individuals. It is appropriate then, to pause from time to time to reflect that the meaning of the 'C' in ICTMA encompasses both Conference and Community, and the progress, challenges and implications that this metamorphosis implies.

Ken Houston, whose name is associated with this presentation, is a former ICTMA President, one of the founding members of the ICTMA community, and the main author of the history of ICTMA: See Houston et al. (2008). This chapter will first draw selectively from that history to illustrate how successive conferences fashioned the agenda that now characterises the community. It will then proceed to reflect on priorities, issues and foci, that stand to challenge the ICTMA community as we move forward.

2 Beginnings

Rumblings in the United Kingdom that led to the forerunner of ICTMA might be traced to the McLone Report (McLone 1973) which described mathematics graduates as:

> Good at solving problems, not so good at formulating them, having a reasonable knowledge of mathematical literature and technique; some ingenuity and capable of seeking out further knowledge. However…not particularly good at planning work, nor at making a critical evaluation of it when completed; and in any event he has … apparently little idea of how to communicate it to others (p. 33)

An early seminal influence was injected by David Burghes, arguably the Father of ICTMA, and organiser of the first two conferences in 1983 and 1985, who sought to enliven the school mathematics curriculum by working with teachers to produce interesting modelling investigations for students at secondary level. In 1978 he began the *Journal of Mathematical Modelling for Teachers*, which in 1981 metamorphosed into *Teaching Mathematics and its Applications* under which name it continues today.

Whilst the tradition of ICTMA conferences may be fairly associated with events in the UK, it should be emphasised that not all initiatives in modelling related activity in education emerged from there, with parallel initiatives emerging at ICME Congresses and elsewhere. At ICME-3 in 1976, Henry Pollak brought applications and modelling to the fore through his lecture on "The Interaction between Mathematics and Other School Subjects" (Pollak 1979). In 1968 Hans Freudenthal began a conference with the theme of "*How* to teach mathematics so as to be useful?" with an opening address "*Why* to teach mathematics so as to be useful?" (Freudenthal 1968). Thus the scene was set in many ways.

3 The Conferences

It is not intended to elaborate at length on the 'C' in ICTMA as representing conference, for the aforementioned comprehensive history is available on the website. Rather it is significant to observe changing trends and emphases, but also issues from earlier days which, together with their relatives and descendants, continue to engage the community. We note the changing emphasis within ICTMA conferences, which began in 1983 with most presentations addressing issues in higher education as a consequence of delegates drawn heavily from UK Polytechnics, many of whom had experience of modelling in industry. A greater emphasis on secondary education, together with the inclusion of 'applications' in the conference title, followed from the increasing presence of delegates from continental Europe, and later the full spectrum of educational levels became represented in conference programs. Among conference themes from the early years that continue to characterise and challenge the ICTMA community today we find:

- A practice of renewing the energy and authenticity in modelling activity by involving practising modellers from within and outside education as plenary speakers.
- Descriptions of innovative attempts to change what were seen as conservative national curricula, and assessment practices, that appeared first at ICTM2 in 1985.
- Identification of *model formulation* as a crucial skill that students do not do very well, and the associated proposing of new ideas in *formulating models.*
- Using modelling as an agent (e.g., via "Peoples' mathematics") for emancipatory purposes – modelling within mathematics education as not just about mathematical empowerment but also about political and intellectual empowerment (Julie 1993).
- The emergence of *technology* as a major theme as evidenced by its appearance in the Proceedings title of ICTMA 8, later to be augmented with consideration of the rapidly changing mix of *techno-mathematical literacies.*
- A widened international participation to embrace every continent except Antarctica, and in particular connection with colleagues from China, where both secondary and tertiary institutions are teaching mathematical modelling which is increasingly featured in secondary and tertiary institutions.

While not a conference publication, a paper reviewing the then situation regarding the place of applications and modelling in mathematics education was published following the fourth conference by two ICTMA members who remain active today (Blum and Niss 1991).

4 ICTMA as Community

So how does ICTMA continue to serve its purpose, as it evolves within and from the traditions established by its history, including its conference legacy? We may approach such an analysis usefully from the viewpoint of ICTMA as a community of practice (Lave and Wenger 1991). The defining characteristics of a community of practice are summarised in (Wenger 2006): "Communities of practice are groups of people who share a concern or a passion for something they do, and learn how to do it better as they interact regularly."

In his terms three characteristics are central to a community's structure and purpose.

The Domain: Identity is defined by a shared domain of interest, and therefore commitment to a shared competence that is a distinguishing feature of the group.

The Community: To pursue interests in their domain, members engage in joint activities and discussions, help each other, share information, and build relationships that enable them to learn from each other.

The Practice: Members of a community of practice are practitioners – they develop a shared repertoire of resources, experiences, stories, insights, and ways of addressing characteristic problems that arise in their domain.

The *community* is constituted by these components in combination, and is cultivated by developing them in parallel.

5 The Domain

Identity is defined by a shared domain of interest, and therefore commitment to a shared competence that is a distinguishing feature of the group. It is too simplistic to just say that our domain of interest is in the teaching and learning of mathematical modelling and applications, for that is a broad church that harbours too many ambiguities if left unqualified. Seven (at least) approaches to the use of mathematics with connections to the real world can be identified within mathematics education literature, that claim links to modelling in some way, and these vary from changes in emphasis to different genres. Only indicative identifying references can be afforded here.

1. *Using contextualised examples to motivate the study of mathematics* (Pierce and Stacey 2006);
2. *Emergent Modelling* (Doorman and Gravemeijer 2009; Gravemeijer 2007);
3. *Modelling as curve fitting* (Riede 2003);
4. *Word problems that use practical settings* (Verschaffel et al. 2010)
5. *Modelling as a vehicle for teaching other mathematical material* (Zbiek and Connor 2006);
6. *Ethnomathematics and street mathematics* (Barbosa 2006; D'Ambrosio 1985).

All of these, with possibly one exception, are equipped to make significant contributions to the teaching of mathematics, and would be at home in any conference with the teaching of applications as a focus. For various reasons, if taken alone, they all lack one, or more, of the aspects necessary for inclusion if the full power of mathematical modelling as a source of real world problem solving expertise in the

ICTMA tradition, and individual empowerment is to be realised. This is not *per se* a criticism for they set out to do different things. The following additional category is the one that most uniquely represents the ICTMA domain of practice.

5.1 Modelling as Real World Problem Solving

This approach to modelling has particular relevance when the legacy and continued work of ICTMA is being represented. It differs in some important respects from those mentioned so far, notably because its origins lie in substantial part with those who have used mathematics to model problems in professional fields such as science and industry, as well as for addressing problems of community or personal interest. Some, including Henry Pollak, and a variety of early ICTMA contributors, took their experience and insights directly across into modelling initiatives in education. Others, such as Pedley (2005) through his Presidential address to the Institute of Mathematics and its Applications, have incidentally provided continuing support, by promoting this concept of modelling in public forums. The example below has been distilled from a problem discussed in his address and published in the given source. It is included here as representative of the approach to modelling that has characterised the domain of the ICTMA community.

5.1.1 New Mexico Atomic Test

Professor Geoffrey Taylor's analysis of the 1945 atomic bomb test in New Mexico, in which he estimated the energy released in the blast, followed the publication in *Life* magazine in 1947, of photos of the expanding blast wave (Fig. 2.1), taken over

Fig. 2.1 Blast wave photo (Source: http://en.wikipedia.org/wiki/Nuclear_weapon_yield)

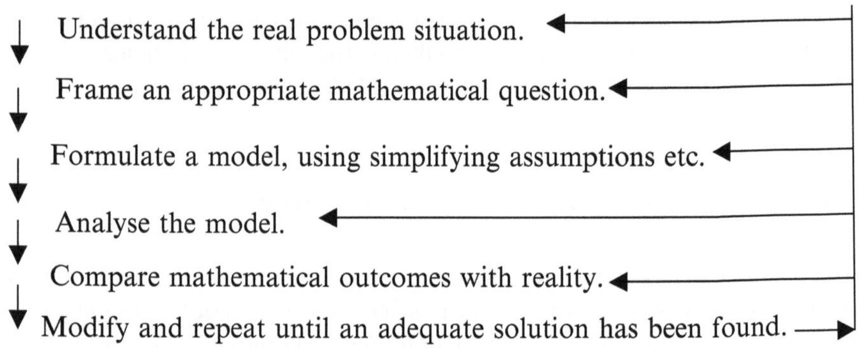

Fig. 2.2 Modelling process (after Pedley 2005)

a succession of small time intervals. The analysis requires no more than senior school mathematics, but is founded on an ability to formulate a mathematical problem from a real life situation.

5.1.2 Reflection on Modelling Issues

As an example of a modelling problem it illustrates the power that modelling competence provides – an empowerment provided by the ability to apply mathematical knowledge to address real world problems. Characteristic of the approach is a cyclical modelling process, typically shown as a chart – the version in Fig. 2.2 is inferred from Pedley's presentation.

The arrows on the right indicate that iterative back tracking may occur between any phases of the modelling cycle as required. This is a compact version of the modelling chart that over the years has appeared in various forms in many sources. Modelling diagrams serve several purposes. In their simplest form (as here) they identify basic stages in the modelling process and provide mental scaffolding for a solution path. In more elaborated forms (see later) they also act as research or design tools. Modelling in this tradition has two purposes: to solve a particular problem at hand, but over time to develop modelling skills, that individuals can apply to problems in their world. This focus has typified the ICTMA culture from the earliest conferences and remains a central interpretation of the modelling process within the community (e.g., Blum and Leiß 2007; Galbraith and Stillman 2006; Niss et al. 2007).

5.2 Authenticity

Together with a commitment to modelling as real world problem solving, 'authenticity' has loomed large as a *domain* issue. A section in the ICMI Study 14 proceedings (Blum et al. 2007) devoted to the theme of authenticity underlines the importance with which it is viewed within the modelling community. It is suggested here that one of the reasons creating potential confusion, is that the sense in which authenticity has been considered is too global. Here it is proposed that authenticity be viewed in terms of four dimensions.

1. Content authenticity
2. Process authenticity
3. Situation authenticity
4. Product authenticity

5.2.1 Content Authenticity

Content authenticity has two aspects. Firstly a problem itself needs to satisfy realistic criteria (involve genuine real world connections), and secondly the individuals addressing it need to possess mathematical knowledge sufficient to support a viable solution attempt. With respect to the latter (Jablonka 1996) reviewed modelling examples from a range of teaching materials, and identified models from a variety of fields, with which students lacked familiarity. She reasonably challenged the value of models if students do not know the techniques needed for their solution. Jablonka thus identifies two issues of importance, one to do with the use of models that are 'black boxes' for students – the other implications of models requiring out-of-reach mathematics. Modelling attempted with attributes like these would certainly fail a test of 'content authenticity'.

In searching for viable problems, those with genuine real world connections con sist broadly of two genres:

Targeted real problems: Problems set with a specific goal. Example – decide the optimum number and spacing of speed bumps along the new stretch of road in front of the College.

Life like problems: The context is real, but there is freedom in choosing the precise problem to be addressed. For example an intention might be to investigate the differential effects on families from proposed changes in a taxation system. A number of possible modelling problems might be developed such as: What are the relative benefits for families in different income brackets *or* what will be the effect on families with school age children *or* how will the new provisions impact on couples versus individuals? Recognition of cultural and social issues and the significance of personal values are central factors here.

5.2.2 Process Authenticity

Process authenticity refers to the conduct of a modelling process that results in solutions that are defensible and robust in terms of the outcomes sought. Kaiser et al. (2006, p. 83) remind us that while variations exist "the important thing is the commonly accepted idea about a general mathematical modelling process." Why is this important? Because it provides an agreed and tested basis from which to apply the process of modelling, teach the process of modelling, and analyse and improve the outcomes of modelling. As a community we need to have confidence in such a process, that can (in a Popperian sense) survive tests of integrity and power that can be asked of it.

Ärlebäck (2009), when introducing modelling to upper secondary students, indicated he found no evidence of the cyclic process widely described in modelling research as central to modelling as real world problem solving. However, earlier in the paper sub-processes that characterised the students' work were identified as follows: reading, making a model (structuring and mathematising), calculating, validating, and writing. All of these are essential components of the modelling process, and validating cannot occur without reviewing a solution in terms of the original problem statement (reading). This alone completes a cycle, without considering further cycles introduced through the checking and reviewing that inevitably takes a solver back through earlier phases in producing a defensible solution. One possibility is that our use of the term 'modelling cycle' may not be as unambiguous as we have assumed, and there may be confusion at times with student activity on modelling problems which is anything but smoothly cyclic. Any such confusion stands to compromise efforts to make authenticity more transparent.

Other difficulties emerge if we rely on assumptions that are not applicable to support the range of conclusions drawn (e.g., Jablonka and Gellert 2007, p. 5):

> The symbolic technology at hand, as well as measuring devices, and domain specific constraints and theorems, all influence the form the real world model will take. So the real world model is by no means a description of the real world problem.

The first sentence is obviously true, and any consideration of process authenticity acknowledges this. On the other hand, the second statement is not a logical consequence, and appropriate evaluation procedures, integral to the modelling process, will provide criteria by which the authenticity of the ensuing models as 'descriptions' of the problem will be judged.

A generic but adaptable approach that empowers individuals as problem solvers stands to benefit users across different areas of content and purpose. This is the position that has enabled the teaching of modelling to benefit from contributions from modellers in many fields of application. It was influential at the birth of the community, continues to be valued, and provides a strong argument why modelling as real world problem solving can never be absorbed entirely into systems that value only prescribed curricular mathematical content.

5.2.3 Situation Authenticity

This critical dimension brings conditions necessary for a valid modelling exercise into direct contact with the workplace, classroom, or other environment within which the modelling enterprise is conducted. The importance of 'situation' is brought home by comments such as Sfard (2008) who claimed that, the minute an 'out of school' problem is treated in school it is no longer an 'out of school problem', and hence the search for authentic real world problems is necessarily in vain. What this does is to make the conception of what it means to be "in school", or "out of school" the definitive construct, by privileging a particular conception of what school mathematics is about, and what mathematics teaching and classrooms are allowed to be – then requiring that modelling fit the stereotype, be subject to associated practices, and hence compromise its integrity. In a similar vein, Jablonka and Gellert (2007, p. 5) argue that in classrooms there is no validation, because the result is not put back into a "real" real situation and that "since within a classroom activity the results are never put into operation, there is no real problem of validation". These are generalisations that ascribe to the field of modelling, weaknesses that may exist in particular implementations – the latter are important to document, but to imply their universal application to classrooms is both inaccurate and unfortunate.

By contrast, what modelling properly conducted can do, is to challenge some of those norms, assumptions, and stereotypes – mathematical, situational, and pedagogical. The essential characteristic, for situational authenticity, is that the requirements of the *modelling task* drive the problem solving process, and for this purpose carry greater authority than beliefs or traditional teaching practices should these compromise the goal. Typically, modelling activity will take place both inside and outside classrooms. Osawa (2002), for example, describes a project in which the goal was to optimise the baton changing practice of relay teams. Activity took place alternately in a classroom, and on the running track which acted as the laboratory within which results were tested. Neither place alone would have sufficed to carry out the complementary theoretical and experimental activities that successive improvements demanded. Equally striking are the actions of individuals who apply their school learning in modelling to address problems particular to them. Examples include the successful modelling of a 12 year old girl to convince her parents that she could both care for and provide for a much wanted pony, and a mature age student who used the cyclic modelling process to redesign the culture he used for growing tomatoes hydroponically.

A recent paper (Jablonka and Gellert 2011) contains several other assertions that require qualification; such as modelling in mathematics education aims at creating a flexible workforce; that modelling conceptions do not see associated competencies as 'culture bound and value driven'; that contextuality of all knowledge is (mis)interpreted in a way that leads to the contention that mathematical concepts can be meaningfully learned only within a 'real life' context. As generalisations the assertions are collectively and specifically refuted within sources such as the following (Niss et al. 2007; Stillman et al. 2008, 2010). When suitably qualified such criticisms become productive warnings against extremism in the parts of a community, but without qualification they suggest a broad fanaticism with which the membership of ICTMA would not be comfortable.

5.2.4 Product Authenticity

As one of the most superficially obvious ideas, product authenticity is both an important and an elusive concept. It is elusive, because it is not always clear when an appropriate 'product' has been achieved. When is a problem solution good enough to warrant the effort expended given that money, and/or time has run out? In the classroom, while money may not be a tangible constraint, time most certainly is. Hence assessing product authenticity involves asking how well an end-point achieved by modelling informs the question asked. So assessing product authenticity involves looking at mathematical outcomes in two ways. Firstly to check that there are no obvious mathematical anomalies which have not been addressed, and secondly to verify that mathematical outcomes have been appropriately absorbed into implications for the real world problem being addressed. If modelling has been curtailed, there needs to be an appraisal of the insights generated, and an indication of what aspects of the problem remain unresolved. This will apply irrespective of whether a project has been curtailed by cost, time, or other factors such as loss of key personnel.

In summary continuing to develop the characteristics of mathematical modelling as real world problem solving, and addressing associated matters of authenticity, continue to occupy a central place in the domain of the ICTMA community.

6 The Community

To pursue interests in their domain, members engage in joint activities and discussions, help each other, share information, and build relationships that enable them to learn from each other.

What does it take to be a *learning community*, which will enable and enhance the interests and activities of its members who want to do more than 'just belong'? The social nature of 'community' suggests Vygotskian lines of thought, and in consequence points to possibilities provided by expanded notions of the zone of proximal development (ZPD).

6.1 The ZPD as Scaffolding

The most widely known definition of the ZPD as the 'distance' between what a learner can achieve alone, and with the assistance of a more advanced teacher or mentor, places the teacher in a pivotal classroom role in supporting students to become more self-regulating participants in learning. In modelling courses some physical embodiment of the modelling process (e.g., a diagram) has been a typical scaffolding prop, which becomes redundant as students internalise the process, and become independently proficient through collaborative practice (Galbraith and Clatworthy 1990).

6.2 The ZPD in Egalitarian Partnerships – Distributed Complementary Competence

Deriving originally from observations of collaborative children's play, this view of the ZPD involves equal status relationships, developing the learning potential in peer groups where students have incomplete but relatively equal expertise – each partner possessing some knowledge but requiring the others' contribution in order to make progress. Compared to the expert-novice situation, the co-production of the task typically involves contest and trial-and-error as the partners begin to appreciate the perspectives of others and coordinate their incomplete competence (Forman and McPhail 1993). Such understanding provides theoretical underpinnings for the enhancement of team problem solving through collaboration – a characteristic of modelling activity.

6.3 The ZPD – In Practitioner- Researcher Relationships

This view applies the concept of a ZPD to whole groups, where participants with partially overlapping ZPDs provide a changing mix of levels of expertise, so enabling many different productive partnerships to be orchestrated (Brown et al. 1993). In professional collaboration, overlapping individual ZPDs can create a combined ZPD which promotes a higher vision of possibilities than either separately could provide. Here partnerships are located in the community itself, where the participants are professionally linked, typically as researchers and/or practitioners. A motivation behind the theorising of this construct has been the experience of the author stimulated, by contributions within teacher – researcher partnerships. Increased possibilities for research were generated through equal-status collaboration, using complementary expertise, which neither alone could have achieved.

Within the ICTMA community of practice both practitioners and researchers, experienced and inexperienced, are represented. How can such a community, use the diverse expertise of its members, to generate a group ZPD that inspires visions that none of us alone could produce? This I think is the spirit intended by Wenger in writing his summary of community.

7 The Practice

Members of a community of practice are practitioners – they develop a shared repertoire of resources, experiences, stories, insights, and ways of addressing characteristic problems that arise in their domain.

The significance of *resources*, *experiences* and *stories* comes to mind when we consider the legacy from ICTMA conferences, where initially for example, many of the talks were of the war story variety: "This is how we teach modelling" or "This is our curriculum," and many were useful descriptions of models that the author(s) had used in their teaching. As Wenger's conception emphasises, there is an honoured place for anecdotes and stories, as those in a field, new and experienced, seek and share ideas, trade experiences, and ask and provide advice from and to each other. This is a highly legitimate activity, necessarily augmented by others that aim to deepen *insight* into the components of the domain of interest of the community, and how its practice is conducted – more usually called research.

7.1 Research in the Domain of Practice

A comprehensive appraisal of research challenges in the field of applications and modelling is contained in Niss (2001). In this section we consider some selective foci that characterise the ICTMA culture, with directions influenced by the author's own activities and priorities. Figure 2.3, after Galbraith and Stillman (2006), will be used to locate some of these directions. The diagram is designed to support research description, analysis, and discussion, rather than serve a teaching purpose – although it does contain an embedded version of the traditional simpler modelling cycle – the stages A to G (with the transitions 1–7), linked clockwise by the heavy single headed arrows.

For present purposes Fig. 2.3 serves to define key foci for research with respect to individuals learning mathematical modelling, and pressure points for those teaching within the field, for example, the kinds of mental activity that individuals engage in as modellers attempting to make the transition from one modelling stage to the next, and which provide key foci for research, are given by the broad descriptors of cognitive activity, (boxed) 1–7 in Fig. 2.3. The light double-headed arrows emphasise that thinking within the modelling process is far from linear (e.g., Borromeo Ferri 2006),

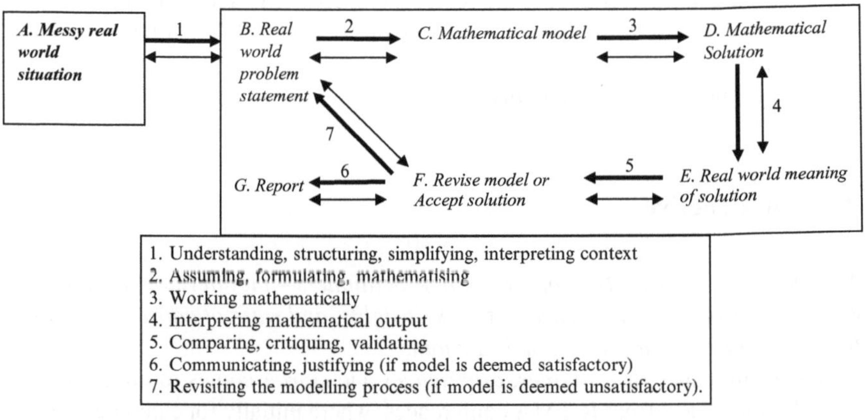

Fig. 2.3 Modelling process (after Galbraith and Stillman 2006)

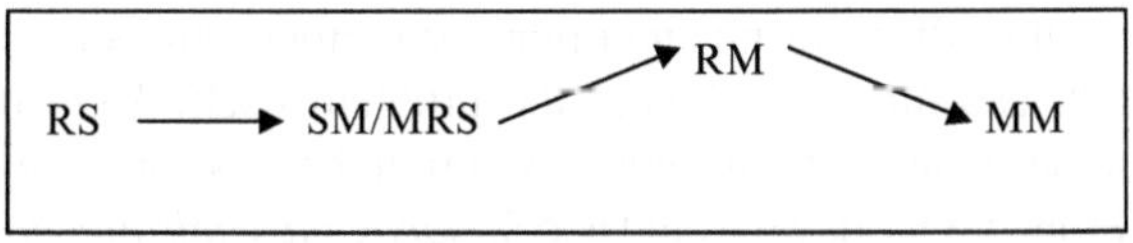

Fig. 2.4 Sample mental infrastructure (after Borromeo Ferri 2006)

and indicate the presence of reflective metacognitive activity as articulated by many researchers (e.g., Maaß 2007). Such reflective activity can look both forwards and backwards with respect to stages in the modelling process. Thus while the stages of problem solution follow a cyclic path, the route taken by an individual modeller is usually anything but smoothly cyclic.

7.1.1 Issues in Formulating Mathematical Models (Modelling in Stages A–C)

As history tells us, challenges in this area have engaged ICTMA participants from the earliest days. Its presence as a *prime focus* separates modelling as real world problem solving, from other educational approaches that use the term 'modelling', and it is both important and exciting that current work is directed within this area. Some of the most systematic research is being conducted by Rita Borromeo Ferri, as reported in the article referenced above. This approach adds formal mental infrastructure (see Fig. 2.4), in terms of a situation model (SM), its real representation (MRS), and a real model (RM), between the real situation (A in Fig. 2.3), and the mathematical model (C in Fig. 2.3).

So we have the question whether a 'real model' represents a formal stage in the modelling process that should be shown in its structure, or whether it is a useful heuristic device that provides substantive help to modellers in moving from a real world problem statement to a mathematical model. What can be agreed is the importance of scaffolding this demanding transition in the modelling process, and the value of researching and applying any means that enhances the ability to achieve this end, whichever of the positions we currently favour.

7.1.2 The Role of Metacognitive Activity

Galbraith and Stillman (2006) developed an approach in which Fig. 2.3 was used as a template to assist in the identification and release of blockages encountered by students in moving between stages of the modelling process. Central to this focus, is the way in which metacognitive activity can enhance or retard a modelling venture. Goos (2002) identified three generic types of situation (*red flag situations*) that should elicit metacognitive monitoring within any stage of a problem solving process – labelled respectively (a) *lack of progress*, (b) *error detection*, and (c) *anomalous results*. She identified three prevalent forms of metacognitive failure. *Metacognitive blindness* occurs when no required action is recognised; *Metacognitive vandalism* occurs when drastic and often destructive actions not only fail to address the issue, but also alter (invalidate) the problem itself. *Metacognitive mirage* occurs when unnecessary actions derail a solution, because an 'imaginary' difficulty is identified. From our work we would add *Metacognitive misdirection* as an inappropriate response that represents inadequacy rather than vandalism; and *Metacognitive impasse* that occurs when no amount of reflective thinking or strategic effort is successful in releasing a blockage. Much more remains to be done in learning to identify and release blockages that inhibit the ability of students in becoming effective modellers.

Furthermore, given that metacognitive activity is located heavily at the transitions indicated in Fig. 2.3, how pedagogy addresses the fostering of metacognitive competencies is crucial to the goal of producing students who are consistently able modellers. The notion of *meta-metacognition,* as discussed for example in Stillman (2011), has come to influence our thinking in this respect. In considering whether given metacognitive activity on the part of students is appropriate, or being properly conducted, a teacher is reflecting on metacognitive activity itself – that is undertaking mental activity that is *meta–metacognitive* in nature. At the macro level how a teacher acts on such reflection will be crucial to the way mathematical modelling is nurtured or stifled in their classroom. At the micro level the capacity of students to make transitions between phases in the modelling cycle, and to release blockages, depends critically upon how they are facilitated in applying the modelling process, and metacognitive strategies central to it. "What should this student be asking her/himself at this point in the modelling process?" is a *meta-metacognitive* self-prompt, that more is required than a suggestion about how to progress past a problem-specific obstacle. It is centrally to do with nurturing effective problem solvers by providing a modelling culture.

7.1.3 Other Research Themes

Here we briefly note some other contemporary research themes that have characterised activity over time within the ICTMA community. Originally *assumptions* were largely considered to inhabit the process of setting up a mathematical model in the first place. While they continue to play a major role in formulation, there is now realisation that they permeate the whole of the modelling process, and are not something to be merely ticked off in an early phase. Galbraith and Stillman (2001), for example, identified three different classes of assumption; those associated with *model formulation*; those associated with *mathematical processes;* and those associated with *strategic choices in the solution process.*

A second area, in which reflection on the modelling process has resulted in a changing understanding of its role over time, is the use of technology in modelling. Geiger et al. (2010) point out that technology related activity takes place during all phases of the modelling cycle, rather than as previously theorised, only at the solving stage. This is relevant not only within the mathematical dimensions of a modelling task, but for the contextual settings within which a problem is addressed, such as collaborative teamwork. It is important that we continue to identify and research aspects once compartmentalised, that represent fluid and dynamic influences across stages of the modelling process.

Other significant continuing research foci with links to ICTMA values and traditions include the identification and assessment of modelling competencies (e.g., Haines and Crouch 2007; Houston and Neill 2003); and the development and researching of modelling programs at all levels of education – undergraduate, primary, and secondary, including teacher education (e.g., Alpers 2011; English 2011; Maaß 2007). We would also acknowledge the parallel work of modelling colleagues in developing and disseminating theoretical and practical dimensions of modelling programs through the medium of model eliciting activities (e.g., Lesh and Doerr 2003), and the development of challenging real word modelling problems through the UMAP and COMAP programs (Garfunkel 2004).

Finally, the classification of modelling perspectives (Kaiser and Sriraman 2006) summarises different purposes adopted within educational settings. In identifying alternative research foci it incidentally draws attention to important distinctions between different 'models of modelling', and alternative perspectives and emphases that they serve. Perspectives represent the interests of those engaging in the modelling activities. For example *socio-critical purposes* (Barbosa 2006) involve making certain assumptions (rather than others) regarding which mathematical questions should emerge from a situational context, and how they are treated. It does not require a separate modelling genre, as has been exemplified through the work of Cyril Julie in his work with students and teachers in South African townships (see: Julie 1993; Julie and Mudaly 2007). Research in all areas identified in the paper continues to engage members of the ICTMA community.

8 Addressing Issues in the Domain and Community – Advocacy

As Wenger (2006) reminds us, members of a community of practice need to develop ways of: "addressing characteristic problems that arise in their domain". One issue confronting a community like ICTMA is how best to promote the growth and impact of the community – what kinds of *advocacy* should we engage in?

As a community ICTMA is committed to enhancing the impact that applications and modelling can make in equipping individuals to use mathematical knowledge in personal, work, and civic contexts. However, in publicly endorsing this goal for mathematics learning do education authorities pay more than lip service to the ideal? What specific provisions ensure that the implemented curriculum contains the time and the resources to make such goals attainable? It would be of great benefit if ICTMA, as an affiliated study group of ICMI, could provide members with support when curriculum statements and intentions open the door to our field of interest in their particular region. Statements representing ICTMA's best advice would be welcome external support for those trying to engender change at local or national levels.

A second important forum is afforded by the Affiliated Study Group time slots at ICME Congresses, where there is opportunity to do two important things. One is to introduce the community to new contacts, for whom a statement of core values, variety of perspectives, central purposes, summary achievements, and future vision is important. The second focus relates to those who already know about ICTMA and its past work. For them an essential message is that the community is on the move, new challenges are being set, new insights achieved, and an increasingly technological and socially challenged world is the context within which problems are being identified and addressed.

9 Conclusion

In concluding it is appropriate to return to ICTMA's roots, and reflect again on how this conference series came to be. We cannot help but note the decline in representation of the group most instrumental in its creation – tertiary mathematics staff with industry experience, concerned with helping students learn to solve real problems. We need to find ways to rejuvenate contact with those, who while primarily involved in solving mathematical problems, have a vital educational interest in helping students to learn these abilities. A strong link between the domains of mathematics and mathematics education needs to be maintained, in order that each can inform the other, and that together we can work to ensure that our community does not become absorbed by either purely mathematical interests, or conservative elements of the education industry.

References

Alpers, B. (2011). The mathematical expertise of mechanical engineers: The case of mechanism design. In G. Kaiser, W. Blum, R. Borromeo Ferri, & G. Stillman (Eds.), *Trends in teaching and learning of mathematical modelling: The 14th ICMI study* (New ICMI studies series, pp. 99–108). New York: Springer.

Ärlebäck, J. (2009). On the use of realistic Fermi problems for introducing mathematical modelling in school. *The Montana Mathematics Enthusiast, 6*(3), 331–364.

Australian Curriculum, Assessment and Reporting Authority. (2010). *Mathematics: Draft consultation version 1.1.0 Australian Curriculum.* Available from www.australiancurriculum.edu.au/Documents/Mathematicscurriculum.pdf. Accessed 30 April 2011.

Barbosa, J. (2006). Mathematical modelling in classrooms: A socio-critical and discursive perspective. *Zentralblatt für Didaktik der Mathematik, 38*(3), 293–301.

Blum, W., & Niss, M. (1991). Applied mathematical problem solving, modelling, applications, and links to other subjects – State, trends and issues in mathematics instruction. *Educational Studies in Mathematics, 22*(1), 37–68.

Blum, W., & Leiß, D. (2007). How do students and teachers deal with modeling problems? In C. Haines, P. Galbraith, W. Blum, & S. Khan (Eds.), *Mathematical modelling (ICTMA 12): Education, engineering and economics* (pp. 222–231). Chichester: Horwood.

Blum, W., Galbraith, P. L., Henn, H.-W., & Niss, M. (Eds.). (2007). *Modelling and applications in mathematics education* (New ICMI studies series No. 10). New York: Springer.

Borromeo Ferri, R. (2006). Theoretical and empirical differentiations of phases in the modeling process. *Zentralblatt für Didaktik der Mathematik, 38*(2), 86–95.

Brown, A. L., Ash, D., Rutherford, M., Nakagawa, K., Gordon, A., & Campione, J. (1993). Distributed expertise in the classroom. In G. Salomon (Ed.), *Distributed cognitions* (pp. 188–228). Cambridge: Cambridge University Press.

D'Ambrosio, U. (1985). Ethnomathematics and its place in the history and pedagogy of mathematics. *For the Learning of Mathematics, 5*, 44–48.

Doorman, L., & Gravemeijer, K. (2009). Emergent modeling: Discrete graphs to support the understanding of change and velocity. *Zentralblatt für Didaktik der Mathematik, 41*, 199–211.

English, L. D. (2011). Modelling with complex data in the primary school. In G. Kaiser, W. Blum, R. Borromeo Ferri, & G. Stillman (Eds.), *Trends in teaching and learning of mathematical modelling: ICTMA 14* (pp. 287–299). New York: Springer.

Forman, E. A., & McPhail, J. (1993). Vygotskian perspective on children's collaborative problem-solving activities. In E. A. Forman, N. Minick, & C. A. Stone (Eds.), *Contexts for learning: Sociocultural dynamics in children's development* (pp. 213–229). New York: Oxford University Press.

Freudenthal, H. (1968). Why to teach mathematics so as to be useful? *Educational Studies in Mathematics, 1*, 3–8.

Galbraith, P. L., & Clatworthy, N. J. (1990). Beyond standard models: Meeting the challenge of modelling. *Educational Studies in Mathematics, 21*(2), 137–163.

Galbraith, P., & Stillman, G. (2001). Assumptions and context: Pursuing their role in modelling activity. In J. Matos, S. Houston, W. Blum, & S. Carreira (Eds.), *Modelling and mathematics education: Applications in science and technology* (pp. 317–327). Chichester: Horwood.

Galbraith, P., & Stillman, G. (2006). A framework for identifying student blockages during transitions in the modelling process. *Zentralblatt für Didaktik der Mathematik, 38*(2), 143–162.

Garfunkel, S. (2004). Research on the teaching and learning of mathematical modelling. In H.-W. Henn & W. Blum (Eds.), *ICMI study 14: Applications and modelling in mathematics education: Pre-conference volume* (pp. 89–96). Kassel: University of Kassel.

Geiger, V., Faragher, R., & Goos, M. (2010). CAS-enabled technologies as 'agents provocateur, in teaching and learning mathematical modelling in secondary classrooms'. *Mathematics Education Research Journal, 22*(2), 48–68.

Goos, M. (2002). Understanding metacognitive failure. *The Journal of Mathematical Behavior, 21*(3), 283–302.

Gravemeijer, K. (2007). Emergent modelling as a precursor to mathematical modelling. In W. Blum, P. Galbraith, M. Niss, & H.-W. Henn (Eds.), *Modelling and applications in mathematics education: The 14th ICMI study* (pp. 137–144). New York: Springer.

Haines, C., & Crouch, R. (2007). Mathematical modelling and applications: Ability and competence frameworks. In W. Blum, P. Galbraith, M. Niss, & H.-W. Henn (Eds.), *Modelling and applications in mathematics education: The 14th ICMI study* (pp. 417–424). New York: Springer.

Houston, S. K., & Neill, N. (2003). Assessding modelling skills. In S. J. Lamon, W. A. Parker, & S. K. Houston (Eds.), *Mathematical modelling: A way of life* (pp. 155–164). Chichester: Horwood.

Houston, K., Galbraith, P., & Kaiser, G. (2008). *ICTMA: The International Community of Teachers of Mathematical Modelling and Applications – The first twenty-five years*. http://www.icmihistory.unito.it/ictma.php. Accessed 25 Sept 2011.

Jablonka, E. (1996). *Meta-Analyse von Zugängen zur mathematischen modellbildung und Konsequenzen für den Unterricht*. Dissertation, Transparent -Verl. Preuss, Berlin.

Jablonka, E., & Gellert, U. (2007). Mathematisation – Demathematisation. In U. Gellert & E. Jablonka (Eds.), *Mathematisation and demathematisation: Social, philosophical and educational ramifications* (pp. 1–18). Rotterdam: Sense Publishers.

Jablonka, E., & Gellert, U. (2011). Equity concerns about mathematical modelling. In B. Atweh, M. Graven, & W. Secada (Eds.), *Mapping equity and quality in mathematics education. Part 2* (pp. 223–236). New York: Springer.

Julie, C. (1993). People's mathematics and the applications of mathematics. In J. de Lange, C. Keitel, I. D. Huntley, & M. Niss (Eds.), *Innovation in maths education by modelling and applications* (pp. 31–40). Chichester: Ellis Horwood.

Julie, C., & Mudaly, V. (2007). Mathematical modelling of social issues in school mathematics in South Africa. In W. Blum, P. Galbraith, M. Niss, & H.-W. Henn (Eds.), *Modelling and applications in mathematics education: The 14th ICMI study* (pp. 503–510). New York: Springer.

Kaiser, G., & Sriraman, B. (2006). A global survey of international perspectives on modeling in mathematics education. *Zentralblatt für Didaktik der Mathematik, 38*(3), 302–310.

Kaiser, G., Blomhoj, M., & Sriraman, B. (2006). Towards a didactical theory for mathematical modeling. *Zentralblatt für Didaktik der Mathematik, 38*(2), 82–85.

Lave, J., & Wenger, E. (1991). *Situated learning: Legitimate peripheral participation*. Cambridge: Cambridge University Press.

Lesh, R., & Doerr, H. M. (Eds.). (2003). *Beyond constructivism: Models and modeling perspectives on mathematics problem solving, learning, and teaching*. Mahwah: Lawrence Erlbaum Associates.

Maaß, K. (2007). Modelling in class: What do we want the students to learn? In C. Haines, P. Galbraith, W. Blum, & S. Khan (Eds.), *Mathematical modelling (ICTMA12): Education, engineering and economics* (pp. 63–78). Chichester: Horwood.

McLone, R. R. (1973). *The training of mathematicians Social Sciences Research Council Report*. London: SSRC.

Niss, M. (2001). Issues and problems of research on the teaching and learning of applications and modelling. In J. F. Matos, W. Blum, S. K. Houston, & S. P. Carreira (Eds.), *Modelling and mathematics education: ICTMA 9: Applications in science and technology* (pp. 72–88). Chichester: Horwood.

Niss, M., Blum, W., & Galbraith, P. (2007). Introduction. In W. Blum, P. Galbraith, M. Niss, & H.-W. Henn (Eds.), *Modelling and applications in mathematics education: The 14th ICMI study* (pp. 3–33). New York: Springer.

OECD. (2001). *Knowledge and skills for life: First results from PISA 2000*. Paris: OECD.

Osawa, H. (2002). Mathematics of a relay – Problem solving in the real world. *Teaching Mathematics and Applications, 21*(2), 85–93.

Pedley, T. J. (2005). Applying mathematics. *Mathematics Today, 41*(3), 79–83.

Pierce, R., & Stacey, K. (2006). Enhancing the image of mathematics by association with simple pleasures from real world contexts. *Zentralblatt für Didaktik der Mathematik, 38*(3), 214–225.

Pollak, H. O. (1979). The interaction between mathematics and other school subjects. In *New trends in mathematics teaching IV* (pp. 232–248). Paris: UNESCO.

Pollak, H. O. (1984). Applications and teaching of mathematics. In J. S. Berry, D. N. Burghes, I. D. Huntley, D. J. D. James, & A. O. Moscardini (Eds.), *Teaching and applying mathematical modelling* (pp. xv–xvi). Chichester: Ellis Horwood.

Riede, A. (2003). Two modelling topics in teacher education and training. In Q.-X. Ye, W. Blum, K. Houston, & Q.-Y. Jiang (Eds.), *Mathematical modelling in education and culture (ICTMA 10)* (pp. 209–222). Chichester: Horwood.

Sfard, A. (2008). *Thinking as communicating: Human development, the growth of discourses, and mathematizing*. Cambridge: Cambridge University Press.

Stillman, G. (2011). Applying metacognitive knowledge and strategies to applications and modelling tasks at secondary school. In G. Kaiser, W. Blum, R. Borromeo Ferri, & G. Stillman (Eds.), *Trends in teaching and learning of mathematical modelling: ICTMA 14* (pp. 165–180). New York: Springer.

Stillman, G., Brown, J., & Galbraith, P. (2008). Research into the teaching and learning of applications and modelling in Australasia. In H. Forgasz, A. Barkatsas, A. Bishop, B. Clarke, S. Keast, W. T. Seah, & P. Sullivan (Eds.), *Research in mathematics education in Australasia 2004–2007* (pp. 141–164). Rotterdam: Sense Publishers.

Stillman, G., Brown, J., & Galbraith, P. (2010). Researching applications and mathematical modelling in mathematics learning and teaching [Editorial]. *Mathematics Education Research Journal, 22*(2), 1–6.

Verschaffel, L., van Dooren, W., Greer, B., & Mukhopadhyay, S. (2010). Reconceptualising word problems as exercises in mathematical modelling. *Journal of Mathematical Didaktics, 31*, 9–29.

Wenger, E. (2006). *Communities of practice: A brief introduction*. http://www.ewenger.com/theory/communities_of_practice_intro.htm. Accessed 25 Sept 2011.

Zbiek, R., & Connor, A. (2006). Beyond motivation: Exploring mathematical modeling as a context for deepening students' understandings of curricular mathematics. *Educational Studies in Mathematics, 63*(1), 89–112.

Chapter 3
Modelling from the Perspective of Commognition – An Emerging Framework

Jonas Bergman Ärlebäck and Peter Frejd

Abstract This chapter explores an emerging framework on mathematical models and modelling using the theoretical perspective of *commognition* to analyse and discern discursive objects in a dialogue between two students engaged in a modelling activity. The results, partly presented as *realization trees*, show a variety of *signifiers* from different discourses coming into play during the modelling, and examples are given of the *activity of negotiation*, which plays an important role in any modelling activity. In addition, it is argued that the framework has potential to bridge different research perspectives on mathematical models and modelling.

1 Introduction

In the research literature in mathematics education, there are many different views taken on, or theoretical perspectives adapted for, mathematical modelling (Frejd 2010; Garcia et al. 2006; Jablonka and Gellert 2007; Kaiser et al. 2011; Kaiser and Sriraman 2006). This plurality is natural considering the different social and cultural realities in which research is being carried out with different objectives and in different traditions. This diversity of perspectives brings a variety of relevant aspects of learning and teaching mathematical models and modelling to the fore, and together they contribute to a rich and many facetted picture of mathematical models and modelling as well as of mathematics in general. However, it also renders the obvious, but easily overlooked risk, that unless interlocutors engaged in discussion about educational aspects of mathematical models and modelling are very explicit

J.B. Ärlebäck (✉) • P. Frejd
Department of Mathematics, Linköping University, Linköping, Sweden
e-mail: jonas.bergman.arleback@liu.se; peter.frejd@liu.se

G.A. Stillman et al. (eds.), *Teaching Mathematical Modelling: Connecting to Research and Practice*, International Perspectives on the Teaching and Learning of Mathematical Modelling, DOI 10.1007/978-94-007-6540-5_3,

and clear about their theoretical underpinnings and intentions, misunderstandings or conversational breakdowns are likely to happen. This happens simply because the parties are talking about different things.

This latter point, considered more generally and in a broader context, is one of the incentives of Sfard (2008) for developing the framework of *commognition*, and it seems a promising and rewarding task to try to formulate a commognitive perspective of mathematical models and modelling in at least two respects. Firstly, and related to the issue just mentioned, commognition presents a *meta-perspective* bridging, contrasting and synthesising the different theories about modelling in the mathematics education community. Secondly, a conceptualisation of mathematical models and modelling in terms of commognition would provide a framework for analysing inter- and intrapersonal communication complementary to the present prevailing perspectives, attending to both the social and cognitive dimensions of modelling. The aim of this chapter is to start exploring mathematical models and modelling from a commoginitive perspective with this latter focus.

2 Theoretical Framework

Commognition is a theoretical framework developed by Sfard (2008) that focuses on both social and individual aspects of thinking and learning; it merges and combines tenets from theories of communication and cognition. The framework uses a set of principles and notions that discuss thinking as a special type of interpersonal communication and learning as a change in discourses. One of the basic principles of commognition is that "discourses permeate and shape all human activities, [and] the change in discourse goes hand in hand with the change in all other human doings" (p. 118). In this framework *discourse* is defined as a "special type of communication made distinct by its repertoire of admissible actions and the way these actions are paired with re-actions; every discourse defines its own community of discourse; discourses in language are distinguishable by their vocabularies, visual mediators, routines, and endorsed narratives" (p. 297). In other words, a discourse is characterised by the meaning and use of language (in a general sense including written and spoken words, symbols, figures, graphs, etc.), not only including consensus on the interpretation of language, but also established conventions for communication and interaction between members of the discourse as well as formalised ways on how to determine what is regarded as 'true' within the discourse. Hence, a discourse functions as delimiter in that it includes or excludes persons from a given discourse. Commonly figuring discourses when a commognitive approach is employed are *colloquial discourses* (or *everyday discourses*) permeated by personal experiences; the complementary *literate discourses* in which communication often is characterised by the use of specialised symbolic artefacts; and, *classroom discourses* capturing school norms and rules (Sfard 2008).

According to Sfard (2008) *discursive objects* arise in a recursive manner in a given discourse to enhance the effectiveness of communication through one or a combination of the three processes of *saming*, *encapsulating*, and *reifying*. Briefly, the act of saming means assigning one name for a collection of different objects (concrete as well as abstract); encapsulating refers to starting to talk about things in singular, although these previously have been considered in plural; and reification is about objectifying so that talk about processes is replaced with talk about objects. The *activity of negotiations*, which can be seen as a recursive pattern of *conjecture-test-evaluation* carried out by the discourse participants, provides a notion for framing these processes in more detail. A discursive object is manifested through its *signifier* and its *realization tree*, where the signifier can be, for example, a word or an algebraic symbol, and the realization tree is the tree-like structure (in the sense of graph theory) of successive meaningful realizations of the signifier. A realization tree can also productively be thought of as a connected graph, possibly containing loops, to stress the "symmetric nature of many signifier-realization relations" (Sfard 2012). Due to the recursive nature inherent in the production of discursive objects, there is often a dual relation between a signifier and its realization making the two notions in different contexts interchangeable.

In terms of commognition, one can interpret a mathematical model as a *discursive object* in a *subsumed discourse*, which means that in a given situation and context a particular mathematical model is a *signifier* with a *realization tree* where the successive realizations branch over and connect multiple relevant discourses. To be engaged in the activity of modelling means to be participating in a modelling discourse, and thus involves singling out the relevant discourses for the problem by finding and making meaningful and productive connections between signifiers/realizations in realization trees belonging to different discourses, subsuming these into the new discourse.

3 Methodology and Method

Taking a commognitive approach on mathematical models and modelling means focusing on the special discourses where mathematical models are used, created, developed and modified. According to Sfard (2008), discourses are characterised and distinguishable in different ways (*vocabulary*, *visual mediators*, *routines*, and *endorsed narratives*), which suggest a number of complementary approaches could be taken to investigate a modelling discourse. This first exploratory attempt to use the commognitive framework analyses the collected and transcribed video data of two students engaged in a modelling activity with a focus on the different discourses evoked, the vocabulary (signifiers) used and the relationship between different signifiers as these enfold and connect (successively constituting realization trees) during the activity.

The data analysed comes from work on the second of two modelling tasks that were given sequentially to three groups of two to three upper secondary students. The students were taking part in a study aiming to investigate the potential use of *realistic Fermi problems* for introducing the notion of mathematical modelling at the upper secondary level (Ärlebäck 2009). These types of problems are open and discussion promoting with clear real-world connections. The decision to use this piece of data was partly based on convenience since the data already were available, and partly on the fact that the chosen group, consisting of two male students, was a known well functioning, dynamic and highly verbal duo when engaged in joint collaborative work. In addition, compared to the first of the two modelling tasks, we also regarded the second task to be more open in terms of relevant contexts coming into play.

The modelling task, *The Snow Clearance Problem*, that the students engaged in, is about snow clearance of a soccer field, a problem motivated in connection to the 2010 opening soccer game in Sweden, when a local soccer club had to ask their supporters for help to clear away all the snow from the soccer field in order for the opening game to take place. The actual task from which the data in this chapter comes was formulated as follows:

The Snow Clearance Problem

A firm gets the job to clear a soccer field from a 2 dm thick layer of snow fallen during the night. If the snow is shovelled into two piles on the two respective shorter sides of the field, how big will the piles be? What would each pile weigh?

In terms of commognition, the specific research question investigated in this chapter is: *What discursive objects can be discerned in the modelling activity of the students working on The Snow Clearance Problem?*

4 Analysis and Results

To illustrate how the data have been analysed, two excerpts from the 30 min conversation between the students are presented in Tables 3.1 and 3.2 respectively. Both excerpts were analysed with focus on identifying the discourses in terms of what signifiers the students use as well as trying to map the students' realization trees created during the conversation. To set the scene, the two students, Markus and Viktor, started the modelling activity by making assumptions about the size of the soccer field. Their estimates are based on Viktor's personal experience having visited a soccer field that was longer than the 100 m racetrack situated next to the field. They calculated the total snow volume as 1,200 m^3 (600 m^3 at each end of the soccer field) and expressed surprise that it was not a large amount. A discussion followed about what happens to snow when it is shovelled together and arguments were put forward that the snow volume becomes compacted to 200 m^3 at each end. In the excerpt in Table 3.1, the students are engaged in a conversation about what the actual question in the problem formulation means.

Table 3.1 Excerpt 1 from the transcripts

Student		Utterance
Markus	[39:10]	What is asked is how big they [the snow piles] will be, then you are searching for the height aren't you?
Viktor	[39:14]	Well, is it? I don't know?
Viktor	[39:19]	If he [the teacher] wanted to know the height, he would have written 'the height' [in the problem formulation].
Markus	[39:22]	Hmm, he probably wants to have the height and the width and maybe the length.
Viktor	[39:25]	Yeah.
Markus	[39:25]	They [he refers to the imaginary people that do the shovelling] are stupid if they don't spread out the snow evenly.
Viktor	[39:27]	If we give the height, the width and the length, then we give the volume… so that the height will be given. We can draw it here too [points at a sheet of paper where the students are writing their solution], that we can do.
Markus	[39:39]	We can say, well say that it will be one and a half metres high.
Viktor	[39:43]	Hmm, I will do it here [writes and draws on the paper] one point five.
Markus	[39:50]	[calculating on his pocket calculator] It will be two point six metres wide if it is square-shaped, so like three metres. However, it will not be square-shaped [cuboid], it will be more rounded.
Viktor	[39:58]	Yeah… That's right if it is rounded, then we have a half circle.

Note: Original in Swedish, translated by the authors

Table 3.2 Excerpt 2 from the transcripts

Student		Utterance
Markus	[46:34]	When someone melts snow then it will be quite little left.
Viktor	[46:36]	It gets VERY little water left.
Markus	[46:38]	One third [of the snow volume remains], because the snow is compacted together. A third!
Viktor	[46:40]	If you bring a snowball inside and let the snowball melt in a glass it will not be this much left [Viktor shows with his hands]. Be serious!
Markus	[46:48]	How about a fifth?
Viktor	[46:50]	I can go with a fifth, but I think it is just a good approximation.
Markus	[46:53]	A fourth?
Viktor	[46:54]	No.
Markus	[46:55]	I still think it is quite, it gets damn compacted when you shovel with a machine. It is not that you shovel little and snow compacts on snow.
Viktor	[47:02]	However, the snow is still frozen. Were you not in the chemistry class and saw that water that turns to ice expands four times. Isn't that so?
Markus	[47:18]	Yes, maybe so, divide it by four then.
Viktor	[47:22]	Hey, we do not have ice, it is far from ice. We have snow and snow is crystals and crystals are much more,… there is much air between [the crystals].
Markus	[47:32]	Ice does not expand four times, I know that.
Viktor	[47:34]	Ok.
Markus	[47:34]	It does not.
Viktor	[47:36]	No, it cannot, because when you put it in the freezer…
Markus	[47:38]	I think snow can expand four times, four-five times if the snow is compacted together.
Viktor	[47:41]	Well, say that snow expands four times.
Markus	[47:43]	Yes, then it is 50 tons on each side that, that is not little. Damn that's lots of snow.

Note: Original in Swedish, translated by the authors

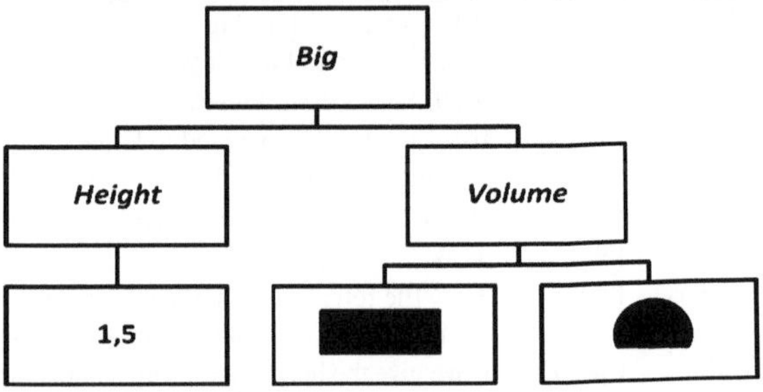

Fig. 3.1 A realization tree from excerpt 1

In the beginning of this excerpt, Viktor and Markus try to interpret the word *big*, which here is analysed as the root signifier in the corresponding realization tree. Any given signifier, for instance a specific word, may mean different things in different discourses. In excerpt 1, the signifier *big* can have several meanings in colloquial discourses such as referring to size, height or weight. Depending on what the signifier means or signifies, it is associated with different realizations. The students try to make the signifier operational by shifting from a colloquial discourse to a literate mathematical discourse by trying to employ a classroom discourse by speculating on what is required from them by their teacher. They assume that the teacher wants the signifier to be realized as a *volume*, but they are not totally sure so, just in case, they also explicitly include the height to please the teacher.

Markus [Table 3.1, 39:25] puts himself in a position of a snow shoveler and uses a colloquial discourse to describe how he assumes a snow shoveler would shovel. Markus assumes the shovelers would make even piles behind each goal and this idea is picked up by Viktor. Using another colloquial discourse argument, Markus emphasises *height* as the relevant realization of *big*, probably based on his own experience of snow piles. Markus's introduction of the new signifier *height* meets no complaints or counter-arguments from Viktor. Finally, the shape of the pile or how the signifier *volume* should be realized is discussed. Markus now uses a colloquial discourse and realizes the signifier *volume* as square-shaped piles with rounded edges. Viktor takes this idea into a more literate mathematical discourse and concludes that the cross sections of the piles are half circles.

The realization tree in Fig. 3.1 summarises and illustrates the identified signifiers together with their realizations. The root signifier *big* was realized as *height* and *volume*, realizations which as signifiers were unpacked even further; the height realized as 1.5 m; and, the volume (the mathematical model), the wanted signifier, was realized first as a cuboid and finally described in terms of a half cylinder.

The second excerpt, see Table 3.2, illustrates the frequently occurring *activity of negotiation* in the transcript. The episode depicts a negotiation about how to realize the relationship between a given volume of snow and the corresponding volume of melted water. This relationship is the key component of the model that describes the change in volume due to phase transition. This model (signifier) is then used to calculate the weight of the snow in the piles.

In the beginning of excerpt 2, Viktor and Markus are engaged in a colloquial discourse on *snow* and *water* resulting in their agreement on the volume of a snowball

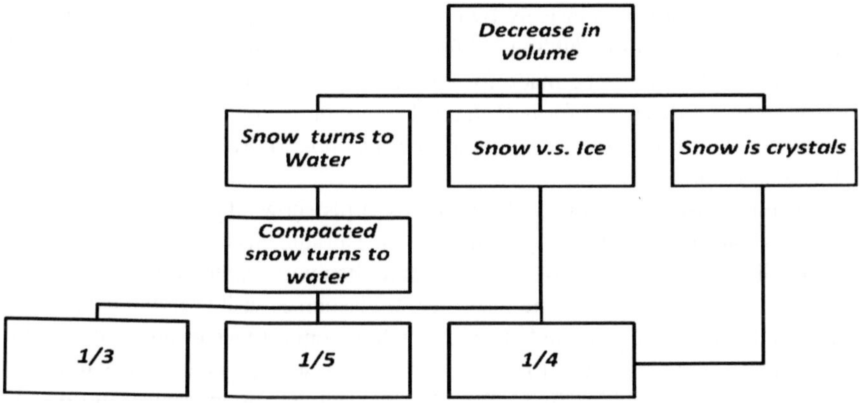

Fig. 3.2 A realization tree from excerpt 2

being bigger than the volume of the liquid water constituting it. However, they do not agree on how big this difference is; how much the *decrease in volume* is, which here is what is analysed as a root signifier. Markus [46:38] gives a conjecture, an idea how the signifier might be realized, presumably drawing on a classroom discourse, that the volume decreases three times. This conjecture is tested by Viktor, based on his own experience of melting snowballs in glasses (a common experiment in Swedish pre-school). Viktor's evaluation results in his dismissing of the conjecture, which is manifested in his utterance to Markus to "be serious". The negotiation continues by conjectures from Markus that the volume first decreases five times and then four times. Viktor tests these conjectures implicitly and evaluates that the volume might decrease five times but not four. However, Markus is persistent and continues to use arguments about the realization of the root signifier from a colloquial discourse and claims that the snowplough compacts the snow together so much that the volume should not decrease as much as five times. Viktor then turns to the literate chemistry discourse in order to convince Markus to agree with his new suggested realization. His argument is that ice expands four times (which is an estimate way above the correct value). This indicates that he is considering *ice* rather than *compacted snow* to be the relevant realization to consider. This transfer from a colloquial discourse to a more literate discourse convinces Markus. Viktor, who is still in the more literate chemistry (or physics) discourse, continues questioning Markus's estimate, and realizes that *snow* and *ice* are not the same thing based on reasoning about properties of crystals. Markus makes a new conjecture and claims that he is sure that the *ice* does not expand four times. Viktor realizes, based on personal experience in the colloquial discourse of putting water into the freezer, that he must have made a mistake trying to invoke the chemistry discourse. Finally, they settle that the snow's volume decreases four times when turned into water. Markus moves to the mathematical discourse and divides the total amount of snow measured in water by four. He obtains 50 tons and is surprised that the result became that much.

Figure 3.2 summarises the activity of negotiation in the realization of the root signifier *decrease in volume* as this enfolded in excerpt 2. Note how the realizations

1/3, 1/4 and 1/5 arise reclusively by conjecture-test-evaluation processes drawing on different discourses in the 'search' for relevant discourses to be subsumed so that a usable realization can be made for the desirable signifier (the model). The discourses that are discussed for potential subsumption in constructing the model are discerned by the talk of *water*, *snow*, *ice*, *melting*, *shovelling/packing*, and *ice expansion* respectively.

To summarise, the analysis of the two excerpts shows that the students in this study use and combine a variety of discourses. The students mainly use several colloquial discourses to realize the signifiers and often with explicit references to their personal experiences. There are also literate discourses visible in the data arising from mathematics, physics and chemistry. Classroom discourses are manifested in the values and norms on how to cooperate in school settings as well as assumptions about what is expected by the teacher. The realization trees presented in Figs. 3.1 and 3.2 illustrate what type of signifiers the students used in working on the problem within the particular context and how those signifiers have been realized. Overall, the discerned discursive objects in this study are manifested by a variety of different signifiers that come into play and were used during the modelling process. The links between these signifiers constructed during the modelling activity, the realization tree, is a result of the interplay between what the two students brought with them cognitively, the communication between the two, and, the social setting in which the modelling activity took place. A commognitive perspective allows all these aspects to be addressed using one framework.

5 Discussion and Conclusion

The students' frequent use of colloquial discourses to identify and realise signifiers may have several different explanations. Firstly, *The Snow Clearance Problem* was part of Markus's and Viktor's first experience of modelling activities. According to Sfard (2008), 'newcomers' need help from 'oldtimers' to make the transition from colloquial discourses to be able to more fully participate in literate discourses. Secondly, realistic Fermi problems in themselves have inherent characteristics inviting the use of colloquial discourses, especially when only a pocket calculator was allowed and not, for instance, the Internet or library resources. Therefore, it was not a surprise that the students use personal experiences as a base for their assumptions. This was found also by Ärlebäck (2009). Using Fermi problems for an introduction to modelling activities suggests the need for follow up lessons with more advanced modelling problems where students have access to more powerful tools and are offered opportunities to validate their work in order for the students to gain more knowledge in mathematical modelling.

The students' use of classroom discourses may also be an effect of the students' lack of modelling experiences; they did not know what was expected and required from them. However, the students seem to be somewhat constrained by the classroom discourses when it comes to validation of their mathematical models. For instance, in excerpt 1, there are hardly any reflections made about whether the shape of the snow piles as symmetric half cylinders is 'realistic' or not. In a classroom or tutoring situation, the lack of validation in the students' conversation suggests a natural point for the teacher to discuss the important role and function of validation in a modelling activity.

The realization trees presented in the chapter depicting signifiers and realizations, also indicate transitions between different discourses as well as how signifiers from different discourses become associated and connected. An example of such a transition is the realization of the signifier *decrease in volume*. The move is made from colloquial discourses towards literate discourses in chemistry or physics and finally realized in a mathematical discourse as a mathematical relationship. These results are similar to findings by Carreira et al. (2002).

An interesting detail in the data is how the signifier *volume* (of snow) initially was realized as 'not much' snow in the beginning of the conversation in excerpt 1, but later, while arguing about the signifier weight, it was realized as 'lots of snow'. This example emphasises that specific signifiers play an important role for how students experience potential realizations depending on measuring unit (m^3 and ton). In addition, it is notable what realistic assumptions the students make when they realize their model of the situation. For example, they implicitly make the assumption to shovel only the inside of the soccer field without thinking about the consequences of this decision. This, again, might be a consequence of the classroom discourse.

From a researcher's perspective, this first attempt to use the commognitive perspective on mathematical models and modelling is challenging since we are all 'newcomers' to this particular research discourse. Exploiting the potential benefits and pitfalls of this framework is in its infancy. Much work remains to be done. Commognition may also bridge and enhance understanding of different research perspectives on mathematical models and modelling by making differences and commonalities explicit and simultaneously providing a framework to relate these. In addition, commognition naturally facilitates conceiving mathematical modelling as an *interdisciplinary subject* including a number of different disciplines. For example, the transcripts of *The Snow Clearance Problem* facilitate discussions between the subjects of mathematics, physics and chemistry, respectively. By using a commognitive perspective on modelling, our analysis highlights how cognitive and social aspects simultaneously are manifested in modelling activities. Conceptualising modelling in terms of singling out relevant discourses in order to construct and develop a *subsumed discourse*, manifested in related signifiers and their realizations, provides a operationalisable approach to analysing the complexity involved in mathematical modelling by explicitly addressing these aspects.

References

Ärlebäck, J. B. (2009). On the use of realistic Fermi problems for introducing mathematical modelling in school. *The Montana Mathematics Enthusiast, 6*(3), 331–364.

Carreira, S., Evans, J., Lerman, S., & Morgan, S. (2002). Mathematical thinking: Studying the notion of 'transfer'. In A. Cockburn & E. Nardi (Eds.), *Proceedings of the 26th international conference of psychology of mathematics education group* (Vol. 2, pp. 185–192). Norwich: PME.

Frejd, P. (2010). Revisiting perspectives on mathematical models and modeling. In C. Bergsten, E. Jablonka, & T. Wedege (Eds.), *Mathematics and mathematics education: Cultural and social dimensions: Proceedings of Madif 7, Stockholm* (pp. 80–90). Linköping: SMDF.

Garcia, F., Gascón, J., Higueras, L., & Bosch, M. (2006). Mathematical modelling as a tool for the connection of school mathematics. *ZDM – The International Journal on Mathematics Education, 38*(3), 226–246.

Jablonka, E., & Gellert, U. (2007). Mathematisation – Demathematisation. In U. Gellert & E. Jablonka (Eds.), *Mathematisation and demathematisation: Social, philosophical and educational ramifications* (pp. 1–18). Rotterdam: Sense.

Kaiser, G., Blum, W., Borromeo Ferri, R., & Stillman, G. (Eds.). (2011). *Trends in teaching and learning of mathematical modelling*. New York: Springer.

Kaiser, G., & Sriraman, B. (2006). A global survey of international perspectives on modelling in mathematics education. *ZDM – The International Journal on Mathematics Education, 38*(3), 302–310.

Sfard, A. (2008). *Thinking as communicating: Human development, the growth of discourses, and mathematizing*. Cambridge: Cambridge University Press.

Sfard, A. (2012). Introduction: Developing mathematical discourse – Some insights from communicational research. *International Journal of Educational Research, 51–52*, 1–9.

Chapter 4
Should Interpretation Systems Be Considered to Be Models if They Only Function Implicitly?

Rita Borromeo Ferri and Richard Lesh

> *Though unconsciousness is, strictly speaking, a business of professional psychologist, it is so closely connected with my main subject that I cannot help dealing scantily with it.*
> *(Hadamard 1945, p. 21)*

Abstract The term "mathematical model" or just "model" is interpreted differently by different people in current international discussions about mathematical modelling. For many, the term "model" is restricted to interpretation systems which are explicit objects of thought. In this chapter we ask the question, if interpretation systems should be considered to be models if they only function implicitly. Furthermore we describe characteristics of what we mean by "implicit models" – as well as possible transitions from implicit to explicit models, and what these transitions look like from a cognitive-psychological perspective.

1 Introduction

There are a variety of reasons why many people have chosen to restrict the term "model" to interpretation systems which are explicit objects of thought. One reason is to encourage students to pay explicit attention to mismatches between models and

R. Borromeo Ferri (✉)
Institute of Mathematics, University of Kassel, Kassel, Germany
e-mail: borromeo@mathematik.uni-kassel.de

R. Lesh
Counseling and Educational Psychology, Indiana University, Bloomington, IN, USA
e-mail: ralesh@indiana.edu

G.A. Stillman et al. (eds.), *Teaching Mathematical Modelling: Connecting to Research and Practice*, International Perspectives on the Teaching and Learning of Mathematical Modelling, DOI 10.1007/978-94-007-6540-5_4,

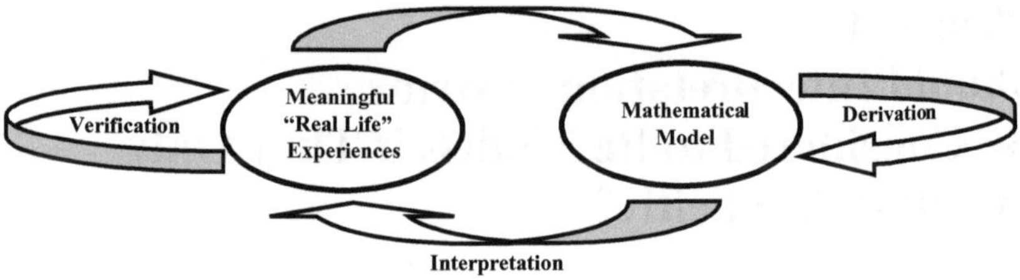

Fig. 4.1 Solutions to modelling problems involve multiple modelling cycle (Note: This diagram is not intended to imply that model development occurs along lockstep paths that always lead from mathematization, to derivation, to interpretation, to verification – and back again for a second, third, or ninth modelling cycle)

modelled systems, and to strengths and weaknesses of alternative models. Another reason is the main goal for many people is to teach modelling using the diagram (Fig. 4.1) or other cyclic diagram as a model of the processes through which models are developed.

But, our own research agenda is about concept development at least as much as it is about modelling for its own sake and the following underlying questions – listed in order of priority – are of interest in both countries: United States of America and Germany. (1) What does it mean to "understand" any one of the most important concepts in the K-16 mathematics curriculum in the United States of America or in the Educational Standards (Mathematics) in Germany? (2) How do these understandings develop? (3) How can development be cultivated, documented, and assessed? Modelling enters into this research agenda because our theoretical perspective suggests that mathematical "understanding" is likely to depend on reasoning that is based on some kind of model (or interpretation system).

According to our *models and modelling perspective* (MMP) on mathematical learning and problem solving, *a model is defined to be a system that is used to interpret (e.g., describe, design, or develop) some other system – for some specific purpose.* The main characteristic that distinguishes mathematical models from (for example) physical, chemical, or historical models is that *mathematical models focus on the structural (or systemic) properties of the system – rather than focusing on the physical, chemical, or historical properties- of the systems they are used to describe, design, or develop.*

Notice that there is nothing in the preceding definition which dictates that models must be explicit objects of attention – nor that they must function consciously rather than subconsciously, explicitly rather than implicitly, or formally or analytically rather than intuitively. In fact, our research has shown that: (a) models often develop along a variety of dimensions – such as concrete-abstract, particular-general, situated-decontextualized, simple-complex, intuitive-formal (Lesh and Harel 2003; Lesh and Yoon 2004), and (b) the models that are most useful in given situations are not necessarily those that are most abstract, most general, most decontextualized, most complex, or most formal (Lesh and Caylor 2007). For example, in contexts ranging from chess to mathematics teaching to business management, highly successful decision makers often develop powerful sense-making systems which function much more rapidly than is possible using formal, analytic, or completely conscious thinking. In fact, in fields ranging from sports (such as basketball or tennis) to performing arts (such as ballroom dancing), it is well known that too much conscious analysis often derails outstanding performances.

> The unconscious self 'is not purely automatic; it is capable of discernment; it has tact, delicacy; it knows how to choose, to divine. What do I say? It knows better how to divine than the conscious self, since it succeeds where that has failed. (Hadamard 1945, p. 40)

It is a common experience to "take a break" during the solution of a difficult problem; and, without conscious thought, to return with a solution in hand. Or, people often find themselves having "day dreamed" while simultaneously carrying out some complex decision-making task – such as driving an automobile or a bicycle. So, it is clear that complex thinking and decision-making can take place without conscious thought. On the other hand, one effective technique that often is used to promote the development of teachers, basketball players, dancers, or decision makers in other fields, is to analyse videotapes (or other examinable records) of past performances. However, when such reflection activities are used, the goal may not be to reduce decision making to an algorithmic process that involves executing formal rules. Instead, even though skill development tends to be an important accompaniment to reflections about complex performances, the most important purpose of reflection activities tends to be to develop increasingly powerful ways of thinking about (or interpreting) the situations.

So, even though increasing formalization and analysis may contribute to the development of more powerful models (or interpretation systems), the explicitly functioning aspects of these models tend to be like the visible tips of icebergs; that is, large parts of the models forever function intuitively and subconsciously. Furthermore, because MMP research tends to focus on situations that involve mathematical models-in-the-making, it has become clear that students' early interpretations of problem solving situations often function without being explicit objects of thought. They are more like windows that students look through (rather than paintings, or diagrams, or written notations that students look at) to make sense of external systems. So, for the purposes of this paper:

> Intuitively functioning models are defined to be those in which decisions are made without relying on explicit, formal, and consciously functioning tools associated with relevant domains of knowledge.

2 Theoretical Reflections About "Implicit or Intuitive Models"

Starting with the metaphor that "the explicitly functioning aspects of these [implicit] models tend to be like the visible tips of icebergs", this section stresses several aspects related to reflections on implicit models and their importance for the teaching and learning of mathematical modelling. In particular, we focus on: (a) the nature of "implicit models" within the modelling process (cycle), (b) the nature of transitions from implicit to explicit models, (c) the nature of situations where mature models might need to function intuitively and (d) the nature of beyond-cognitive aspects of mathematical models.

2.1 *The Nature of "Implicit Models" Within the Modelling Process (Cycle)*

Just as the term "mathematical modelling" is discussed in ways that are significantly different around the world (Kaiser and Sriraman 2006), notions about the concept of "model" itself also are often not easy to combine. In our sense, the concept of "model" is often interpreted or defined as a required (written, explicit, verbalised) product of an individual while solving a modelling problem. Looking at the several existing modelling cycles (see e.g., Borromeo Ferri 2006), one can see that there is a strong consensus using terms like "real model" and "mathematical model", because of its embedding in the history of mathematical modelling for educational purposes. But what, for example, about the term "situation model"? Already in Germany, there has been a discussion about this term – in particular labelling it with "model", because Blum and Leiß (2007) used the term "situation model" in their descriptions of modelling cycles. So, this term has gained more attention in discussions about modelling in mathematics education. In particular, the term "situation model" has been used in connection with non-complex modelling problems – to be precise, with word problems (see Kintsch and Greeno 1985; Nesher 1982; Verschaffel et al. 2000) and has its origin in linguistics.

A "situation model" can be described as a mental representation of the situation, which is given in a word problem and that is why the first author did not use "situation model" but "mental representation" in her modelling cycle (see Borromeo Ferri 2006). One reason for this was that a "mental representation" is a very distinct way of thinking about a given situation, and is affected by a variety of other personal attributes and experiences and is consequently more difficult to share with others. Whereas explicit mathematical models in written or verbalised forms can be communicated much better. Furthermore, it often includes phases of building implicit models (see later). So, certain aspects of a "situation model" also can function as an implicit model. For example, decisions often are made without relying on explicit, formal, and consciously functioning tools associated with relevant domains of knowledge.

We believe that there are many open questions about how the term "model" should be interpreted in the context of modelling processes. We also think that studies investigating strengths and weaknesses of alternative perspectives are likely to provide productive areas of research on modelling. In particular, it may be useful to investigate similarities and differences between "implicit models" and "mental models" which have been investigated at a variety of stages of the modelling process as depicted in Fig. 4.2.

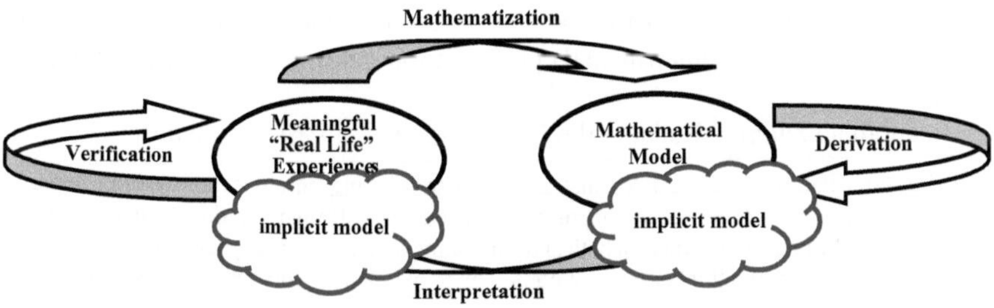

Fig. 4.2 Solutions to modelling problems include "implicit models"

2.2 *The Nature of Transitions from Implicit to Explicit Models*

Considerations about transitions from implicit to explicit models raise a variety of issues that warrant investigation. For example, if we assume implicit models are on an unconscious level and that explicit models are more conscious and can be better communicated to others, then, taking a cognitive development perspective (see e.g., Vygotsky, Piaget, Aebli), it may be true that implicit models are especially important for younger children, or for older students who are at early stages in the development of specific models, or for students who are functioning at lower cognitive levels. For example, for problem solvers and decision makers ranging from primary school children through economists in business communities, it is clear that early interpretations of situations often function implicitly. So, how and why do transitions from implicit to explicit models occur? What enables problem solvers to recognise the need for such transitions? How can students become aware of model-reality mismatches if the models themselves are not functioning as explicit objects of thought? Similarly, how can students learn about modelling processes if relevant models are not explicit objects of thought? Or, can several partial models function intuitively and at the same time? And, if so, how can model-model mismatches be detected?

Before discussing such questions, it is useful to reconsider the two quotes from Hadamard that we cited earlier. Hadamard's intensive work and observations about unconsciousness and "fringe-consciousness" in the field of mathematics is informative. He was asking his mathematics colleagues how they form new "creative ideas" (see Hadamard 1945, p. 34f). In his thinking about how mathematicians discover mathematics he was especially interested in how these ideas or thoughts become conscious – and how they came to be communicated. So he, like us, was interested in how mathematical concepts are developed for any age group. He was especially interested in how they begin to emerge at early stages – and at unconscious levels.

We agree with Hadamard when he said that "the very fact that it is unknown to the usual self gives to it such an appearance of mystery" Hadamard 1945, p. 21. "Fringe-consciousness" and consciousness, speaking in the sense of Hadamard (1945, p. 38), "are so close to each other, exchanges between them are so continuous and so rapid, that it seems hardly possible to see how they divide their roles" – we think it could be the same (mental) phenomenon getting from implicit to explicit models.

In another example, Hadamard refers (1945, pp. 35–36) to a chemist

> who, after half an hour, realized that he had been working on a question without being aware that he was doing so, and in such abstracted state of mind that during that time he forgot that he had already taken a bath and was taking a second one: a special case of unconscious process, as the thinker was not conscious of his mental work while it was going on, but perceived it when ended.

This example again illustrates an important way that thought sometimes functions; and, it also is relevant to transition questions that arise when we consider the stage of the so-called "fringe-consciousness" in model development. As an interpretation system becomes more explicit:

> At a first glance ideas are never in a more positively conscious state than when we express them in speaking. However, when I pronounce one sentence, where is the following one? Certainly not in the field of my consciousness, which is occupied by sentence number one; and nevertheless, I do think of it, and it is ready to appear the next instant, which cannot occur if I do not think of it unconsciously. (Hadamard 1945, p. 24)

Already, that could be the stage when an individual developed a "real model" or a "mathematical model" – because he or she is aware of his or her actions. From a cognitive-psychological point of view the mentioned thoughts can be very helpful, although we do not know how the interpretation systems function implicitly.

2.3 *The Nature of Situations Where Mature Models Might Need to Function Intuitively*

In what kind of problem solving situations are intuitively functioning models likely to be needed? Answers to this question include: (a) situations in which sense-making systems must function much more rapidly than is possible using formal, analytic, or completely conscious thinking, or (b) situations in which the amount of information that needs to be taken into account exceeds the processing power of available models. These characteristics often occur for problems in which there is more than a single "agent" or situations that cannot be described adequately using only a single function (or input-output rule). For example, the problem may involve feedback loops in which A sends data or information to B, B sends data or information to C, and C sends data or information back to A. Or, it may involve even simpler feedback loops in which A sends data or information to B, and B sends data or information back to A. Or, it may involve situations in which many "agents" interact – like cars on streets through a city. (see also Lesh and Doerr 2003; Lesh and Zawojewski 2007)

One such situation occurs in the "beer game" which is a famous simulation or "case study" that in often used in graduate schools offering degrees in business management. The beer game involves a producer, a distributor, and a seller; and, one characteristic that makes it interesting in that, if each of the "agents" acts in a way that is greedy (by maximizing personal profits without regard for others), then everybody loses. But, if everybody thinks of the system-as-a-whole, then everybody can win – and the system-as-a-whole can function in a way that is stable (with no "crashes" in which some key player loses, as in predator-prey situations in which one extinct species leads to another and another).

Unfortunately, mathematics problems with the preceding kinds of characteristics almost never occur in school textbooks – where: (a) most of the problems can be solved using only a single one-way function, and (b) issues involving maximisation, minimisation, or stabilisation never arise (perhaps because of the obsolete belief that such problems cannot be solved without the use of calculus). Yet, today, a wide range of such problems can be solved quite readily using a few basic mathematical concepts and a modern graphing calculator. Furthermore, many elegant simulations of such systems can be downloaded from reliable internet sites.

2.4 *The Nature of Beyond-Cognitive Aspects of Mathematical Models*

When students develop mathematical interpretations (or models) of problem solving or decision making situations, they do not simply engage conceptual systems that are purely logical or mathematical. They also engage feelings, dispositions, attitudes, beliefs, and a variety of metacognitive functions. Many of these beyond-cognitive attributes function without conscious thought – but function whenever the logical-mathematical aspects of the model are functioning. So, what is the instructional value of treating the development of beyond-cognitive attributes as part of mathematical model-development?

Like Skemp (1987, p. 35f), we believe in a high instructional value of emphasizing the development of beyond-cognitive attributes. Due to the fact that there are a great many beyond-cognitive attributes, we mention here for example (mathematical) beliefs (see e.g., Opt' Eynde et al. 2002; Pekhonen and Törner 1996; Schoenfeld 1992) as an important influencing factor on modelling. Beliefs are, roughly characterised, an affective domain. In the meantime the research on beliefs has changed from broadly defined attitudes to more specific sub-categories, such as beliefs about mathematics education (mathematics as a subject, mathematical learning and problem solving, mathematics teaching in general), beliefs about self or beliefs about the social context. Maaß (2004; see also Kaiser and Maaß 2007) demonstrated in her empirical year-long study, that many students of grade seven changed their mathematical beliefs. In particular the usefulness of mathematics was an important aspect for the students, which they did not recognise before and which caused more motivation for learning mathematics. Also, a lot of other studies in the last years came to these results. In this section our hypothesis is that encouraging students in their beyond-cognitive attributes could enhance the general "willingness for mathematical

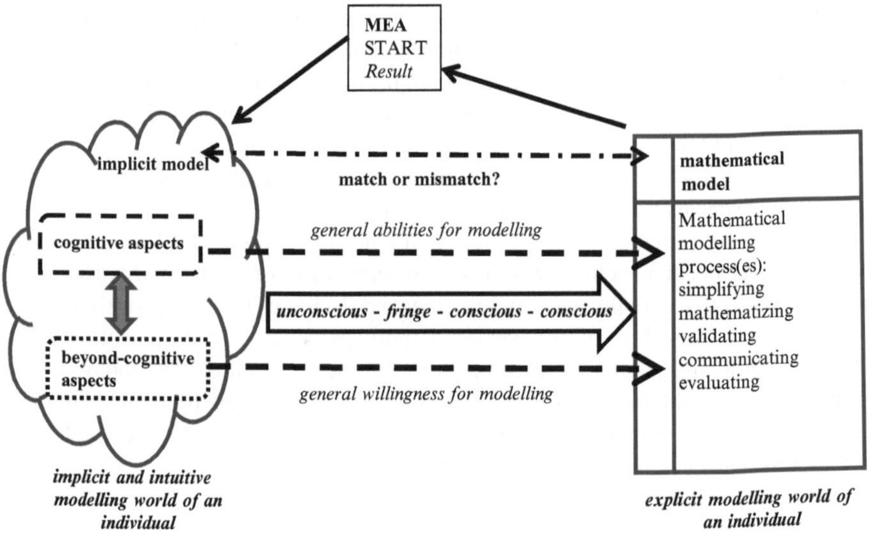

Fig. 4.3 Implicit and explicit modelling world of an individual

modelling activities" (see Fig. 4.3). In our view the following aspects could be a strong basis for beyond-cognitive attributes being a part during model development:

- If individuals should experience the diversity of mathematics (this also means different kinds of mathematical problems), then conceptual systems could be handled more *flexibly* for building mathematical models.
- If individuals have learned to talk and reflect about mathematics and their own way of understanding mathematics, then the modelling process could be handled more *goal-oriented.*
- If individuals were aware of the first mentioned points and then trust in their intuition when they approach a (modelling) problem, then interpretation systems could function as implicit models with a *match* to a successfully built mathematical model.

In such studies, it has become clear that theoretical reflections about "implicit" or "intuitive" models have a broad spectrum, but have a strong focus on psychological theories concerning unconsciousness. We pointed out four aspects about the nature of "implicit models" in order to give some more food for thought in this direction.

3 Summary and Conclusion

In this chapter, our aim has been to sort out a variety of issues related to the general question of whether interpretation systems should be considered to be models if they only function implicitly. Like the process of model development in general, we believe that there is no single correct answer to our question. There are both strengths and weaknesses associated with a variety of perspectives. But, when considering "implicit models", four issues that appear to be especially significant involve: (a) the nature of "implicit models" within the modelling process (cycle), (b) the nature of transitions from implicit to explicit models, (c) the nature of situations where mature models might need to function intuitively and (d) the nature of beyond-cognitive aspects of mathematical models.

We illustrate our conclusion in Fig. 4.3:

In the context of model-eliciting activities (MEAs) that we have investigated in our research, the diagram emphasises parallel and interacting developments between the "explicit modelling world of an individual" (right side of the figure) and on the other hand a fuzzy "implicit and intuitive modelling world of an individual" (left side of the figure). The first stage when working on a modelling eliciting activity or another complex modelling problem in every kind of area is this fuzzy "implicit modelling world" including cognitive aspects and beyond-cognitive attributes which influence each other in several (unconscious) ways. Cognitive aspects comprise general abilities for modelling, which means in particular mathematical abilities and modelling competencies and thus are necessary for developing an adequate mathematical model. At the same time these cognitive aspects are influenced by beyond-cognitive attributes such as beliefs and feelings, which in turn build a basis for general willingness dealing with the MEA at all. It is impossible to reconstruct these mental actions of an individual. But these interpretation systems within this fuzzy unconscious world are the bricks for the upcoming mathematical model in the "explicit modelling world" through the fringe-consciousness and finally consciousness. The match or the mismatch of the interpretation systems developed in the implicit world as an implicit model with the mathematical model in the explicit world cannot be investigated, because it disappeared in the mystery of the unconsciousness:

> The unconscious self 'is not purely automatic; it is capable of discernment; it has tact, delicacy; it knows how to choose, to divine. What do I say? It knows better how to divine than the conscious self, since it succeeds where that has failed. (Hadamard 1945, p. 40)

References

Blum, W., & Leiß, D. (2007). How do students and teachers deal with modelling problems? In C. Haines, P. Galbraith, W. Blum, & S. Khan (Eds.), *Mathematical modelling: Education, engineering and economics* (pp. 222–231). Chichester: Horwood.

Borromeo Ferri, R. (2006). Theoretical and empirical differentiations of phases in the modelling process. *Zentralblatt für Didaktik der Mathematik, 38*(2), 86–95.

Hadamard, J. (1945). *The psychology of invention in the mathematical field.* Toronto: Princeton University Press.

Kaiser, G., & Maaß, K. (2007). Modelling in lower secondary mathematics classrooms – Problems and opportunities. In W. Blum, P. Galbraith, H.-W. Henn, & M. Niss (Eds.), *Applications and modelling in mathematics education* (pp. 99–108). New York: Springer.

Kaiser, G., & Sriraman, B. (2006). A global survey of international perspectives on modelling in mathematics education. *Zentralblatt für Didaktik der Mathematik, 38*(3), 302–310.

Kintsch, W., & Greeno, J. (1985). Understanding word arithmetic problems. *Psychological Review, 92*(1), 109–129.

Lesh, R., & Caylor, E. (2007). Introduction to the special issue: Modeling as application versus modeling as a way to create mathematics. *International Journal of Computers for Mathematical Learning, 12*(3), 173–194.

Lesh, R., & Doerr, H. (2003). *Beyond constructivism: Models & modeling perspectives on mathematics teaching, learning, and problems solving*. Hillsdale: Lawrence Erlbaum.

Lesh, R., & Harel, G. (2003). Problem solving, modeling, and local conceptual development. *Mathematical Thinking and Learning, 5*(2 & 3), 157–190.

Lesh, R., & Yoon, C. (2004). Evolving communities of mind – In which development involves several interacting & simultaneously developing strands. *Mathematical Thinking and Learning, 6*(2), 205–226.

Lesh, R., & Zawojewski, J. (2007). Problem solving and modeling. In F. Lester (Ed.), *The second handbook of research on mathematics teaching and learning* (pp. 763–804). Charlotte: Information Age Publishing.

Maaß, K. (2004). *Mathematisches modellieren im Unterricht. Ergebnisse einer empirischen Studie*. Hildesheim: Franzbecker.

Nesher, P. (1982). Levels of description in the analysis of addition and subtraction. In T. P. Carpenter, J. M. Moser, & T. Romberg (Eds.), *Addition and subtraction: A cognitive perspective* (pp. 25–38). Hillsdale: Erlbaum.

Opt' Eynde, P., De Corte, E., & Verschaffel, L. (2002). Framing students' mathematical-related beliefs: A quest for conceptual clarity and comprehensive categorization. In G. Leder, E. Pekhonen, & G. Törner (Eds.), *Beliefs: A hidden variable in mathematics education* (pp. 13–37). Dordrecht: Kluwer.

Pekhonen, E., & Törner, G. (1996). Mathematical beliefs and different aspects of their meaning. *Zentralblatt für Didaktik der Mathematik, 28*(4), 101–108.

Schoenfeld, A. H. (1992). Learning to think mathematically: Problem solving, metacognition, and sense making in mathematics. In D. Grouws (Ed.), *Handbook of research on mathematics teaching and learning* (pp. 334–370). New York: Macmillan.

Skemp, R. (1987). *The psychology of learning mathematics*. Hillsdale: Erlbaum.

Verschaffel, L., Greer, B., & De Corte, E. (2000). *Making sense of word problems*. Lisse: Swets & Zeitlinger.

Chapter 5
Mathematical Modelling, Mathematical Content and Tensions in Discourses

Andréia Maria Pereira de Oliveira and Jonei Cerqueira Barbosa

Abstract In this chapter, we present part results of an empirical study on tensions in discourses manifested by teachers when they implemented mathematical modelling in the pedagogic practices. The focus is on analysing the tension of students' mathematical performance. Using Bernstein's theoretical frame, we followed three teachers from the lower secondary school level from Brazilian public schools. These teachers were videotaped during their modelling-based lessons. The nature of the research analysis is qualitative. The procedures used for collecting data were observations accomplished through recordings of lessons, interviews after each lesson and teachers' narratives on their lessons. The results have shown that the tension of students' mathematical performance is related to what and how to teach mathematical content in the modelling environment, when students do not have a mathematical performance to solve problems from daily life situations.

1 Introduction

Mathematical modelling has been one of the ways to promote the connections of everyday life in the classroom. We define *mathematical modelling* as a learning environment where students are invited to solve problems from daily life, professional areas or situations in scientific disciplines, through mathematics (Barbosa 2006). By *learning environment*, we mean the social conditions provided to students for the

A.M.P. de Oliveira (✉)
Dipartimento de Ciencias Exatas, State University of Feira de Santana, Feira de Santana, BA, Brazil
e-mail: ampodeinha@gmail.com

J.C. Barbosa
Faculdade de Educação, Federal University of Bahia, Salvador, BA, Brazil
e-mail: joneicerqueira@gmail.com

G.A. Stillman et al. (eds.), *Teaching Mathematical Modelling: Connecting to Research and Practice*, International Perspectives on the Teaching and Learning of Mathematical Modelling, DOI 10.1007/978-94-007-6540-5_5,

development of some activities (Skovsmose 2001). The problem that students are asked to formulate and/or solve about daily life, professional areas or situations in other disciplines is a *mathematical modelling task*.

Recent studies have discussed how mathematical modelling demands specific actions from students and teachers to address daily situations in the classroom: interventions in the students' modelling process (Leiß 2005), the pedagogical knowledge for teaching modelling (Doerr 2006, 2007; Doerr and English 2006), classroom activities and type of teaching patterns through modelling (Antonius et al. 2007) and mathematical thinking styles (Borromeo Ferri and Blum 2010). Regarding these discussions questions arise: How have teachers engaged their students to work on mathematical modelling tasks in mathematics lessons? How have teachers worked mathematical content while engaging their students in modelling tasks? Empirical evidence on teachers' pedagogic practices in mathematical modelling is still scarce, especially regarding studies that examine what happens when using mathematical modelling in school settings.

In this chapter, we will examine *tensions in discourses* when teachers engaged learners in mathematical modelling tasks in their classrooms through an analysis of teachers' discourses based on Bernstein's theoretical framework. We define *discourse* as an oral or written text produced by an individual in a specific social context. In particular, we focus on *the tension of students' mathematical performance* when teachers implemented mathematical modelling in the pedagogic practices.

2 Tensions in Discourses

In this study, Bernstein's theory (1990, 2000) is employed to highlight the tensions in teachers' discourses when they engaged learners in mathematical modelling tasks in the pedagogic practices. Empirical studies have suggested that teachers play a crucial role in ensuring the implementation of modelling in the pedagogic practices (Doerr 2006, 2007; Doerr and English 2006). Bernstein (1990, 2000) uses the term *pedagogic recontextualising* for the movement of a discourse from its original site to a pedagogic site. To understand the process of pedagogic recontextualising, we will use Jablonka's (2007) description of mathematics, as a school subject:

> mathematics is a highly specialised activity that consists of a range of practices, some of which employ sophisticated tools and sign systems. The recontextualisation of parts of those practices establishes the school subject mathematics as it is defined in curriculum documents (p. 194).

This recontextualising process involves the selection of those practices and their relocation into school mathematics that it is operated by a *pedagogic discourse*.

Bernstein (1990) defined *pedagogic discourse* as a principle for the selection of discourses that are relocated according to their own order. It is "a principle for appropriating other discourses and bringing them into a special relation with each other for the purpose of their selective transmission and acquisition" (pp. 183–184). The notion of discourse as text, presented in the introduction, is different from the concept of pedagogic discourse which is a principle, because pedagogic discourse "cannot be identified with any of the discourses it has recontextualized" (p. 184). In this sense, once everyday life situations are moved to the classroom by teachers through pedagogic recontextualising, the pedagogic discourse selectively relocates and refocuses them in agreement with the rules present in the pedagogic practice. In addition, pedagogic discourse places them in a special relationship with other discourses to constitute its own order. Thus, the movement of everyday life situations for classroom practice is regulated by rules that have already been socially established and legitimated in this pedagogic context.

Doerr (2006) examined the ways in which teachers identified, interpreted and responded to students' work with modelling tasks. The results suggested that teachers develop sophisticated schemas to understand the diversity of students' ways of thinking. Teachers' actions supported students' engagement in the task and led them to review and refine their own mathematical thinking. Similarly, the results in Leiß's study (2005) pointed out that teachers' interventions were important in terms of facilitating student understanding of the problem, finding an appropriate model for the situation and reflecting on the model. These results mean that promoting modelling in the classroom has provoked some change in the pedagogical relationship between teachers and students.

In Brazil and in many countries, teachers have had contact with mathematical modelling through teacher education programmes (in-service or pre-service), results from research in the community of mathematics education, conferences, which are pedagogic recontextualising fields, as named by Bernstein (2000). The discourse on modelling, that is moved by teachers from a pedagogic recontextualising field to the classroom (defined as a field of reproduction) is a specialised discourse, named by Barbosa (2006) as *school mathematical modelling.*

Bernstein (1990, 2000) uses two concepts to stress relations of power (classification) and control (framing) in the pedagogic practice. Classification embodies power relations between different categories, as for example, kinds of discourses. It is defined by the space between categories that are maintained by power relations. This space is defined by Bernstein (2000) as *insulation*. On the other hand, framing regulates relations within a context. It refers to different forms of legitimating communication in any pedagogic practice. Framing refers to control of communication (selection, sequencing, pacing and criteria) in pedagogic relations, as for example, between teachers and students.

According to Bernstein (2000), classification establishes *recognition rules* that regulate what meanings are relevant in a context, and framing establishes *realisation rules* that regulate how the meanings are to be put together to create the legitimate text according to this context. Bernstein (2000) defines *legitimate text* as any realisation on the part of the acquirer which is evaluated. In short, "classification provides us with the limits of any discourse, whereas framing provides us with the form of the realisation of that discourse" (Bernstein 2000, p. 12). Lerman and Zevenbergen (2004) argue that school mathematics is a specialised discourse with strong classification and framing, because "it is often taught as a discipline quite distinct from others, and taught in a way where there is an emphasis on specialised skills" (p. 29).

Regarding Bernstein's theoretical frame, what happens when mathematical modelling is moved to the classroom? How have teachers moved it to the school context? We used the aforementioned concepts to understand how teachers have engaged learners in mathematical modelling tasks in the pedagogic practices. Doerr (2006) and Doerr and English (2006) have argued that understanding teachers' knowledge means knowing how teachers interpret their pedagogic practices in the classroom, and how and when those interpretations influence decisions and actions in the classroom. "It is precisely teacher's interpretations of a situation that influence when and why as well as what it is that the teacher does" (Doerr 2006, p. 5).

Empirical studies have discussed teachers' dilemmas and uncertainties when using modelling in the classroom (Blomhøj and Kjeldsen 2006; Doerr and English 2006). Some teachers' dilemmas pointed out by Blomhøj and Kjeldsen (2006) are: understanding of the phases in the modelling process from a holistic point of view or as an inner part of the modelling process to work the mathematical content; the goal of modelling as an educational one or as a mean for motivating and supporting the students' learning of mathematics; and how to develop students' autonomy when working with projects. Doerr and English (2006) identified teacher uncertainty with how students can develop mathematically viable solutions. This uncertainty is related to the teachers' legitimate action in promoting modelling in the pedagogic practice. It constituted a kind of tension between how to develop the task and the uncertainty of which solutions the students might develop to solve the problem. These dilemmas and uncertainties might be seen as results of trying to place a new discourse in the pedagogic practice.

Based on Bernstein's theory, we use the expression *tensions in discourses* to understand how teachers engage learners in mathematical modelling tasks in the pedagogic practices. *Tensions in discourses* are manifested by teachers through contradictions, cleavages and dilemmas that are constituted because of the space between categories (present discourses in the pedagogic practice and a new discourse in the pedagogic practice). "The classificatory principle creates order, and the contradictions, cleavages and dilemmas which necessarily inhere in the principle of a classification are suppressed by the insulation" (Bernstein 2000, p. 7). In this sense, discourse on modelling is moved to the classroom through a pedagogic recontextualising process. As a result, it might change the values of classification (what can be said) and framing (how it can be said) in a mathematics lesson.

It means that *tensions in discourses* might be interpreted in terms of a recontextualising process, because taking a new discourse into the classroom involves crossing the insulation between its original site and a pedagogic site. This new discourse is positioned by the pedagogic discourse, presenting a discontinuity in relation to the present discourses in the pedagogic practice. This discontinuity is justified by the insulation among discourses that are positioned in the pedagogic practice. In this sense, *tensions in discourses* can be identified when there are characteristics of discourses that had been consolidated and legitimated in the pedagogic practice and discourses that have been brought into it. The expression *tensions in discourses* has its origin in the discontinuity among legitimate discourses and a new discourse positioned by the pedagogic discourse, when teachers decide what can be said in the pedagogic practice when using mathematical modelling and how it can be said. This discontinuity is manifested in the teachers' discourses through contradictions, cleavages and dilemmas which occur in specific moments in the pedagogic practice and shows the tensions in discourses that are denominated *situations of tension*.

3 Context and Methodology

The research context was the first modelling experience of three lower secondary school level teachers from public schools in the Northeast of Brazil. The teachers who developed modelling-based lessons were Boli, Maria and Vitoria (pseudonyms). Each of them has been teaching for more than 14 years in public schools with classes of disadvantaged students. The teachers organised the modelling environment according to what Barbosa (2003) calls Case 2; in other words, teachers present a problem and students are required to collect and investigate data. The teachers elaborated some tasks during the development of the modelling environment. They then presented some problems with quantitative and qualitative data and students solved them (framed in Case 1 Barbosa 2003). Each teacher's lesson was organised in small student groups, who solved the teacher assigned tasks.

At the time of data collection, the teachers were finishing a training program for non-certified teachers at the State University of Feira de Santana. The first author of this chapter was a lecturer for them in two semesters devoted to mathematical modelling. They, and their colleagues, engaged in the approach of using problems from daily situations and in the development of modelling projects by themselves, as well as in the implementation of modelling in their classrooms.

The research was framed according to the qualitative perspective (Denzin and Lincoln 2005), because its purpose was the analysis of the tensions in discourses when teachers were using modelling in pedagogic practices. Each teacher was videotaped during the modelling-based lessons. The videotaping focused on the teacher and the interactions between teacher and students and these videotapes were transcribed. After each lesson, interviews were conducted with each teacher that described how the modelling task was developed. The interviews were also recorded and transcribed. The teachers wrote narratives about each lesson and these were analysed. With the purpose of producing theoretical understandings, based on the collected data and guided by a research question, the teachers' pedagogic practices were analysed in terms of the relationships between agents (teachers and students) and discourses that were present in the pedagogic practice as well as discourses on mathematical modelling moved into the pedagogic practice.

The analysis of data had some inspiration in the analytical procedures of grounded theory (Charmaz 2006), namely, the elaboration of codes and categories of the transcribed data. The analysis of data occured in three phases. In the first phase, the transcripts of the videotapes and interviews, as well as teacher narratives on their lessons were read. The second phase was identification of extracts. These extracts are pieces of data from the interviews, class videotapes and teacher narratives. Following this, each extract was read, line by line, and we used an open-ended coding of the tensions in discourses that we identified in the teachers' pedagogic practice. In the third phase, we classified the codes into general categories and we were able to understand the focus of the study by integrating the results from literature and Bernstein's theory.

4 The Tension of Students' Mathematical Performance

To investigate the focus of this study, we examined the pedagogic practices of three teachers (one male and two female) through an analysis of their discourses. The study focuses on analysis of the *tension of students' mathematical performance* when teachers have implemented mathematical modelling into their classroom practice. In this section, we will present this tension in discourses that we identified in teachers' pedagogic practices.

The teachers introduced real problems[1] in their lessons and the students had difficulties in solving them, because they did not understand how the mathematical content could be used, that is, they were not able to produce a legitimate text to make the modelling task. The teachers did not expect that their students would show difficulties in using previous mathematical content to solve the problems. As well they did not expect that the students would not use other content to solve these problems. In other words, teachers wanted to know what can be done and how it can be done when the students are not able to solve the modelling task, because they have difficulties in using the mathematical content to solve it. This tension was identified when the teachers had decided how to approach (previous and new) mathematical content when modelling was implemented in the pedagogic practices. In this section, we present three *situations of tension* to discuss how each teacher understood and dealt with the *tension of students' mathematical performance* during the modelling task.

4.1 Situation of Tension 1: Students Making Mistakes in Mathematical Procedures

Teacher Boli implemented a modelling task entitled *Basic Basket of Goods* in two classes (Year 9) in the lower secondary level (see de Oliveira and Barbosa 2010 for further detail). He organised modelling in several phases, such as discussing the theme, introducing a problem, defining the products and quantities of a basic basket, students working in groups, students getting information to solve the problem, defining families' expenses, making calculations and comparisons, and drawing graphs. He used part of the lessons that were to be dedicated for the modelling task to deal with student difficulties in relation to previous mathematical content required to solve the problem. This situation of tension happened in Boli's class

[1] We use "real problem" to mean a problem from daily life or situations in scientific disciplines.

when he asked the students who were organised in groups to present their calculations of the costs of a family, having as parameter the minimum wage. He asked that each group write their table on the whiteboard with the expenses and their percentages: "What could you see here for each percentage calculation in relation to the expenses of a family with these expenses?" Then, Boli discussed the values found by the groups and asked students to explain the observed values. At this point, he noted that some calculations were incorrect. In the interview, he commented on the incorrect calculations of the students:

Boli: I observed that the calculations were wrong and they remade them again. I was concerned with the high values that they had found. [...] When students had errors, I had to stop everything and they did the calculations again. I guided each group how to do the calculations. Then, I asked that they calculate the percentage of each item relative to the minimum wage. So, I noticed that almost all students did not know rule of three and percentage.

Boli was concerned about the incorrect calculations of the students. This situation indicated that students could not solve the task because they did not know the mathematical content required in the proposed problem. Boli requested that the students do the calculations again. After that, he followed the groups doing the calculations and explained the required content in the solution of the problem.

4.2 *Situation of Tension 2: Students Unable to Solve the Problems*

Teacher Maria implemented a modelling task entitled *Analyzing Water Bills* in a Year 6 class at the lower secondary level. She organised it in some phases, such as discussing the theme, introducing a problem, students working in groups, students getting information, analysing tables, making calculations and comparisons. She noticed that her students had stopped developing the task, because they did not know how to use the mathematical content to solve the problem. This situation of tension arose in Maria's class when her students were solving problems about the waste of water for household chores: washing clothes, watering the garden, washing dishes, washing the sidewalk, washing the car. She noted her students were not solving the problems. She asked: "Why are you not doing something?" A student answered: "Teacher, we have no idea how to start doing something". In the interview, Maria commented on students' difficulties in solving the problems proposed:

Maria: [...] I was worried because most of the students stopped doing the task. What will I do now? Can I explain it to them? I was observing the groups answer the questions and I noticed the big problem they had when making the calculations. Then, I decided to show them, on the whiteboard, how to solve one of the problems. After that, I thought they started working more easily and they made the calculations without great efforts.

Maria was concerned about how she could intervene to help her students when they had difficulties in using mathematical content to solve the modelling task. Her students did not solve the problem, because they did not know which mathematical content could be used to solve it. As she did not know how to intervene to help her students, she asked the lecturer for help to work with the students' difficulties. After the lecturer explained that she could intervene and help the students, Maria explained to the students how to solve one of the problems.

4.3 Situation of Tension 3: Students Not Using New Mathematical Content Taught by Teacher

Vitoria implemented a modelling task entitled *The Minimum Wage and a Family's Cost of Living* in a Year 8 class at the lower secondary level. She organised it as follows: discussing the theme, introducing a problem, defining families' expenses, students getting information, establishing the products and quantities of a basic basket of goods, collecting data, making calculations and comparisons, and elaborating tables. A tension situation arose in Vitoria's class when she tried to approach new mathematical content required in the resolution of the proposed task. She tried to work with other mathematical content, but her students had difficulties to use this other content to solve the problem. She explained that they had difficulties, because they had no contact with the content before the modelling task. Thus, she wanted to know how to work on new mathematical content in the modelling task. The following extract shows this:

Vitoria: I don't know what contents we will work with the modelling task. Students have many difficulties, as for example, with graphs. They do not know what a graph is. I wanted to work with graphs and other mathematical contents. I have worked until this moment with operations and percentage. I have difficulties to explain other contents with students. I don't know how to work them, because students do not know the other contents. I imagine this problem will request "function" and I think I will be able to work with it. […] I tried to work with graph, but they have no notion of what a graph is. I tried to make a graph on the expenses but… I explained [to] them how to make a graph, but they had many difficulties.

We noticed that Vitoria admitted that she did not know how to approach new content in the modelling task. Nevertheless, she tried to approach new content, but her students had difficulties to understand it and to use it to solve the problem. Due to this situation, she argued that the students need to have previous knowledge of the content that will be used in the modelling task. She was concerned about how to approach new mathematical content in a modelling task.

5 Conclusion and Implications

In this study, we present extracts that refer to a tension in teachers' discourses: the *tension of students' mathematical performance*. We now discuss how this tension has been constituted while the teachers developed mathematical modelling in pedagogic practices.

Students making mistakes in mathematical procedures; *students unable to solve the problems* and *students not using new mathematical content taught* were three situations of tension that represented students' actions in the modeling environment. These situations have blocked following the task at the time that teachers accompanied the students to solve problems. In relation to Boli, the students used incorrect procedures to solve the problem and he had to explain the content required in the task and already studied by students in previous lessons. Maria's students were unable to mobilise any mathematical content already studied in previous years to solve the problems. Vitoria's students were unable to use new mathematical content worked during the modelling task. They failed to mobilise this new content in solving the problem. These students' actions refer to their *mathematical performance*. Students' *mathematical performance* is related to mathematical procedures and use of mathematical content to deal with the problem.

The teachers introduced real problems in the pedagogic practices and the students had difficulties to solve them, because they did not understand how to use the mathematical content to solve the problems and they were not able to produce a *legitimate text* (Bernstein 2000) for the development of the modelling task. They had encountered difficulties in solving problems from daily life situations. How do we address the students' difficulties to solve these problems? How do we teach (previous and new) mathematical content? These questions represent discontinuities in relation to the present discourses and a new discourse (the discourse of mathematical modelling) in the pedagogic practice. These discontinuities refer to *the tension of students' mathematical performance*. This tension refers to the discontinuity in relation to what and how to deal with mathematical performance in the modelling environment. Due to that tension, the teachers wanted to know how to work the mathematical content, before or during the modelling task; how to work new mathematical content; how to work the students' difficulties with mathematical content, in other words, what to do and how to approach the mathematical content in the modelling environment. To deal with this tension they had worked the previous and new mathematical content in the development of the modelling task in their lessons. Antonius et al. (2007) point out that it is a timing dilemma as to when teachers work on mathematical content related to the theme of the modelling task, that is, to decide to approach the mathematical topics before, during or after the task.

Implications of the research are firstly understanding how *tensions in discourses* can contribute to the teachers' professional development when teachers produce actions and strategies to deal with them; and secondly for practice teaching education programmes in mathematical modelling need to discuss tensions in discourses to support teachers in implementing this learning environment in the classroom. Thus, tensions in discourses can provide opportunities for teachers to realize actions and strategies to implement mathematical modelling in the pedagogic practice.

References

Antonius, S., Haines, C., Jensen, T. H., & Niss, M. (with Burkhardt, H.). (2007). Classroom activities and the teacher. In W. Blum, P. Galbraith, H. Henn, & M. Niss (Eds.), *Modelling and applications in mathematics education: The 14th ICMI study* (pp. 295–308). New York: Springer.

Barbosa, J. C. (2003). What is mathematical modeling? In S. Lamon, W. Parker, & S. Houston (Eds.), *Mathematical modeling: A way of life ICTMA 11* (pp. 227–234). Chichester: Horwoord.

Barbosa, J. C. (2006). Mathematical modelling in classroom: A critical and discursive perspective. *ZDM – The International Journal on Mathematics Education, 38*(3), 293–301.

Bernstein, B. (1990). *Class, codes and control: Vol. 4. The structuring of pedagogic discourse.* London: Routledge.

Bernstein, B. (2000). *Pedagogy, symbolic control and identify: Theory, research, critique*. Lanham: Rowman & Littlefield.

Blomhøj, M., & Kjeldsen, T. H. (2006). Teaching mathematical modeling through project work. *ZDM – The International Journal on Mathematics Education, 38*(2), 163–177.

Borromeo Ferri, R., & Blum, W. (2010). Insights into teachers' unconscious behavior in modeling contexts. In R. Lesh, P. Galbraith, C. Haines, & A. Hurford (Eds.), *Modelling students' mathematical modeling competencies* (pp. 423–432). New York: Springer.

Charmaz, K. (2006). *Constructing grounded theory: A practical guide through qualitative analysis*. Thousand Oaks: Sage.

de Oliveira, A. M. P., & Barbosa, J. C. (2010). Mathematical modeling and the teachers' tensions. In R. Lesh, P. L. Galbraith, C. R. Haines, & A. Hurford (Eds.), *Modeling students' mathematical modeling competencies: ICTMA 13* (pp. 511–517). New York: Springer.

Denzin, N., & Lincoln, Y. (2005). Introduction: The discipline and practice of qualitative research. In N. Denzin & Y. Lincoln (Eds.), *Handbook of qualitative research* (3rd ed., pp. 1–32). Thousand Oaks: Sage.

Doerr, H. M. (2006). Examining the tasks of teaching when using students' mathematical thinking. *Educational Studies in Mathematics, 62*(1), 3–24. doi:10.1007/s10649-006-4437-9.

Doerr, H. M. (2007). What knowledge do teachers need for teaching mathematics through applications and modelling? In W. Blum, P. Galbraith, H. Henn, & M. Niss (Eds.), *Modelling and applications in mathematics education: The 14th ICMI study* (pp. 69–78). New York: Springer.

Doerr, H. M., & English, L. D. (2006). Middle grade teachers' learning through students' engagement with modeling tasks. *Journal of Mathematics Teacher Education, 9*, 5–32. doi:10.1007/s10857-006-9004-x.

Jablonka, E. (2007). The relevance of modelling and applications: Relevant to whom and for what purpose? In W. Blum, P. Galbraith, H. Henn, & M. Niss (Eds.), *Modelling and applications in mathematics education: The 14th ICMI study* (pp. 193–200). New York: Springer.

Leiß, D. (2005). Teacher intervention versus self-regulated learning? *Teaching Mathematics and its Applications, 24*(2–3), 75–89.

Lerman, S., & Zevenbergen, R. (2004). The socio-political context of the mathematics classroom: Using Bernstein's theoretical framework to understand classroom communications. In P. Valero & R. Zevenbergen (Eds.), *Researching the socio-political dimensions of mathematics education: Issues of power in theory and methodology* (pp. 27–42). Dordrecht: Kluwer.

Skovsmose, O. (2001). Landscapes of investigation. *ZDM – The International Journal on Mathematics Education, 33*(4), 123–132.

Chapter 6
Ethnomodelling as a Methodology for Ethnomathematics

Milton Rosa and Daniel Clark Orey

Abstract The application of ethnomathematical techniques and tools of modelling allows us to examine systems taken from reality and offers insight into forms of mathematics done in a holistic way. The pedagogical approach that respects a diversity of cultural forms of mathematics is best represented through ethnomodelling, which is a process of translation and elaboration of problems and the questions taken from reality. We would like to broaden the discussion of possibilities for the inclusion of ethnomathematics and associated ethnomodelling perspectives that respect the social diversity of distinct cultural groups with guarantees for the development of understanding different ways of doing mathematics through dialogue and respect.

1 Introduction

Culture and society considerably affect the way individuals understand mathematical ideas and concepts. Research in ethnomathematics has demonstrated how mathematics has grown from the many diverse and distinct cultural traditions that comprise human activity over time. In this regard, all cultural groups have developed unique ways of incorporating mathematical knowledge and have often come to represent given cultural systems, especially in ways that cultural groups quantify and use numbers, incorporate geometric forms and relationships, and measure and classify objects (D'Ambrosio 1990).

Each cultural group has developed unique and distinct ways to *mathematise* their own realities (Rosa and Orey 2006). Mathematisation is a process in which individuals from different cultural groups have developed mathematical tools that can help

M. Rosa (✉) • D.C. Orey
Centro de Educação Aberta e à Distância (CEAD), Universidade Federal de Ouro Preto (UFOP), Ouro Preto, MG, Brazil
e-mail: milton@cead.ufop.br; oreydc@cead.ufop.br

G.A. Stillman et al. (eds.), *Teaching Mathematical Modelling: Connecting to Research and Practice*, International Perspectives on the Teaching and Learning of Mathematical Modelling, DOI 10.1007/978-94-007-6540-5_6,

them to organise, analyse, comprehend, understand, model, and solve problems located in real-life contexts. These tools allow for identification and description of mathematical ideas, procedures, and practices by schematising, formulating, and visualising a problem in different ways, discovering relations and regularities, and translating real world problems to mathematical ideas through mathematisation.

Inclusion of a diversity of ideas brought by students from other cultural groups can give confidence and dignity to students, while allowing them to see a variety of perspectives and provide a base for learning academic-Western mathematics (Bassanezi 2002). Equally important is the search for alternative methodological approaches. One alternative methodological approach is *ethnomodelling* (Rosa and Orey 2010), a practical application of ethnomathematics, and which adds the cultural perspective to modelling concepts.

When justifying the need for a culturally bound view on mathematical modelling, our sources are rooted in the theory of ethnomathematics and modelling (Bassanezi 2002; D'Ambrosio 1990; Rosa and Orey 2003). Research on culturally bound modelling ideas seeks to address the problem of mathematics education in non-Western cultures by bringing the cultural background of students into the traditional curriculum (Rosa and Orey 2010).

2 Ethnomathematics

Ethnomathematics as a research paradigm is wider than traditional concepts of mathematics, ethnicity or any current sense of multiculturalism. Ethnomathematics is described as the arts and techniques (*tics*) developed by individuals from diverse cultural and linguistic backgrounds (*ethno*) to explain, to understand, and to cope with their own social, cultural, environmental, political, and economic environments (*mathema*) (D'Ambrosio 1990). *Ethno* refers to distinct groups identified by cultural traditions, codes, symbols, myths, and specific ways of reasoning and inferring. Detailed studies of mathematical procedures and practices of distinct cultural groups most certainly allow us to further our understanding of the internal logic and mathematical ideas of diverse groups of students.

Ethnomathematics is the intersection of cultural anthropology, mathematics, and mathematical modelling, which is used to help students understand and connect diverse mathematical ideas and practices found in their communities to traditional-academic mathematics (Fig. 6.1).

Ethnomathematics, as well, is a program that seeks to study how students have come to understand, comprehend, articulate, process, and ultimately use mathematical ideas, procedures, and practices that enable them to solve problems related to their daily activities. This holistic context helps students to reflect, understand, and comprehend extant relations among all components of systems under study. In this regard, educators should be empowered to analyse the role of students' *ethnoknowledge* in the mathematics classroom (Borba 1990), which is acquired by students in the process of pedagogical action of learning mathematics in culturally relevant educational systems.

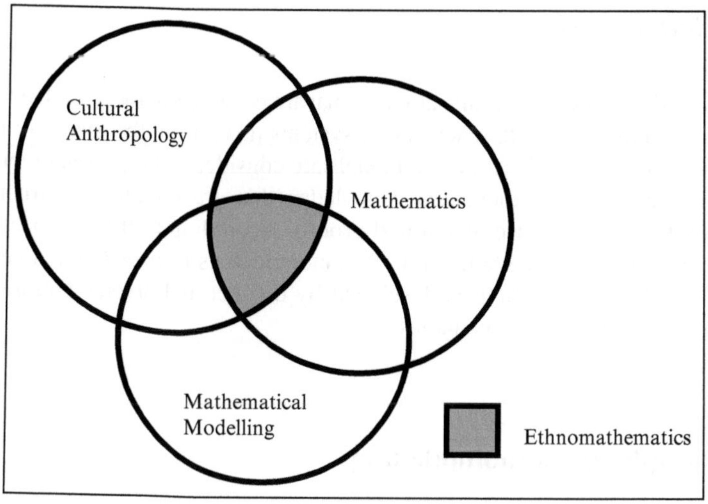

Fig. 6.1 Ethnomathematics as an intersection of three research fields (Rosa 2000)

3 Ethnomodelling

Ethnomodelling is the process of the elaboration of problems and questions that grow from real situations, which form an image or sense of an idealised version of the *mathema*. According to Rosa and Orey (2006), the focus of this perspective essentially forms a critical analysis of the generation and production of knowledge (creativity), and forms an intellectual process for its production, the social mechanisms of institutionalisation of knowledge (academics), and its transmission (education). D'Ambrosio (2000) affirmed that "this process is [called] modelling" (p. 142). By analysing reality as a whole, ethnomodelling allows those engaged in the modelling process to study systems of reality in which there is an equal effort to create an understanding for all components of the system as well as the interrelationships among them (D'Ambrosio 1993).

The use of modelling as pedagogical action for an ethnomathematics program values students' previous knowledge and traditions (Rosa and Orey 2007). This is done by developing student capacity to assess and translate daily phenomena and by elaborating mathematical models in their different applications. The ethnomodelling process starts with the social context, reality, and interests of students and not by enforcing a set of external values and decontextualised curricular activities without meaning for the students. This process is defined as "the mathematics practiced and elaborated by different cultural groups, which involves the mathematical practices present in diverse situations in the daily lives of diverse group members" (Bassanezi 2002, p. 208). In this regard, ethnomodelling uses mathematics as a language for understanding, simplification, and resolution of problems.

3.1 Ethnomodels

We define ethnomodels as cultural artefacts that are pedagogical tools used to facilitate the understanding and comprehension of systems taken from the reality of cultural groups (Rosa and Orey 2009). Ethnomodels are considered the external representations consistent with mathematical knowledge that are socially constructed and shared by members of specific cultural groups. According to this perspective, the primary objective for the elaboration of ethnomodels is to *translate* mathematical ideas, procedures, and practices developed by distinct and diverse cultural group members into academic mathematics.

4 Examples of Ethnomodelling

Many interesting models are formulated by using information and data obtained from studies and research related to ethnomathematics, which proposes the examination of knowledge systems adopted by distinct cultural groups. When some of this knowledge consists of mathematical ideas and procedures found via ethnomodels, we are better able to reach the origin of mathematical practices.

4.1 Measuring Land

Knijnik (1996) proposed activities related to the demarcation of land with the participants of the Landless Peoples' Movement (Movimento dos Sem Terra – MST) in Southern Brazil. The demarcation activity examined the method of *cubação* of the plots, which is a traditional mathematical practice applied by the participants of this movement. Flemming et al. (2005) defined *cubação* of the land as the solution of "problems of the measurement of land using diverse shapes" (p. 41). The use of the *cubação* land practice as a pedagogical proposal to elaborate activities for teaching and learning mathematics shows the importance of the contextualisation of problems in the learning environment of ethnomodelling through the elaboration of ethnomodels.

4.1.1 An Ethnomodel to Calculate the Area of Land

The landless people needed to calculate the area of figures with irregular quadrilateral shapes (Fig. 6.2a). One of these problems states that it is necessary to *calculate the area of land, which has a quadrilateral shape that measures 114 metres x 152*

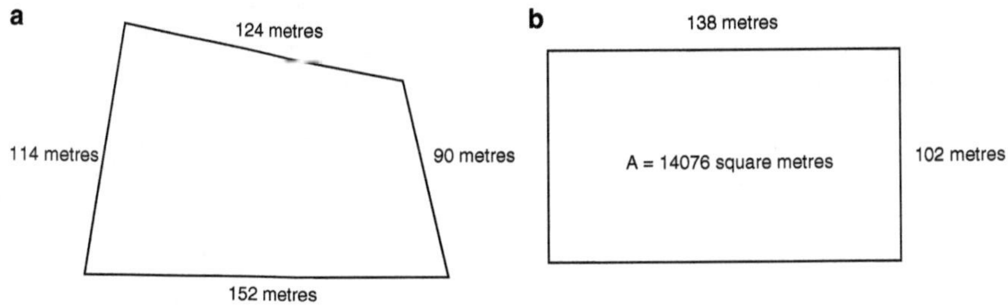

Fig. 6.2 (**a**) A problem to calculate the area of a figure with irregular quadrilateral shape. (**b**) Representation of a model that transforms the irregular quadrilateral in a rectangle

metres x 90 metres x 124 metres" (Fleming et al. 2005, p. 42). Thus, the mathematical knowledge of the landless can be represented by a model that transforms the shape of the given land into a rectangle of 138 m × 102 m with an area of 14,076 m^2 (Fig. 6.2b).

The model of this mathematical practice can be explained by the following ethnomodel:

- Transform the shape of the irregular quadrilateral into a rectangle whose area can be easily determined through the application of the formula $A = b \cdot h$.
- Determine the dimensions of the rectangle by calculating the mean of the two opposite sides of the irregular quadrilateral.

$$Base = \frac{152 + 124}{2} = 138 \text{ m}$$

$$Height = \frac{114 + 90}{2} = 102 \text{ m}$$

- In order to determine the area of this irregular quadrilateral, it is necessary to determine the area of the rectangle.

$$A = b \cdot h$$
$$A = 138 \cdot 102$$
$$A = 14,076 \text{ m}^2$$

Regarding this problem, there is another procedure from the mathematical knowledge of the landless that can be explained through another ethnomodel. The irregular shaped quadrilateral parcel presented in this example can also be transformed into "a square with sides of 120 metres, therefore with an area of 14400 square metres" (Flemming et al. 2005, p. 42). It is possible to observe

Fig. 6.3 Geometric scheme used by wine producers in the construction of wine barrels (Bassanezi 2002, p. 47)

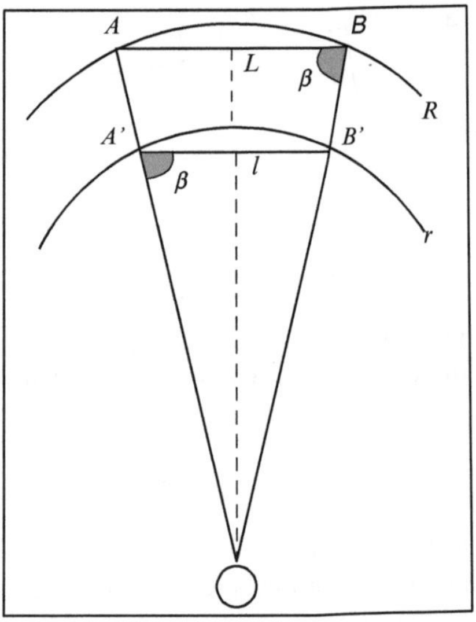

that the value of 120 was calculated by adding the dimensions of the quadrilateral and then dividing it by four, which is the number of sides of the irregular quadrilateral.

A model is efficient if we realise that it is only an approximation of reality (Bassanezi 2002). Thus, both methods present an approximated calculation of the area of the irregular quadrilateral that satisfies the needs of this specific group (Flemming et al. 2005).

4.2 The Wine Barrel

D'Ambrosio (2002) noted an ethnomathematics example that offers us a mathematical modelling methodology, with a group of Brazilian teachers who studied wine production. The motivation was to find the volume of wine barrels and to apply the techniques learned by ancestors of the wine producers who came to Southern Brazil as Italian immigrants in the early twentieth century.

In order to construct a wooden wine barrel with pre-established volume, it is necessary for wine producers to cut wooden strips to fit perfectly. This process drew the attention of the students who were interested in knowing what kind of inherited mathematics the wine producers were using in their geometric schemes. In Fig. 6.3, the larger circle (R) represents the base of the barrel while the smaller circle (r) represents its cover. The wine barrels are shaped like a truncated cone and are constructed by interlocking wooden strips (Fig. 6.4).

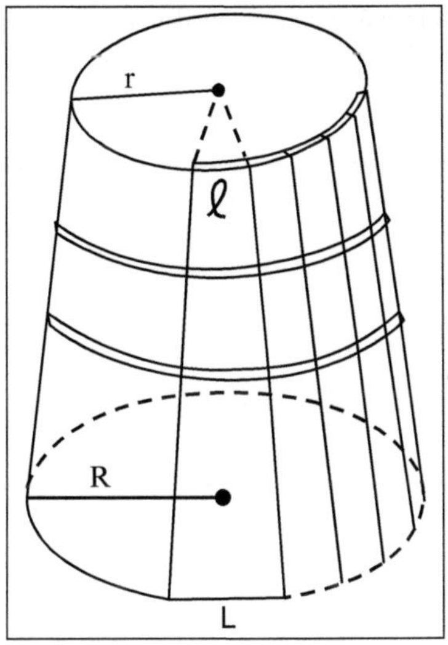

Fig. 6.4 Wine barrel shaped like a truncated cone (Bassanezi, 2002, p. 48)

This process was investigated from an ethnomathematics perspective as the cultivation of vines and production of wine barrels are linked strongly to the history and culture of people in that particular region of Brazil. The wine case study is an excellent example that typifies the connection between ethnomathematics and mathematical modelling through ethnomodelling (D'Ambrosio 2002; Rosa and Orey 2010).

4.3 Modelling the Tipi

The word *tipi* from the Sioux language refers to a conical dwelling common among the prairie peoples of North America. Spatial geometry is inherent in the shape of the tipi which was used to symbolise the universe in which the people lived. The nomadic prairie people observed that the tripodal foundation appeared to be perfectly adapted for the harsh environment in which they lived (Orey 2000). In this regard, if we look at variations between tripodal and quadripodal tipis, there is some evidence that they have an understanding of geometry concepts such as triangles and their geometrical characteristics and properties, which show manifestations of mathematical knowledge. The majority of Sioux tipis use the tripod foundation or three-pole foundation because it is stronger and offers a firmer foundation than a quadripodal or four-pole tip foundation (Orey 2000).

Fig. 6.5 Tripodal foundation of the Tipi

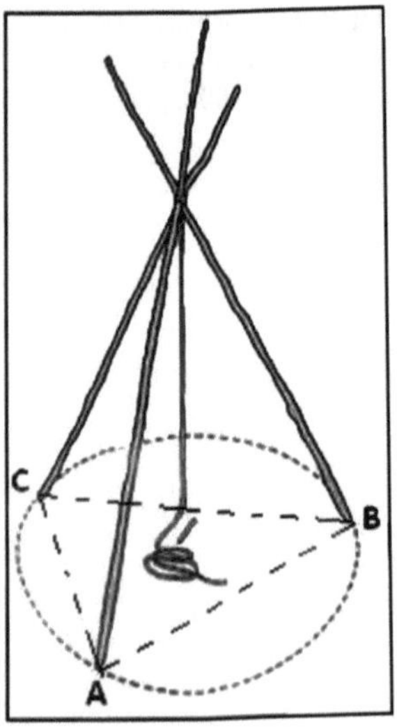

4.3.1 Tripodal Versus Quadripodal Foundations of the Tipi

Mathematisation helps us to explain why a tripod is more stable than a quadripodal or four-legged structure. Imagine three points, A, B, and C, which are not collinear. There are an infinite number of planes that pass through points A and B that contain line AB. Only one of these planes also passes through point C. Three points determine a plane if, and only if, they are non-collinear. In terms of geometry, this can be explained by using the plane postulate, which states that through three non-collinear points, there is exactly one plane.

For example, in a four-legged table, there is the possibility of the extremity of one of the legs not belonging to the same plane. A table that has three legs, however, is always balanced. Similar to a three-legged table, the tripod foundation of the tipi (Fig. 6.5) appears to be perfectly adapted for the harsh environment in which it was used as it provided a stable structure that was lightweight and portable. At the same time, it withstood the prevailing winds and extremely variable weather of this region. Let us look at this information mathematically.

The base formed by the tripod is ΔABC in which the midpoints of each of the sides are points M, N, and P (Fig. 6.6a). In this regard, it is possible to connect the midpoint of each opposite side of ΔABC to each of its vertices, which form line segments *AM*, *BN*, and *CP.* These line segments form three medians that intersect at only one point called the *centroid*, which is the balance point or centre of gravity of ΔABC (Fig. 6.6b).

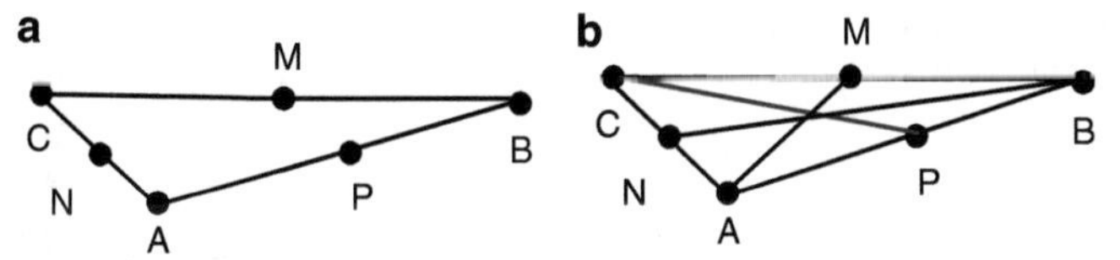

Fig. 6.6 (**a**) The tripod base of the Tipi. (**b**) The three medians of ΔABC

Fig. 6.7 The circumcentre of ΔABC

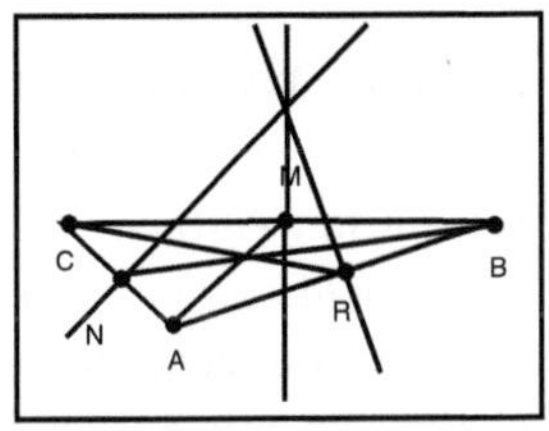

4.3.2 Determining the Centre of the Circular Base of the Tipi

Tipi dwellers placed the altar and fire at the centroid of the tipi because it "holds a definite power and holiness" (Orcy 2000, p. 246). It is possible to use geometric concepts to determine how tipi dwellers determine the centre of the circular base of the Tipi. An ethnomodel can explain how to determine the centre, height and lateral area of the Tipi cover. By using the triangle formed by the tripod, it is possible to trace each one of the sides of the triangle by using lines passing perpendicularly through the midpoints of each of its sides. It can be shown that these lines are perpendicular bisectors. These bisectors are intercepted at the same point, which is an equal distance from the vertices of the triangle. This point is called the *circumcentre* (Fig. 6.7).

The Tipi scaffolding itself is constructed by placing poles around its tripod foundation, thus assisting in the calculation of the area of its base. This means that the poles are placed as if they were points on a circumference. However, it is important to highlight that the base of the Tipi is slightly oval because it is not a perfect cone (Fig. 6.8).

The centre of the Tipi holds a definite power and holiness, which is more than just necessity or aesthetics that went into finding the centre of the Sioux home. The mathematical ideas implicit in this mathematical knowledge was passed on to the members of the Sioux people across generations by the women, who were responsible for the construction, and upkeep of these unique conical dwellings. From this, it is possible to conclude that Sioux understanding of the strength of tripodal construction is valid general and universal knowledge compatible with Western mathematical knowledge.

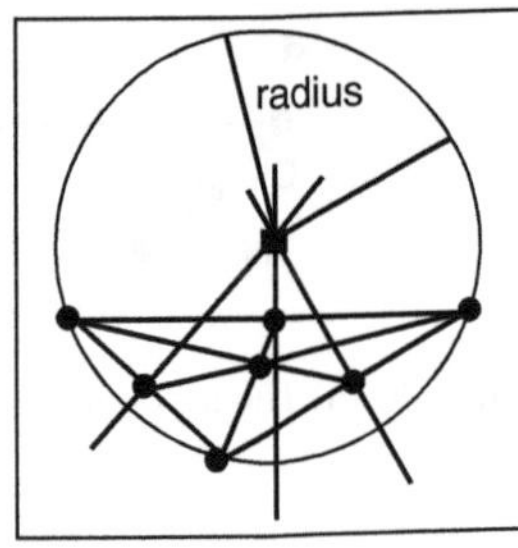

Fig. 6.8 The circular area of the base of the Tipi

5 The Methodology of Ethnomodelling

What is most difficult for researchers and educators is to learn how to connect what they would consider fundamental in academic mathematical ideas and procedures of the school community to what concepts have become almost universal in their mathematical practices. Most importantly, it is crucial to understand how to translate this knowledge into formalised aspects of academic mathematics through ethnomodelling (Rosa and Orey 2007).

There are three reasons for the application of ethnomodelling as a methodology for ethnomathematics:

1. Ethnomodelling is an effective path that can be used to reach traditional mathematical concepts.
2. Ethnomodelling can be used to develop intercultural classroom activities.
3. Ethnomodelling is a pedagogical action that can be used to transform the relationship between mathematics and society.

This paradigm suggests that developing a curricular praxis of ethnomodelling by investigating the ethnomathematics of a culture in constructing a mathematics curriculum values contributions of other mathematical knowledge traditions. One way to achieve this goal is for teachers to interpret alternative ethnomathematical approaches by starting with the outside sociocultural reality of their students. However, students may refuse to study their reality because it may be oppressive, and this may mean that students may not be able to identify their reality as contextualised mathematics.

On the other hand, they have a grounded mathematical knowledge based on previous experiences. In this educational process, perhaps teachers should not start with the students' own realities, but start with their own conceptions of mathematics, even though they might be traditional. Further, teachers should be encouraged to explore student ethnomathematical knowledge by applying contextualised mathematics activities to uncover important mathematical ideas and concepts.

It is beneficial to apply an ethnomathematical ethnographic perspective in order to come to a good understanding of mathematical aspects of a given cultural group, and having a clear purpose of this educational activity. For best practices, implementing an ethnomodelling perspective must be preceded by an inventory of previous knowledge of the student. Coming to understand student context, simply put is good pedagogy, and provides for the construction of relevant mathematical experiences that allow for a deeper appreciation of mathematical beauty and utility. It is important to understand what mathematical ideas, procedures, and practices are important to their particular cultural environment and historical contexts.

6 Final Considerations

Any study of ethnomathematics using modelling represents a powerful means for validating students' real life experiences and allows students to become familiar with tools that may enable them to become full participants in society. In this process, the discussion between teachers and students about the efficiency and relevance of mathematics in different contexts should permeate instructional activities. The role of teachers is to help students to develop a critical view of the world by using mathematics.

It is necessary that current researchers continue to investigate ethnomodelling in terms of non-Western cultural contexts and consider implementing new views into old themes. In this regard, we hope that this discussion broadens the discussion of possibilities and potentialities for the inclusion of ethnomathematics and mathematical modelling perspectives that respect social and cultural diversity of all students with guarantees for the understanding of our differences through dialogue and respect.

References

Bassanezi, R. C. (2002). *Ensino-aprendizagem com modelagem matemática* [Teaching and learning with mathematical modelling]. São Paulo: Editora Contexto.

Borba, M. C. (1990). Ethnomathematics and education. *For the Learning of Mathematics, 10*(1), 39–43.

D'Ambrosio, U. (1990). *Etnomatemática* [Ethnomathematics]. São Paulo: Editora Ática.

D'Ambrosio, U. (1993). Etnomatemática: um programa [Ethnomathematics: A program]. *A Educação Matemática em Revista, 1*(1), 5–11.

D'Ambrosio, U. (2000). Etnomatemática e modelagem [Ethnomathematics and modelling]. In M. C. Domite (Ed.), *Anais do Primeiro Congresso Brasileiro de Etnomatemática – CBEm-1* (p. 142). São Paulo: FE-USP.

D'Ambrosio, U. (2002). *Etnomatemática*: elo entre as tradições e a modernidade [Ethnomathematics: a link between traditions and modernity]. Belo Horizonte, MG: Autêntica.

Flemming, D. M., Flemming Luz, E., & Collaço de Mello, A. C. (2005). *Tendências em educação matemática* [Tendencies in mathematics education]. Pallhoça: UNISUL.

Knijnik, G. (1996). *Exclusão e resistência: educação matemática e legitimidade cultural* [Exclusion and resistance: Mathematics education and cultural legitimacy]. Porto Alegre: Ed. Artes Médicas.

Orey, D. C. (2000). The ethnomathematics of the Sioux tipi and cone. In H. Selin (Ed.), *Mathematics across culture: The history of non-western mathematics* (pp. 239–252). Dordrecht: Kulwer Academic Publishers.

Rosa, M. (2000). *From reality to mathematical modelling: A proposal for using ethnomathematical knowledge*. Unpublished master thesis, California State University, Sacramento.

Rosa, M., & Orey, D. C. (2003). Vinho e queijo: etnomatemática e modelagem! [Wine and cheese: Ethnomathematics and modelling!]. *Bolema, 16*(20), 1–16.
Rosa, M., & Orey, D. C. (2006). Abordagens atuais do programa etnomatemática: delinenando-se um caminho para a ação pedagógica [Current approaches in ethnomathematics as a program: Delineating a path toward pedagogical action]. *Bolema, 19*(26), 19–48.
Rosa, M., & Orey, D. C. (2007). Cultural assertions and challenges towards pedagogical action of an ethnomathematics program. *For the Learning of Mathematics, 27*(1), 10–16.
Rosa, M., & Orey, D. C. (2009). Symmetrical freedom quilts: The ethnomathematics of ways of communication, liberation, and arts. *Revista Latinoamericana de Etnomatemática, 2*(2), 52–75.
Rosa, M., & Orey, D. C. (2010). Ethnomodeling: A pedagogical action for uncovering ethnomathematical practices. *Journal of Mathematical Modelling and Application, 1*(3), 58–67.

Chapter 7
Dual Modelling Cycle Framework for Responding to the Diversities of Modellers

Akihiko Saeki and Akio Matsuzaki

Abstract The modelling cycle (e.g., Blum and Leiß How do students and teachers deal with modelling problems? In: Haines C, Galbraith P, Blum W, Khan S (eds) Mathematical modelling (ICTMA12): education, engineering and economics. Horwood, Chichester, pp 222–231, 2007) contains the repeated processes that modellers are asked for in deepening their thinking. When we investigate the modelling cycle of modellers, we have to consider the diversities of various modellers in their modelling progress (Blum and Borromeo Ferri J Math Model Appl 1(1):45–58, 2009). In addition, problems or tasks can be changed from the initial real situation and problem with the aim of meeting the need of modellers. In this chapter, we make two modelling cycles parallel and focus on interactions between cycles, and call it the dual modelling cycle. We show three types of modelling cycles based on the dual modelling cycle framework through the example of one modelling cycle based on an *Oil Tank Task*, and another modelling cycle based on a *Toilet Paper Tube Task*.

1 Modelling Cycles for Responding to Diversities of Modellers

We use the Blum and Leiß (2007, p. 225) modelling cycle (Fig. 7.1) because this cycle indicates as its first step the "constructing" process from "real situation & problem" to "situation model" in the "rest of the world". Especially, progression of

A. Saeki (✉)
Graduate School of Education, Naruto University of Education, Naruto, Japan
e-mail: asaeki@naruto-u.ac.jp

A. Matsuzaki
Faculty of Education, Saitama University, Saitama, Japan
e-mail: makio@mail.saitama-u.ac.jp

G.A. Stillman et al. (eds.), *Teaching Mathematical Modelling: Connecting to Research and Practice*, International Perspectives on the Teaching and Learning of Mathematical Modelling, DOI 10.1007/978-94-007-6540-5_7,

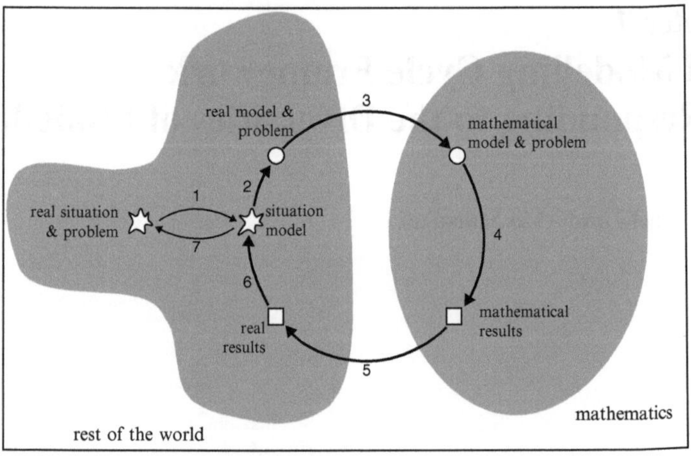

Fig. 7.1 Modelling cycle (Blum and Leiß 2007)

this process depends on the characteristics of each modeller and is the first cognitive barrier in tackling modelling tasks (Blum and Borromeo Ferri 2009). Borromeo Ferri (2007) focused on individual modelling progress and called such modelling progress a "modelling route" because each modeller's modelling progress differs. When we investigate the modelling cycle of modellers, we have to consider the diversities of various modellers in their modelling progress.

2 Background of Dual Modelling Cycle Framework

Polya (1988) advised problem solvers in the second step "devising a plan (responding to what is asked for)" of his problem solving model, "If you cannot solve the proposed problem try to solve first some related problem" (p. 10). Thus, variation between problems is performed by generalisation, specialisation, an analogy, and various decompositions and combinations. This idea is useful to understand and solve a problem, and provides the possibility of making the union which is helpful to solve the original problem.

2.1 Historical Background

Such variation of a problem is seen also in the history of the field of natural science. Galileo Galilei, for example, discovered the law of free-fall motion from the results of research of slope motion. In research of slope motion, he experimented by lessening friction on a slope and a ball, air resistance of the ball, and influence of rotation. Thus, when an experiment is directly impossible, experiments on the basis of

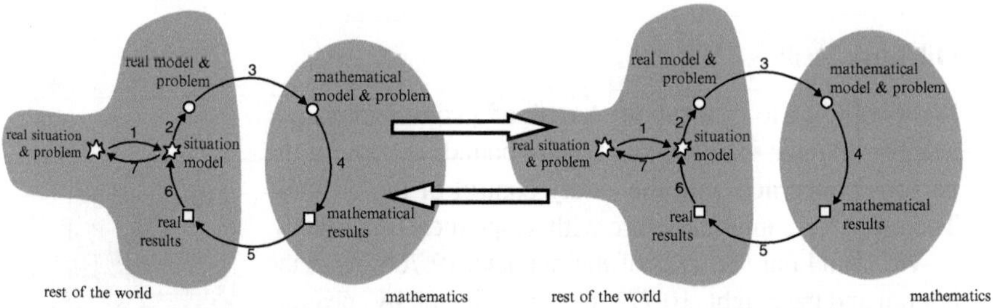

Fig. 7.2 Dual modelling cycle framework

resemblance/approximation conditions are one approach of the scientific method which measures or observes the result and finds out a rule and a law indirectly. For solving problems through such scientific methods, we built a framework using the modelling cycle for variation between problems.

2.2 *Framework for Corresponding Changes in the Modelling Tasks*

One of the important characteristics of modelling is deepening of the models constructed through repeating progress, and the conventional modelling cycles are also intended to represent this. Models and connections between variables which construct models change during modelling progression (Matsuzaki 2007, 2011; Stillman 1996; Stillman and Galbraith 1998), and a modeller might tackle tasks or problems that are different from the initial modelling task in this process. In this chapter, we make two modelling cycles parallel and focus on switching between these cycles, which we call the *dual modelling cycle*. One of the reasons for creating a dual modelling cycle is that there might be cases for changing from an initial modelling task to a similar modelling task through both modelling progressions. Furthermore, we hope that distinguishing of each modeller's "modelling route" becomes clear by making two modelling cycles parallel and focusing on switching between these cycles as in Fig. 7.2.

3 Three Types of Modelling Cycle from the Perspective of Dual Modelling Cycles

In this chapter, we use the *Oil Tank Task* as the initial modelling task and explain three types of modelling cycles.

Oil Tank Task

We would like to measure the length of a spiral banister of a heavy oil tank. However, it is out of bounds except for the persons concerned with the safety control of the oil tank. Therefore we cannot measure with a tape measure directly.

We found out the sizes of the diameter (9.766 m) at the bottom and the height (10.772 m) of the oil tank by speaking to the persons concerned. Based on these data, we decided to find the length of the spiral banister of the oil tank.

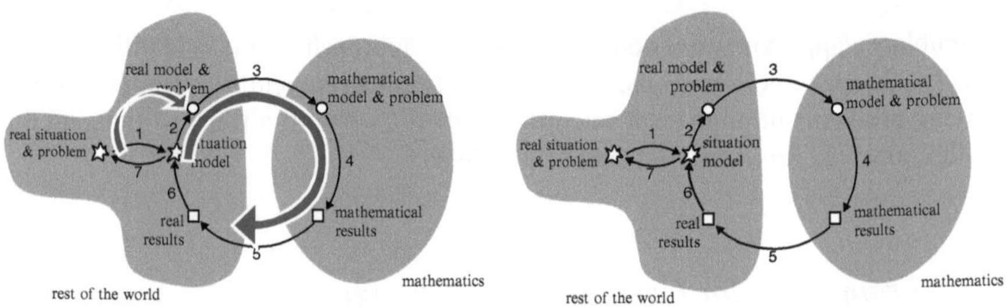

[Modelling cycle of oil tank] [The second modelling cycle]

Fig. 7.3 Single modelling cycle diagram

Though only two pieces of data (i.e., sizes of a diameter and the height) are available from the description of this task, we prepared other data (e.g., beginning height of a spiral banister from the ground 1.050 m) and would present these necessary data to each modeller only on demand during solving of the task.

3.1 *Type 1: Single Modelling Cycle*

If a modeller thinks of only the initial modelling task, we can explain the modelling by a typical single modelling cycle (see Fig. 7.1). In this instance, the second modelling cycle is not necessary for a modeller to solve the initial modelling task. As prior knowledge or experiences of each modeller are scaffolding the progress of the modelling cycle (Matsuzaki 2007, 2011), the first modelling becomes a basis for progression. In other words, we can explain this modelling by a single modelling cycle from the perspective of dual modelling cycles (see Fig. 7.3). In this case, we assume that a modeller will tackle the *Oil Tank Task* based on only an oil tank situation. For example, a modeller could make a parallelogram or a rectangle as a developing image model of an oil tank. This model is located in "real model & problem" stage because this figure is cut and developed along with a spiral banister.

Fig. 7.4 A toilet paper tube as a similar model for the spiral banister of an oil tank

3.2 *Type 2: Double Modelling Cycle*

When a modeller cannot forecast solutions of the initial task, he/she will be able to imagine other models based on personal prior knowledge or experiences (Matsuzaki 2007, 2011); otherwise his/her modelling progress is stopped at this point. In this case, some modellers might think that an oil tank looks like a toilet paper tube (see Fig. 7.4). At that time, transition from the modelling cycle for the oil tank to a modelling cycle for a toilet paper tube is implemented. Other modellers might imagine a signpole which can be seen in the front of some barber shops or a screw from the shape of the spiral banister of an oil tank. These images could be useful models for solving the *Oil Tank Task*. In this chapter we explain use of a toilet paper tube as a similar model for a spiral banister of the oil tank.

A slit of a toilet tube looks like the shape of a spiral banister. Additionally, it is impossible to open along the actual spiral banister of an oil tank, but it is easy to open along the slit of an actual toilet paper tube. So we assume a modelling cycle for a toilet paper tube as the second modelling cycle (Fig. 7.5) can be used because a toilet paper tube might be a similar model which we can use instead of an oil tank.

Thus, the initial modelling task is the *Oil Tank Task* and another modelling task is located in a second modelling cycle in this double modelling cycle. Some modellers might tackle the question as the new task located in the modelling cycle of a toilet paper tube: "How long is a spiral of a toilet paper tube?" They could imagine parallelograms as a developing model of the toilet paper tube. In this type of modelling, the two modelling cycles are separate modelling cycles, even if each modelling task is solved successfully or unsuccessfully. In other

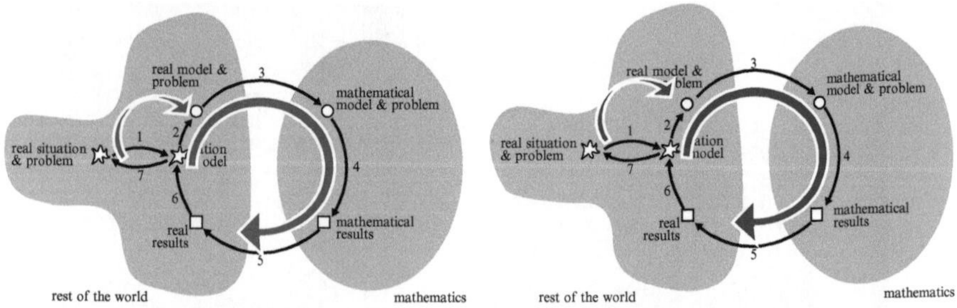

[Modelling cycle of oil tank] [Modelling cycle of toilet paper tube]

Fig. 7.5 Double modelling cycle diagram

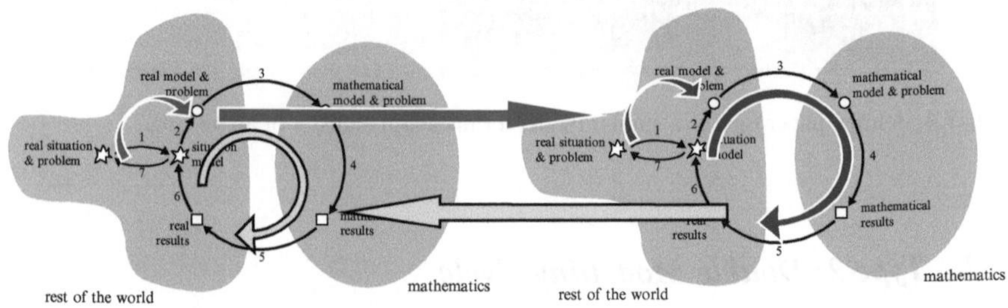

[Modelling cycle of oil tank] [Modelling cycle of toilet paper tube]

Fig. 7.6 Dual modelling cycle diagram

words, switching between the modelling cycle of the oil tank and modelling cycle of the toilet paper tube is not implemented.

3.3 Type 3: Dual Modelling Cycle

This type is advanced from type 2. In this type, a modeller can apply the results obtained through the second modelling cycle to the first modelling cycle (see Fig. 7.6). The differences between the double modelling cycle and the dual modelling cycle are (1) there is switching between two modelling cycles, and (2) progress in the first modelling cycle based on results or limitations obtained from the second modelling cycle is deliberate.

In this case, the initial modelling task is the *Oil Tank Task* and the second modelling task is the *Toilet Paper Tube Task*. One thing that our modelling teachers have to pay attention to in dual modelling cycles is that it is different for each modeller

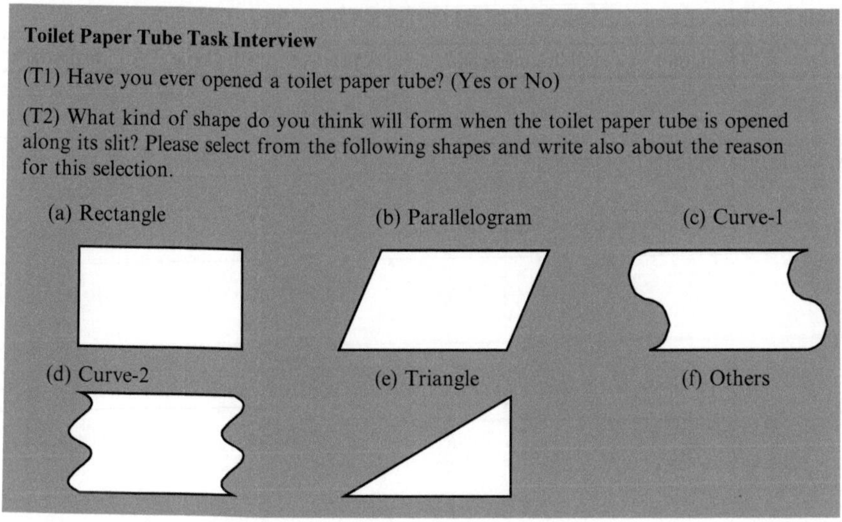

Fig. 7.7 *Toilet Paper Tube Task* interview

as he/she will apply answers from the second modelling cycle to the *Oil Tank Task*. This implies we have to teach bearing in mind the diversities of modellers. In addition, a modeller is not necessarily able to solve the *Oil Tank Task* even if he/she can solve the task located in the second modelling cycle.

4 Example of Problems in the Second Modelling Cycle

During the explanation of three types of modelling cycles, we have set the modelling cycle of a toilet paper tube as the second modelling cycle of the dual modelling cycle framework.

4.1 Street Interview regarding the Toilet Paper Tube Task

To gain a sense of the range of likely responses for the *Toilet Paper Tube Task* we conducted an interview (see Fig. 7.7) on the street on 2nd and 3rd September 2010. Collected results of responses ($N=41$) for each question are shown in Table 7.1.

There were a few responses for (a) rectangle for question (T2) even though the development of a cylinder is learnt in school mathematics. There was no characteristic relationship between the responses to questions (T1) and (T2) and experience opening a toilet paper tube.

Table 7.1 Responses for questions (T1) and (T2) of the *Toilet Paper Tube Task*

(T2)	(a) Rectangle		(b) Parallelogram		(c) Curve-1		(d) Curve-2		(e) Triangle	
(T1)	Yes	No	Yes	No	Yes	No	Yes	No	Yes	No
	0	2	5	9	4	6	1	0	5	9

Note. No one chose (f) for question (T2)

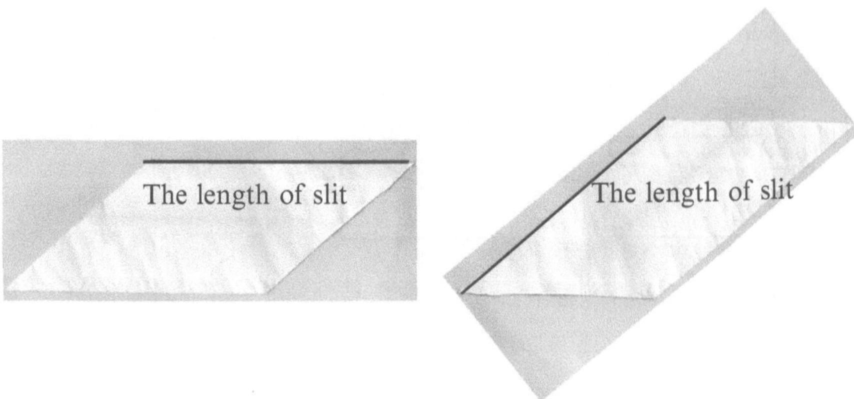

Fig. 7.8 Two cases for the base of parallelogram formed from opened toilet paper tube

5 Designing Lesson Practices

5.1 To Explore the Length of a Spiral Banister of the Heavy Oil Tank

For solving the *Oil Tank Task*, an advantage of using a toilet paper tube which is similar to the oil tank is that we can cut and open it easily. After opening it, modellers have to measure data for the resulting parallelogram. Thus, the modellers have to decide on the necessary variables for solving the *Oil Tank Task*. Variables for solving the *Toilet Paper Tube Task* are not necessarily those for solving the *Oil Tank Task*. There are two cases to decide between for a base of the parallelogram from the opened toilet paper tube. At first, we would start to consider the left hand parallelogram in Fig. 7.8.

We can measure the length of slit as 183 mm directly. Angle θ (see Fig. 7.9) is needed to explore the following relationship: $\sin\theta = \frac{80}{128} = 0.625$. From this value, angle θ is nearly 38°. Alternatively, we can measure this angle by using a protractor directly. In addition, the length of the diameter at the bottom of a toilet paper tube is 41 mm and the height of the tube is 112 mm.

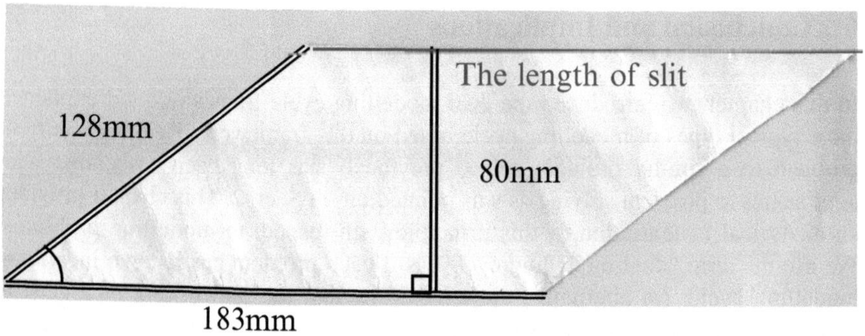

Fig. 7.9 Dimensions of a toilet paper tube

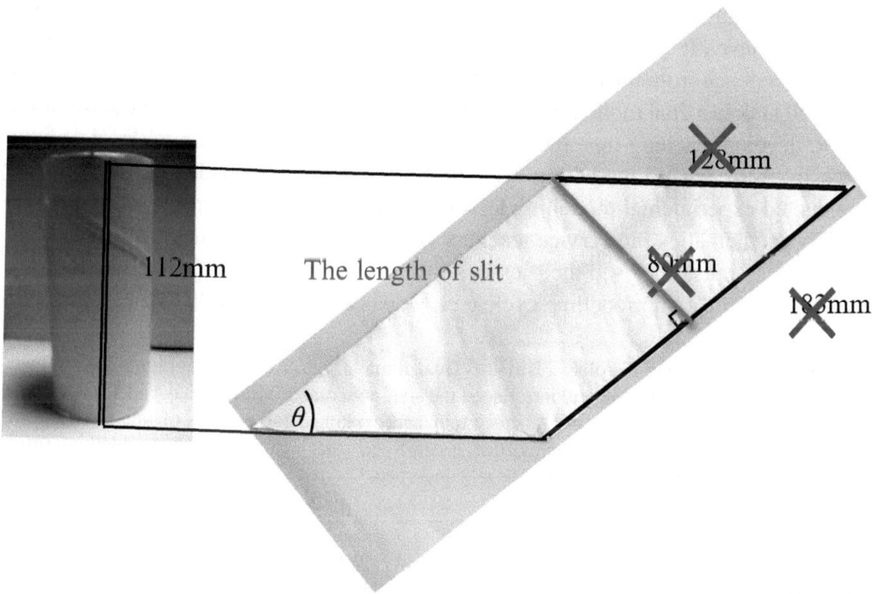

Fig. 7.10 Necessary measuring data to lead to the answer for the length of a slit

We can find the length of the slit of the toilet paper tube through measuring several lengths but necessary measuring data are only angle θ and the height of the toilet paper tube (Fig. 7.10). These data are necessary to lead to finding the length of a spiral banister of an oil tank through exploration based on a toilet paper tube as a model. Of course, modellers have to transfer the relationship of the data between the toilet paper tube and the oil tank.

6 Conclusion and Implications

In this chapter, we introduced the dual modelling cycle framework and explained three typical types of modelling cycle based on this framework. Relating a current problem to a similar problem solved previously has long been associated with approaches to problem solving as was pointed out in Sect. 2. This chapter provides an analytical instantiation of this principle with respect to modelling problems. We use the term "dual modelling cycle" as if each problem has its own individual modelling cycle. An alternative view would be that the outworking of a generic modelling cycle has characteristics that differ according to the specifics of a given problem. However, our purpose in using the dual modelling cycle has its basis in its utility in facilitating teaching when a current modelling task is able to be related to a similar modelling task in order to solve the initial task through feedback from the solution of the second.

We have implemented experimental modelling lessons for students at Year 5 primary level (Kawakami et al. 2012) and plan to design practice lessons for students at secondary level. It is important for implementing our modelling lessons that we teach how to feedback from the modelling cycle of a toilet paper tube to the modelling cycle of an oil tank in a dual modelling cycle. In other words, it is necessary for our researchers to develop teaching materials matched for students of each school level to feedback from the second modelling cycle to the first modelling cycle. We have already conducted experimental modelling lessons for undergraduate primary and lower secondary mathematics pre-service teachers with the intention of supporting this feedback process, and analysed their lessons from the perspective of providing empirical evidence for the dual modelling cycle (see Matsuzaki and Saeki 2013).

Acknowledgement Mr. Kurokawa Hiroshi (Tokushima office, Nisshinbo Holdings Inc.) allowed us to measure a heavy oil tank and introduced the staff concerned. Mr. Kurebayashi Yutaka (Tokai Seishi Kogyo Co., LTD.) gave us a factory tour and explained the process of making the toilet paper tube. Mr. Kainuma Kenji (Anan office, Asahi Synchrotech Corporation) provided important plan data of the heavy oil tank. Thanks to all concerned.

References

Blum, W., & Leiß, D. (2007). How do students and teachers deal with modelling problems? In C. Haines, P. Galbraith, W. Blum, & S. Khan (Eds.), *Mathematical modelling (ICTMA12): Education, engineering and economics* (pp. 222–231). Chichester: Horwood.

Blum, W., & Borromeo Ferri, R. (2009). Mathematical modelling: Can it be taught and learnt? *Journal of Mathematical Modelling and Application, 1*(1), 45–58.

Borromeo Ferri, R. (2007). Modelling problems from a cognitive perspective. In C. Haines, P. Galbraith, W. Blum, & S. Khan (Eds.), *Mathematical modelling (ICTMA12): Education, engineering and economics* (pp. 260–270). Chichester: Horwood.

Kawakami, T., Saeki, A., & Matsuzaki, A. (2012). Necessity for modelling teaching corresponding to diversities: Experimental lessons based on dual modelling cycle framework for the 5th grade pupils. In *ICME-12 pre-proceedings* (pp. 3291–3300). Seoul: ICME.

Matsuzaki, A. (2007). How might we share models through cooperative mathematical modelling? Focus on situations based on individual experiences. In W. Blum, P. Galbraith, H.-W. Henn, & M. Niss (Eds.), *Modelling and applications in mathematics education: The 14th ICMI study* (pp. 357–364). New York: Springer.

Matsuzaki, A. (2011). Using response analysis mapping to display modellers' mathematical modelling progress. In G. Kaiser, W. Blum, R. Borromeo Ferri, & G. Stillman (Eds.), *Trends in teaching and learning of mathematical modelling: ICTMA14* (pp. 499–507). New York: Springer.

Matsuzaki, A., & Saeki, A. (2013). Evidence of a Dual Modelling Cycle: Through a Teaching Practice Example for Pre-service Teachers. In G. Stillman, G. Kaiser, W. Blum, & J. Brown (Eds.), *Teaching mathematical modelling: Connecting to research and practice* (pp. 195–205). New York: Springer.

Polya, G. (1988). *How to solve it: A new aspect of mathematical method* (1st Princeton Science Library ed.). Princeton: Princeton University Press.

Stillman, G. (1996). Mathematical processing and cognitive demand in problem solving. *Mathematics Education Research Journal, 8*(2), 174–197.

Stillman, G., & Galbraith, P. (1998). Applying mathematics with real world connections: Metacognitive characteristics of secondary students. *Educational Studies in Mathematics, 36*(2), 157–195.

Chapter 8
The Eyes to See: Theoretical Lenses for Mathematical Modelling Research

Nils Buchholtz

Abstract This chapter will contribute to the discussion in one of the two symposia held at ICTMA-15, in particular the symposium about the role of theory in the research about mathematical modelling. First, the necessity to discuss theoretical approaches in the research about mathematical modelling and the idea of the symposium will be pointed out. Subsequently in the second part of the chapter the role of theory in research will be described more differentially. Finally, the contributions to the symposium are classified concerning the role of theory in the theoretical approach. The chapter closes with open questions still remaining, which may guide future symposia on the topic.

1 Introduction

In the preface of the first volume of the newly established series "International Perspectives on the Teaching and Learning of Mathematical Modelling" Kaiser et al. (2011) point out the diversity that has evolved over the past 30 years in the discussion of mathematical modelling and applications within the scientific community. The growing interest has previously culminated in an ICMI study on this theme (Blum et al. 2007) and is evidence of the increasing importance of the scientific discussion of this particular field of research within mathematical and mathematics educational research. Increasing diversity, however, in many cases is accompanied with an increasing complexity of the respective methodological, theoretical and practical approaches to research. Not least for this reason, the

N. Buchholtz (✉)
Faculty of Education, Psychology and Human Movement,
University of Hamburg, Hamburg, Germany
e-mail: Nils.Buchholtz@uni-hamburg.de

G.A. Stillman et al. (eds.), *Teaching Mathematical Modelling: Connecting to Research and Practice*, International Perspectives on the Teaching and Learning of Mathematical Modelling, DOI 10.1007/978-94-007-6540-5_8,

diversity in research about mathematical modelling also was one of the basic points of discussion within the Modelling and Applications Working Group at the 5th Congress of the European Society for Research in Mathematics Education (CERME 5) in 2007.

A first classification of the variety of research approaches was formulated by Kaiser and Sriraman previously in 2006, who classified the different approaches on the basis of their understanding of goals of modelling and distinguished various perspectives related to the central educational aims in mathematics modelling (Kaiser and Sriraman 2006). Gabriele Kaiser gave a plenary lecture related to this classification at ICTMA-15 (see Kaiser and Stender, Chap. 23 this volume). Although the classification was considered to be very useful in helping and understanding different approaches, it was also stressed that such a classification is a strong systematising simplification, which might not reflect all research approaches appropriately. Nevertheless, the classification is an important working instrument, but the need to revise this classification was also mentioned. The CERME 5 Working Group revised the classification considering the main point of criticism being that the classification of Kaiser and Sriraman did not distinguish between didactical approaches and research perspectives. "Didactical approaches are characterized by a normative orientation concerning the overall aims of applications and modelling in mathematics education in contrast to research perspectives, which guide studies on special aspects concerning applications and modelling" (Kaiser et al. 2007, p. 2036).

This distinction clarifies in greater degree the differentiation between practice- and research-oriented approaches, but always with the proviso of a better mutual understanding. An essential feature of the diversity of different approaches was seen in the theoretical perspective: "It makes clear, that the various approaches promoting applications and modelling in school or university teaching come from very different theoretical perspectives spanning the debate from ethno-mathematics to problem solving" (Kaiser et al. 2007, p. 2041).

The reference to theory within different research approaches can vary and be less pronounced. For this reason, the question is legitimated, whether the reference to various scientific theories can serve as a distinguishing criterion of research approaches in general. If one wants to classify different research approaches, the necessary abstraction of key criteria of differentiation makes a comparison difficult in general. What criteria are appropriate and how many criteria should be considered, by which research approaches can be classified?

As the integration of theory into the research approach as a classification criterion was mentioned in the previous discussions, at ICTMA-15 for the first time a symposium on the role of theory in various research approaches was conducted. Jonei Barbosa,[1] Vince Geiger and Milton Rosa presented their different research approaches with emphasis on the theoretical reference.

[1] No chapter submitted.

2 Role of Theory in the Research Approach

The symposium was named "The Eyes to See: Theoretical Lenses for Mathematical Modelling Research". This title incorporates a reference to the influence of (a specific) theory on the research process. As research in general is a hermeneutic process, theory can – in a very simplified manner – be considered as a pair of glasses with which the researcher looks at his or her research object. A research object like mathematical modelling therefore appears within the different research approaches exactly the way as described by the underlying theory. More generally, already in the 1960s the American philosophers of science, Norwood Russell Hanson and Thomas Samuel Kuhn, were of the opinion that observations are in principle "theory-laden". Research results in this sense never appear pure, and a fundamental conception of research, which neglects a theoretical context of research results, therefore would be inadequate (Hanson 1958; Kuhn 1962).

This role of the influence of theory was emphasised by the title of the symposium in two ways: firstly by the term "lens", secondly by the word "theory" itself, which has its etymological meaning derived from the Greek θεωρεἰν which means "to view" or "to observe" and therefore refers to the perspective, under which a researcher is investigating a research object. Understanding the integration of theory into the research act as a filter, the resulting diversity of theoretical approaches even in the research on mathematical modelling is not surprising at all.

On the contrary, the diversity of theories in research can be perceived not only as a delimiting criterion in the context of classifications, but also as a rich resource for grasping complex realities. Bikner-Ahsbahs and Prediger (2010), who analysed the role of theory in mathematics educational research in general, highlight two important aspects on this: (1) Theories evolve independently in different regions of the world and different cultural circumstances. (2) The complexity of the topic of research itself, since mathematics learning and teaching is a multi-faceted phenomenon which cannot be described by one monolithic theory alone, means a variety of theories is necessary to do justice to the complexity of the field. Not least because of this, the Theory Working Group of CERME claims that the notion of different intentions behind research effort calls for strategies for comparing and connecting theories or research obtained results using different theoretical approaches (Prediger et al. 2010). However, concerning comparability, the necessary (relatively rough) distinction of Kaiser et al. (2007) between different theoretical backgrounds like the problem-solving debate for contextual modelling or the socio-critical approaches to political sociology for socio-critical and socio-cultural modelling is limited in distinguishing different research approaches as long as the range of theoretical reference within one research approach is not clarified.

If the theoretical perspective should serve as a distinguishing feature of different research approaches, according to Bikner-Ahsbahs and Prediger (2010) it has to become clear, what is meant by using the terms "theory" or "theoretical approach" or "theoretical perspectives". They suggest that there is no unique definition of shared theory and theoretical approach among mathematics education researchers. Theory can influence the research approach in different ways.

Bikner-Ahsbahs and Prediger (2010) distinguish different types of views on theory:

- A normative *more static view*, which regards theory as a human construction to present, organize and systemize a set of results about a piece of the real world, which becomes a tool to be used. In this sense, a theory is given to make sense of something in some kind and some way (for example Bernstein's structuralist perspective…)
- And a *more dynamic view*, which regards a theory as a tool in use, rooted in some kind of philosophical background which has to be developed in a suitable way in order to answer a specific question about an object. In this sense the notion of theory is embedded in the practical work of researchers. It is not ready for use, the theory has to be developed in order to answer a given question (for example…an interpretative approach…). In this context, the term 'theoretical approach' is sometimes preferred to 'theory'. (p. 485).

Bikner-Ahsbahs and Prediger note with this dialectic understanding of theory:

If we approach the notion of theory in this way from its role in research and from research practices, theories can be understood as guiding research practices and at the same time being influenced by or being the aim of research practices. (p. 487)

3 Examples of Theory Guiding Research

Against this background, let us have a closer look at the research featured in the symposium:

Vince Geiger participated with a contribution in which he considers the potential of a tetrahedral model for teaching and learning "to describe and explain mathematical modelling and application activity that incorporates roles of students, teachers, mathematical knowledge and artefacts during engagement with mathematical applications and modelling" (Geiger, Chap. 9 this volume). He refers, from the perspective of research into mathematical modelling, to a more general theoretical model originally formulated by Rudolf Strässer and applies it for the description and explanation of the activity that takes place in the process of mathematical modelling in a holistic way. He illustrates the applicability of Strässer's tetrahedral model to mathematical modelling and application activity by analysing an episode from his research data. "In this episode the teacher was working with a year 11 class on a unit of work which engaged students in the study of a range of mathematical functions." It shows that "consideration must be given to different types of interactions between participants and between participants and artefacts." The result of the application is that Strässer's tetrahedral model seems to work for the interactions between the students and the digital tool, but does not describe the social interactions between the participants of this episode sufficiently well. This is why another sphere is added to the tetrahedral model, which already Strässer had pointed to: Chevallard's "noosphere". "This implies that a socio-cultural perspective must be considered when developing frameworks which incorporate the concept of a noosphere." Geiger combines the two theoretical approaches in order to describe his episode. He summarises "that there is still much research to be done into how the role of social aspects of learning and teaching can be theorised together with the roles of students, teachers, secondary artefacts and mathematical knowledge."

Milton Rosa and Daniel Clark Orey contributed to the symposium with a presentation on ethnomodelling as a research lens (Rosa and Orey, Chap. 10 this volume). Ethomodelling "may be considered as the intersection area of cultural anthropology, ethnomathematics, and mathematical modelling". They see ethnomodelling as an alternative methodological approach to analyse mathematical modelling and mathematical ideas in different cultural contexts and describe the dilemma between an "emic" and an "etic" perspective in this field of research, which occurs when taking ethnomodelling as a research lens. "Using an ethnomodelling perspective, the emic constructs are the accounts, descriptions, and analyses expressed in terms of the conceptual schemes and categories that are regarded as meaningful and appropriate by the members of the cultural group under study." A cultural bias may occur if researchers, who are not embedded in the culture they observe, assume that an emic construct is actually etic. "Thus, according to this context, mathematical ideas and procedures are etic if they can be compared across cultures using common definitions and metrics". Rosa and Orey claim that researchers and investigators should gain an informed sense of distinctions of the cultural-dependent mathematical knowledge used in modelling. This dilemma has a pedagogical inference. They argue that traditional mathematical modelling does not fully take into account the implications of the cultural aspect of human social systems and suggest regarding ethnomodelling as a pedagogical approach which integrates an emic perspective into a mathematical modelling curriculum.

The different contributions presented at the symposium vary in the way theory is embedded in the research approach. With Rosa and Orey, the theoretical approach can be seen in the orientation to whole research paradigms (like the sociological perspective or the ethno-modelling). Geiger's approach, in contrast, refers to local theoretical approaches that relate to specific aspects of the research object (Strässer's tetrahedral model) and therefore can be combined with each other. Common to all contributions was the noting that a specific or broader theory is used to explain the research object or to interpret modelling in mathematics education (more static view). In the interpretive approach of Geiger and the development of Strässer's theory, however, we also find a dynamic view of theory.

4 Open Questions

Bikner-Ahsbahs and Prediger (2010, p. 487) develop a fivefold category schema for theory guiding research, allowing them to distinguish different research approaches concerning incorporation of theory.

> Most theories about and within mathematics education share their *research object* up to a certain point: it is about aspects of mathematics teaching and learning. But they differ in the *situations* that are considered and what exactly in these situations is theoretically

conceptualized, in the *methods* that are used for generating results for theory building and in their *aims*. … [They add] the *sort of questions considered to be relevant* […] to describe more precisely how research practice, background theory and its philosophical base are deeply interwoven.

This raises the question whether this scheme is an appropriate tool to describe the different approaches to research within the research on Mathematical Modelling. The symposium, “The Eyes to See: Theoretical Lenses for Mathematical Modelling”, was a first step to deal with the influence of theory in research on mathematical modelling. There still remain open questions which should be addressed in future discussions on the role of theory in research on Mathematical Modelling:

- How is theory actually influencing the different understanding of mathematical modelling and the research on mathematical modelling?
- Is the role of theory appropriate to distinguish between various research approaches on mathematical modelling?
- How can we combine different theories on mathematical modelling and what happens if different theories contradict each other?

References

Bikner-Ahsbahs, A., & Prediger, S. (2010). Networking of theories – An approach for exploiting the diversity of theoretical approaches. In G. Kaiser & B. Sriraman (Series Eds.) & B. Sriraman & L. English (Vol. Eds.), *Advances in Mathematics Education: Vol. 1. Theories of mathematics education: Seeking new frontiers* (pp. 483–506). New York: Springer.

Blum, W., Galbraith, P., Henn, H.-W., & Niss, M. (Eds.). (2007). *Modelling and applications in mathematics education: The 14th ICMI study*. New York: Springer.

Hanson, N. R. (1958). *Patterns of discovery: An inquiry into the conceptual foundations of science*. Cambridge: Cambridge University Press.

Kaiser, G., & Sriraman, B. (2006). A global survey of international perspectives on modelling in mathematics education. *ZDM – The International Journal on Mathematics Education, 38*(3), 302–310.

Kaiser, G., Sriraman, B., Blomhøj, M., & Garcia, F. J. (2007). Report from the working group modelling and applications: Differentiating perspectives and delineating commonalties. In D. Pitta-Pantazi & G. Philippou (Eds.), *CERME 5, proceedings of the fifth Congress of the European Society for Research in Mathematics Education* (pp. 2035–2041). Larnaca: ERME & University of Cyprus.

Kaiser, G., Blum, W., Borromeo Ferri, R., & Stillman, G. (2011). Trends in teaching and learning of mathematical modelling – Preface. In G. Kaiser, W. Blum, R. Borromeo Ferri, & G. Stillman (Eds.), *Trends in teaching and learning of mathematical modelling: ICTMA 14* (pp. 1–5). New York: Springer.

Kuhn, T. S. (1962). *The structure of scientific revolutions*. Chicago: University of Chicago Press.

Prediger, S., Bosch, M., Kidron, I., Monaghan, J., & Sensevy, G. (2010). Different theoretical perspectives and approaches in mathematics education research: Strategies and difficulties when connecting theories. In V. Durand-Guerrier, S. Soury-Lavergne, & F. Arzarello (Eds.), *CERME 6, proceedings of the sixth Congress of the European Society for Research in Mathematics Education* (pp. 1529–1534). Lyon: Institut National de Recherché Pédagogique.

Chapter 9
Strässer's Didactic Tetrahedron as a Basis for Theorising Mathematical Modelling Activity Within Social Contexts

Vince Geiger

Abstract This chapter considers the potential of Strässer's tetrahedral model for teaching and learning mathematics as a means for examining the activity and/or roles of students, teachers, mathematical knowledge and artefacts during engagement with mathematical applications and modelling. Research data, collected from mathematics classrooms, will be used to illustrate how the interactions and transformations that are part of this activity are played out in authentic classroom settings.

1 Introduction

Since research interest in mathematical modelling and applications began to emerge in the 1970s (Niss 2008) researchers have attempted to understand and describe the use of models and the process of mathematical modelling in mathematics classrooms. Niss (2008) observes that initially this movement focused on how students' made use of pre-existing models, this is, "models were taken for granted and used, but not built, and the activity concentrated on working solely within a mathematical context derived from the model" (p. 80). Over time, the models themselves, the constructions of these models, and students' active involvement in the process of construction became the centre of interest for mathematics education researchers. From this research, different branches of the field have developed including models of mathematical modelling, interest in modelling competencies, the role of modelling in assessment, and the role of technology in enhancing the mathematical modelling process (Stillman 2007). While these sub-fields of research explore important elements of the processes associated with engaging students with authentic

V. Geiger (✉)
Education QLD | Faculty of Education, Australian Catholic University,
Brisbane, QLD, Australia
e-mail: vincent.geiger@acu.edu.au

G.A. Stillman et al. (eds.), *Teaching Mathematical Modelling: Connecting to Research and Practice*, International Perspectives on the Teaching and Learning of Mathematical Modelling, DOI 10.1007/978-94-007-6540-5_9,

applications of mathematics, there appears to have been less attention invested into a holistic understanding of the role of all elements of the mathematical modelling process including teachers, students, mathematical knowledge and artefacts such as physical, symbolic and digital tools. These artefacts include tasks, textbooks, tools of measurement and computers and on-line facilities.

The aim of this chapter is to explore the potential of a model for teaching and learning proposed by Strässer (2009) to describe and explain mathematical modelling and application activity that incorporates roles of students, teachers, mathematical knowledge and artefacts during engagement with mathematical applications and modelling.

2 Theoretical Framework

Verillon and Rabardel's (1995) iconic work on the distinction between an artefact and an instrument has laid the groundwork for a significant corpus of literature within which theories of teaching and learning incorporate a wider range of influences than on students and teachers alone. From Verillon and Rabardel's (1995) perspective, artefacts, which include both physical and sign tools, have no intrinsic meaning of their own but a meaningful relationship develops between an artefact and a user when both combine to work on a specific task. Different tasks will require different relationships between the user and the artefact, and the development of these relationships is referred to as instrumental genesis. This instrumental genesis has two components (as described by Artigue 2002). Firstly, the transformation of the artefact itself into an instrument is known as instrumentalisation. In this process the potentialities of the artefact for performing specific tasks are recognised. Secondly, the process that takes place within the user, in order to use the instrument for a particular task, is known as instrumentation. This process leads to the development of schemas of instrumented action, which are developed either personally or through the appropriation of pre-existing schemas from others. Thus an instrument is composed of the artefact, along with its affordances and constraints, and the user's task specific schemas. In concert, these elements provide direction for the use of the instrument in a given context. Finally, the process of instrumental genesis is bilateral in that the possibilities and constraints shape the conceptual development of the user, while at the same time, the user's conceptualisation of the artefact and thus its instrumentation leads, in some cases, to the user changing the instrument (Drijvers and Gravemeijer 2005).

Prominent within the mathematics education literature that is concerned with the process of instrumental genesis are writers who have attempted to theorise the nature of interaction between user and digital tools. Guin et al. (2005) and Artigue (2002), for example, have attempted to explain how digital artefacts such as Computer Algebra Systems (CAS) are transformed into instruments for learning through interaction with teachers and students. A teacher's activity in promoting a

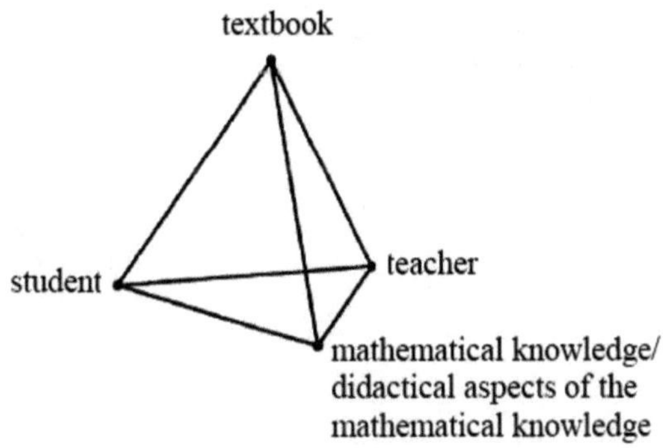

Fig. 9.1 Rezat's (2006) model for the use of textbooks in teaching and learning

student's instrumental genesis is known as instrumental orchestration (Trouche 2003, 2005). Social aspects of learning are recognised within this process and take the form of student activity that makes explicit the schemas that individuals have developed within a small group or a whole class. These schemas are then available for appropriation by other class members through careful and selective questioning by the class teacher, that is, the teacher orchestrates the interaction so that a new individual scheme is shared with others.

More recently, Gueudet and Trouche (2009) have extended the concept of the instrumental approach to include other artefacts, in addition to CAS enabled technologies, when considering the professional work of teachers. They extend our understanding of artefacts by introducing the term *resources* to identify any artefact with the potential to promote semiotic mediation in the process of learning including computer applications, student worksheets or discussions with a colleague. In their view, resources must undergo a process of genesis, parallel to that of the instrumental approach, in which a resource is appropriated and reshaped by a teacher, in a way that reflects their professional experience in relation to the use of resources, to form a scheme of utilisation. The combination of the resource and the scheme of utilisation is called a document. The process of documental genesis is an ongoing one as utilisation schemes will be reshaped as the teacher gains more experience through the use of a resource.

Theorising about the role of artefacts in mediating transformative learning experiences from a different perspective, Rezat (2006) draws on activity theory to assign artefacts, such as textbooks, a position in a tetrahedral model (Fig. 9.1) which uses Steinbring's (2005) didactical triangle as a base. The faces of the tetrahedron represent activity sub-systems of textbook use in which one element mediates between the other two elements of the triangular face. For example, the face that is bounded by the elements student-teacher-textbooks represents activity in which the teacher mediates the use of the textbook in order for students to learn.

In order to extend Rezat's (2006) tetrahedral model to include additional aspects of teaching and learning, Strässer (2009) draws on Wartofsky's (1979) classification of artefacts. In his model, Strässer (2009) includes the theoretical constructs of

Fig. 9.2 Strässer's (2009) tetrahedral model for teaching and learning

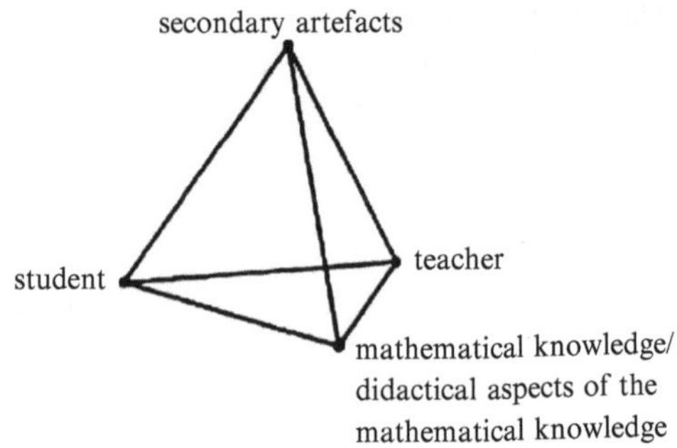

semiotic mediation and instrumental genesis, as well as incorporating Wartofsky's (1979) broader view of arefacts. He argues that Wartofsky's class of "secondary" artefacts, which are "used in the preservation and transmission of the acquired skills or modes of action or praxis by which this production is carried out" (Wartofsky 1979, p. 2005), can replace textbooks in Rezat's tetrahedron to create a more inclusive model for the use of instruments in teaching and learning mathematics. In this new model, the edges which join elements allow for a number of processes. The edges that join student and secondary artefacts and teacher, for example, allow for the process of instrumental genesis. The vertices of secondary artefact and mathematical knowledge are joined by an edge that represents semiotic mediation (Fig. 9.2).

3 Strässer's Tetrahedron as a Model of Mathematical Modelling and Applications Activity

The applicability of Strässer's tetrahedral model to mathematical modelling and application activity is illustrated by using this model to analyse an episode reported by Geiger et al. (2010). In this episode, the teacher worked with a Year 11 class on a unit of work which engaged students in the study of a range of mathematical functions. As part of this unit the teacher challenged students to use linear, quadratic, cubic, exponential and power functions to model authentic life-related data. During one lesson students were asked to work on the *Algal Bloom Problem*, outlined in the shaded section below, using the data in Table 9.1.

Students were expected to build a mathematical model for these data by first inspecting a scatterplot; the characteristics of which were then used to determine the general form of the function that would best fit the data. This general form was then to be adapted to the specific data presented in the task and used to address the final questions. Students had earlier studied strategies for determining if a particular function type was most suited to a data set.

Algal Bloom Problem

The CSIRO has been monitoring the rate at which Carbon Dioxide is produced in a section of the Darling River. Over a 20 day period they recorded the rate of CO_2 production in the river. The averages of these measurements appear in the table below.

The CO_2 concentration $[CO_2]$ of the water is of concern because an excessive difference between the $[CO_2]$ at night and the $[CO_2]$ used during the day through photosynthesis can result in algal blooms which then results in oxygen deprivation and death of the resulting animal population and sunlight deprivation leading to death of the plant life and the subsequent death of that section of the river.

From experience it is known that a difference of greater than 5 % between the $[CO_2]$ of a water sample at night and the $[CO_2]$ during the day can signal an algal bloom is imminent.

Table 9.1 Rate of CO_2 production versus time

Time in hours	0	1	2	3	4	5	6	7	8	9
Rate of CO_2 production	0	−0.042	−0.044	−0.041	−0.039	−0.038	−0.035	−0.03	−0.026	−0.023

Time in hours	10	11	12	13	14	15	16	17	18	19
Rate of CO_2 production	−0.02	−0.008	0	0.054	0.045	0.04	0.035	0.03	0.027	0.023

Time in hours	20	21	22	23	24
Rate of CO_2 production	0.02	0.015	0.012	0.005	0

Is there cause for concern by the CSIRO researchers?
Identify any assumptions and the limitations of your mathematical model.

Prior to this activity, students were introduced to a technique where *ln* versus *ln* plots of data sets were used to determine if a power function was an appropriate basis on which to build a mathematical model. This appears to have influenced the actions of two students as the transcript below indicates.

Researcher: So you are up to building the model are you?
Student 1: Well we worked out a plan of what we are going to do, we are just putting it on paper.
Researcher: So do you want to tell me what the plan is?
Student 1: The plan is to do the Log/Log plot of both the data to see if they are modelled by a power function. We have previously seen that the…
Researcher: So that is something you have learnt to do over time? Whenever you see data look like that, you check if it's a power function by using Log/Log?
Student 1: Yes

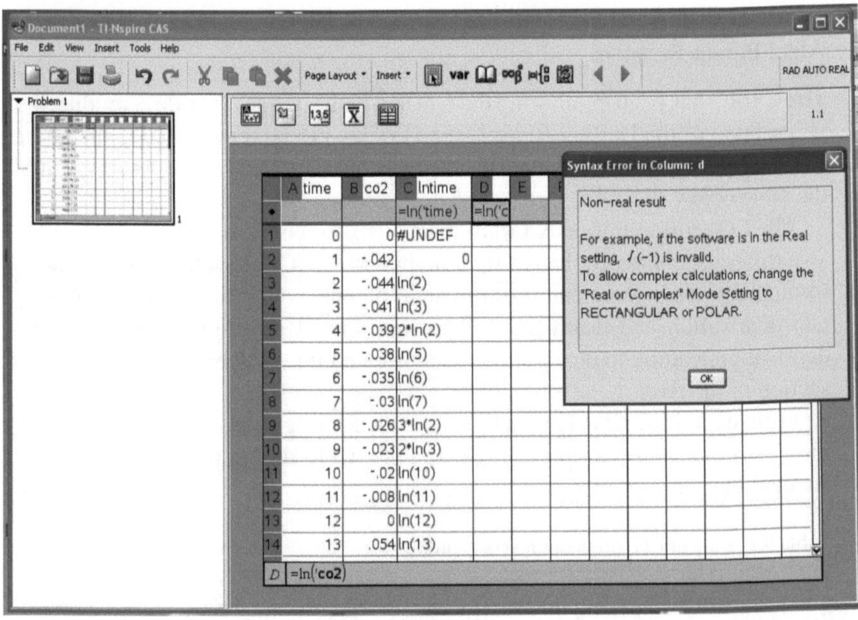

Fig. 9.3 Nspire display of spreadsheet for natural logarithm of time and CO_2 output data

These students experienced problems with this approach, as the technique employed in this case meant the students tried to find the natural logarithm of 0.

Student 1: 0.44 zero… (entering information into the Nspire device). Don't tell me I have done something wrong. Dammit. [Mumbles] Start at zero is it possible to do a power aggression (sic)? I don't think so!

This comment was in response to the display that resulted when the students attempted to find the natural logarithm of both *Time* and *CO_2 output* data using the spreadsheet facility of their handheld device (see Fig. 9.3). Students were surprised by the outputs they received for both sets of calculations, that is, the #UNDEF against the 0 entry in the *Time* column and the lack of any entries in the *CO_2* column. In addition, an error message was produced indicating the results of the students' entries were problematic for the handheld device.

After a little more thought the students realised where the problem lay.

Student 1: *Ln* time is going to be equal to the *Ln* of actually time. Time…. Oh, is that undefined 'cause it's zero?

Student 2: Yep.

Student 1: Right now if I go back to my graph… Enter

Student 2: If you try zero fit, it will just go crazy.

Students eventually identified the problem with their approach and realised their initial assumption, that is, the best model for the whole data set was a power function, was at fault. Eventually, they realised it was best to model the data with two separate functions.

Student 1: So we have fitted a linear model for the top data and then we fitted a power function to the bottom data given we take the absolute value of those the question asks, the difference greater than 5 % we need to look at the actual CO_2 produced, now what we have got is the rate, to go back to the actual CO_2 absorbed we need to integrate the model or both models and then use the percentage difference formula – predicted minus actual divided by actual or, in this case, night minus day divided by day multiplied 100 to look at whether for any x or any t there is any percentage difference greater than 0.05.

Towards the end of the lesson, the teacher asked these students to share their experience with the rest of the class and by doing so, made public what these two students had learned together over the lesson.

This episode begins with the teacher setting a task which is represented on Strässer's tetrahedron by the face defined by student, teacher and secondary artefacts. The edges on this face between student and secondary artefacts and teacher and secondary artefacts allow for the process of documental genesis. The edge between teacher and secondary artefacts represents an interaction within which the teacher is attempting to shape a tool that will mediate the learning of his/her student(s). The intent of this interaction is to facilitate the process of semiotic mediation which takes place in the theoretical space between student, secondary artefacts (the task) and mathematical knowledge. The result of this process, however, is influenced by students' prior learning which has generated schemes of instrumental action which are not valid in the presented situation. Students appear to have chosen these schemes (power functions fitted to real life data) because they were the most recently used in mathematics classes.

The approach chosen by students, however, is disrupted because of a second level of activity that takes place on the student-secondary artefacts-mathematical knowledge face of the tetrahedron. Here, interaction between students, digital tools and mathematical knowledge produces a confounding outcome from which students realise their initial approach is inappropriate (their model does not accommodate zero values). This realisation forces them to reconsider their choice of action scheme and to choose a different direction; to model the data with a piece-wise function. It is not clear from the data whether students have invoked an already existing alternative scheme or that the disruption caused by the unexpected result led to the genesis of a new scheme in which the task, their selection of modelling function and the results displayed on their digital tools, were all accommodated. Finally, the teacher brings these students into the space on the tetrahedron defined by teacher, student and mathematical knowledge.

While this model accommodates most of the activity which takes place within this teaching and learning episode, for example, the interactions between digital tools and students, it does not appear to provide for a theoretical space for the interactions which take place between the students as they work their way forward through the problem. As Strässer points out, one of the shortcomings of this model is that it does not clearly theorise a role for social aspects of teaching and learning mathematics. While he does not attempt to address this gap in theory himself, he points to the concept of Chevallard's (1985/1991) "noosphere"

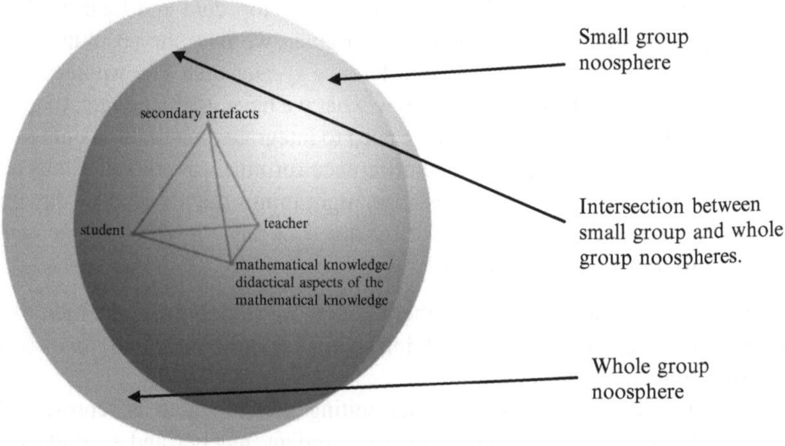

Fig. 9.4 Noospheres surrounding Strässer's tetrahedral model for teaching and learning

– a space that accommodates "all persons and institutions interested in the teaching and learning of mathematics" (Strässer 2009 p. 75) and suggests his model for teaching and learning should also provide for productive interactions between all participants. This means that consideration must be given to different types of interactions between participants and between participants and artefacts. In the case discussed above, it is the interaction between student-student and digit tool which brings about a conflict that leads to new learning through the resolution of this conflict. Further, this new knowledge, which is developed in a small group situation, is eventually shared with the whole class. A possible representation of these interactions, developed from Strässer's tetrahedral model, appears in Fig. 9.4. In this representation his model is nested within two realms based on the concept of a noosphere, one that represents small group interaction and one that represents whole class interaction. The intersection between these two spheres represents the interaction between both social groups. It is important to note that these spheres are not permanent theoretical objects but that they are instantiated as different types of in-class interactions take place and then disappear again when a particular type of interaction comes to a conclusion.

4 Conclusion

How to describe and explain all of the activity that takes place in the process of mathematical modelling in a holistic way is clearly an area in which greater research effort is required. Strässer's extended model highlights the need to develop frameworks which describe, theorise and interpret the ways that teachers and students engage with mathematical knowledge and secondary artefacts within learning communities. This implies that a socio-cultural perspective must be considered when developing frameworks which incorporate the concept of noosphere.

The example described and interpreted in this chapter by no means offers an exhaustive list of possibilities for what might constitute a plethora of different types of mathematical modelling processes within different types of social settings. Strässer's concept of a realm of spaces for social interaction, based on Chevallard's (1985/1991) spheres that surround Strässer's tetrahedral model of learning and teaching, offers one possibility for investigating the roles of cognitive, semiotic and social elements in the process of mathematical modelling within authentic classroom settings. An analysis of the classroom episode presented in this chapter also suggests that these spheres are not concentric nor independent entities, as interaction between spheres can occur, as was the case when the teacher asked two students to describe the results of their interaction in attempting to solve the *Algal Bloom Problem* to the whole class. Such points of intersection may also be sites that provide for dynamic learning possibilities. The evidence presented here suggests that there is still much research to be done into how the role of social aspects of learning and teaching can be theorised together with the roles of students, teachers, secondary artefacts and mathematical knowledge.

References

Artigue, M. (2002). Learning mathematics in a CAS environment: The genesis of a reflection about instrumentation and the dialectics between technical and conceptual work. *International Journal of Computers for Mathematical Learning, 7*(3), 245–274.

Chevallard, Y. (1985/1991). *La transposition didactique* (2nd ed. 1991 incl. a "postface" and a reprint of the case study by Chevallard & Joshua on distance from *Recherches en Didactique des Mathématiques* 1982). Grenoble: Pensées sauvages.

Drijvers, P., & Gravemeijer, K. (2005). Computer algebra as an instrument: Examples of algebraic schemas. In D. Guin, K. Ruthven, & L. Trouche (Eds.), *The didactical challenge of symbolic calculators: Turning a computational device into a mathematical instrument* (pp. 163–196). New York: Springer.

Geiger, V., Faragher, R., & Goos, M. (2010). CAS-enabled technologies as 'agents provocateurs' in teaching and learning mathematical modelling in secondary school classrooms. *Mathematics Education Research Journal, 22*(2), 48–68.

Gueudet, G., & Trouche, L. (2009). Towards new documentation systems for mathematics teachers? *Educational Studies in Mathematics, 71*(3), 199–218.

Guin, D., Ruthven, K., & Trouche, L. (2005). *The didactical challenge of symbolic calculators: Turning a computational device into a mathematical instrument*. New York: Springer.

Niss, M. (2008). Perspectives on the balance between applications and modelling and "pure" mathematics in the teaching and learning of mathematics. In M. Menghini, F. Furinghetti, L. Giacardi, & F. Azarello (Eds.), *The first century of the International Commission on Mathematical Instruction (1908–2008): Reflecting and shaping the world of mathematics* (pp. 69–84). Roma: Instituto della Enciclopedia Italiana.

Rezat, S. (2006). A model for textbook use. In J. Novotna, H. Moraova, M. Kratka, & N. Stehlikova (Eds.), *Proceedings of the 30th conference of the International Group for the Psychology of Mathematics Education* (Vol. 4, pp. 409–416). Prague: PME.

Steinbring, H. (2005). *The construction of new mathematical knowledge in classroom interaction: An epistemological perspective*. New York: Springer.

Stillman, G. (2007). Upper secondary perspectives on applications and modelling. In W. Blum, P. Galbraith, H.-W. Hen, & M. Niss (Eds.), *Modelling and applications in mathematics education: The 14th ICMI study* (pp. 461–468). New York: Springer.

Strässer, R. (2009). Instruments for learning and teaching mathematics: An attempt to theorize about the role of textbooks, computers and other artefacts to teach and learn mathematics. In M. Tzekaki & H. Sakonidis (Eds.), *Proceedings of the 33rd conference of the International Group for the Psychology of Mathematics Education* (Vol. 1, pp. 67–81). Thessaloniki: PME.

Trouche, L. (2003). From artifact to instrument: Mathematics teaching mediated by symbolic calculators. *Interacting with Computers, 15*(6), 783–800.

Trouche, L. (2005). Instrumental genesis, individual and social aspects. In D. Guin, K. Ruthven, & L. Trouche (Eds.), *The didactical challenge of symbolic calculators: Turning a computational device into a mathematical instrument* (pp. 197–230). New York: Springer.

Verillon, P., & Rabardel, P. (1995). Cognition and artifacts: A contribution to the study of thought in relation to instrumental activity. *European Journal of Psychology of Education, 10*, 77–103.

Wartofsky, M. W. (1979). *Models. Representation and the scientific understanding* (Vol. 129). Dordrecht: Reidel.

Chapter 10
Ethnomodelling as a Research Lens on Ethnomathematics and Modelling

Milton Rosa and Daniel Clark Orey

Abstract Mathematics used outside of the school may be considered as a process of modelling rather than a mere process of manipulation of numbers and procedures. The application of ethnomathematical techniques and the tools of modelling allow us to see a different reality and give us a more holistic insight into the kind of mathematics used by specific groups of people. The pedagogical approach that connects the academic to the cultural aspects of mathematics is named ethnomodelling; which is a process of translation and elaboration of problems and questions taken from systems that are part of any given cultural group.

1 Introduction

When researchers investigate the knowledge possessed by members of diverse cultural groups, they may be able to find distinctive characteristics of mathematical ideas that we might label as ethnomathematics. It is important to note that, an outsider's understanding of cultural traits is always an interpretation that may emphasise inessential features and misinterpretations of this mathematical knowledge. The challenge arising from this approach is related to how we can find and understand culturally bound mathematical ideas without letting the culture of the researcher or investigator interfere with the culture of the members of the cultural group under study. This can happen when the members of distinct cultural groups share their own interpretation of their culture (*emic*) as opposed to an outsider's interpretation (*etic*).

M. Rosa (✉) • D.C. Orey (✉)
Centro de Educação Aberta e á Distância (CEAD), Universidade Federal de Ouro Preto (UFOP), Ouro Preto, MG, Brazil
e-mail: milton@cead.ufop.br; oreydc@cead.ufop.br

G.A. Stillman et al. (eds.), *Teaching Mathematical Modelling: Connecting to Research and Practice*, International Perspectives on the Teaching and Learning of Mathematical Modelling, DOI 10.1007/978-94-007-6540-5_10,

Researchers, investigators, educators, and teachers who take on an emic perspective believe that many factors such as cultural and linguistic backgrounds, social and moral values, and lifestyle come into play when mathematical ideas, concepts, procedures, and practices are developed by the members of cultural groups. Different cultural groups have developed different ways of doing mathematics in order to understand and comprehend their own cultural, social, political, economic, and natural environments (Rosa 2010). In other words, each cultural group has developed unique and quite often distinct ways to *mathematise* their own realities (D'Ambrosio 1990).

From this perspective, *mathematisation*[1] is a process in which individuals from different cultural groups come up with different mathematical tools that help them organise, analyse, comprehend, understand, and solve specific problems located in the context of their real-life situation (Rosa and Orey 2006). These tools allow them to identify and describe a specific mathematical idea or practice in a general context by schematising, formulating, and visualising a problem in different ways, discovering relations and regularities, and transferring a real world problem to a mathematical idea through mathematisation.

It is important to search for alternative methodological approaches as Western mathematical practices are accepted worldwide in order to archive, study and record historical forms of mathematical ideas that have occurred in different cultural contexts. One alternative methodological approach is *ethnomodelling*; the practical application of ethnomathematics, which adds a cultural perspective to the mathematical modelling process. When justifying the need for a culturally bound view of mathematical modelling, our sources are derived from the theory of ethnomathematics (D'Ambrosio 1990).

2 Ethnomodelling

Studies conducted by Orey (2000) and Rosa and Orey (2009) "have revealed sophisticated mathematical ideas and practices that include geometric principles in craft work, architectural concepts, and practices in the activities and artefacts of many indigenous, local, and vernacular cultures" (Eglash et al. 2006, p. 347). These concepts are related to numeric relations found in measuring, calculation, games, divination, navigation, astronomy, modelling, and a wide variety of other mathematical procedures and cultural artefacts.

Eglash et al. (2006) and Rosa and Orey (2006) use the term *translation* to describe the process of modelling local cultural systems, which may have a Western academic mathematical representation. However, indigenous designs may be merely analysed from a Western view such as the application of symmetry classifications from crystallography to indigenous textile patterns (Eglash et al. 2006). On the

[1] We acknowledge that there are other meanings of mathematisation in the field of modelling education. For example, in Kaiser's (2006) point of view, mathematisation is considered as the process in which the real world model is translated into mathematics, leading to a model of the original situation, object or context.

Fig. 10.1 Geometric lace patterns

other hand, ethnomathematics makes use of modelling by attempting to apply it in order to establish relations between the local conceptual framework and the mathematical ideas embedded in relation to global designs through translations. In this regard, this relationship may be named as ethnomodelling because "the act of translation is more like mathematical modeling" (Eglash et al. 2006, p. 348).

In some cases, "the translation to Western mathematics is direct and simple such as counting systems and calendars" (Eglash et al. 2006, p. 347). However, there are cases in which mathematical ideas and concepts are "embedded in a process such as iteration in bead work, and in Eulerian paths in sand drawings" (p. 348). This means that mathematical knowledge arises from emic rather than etic origins, which is a reasonable assumption, as ethnomathematics often uses modelling to establish relations between conceptual frameworks and the mathematics embedded in indigenous designs (Eglash et al. 2006). For example, the mathematical knowledge that lace makers in the Northeast part of Brazil use to make geometric lace patterns, as shown in Fig. 10.1, have mathematical concepts arising from emic origins. The lace makers are not able to associate these geometrical principles to the mathematical aspect of this practice, which arises from etic origins.

Ethnomodelling takes into consideration the processes that help in the construction and development of scientific and mathematical knowledge, which includes collectivity, creativity, and inventivity. In so doing, it is impossible to imprison mathematical concepts in registers of univocal designation of reality because there are distinct systems that provide an unambiguous representation of reality as well as universal explanations (Craig 1998). This means that mathematics may not be, as is often thought, a universal language because its principles, concepts, and foundations are not always the same everywhere around the world.

The processes and production of scientific and mathematical ideas, procedures, and practices operate in the register of interpretative singularities regarding the possibilities for the symbolic knowledge construction in different cultural groups. In this regard, we believe that through "using ethnomodelling as a tool towards pedagogical action of the ethnomathematics program, students have been shown to learn how to find and work with authentic situations and real-life problems" (Rosa and Orey 2010, p. 60).

3 The Emic and Etic Constructs of Ethnomodelling

The elaboration of models that represent systems taken from reality are representations that help members of cultural groups to understand and comprehend the world by using small units of information, called *ethnomodels*, which link their cultural heritage with the development of their mathematical practices. It is our understanding that this approach may help the organisation of pedagogical action that occurs in classrooms through the use of emic and etic aspects of these mathematical practices.

In using an ethnomodelling perspective, the emic constructs are the accounts, descriptions, and analyses expressed in terms of conceptual schemes and categories that are regarded as meaningful and appropriate by the members of the cultural group under study. This is in accordance with perceptions and understandings deemed appropriate by the insider's culture (Lett 1996). The validation of emic knowledge comes through consensus; that is the consensus of those who must agree that these constructs match shared perceptions and portray the characteristic of their culture.

In contrast, the etic constructs represent the accounts, descriptions, and analyses of mathematical ideas, concepts, procedures, and practices expressed in terms of conceptual schemes and categories regarded as meaningful and appropriate by the community of scientific observers and researchers (Lett 1996). An etic construct needs to be precise, logical, comprehensive, replicable, theoretical, and observer-researcher independent. Researchers and investigators have to acknowledge and recognise that the people under study possess scientifically valid mathematical knowledge (D'Ambrosio 1990).

4 Ethnomodels

We argue that traditional mathematical modelling does not fully take into account the implications of cultural aspects of human social systems. The cultural component in this process is critical as it emphasises "the unity of culture, viewing culture as a coherent whole, a bundle of [mathematical] practices and values" (Pollak and Watkins 1993, p. 490) that are incompatible with the rationality of the elaboration of traditional mathematical modelling processes. In the context of mathematical knowledge, what is meant by the cultural component varies widely and ranges from viewing mathematical practices as socially learned and transmitted mathematical practices that are viewed as made of abstract symbolic systems with an internal logic that gives a symbolic system its mathematical structure, power and usefulness. This is the process by which knowledge transmission takes place from one person to another, central to elucidating the role of culture in the development of mathematical knowledge (D'Ambrosio 1990). If the latter is considered, then culture plays a far-reaching and constructive role with respect to mathematical practices that cannot be induced simply through observation of these practices.

As mathematical knowledge is developed by the members of a cultural group, it possesses an abstract symbol system whose form is the consequence of an internal logic; then students learn the specific instances of the usage of that symbol system as well as deriving from those instances a cognitive-based understanding of the internal logic of the mathematical symbology system itself.

In the context of constructing ethnomodels of the mathematical practices of sociocultural systems, mathematical knowledge consists of socially learned and transmitted mathematical practices (Rosa and Orey 2010). The cognitive aspects needed in this framework are primarily decision processes by which the ethnomodels are either accepted or rejected as part of their own repertoire of mathematical knowledge. This represents the conjunction of two scenarios and appears to be adequate to encompass the full range of cultural-mathematical phenomena, made up of abstract symbolic systems with an internal logic, which provides its own mathematical structure.

In effect, there are two ways in which we recognise, represent and make sense of mathematical phenomena. First, there is a level of cognition that we share, to varying degrees, with all members of our own and other cultural groups. This level would include the cognitive models that are elaborated at a non-conscious level, which serve to provide an internal organisation of external mathematical phenomena and to provide the basis upon which mathematical practices take place.

Second, there is a culturally constructed representation of external mathematical phenomena that also provide an internal organisation for these phenomena; but where diverse forms of representation arise through formulating an abstract, conceptual structure that provides form and organisation for external phenomena. This must occur in a consistent manner with forms and patterns of those same phenomena as external phenomena; that is, the cultural construct provides a constructed reality.

The implications for mathematical modelling of systems are that models of a cultural construct may be considered a symbol system organised by the internal logic of cultural group members. However a model built without a first-hand sense for the world being modelled should be viewed with suspicion (Eglash et al. 2006). If not influenced by their prior theory and ideology, researchers and investigators will develop an informed sense of the distinctions that make a difference from the point of view of the mathematical knowledge of people being modelled. In the end, they should be able to communicate to outsiders (global-etic) what matters to insiders (local-emic).

According to this perspective, we define *ethnomodels* as culturally-based models that are pedagogical tools used to facilitate the understanding and comprehension of systems that are taken primarily from the common, day to day reality of cultural groups. This means that ethnomodels can be considered as external representations that are precise and consistent with the scientific and mathematical knowledge that is socially constructed and shared by members of specific cultural groups. From this perspective, the primary objective for ethnomodel elaboration is to *translate* mathematical ideas, procedures, and practices developed by the members of distinct cultural groups.

5 An Ethnomodelling Example: Modelling the Mangbetu Ivory Sculpture

To give the reader an example, it may be informative to examine the mathematical ideas found in an ivory hatpin from the Mangbetu people, who occupy the Uele River area in the northeastern part of the Democratic Republic of Congo in Africa, and the geometric algorithm involved in its production, which "gives explicit instructions for generating a particular set of spatial patterns" (Eglash 1999, p. 61).

In this perspective, the creation of a Mangbetu design may reflect the artisans' desire to "make it beautiful and show the intelligence of the creator" (Schildkrout and Keim 1990, p. 100) by adhering to angles that are multiples of 45°. This is only one part of an elaborated geometric esthetic based on multiples of 45° angles, which is used in many Mangbetu designs. However, this also may suggest that if there were no rules to follow, then it would have been difficult to compare designs. On the other hand, by restricting the permissible angles to a small set, the Mangbetu were better able to display their geometric accomplishments (Eglash 1999).

Figure 10.2 shows the decorative end of this ivory hatpin that is composed of four scaled similar heads, which shows a scaling design. According to Eglash (1999), the combination of the 45° angle construction technique with the scaling properties of the ivory carving may reveal its underlying structure, which has three interesting geometric features. First, each head is larger than the one above it, and faces in the opposite direction. Second, each head is framed by two lines, one formed by the jaw and one formed by the hair; these lines intersect at approximately 90°. Third, there is an asymmetry in which the left side shows a distinct angle about 20° from the vertical.

Figure 10.3 shows the geometric analysis of this sculpture in which the sequence of shrinking squares can be constructed by applying an iterative process that bisects

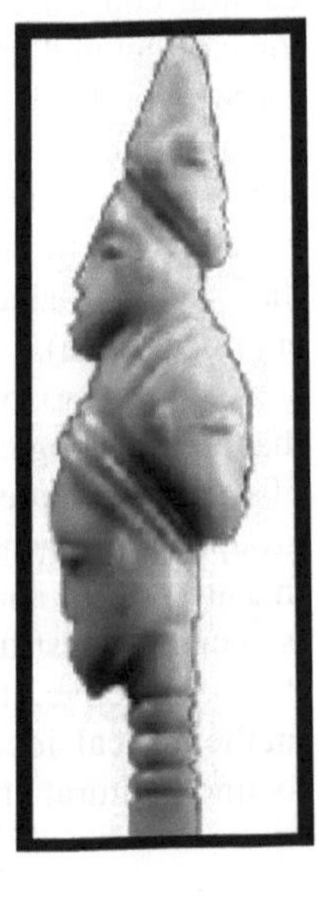

Fig. 10.2 Mangbetu ivory sculpture (Eglash 1999)

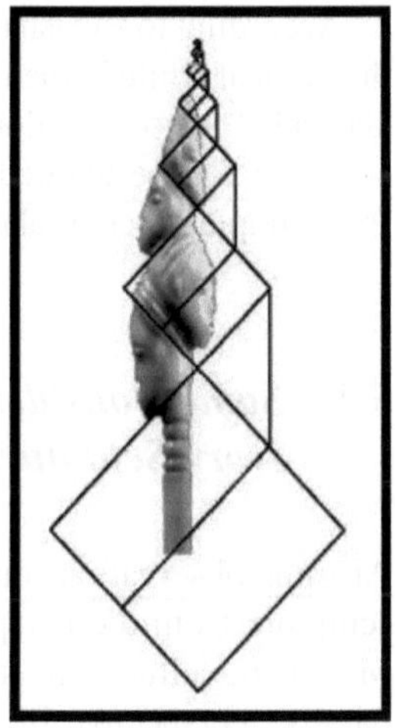

Fig. 10.3 Geometric analysis of a Mangbetu ivory sculpture (Eglash 1999)

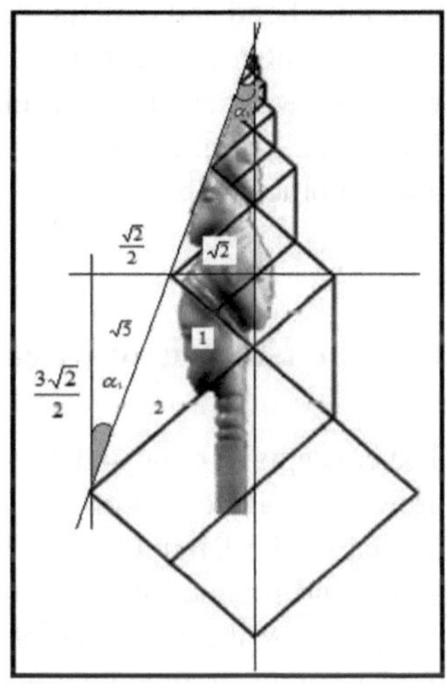

Fig. 10.4 Geometric relations in the Mangbetu ivory sculpture (Eglash 1999)

one square to create the length of the side for the next square. However, Eglash (1999) stated that it is not possible "to know if this iterative squares construction was the concept underlying the sculpture's design, but it does match the features identified in this process" (p. 68).

Figure 10.4 shows the geometric relations in the sculpture iterative square structure. We can see that since α_1 and α_2 are alternate interior angles of a transversal intersecting two parallel lines, then $\alpha_1 = \alpha_2$. In so doing, we have that:

$$\tan\alpha_1 = \frac{\frac{\sqrt{2}}{2}}{\frac{3\sqrt{2}}{2}} = \frac{\sqrt{2}}{3\sqrt{2}} = \frac{1}{3} \text{ and } \alpha_1 = \arctan\frac{1}{3} \cong 18^\circ$$

According to Eglash (1999), in the ivory sculpture, the left side is about 20° from the vertical while in the iterative squares structure, the left side is about 18° from the vertical. The construction algorithm can be continued indefinitely, and the resulting structure can be applied to a wide variety of mathematics teaching applications, from simple procedural construction to trigonometry.

5.1 Some Considerations Regarding the Mangbetu Ivory Sculpture

An emic observation sought to understand this mathematical practice for the ivory sculpture from the perspective of internal dynamics and relationships within the Mangbetu culture. An etic perspective provides a cross-cultural contrast and comparative perspective by using aspects of academic mathematics to translate this phenomenon for understanding by those from different cultural backgrounds so as to comprehend and explain this mathematical practice as a whole from the point of view of outsiders. The emic perspective clarifies intrinsic cultural distinctions while the etic perspective seeks objectivity as an outside observer across cultures. Both emic and etic perspectives are essential to help us to understand this mathematical practice.

6 The Emic-Etic Dilemma in Ethnomodelling Research

The concepts of emic and etic were first introduced by the linguist Pike (1954) who drew on an analogy with two linguistic terms. Phon*emic*, which is the sounds used in a particular language and phon*etic*, which is the general aspects of vocal sounds and sound production in languages. In this regard, all the possible sounds human beings can make constitute *phonetics*. However, when people actually speak a particular language, they may not hear all possible sounds. As modelled by linguists, not all sounds *make a difference* because they are locally significant. This means that they are the *phonemics* of that language.

If we make an analogy with this approach to ethnomodelling, it is possible to state that the emic perspective is about differences that make mathematical practices unique from an *insider's* point of view. We argue that *emic* models are grounded in what matters in the mathematical world of those being modelled. However, many models are *etic* in the sense that they are built on an outsider's view of the world being modelled. In this context, etic models represent how the modeller thinks the world works through systems taken from reality while emic models represent how people who live in such worlds think these systems work in their own reality. The etic perspective always plays an important role in ethnomodelling research, yet the emic perspective should also be taken into consideration in this process because, in this perspective, the emic models sharpen the question of what an agent-based model should include to serve practical goals in modelling research. Thus, according to this context, mathematical ideas and procedures are *etic* if they can be compared across cultures using common definitions and metrics while the focus of the *emic* analysis of these aspects are emic if the mathematical concepts and practices are unique to a subset of cultures that are founded in the diverse ways in which etic activities are carried out in a specific cultural setting.

While emic and etic are often thought to create a conflicting dichotomy, Pike (1954) originally conceptualised them as complementary viewpoints. Berry (1999) argued that rather than posing a dilemma, the use of both approaches can deepen our understanding of important issues in scientific research and investigations. Usually, in ethnomodelling research, an emic analysis focuses on a single culture and employs descriptive and qualitative methods to study a mathematical idea, concept, procedure, or practice of interest. An emic focuses on the study within the cultural group context in which the researcher tries to develop research criteria relative to the internal characteristics or logic of the cultural system. Meaning is gained relative to the context and therefore not transferable to other contextual settings. On the other hand, an etic analysis would be comparative, examining many different cultures by using standardised methods. The primary goal of the emic approach is a descriptive idiographic orientation of the mathematical phenomena because it puts emphasis on the uniqueness of each mathematical idea, concept, procedure, or practice. In contrast, the etic approach tries to identify lawful relationships and causal explanations valid across different cultures. Thus, if researchers and educators wish to make statements about universal or etic aspects of mathematical knowledge, these statements need to be phrased in abstract ways. Conversely, if they wish to highlight the meaning of these generalisations in specific or emic ways, then they need to refer to more precisely specified mathematical events.

We also agree with Pike's (1954) viewpoint that the etics perspective may be a way of gaining insight into emics of the members of cultural groups. In this regard, the etic perspective may be useful for penetrating, discovering, and elucidating emic systems that were developed by members of distinct cultural groups. We would like to highlight that in the dialectical perspective, the etic perspective claims to knowledge of any given cultural group have no necessary priority over its competing emic claims. According to this point of view, Eglash et al. (2006) stated that there is a necessity to depend "on acts of 'translation' between emic and etic perspectives" (p. 347). In this regard, the cultural specificity may be better understood with the background of communality and the universality of theories and methods and vice versa.

The rationale behind the emic-etic dilemma is the argument that mathematical phenomena in their full complexity can only be understood within the context of the culture in which they occur. The emic approach tries to investigate the mathematical phenomena and their interrelationships and structures through the eyes of the people native to a particular cultural group.

7 The Emic-Etic Perspective in an Ethnomodelling Curriculum

The mathematical knowledge of distinct cultural group members combined with Western mathematical knowledge systems results in dialectical perspectives in mathematics education. An emic analysis of a mathematical phenomenon is based on internal structural or functional elements of a particular cultural group. An etic analysis is based on predetermined general concepts external to that cultural group (Lovelace 1984). The emic perspective provides internal conceptions and perceptions of mathematical ideas and concepts while the etic perspective provides the framework for determining the effects of those beliefs on the development of the mathematical knowledge. In this regard, the acquisition of mathematical knowledge is based on the applications of current mathematics curriculum, which is assessed based on multiple instructional methodologies across various cultures.

Reasons for academic failure in every country are complex, but one reason for this seemingly universal problem may be related to educators and curriculum developers who ignore the emic perspectives in the mathematics curriculum that they elaborate for school use. An emic-etic perspective includes a respect for alternative epistemologies, and of holistic and integrated natures of the mathematical knowledge of members of diverse cultural groups found in many urban centres. An ethnomodelling practice provides an ideological basis for learning with, and from, the diverse cultural and linguistic backgrounds of members of distinct cultural groups (Rosa and Orey 2010). However, in this kind of curriculum it is crucial to understand that an etic construct is a mathematical-theoretical idea that is assumed to apply in all cultural groups while an emic construct is one that applies only to members of a specific cultural group. This means that a cultural bias may occur if researchers assume that an emic construct is actually etic, which results in an imposed etic perspective in which a culture-specific idea is wrongly imposed on another culture. In this context, an emic-etic perspective in an ethnomodelling curriculum practice provides the underlying philosophy of knowledge generation and exchange within and between all subsystems of mathematics education.

8 Final Considerations

Currently, many local mathematical practices are disappearing because of the intrusion of often foreign values, technologies and ideas, and as a result of the development of concepts that promise short term gains or solutions to problems faced by cultural groups without being capable of sustaining them; defined in this manner, the usefulness of the emic and etic distinction is evident. In every country, researchers, educators, and teachers have been enculturated to a particular cultural worldview; they therefore need a means of distinguishing between the answers they derive as enculturated individuals and the answers they derive as observers. By utilising the research provided by both emic and etic approaches, it is possible to gain a more complex understanding of the cultural groups under study.

An alternative goal for ethnomodelling research must be the acquisition of both emic and etic knowledge. Emic knowledge is essential for an intuitive and empathic understanding of mathematical ideas of a culture and it is essential for conducting effective ethnographic fieldwork. Etic knowledge, on the other hand, is essential for cross-cultural comparison, the essential components of ethnology, because such comparison necessarily demands standard units and categories. Furthermore, emic knowledge is a valuable source of inspiration for etic hypotheses. Finally, we define *ethnomodelling* as the study of mathematical phenomena within a culture because it is a social construction and culturally bound.

References

Berry, J. W. (1999). Emics and etics: A symbiotic conception. *Culture & Psychology, 5*, 165–171.

Craig, E. (1998). *Routledge encyclopedia of philosophy: Questions to sociobiology* (Vol. 8). New York: Routledge.

D'Ambrosio, U. (1990). *Etnomatemática* [Ethnomathematics]. São Paulo: Editora Ática.

Eglash, R. (1999). *African fractals: Modern computing and indigenous design*. Piscataway: Rutgers University Press.

Eglash, R., Bennett, A., O'Donnell, C., Jennings, S., & Cintorino, M. (2006). Culturally situated designed tools: Ethnocomputing from field site to classroom. *American Anthropologist, 108*(2), 347–362.

Kaiser, G. (2006). Modelling and modelling competencies in school. In C. P. Haines, P. Galbraith, W. Blum, & S. Khan (Eds.), *Mathematical modelling (ICTMA 12): Education, engineering, and economics* (pp. 110–119). Chichester: Horwood.

Lett, J. (1996). Emic-etic distinctions. In D. Levinson & M. Ember (Eds.), *Encyclopedia of cultural anthropology* (pp. 382–383). New York: Henry Holt.

Lovelace, G. (1984). Cultural beliefs and the management of agro-ecosystems. In T. Rambo, P. E. Rambo, & P. E. Sajise (Eds.), *An introduction to human ecology research on agricultural systems in South East Asia* (pp. 194–205). Honolulu: East–West Centre.

Orey, D. C. (2000). The ethnomathematics of the Sioux tipi and cone. In H. Selin (Ed.), *Mathematiics across culture: The history of non-western mathematics* (pp. 239–252). Dordrecht: Kulwer Academic.

Pike, K. L. (1954). *Emic and etic standpoints for the description of behaviours*. Glendale: Summer Institute of Linguistics.

Pollak, R., & Watkins, S. (1993). Cultural and economic approaches to fertility: Proper marriage or mésalliance? *Population and Development Review, 19*, 467–496.

Rosa, M. (2010). *A mixed-methods study to understand the perceptions of high school leader about English language learners (ELL): The case of mathematics*. Unpublished doctoral dissertation, College of Education, California State University, Sacramento.

Rosa, M., & Orey, D. C. (2006). Abordagens atuais do programa etnomatemática: delinenando-se um caminho para a ação pedagógica [Current approaches in ethnomathematics as a program: Delineating a path toward pedagogical action]. *Bolema, 19*(26), 19–48.

Rosa, M., & Orey, D. (2009). Symmetrical freedom quilts: The ethnomathematics of ways of communication, liberation, and art. *Revista Latinoamericana de Etnomatemática, 2*(2), 52–75.

Rosa, M., & Orey, D. C. (2010). Ethnomodeling: A pedagogical action for uncovering ethnomathematical practices. *Journal of Mathematical Modelling and Application, 1*(3), 58–67.

Schildkrout, E., & Keim, C. (1990). *African reflections: Art from northeastern Zaire*. Seattle: University of Washington Press.

Part II
Research into, or Evaluation of, Teaching Practice

Chapter 11
Real-Life Modelling Within a Traditional Curriculum: Lessons from a Singapore Experience

Ang Keng Cheng

Abstract Many mathematics teachers in Singapore are not familiar with designing, implementing and facilitating modelling activities in the classroom. In this chapter, we report an experiment in which a teacher with no experience in mathematical modelling made an attempt to conduct a modelling activity in his mathematics class at a local secondary school. Data in the form of students' work and feedback, lesson videos and interviews were collected. We examine the reactions from students, and discuss the lessons gained from this experience. From the data collected, it appears that while students were generally motivated in carrying out the task, they had some difficulties handling the mathematics involved and the modelling process. The teacher was able to design a reasonable task, but had some difficulty with classroom implementation of the activity.

1 Introduction

In 1990, a new mathematics curriculum based on a framework with mathematical problem solving as a core theme was designed and implemented in Singapore by the Ministry of Education (Lim 2002). The intention was to frame the curriculum as one in which a learner, through different processes, acquires the skills and concepts in various mathematical topics and develops the appropriate attitudes and metacognitive attributes for the purpose of mathematical problem solving. Since then, the Singapore mathematics curriculum has undergone several rounds of refinements although the central theme – mathematical problem

K.C. Ang (✉)
Mathematics and Mathematics Education Academic Group,
National Institute of Education, Nanyang Technological University, Singapore
e-mail: kengcheng.ang@nie.edu.sg

G.A. Stillman et al. (eds.), *Teaching Mathematical Modelling: Connecting to Research and Practice*, International Perspectives on the Teaching and Learning of Mathematical Modelling, DOI 10.1007/978-94-007-6540-5_11,

solving – has remained unchanged. One notable change made in 2007 was the explicit mention of "Applications and Modelling" as one of the areas of development in the latest, revised framework (Ministry of Education 2006a, b). However, not much else has changed in the syllabuses of all the mathematics subjects offered in Singapore schools to promote or include mathematical modelling.

Mathematics teachers in Singapore have very little, if any, training and exposure to mathematical modelling or the teaching of mathematical modelling (Ang 2010). It has been suggested that some form of support or course like those reported by Kaiser (2005) would have been helpful for teachers. In her report, Kaiser concluded that through the kinds of seminars that she and her team had planned and conducted for prospective teachers, "a change of mathematics teaching towards a stronger consideration of modelling and real world context is possible" (pp. 107–108). However, in the current context, given there are other priorities to be addressed, a comprehensive programme in mathematical modelling may not be viewed as the most essential component in the local teacher training curriculum.

Nevertheless, it may be useful for a teacher to be guided by a set of principles when designing modelling tasks. One such framework was proposed by Galbraith (2006, p. 233) who outlined such a set to guide a teacher through problem design. These include ensuring the modelling activity or task (a) has "a genuine link with the real world of the students"; (b) provides opportunities to "identify and specify" appropriate mathematical questions; (c) requires students to make assumptions or collect data in the formulation process; (d) contains a feasible solution accessible by students; and (e) includes "an evaluation procedure" to check the model. Additionally, Galbraith suggested that the problem be structured into "sequential questions that retain the integrity of the real situation".

Currently, a typical mathematics teacher in Singapore will neither have the opportunity to undergo an intensive course on "mathematical modelling in schools" as advocated by Kaiser (2005), nor have heard of the "principles of design" of modelling problems as proposed by Galbraith (2006). It is therefore both interesting and instructive to see how such a teacher would handle a modelling task from design to implementation on his/her own, and how the lesson would turn out. This study attempts to investigate this.

The objective of the present study is twofold. Firstly, we wish to determine whether a typical Singapore mathematics teacher, who has no training or experience in mathematical modelling, is able to design and implement an appropriate modelling task. Secondly, we wish to assess the general performance and response of students undertaking a modelling task given that such activities are not a norm in their mathematics learning experience. We hope that through this study, we can gain insights into what can, or should, be done if we wish to promote the teaching of mathematical modelling amongst local teachers.

2 Method

2.1 *Participants*

The teacher involved in this study is a relatively young teacher with about 5 years of experience teaching secondary mathematics. While he is a trained teacher, he has no formal training in mathematical modelling, whether as a student or a teacher. However, prior to this study, he has attended one 3 hour workshop on mathematical modelling, and been exposed to a few local publications on mathematical modelling. The student participants of this study included 20 Secondary 3 (aged 14–15) students of mixed ability, from a typical secondary school in Singapore. The students were divided into groups of four, and each group was expected to work on the same problem. They had recently been taught basic trigonometry a week before the implementation of the modelling task, and did not have any experience or lessons on mathematical modelling previously.

2.2 *The Task*

The modelling task required students to design and plan the layout of car park spaces within a car park subject to some constraints. It was a relevant and real problem as the school premises had often been used for various events, resulting in a severe shortage of car park space during such times. The problem was stated as follows:

School Car Park Spaces Task

The existing car park in our school cannot accommodate the overwhelming number of teachers, guests and parents who drive to the school whenever the school hosts special events or workshops. The school board is considering the possibility of converting a part of the school field (measuring 110 m by 60 m) to a car park.

Design a possible layout for a car park with at least 100 parking spaces while leaving as much space as possible to the remainder of the field for other activities.

A task sheet was designed by the teacher to guide students through the task. Apart from the problem statement, the task sheet consisted of seven sub-tasks or questions intended to provide some form of scaffolding for successful completion of the task. These are given below:

1. Collect data for the following (*from cars parked in the school*).

Car brand	Length of car	Width of car	Length of front door

2. What is the minimum opening angle of a car door so that the driver/passenger can get in or out of the car comfortably? Describe how you obtained this angle.
3. Determine the size of a parking space: Make a plan view sketch of how two cars can be parked side by side to each other in two parking spaces. Show the possible obstruction when the door is opened and comment on your observation.
4. Using the information gathered in (1), (2) and (3), form a mathematical expression to help you find the width of the car park space.
5. Using the internet as a resource,

   (a) list the different modes of parking a car;

   (b) choose one of the parking angles (except 90°) for your design, find the minimum width of the parking aisle.

6. Find the minimum width of the access way. You may use the internet as a resource or estimate from some site measurements.
7. Design the overall car park layout. You should take into consideration the minimum width of the parking aisle and the access way.

2.3 *Data Collection*

The task was implemented in one 120-min lesson. One video camera was set up in the classroom. As the class size was not large, it was possible for the videographer to move around to capture critical incidents involving student discussion or teacher intervention throughout the lesson. Part of the lesson involved students making actual physical measurements of a typical car, and the space required for someone to get out of the car (see sub-tasks 1 and 2 above). This was done outdoors, and was also video-recorded.

Although students worked and discussed the problem in groups, they had to fill out the task sheets individually. All task sheets and students' work were collected at the end of the activity. In addition, students were asked to write down what they have learnt from the task and their general feelings towards the activity in an open feedback form. The teacher was interviewed after the lesson to gather more information and insights into his actions and decisions made during the lesson, as well as his choice and design of the task.

3 Results and Discussion

3.1 *The Lesson*

To motivate the modelling task, the teacher began the lesson with a video on road rage arising from a dispute over parking space. Students were then asked for their views on what could have prevented the incident. It did not take long for them to conclude that the incident had happened because of poor planning in the parking space of the car park.

Students, in their groups, were then given the task sheet and encouraged to discuss the problem. After about 15 min of discussion, they went outside to an existing car park and took the necessary measurements. Each group was supplied with a measuring tape and large protractor. The data collection phase took about another 20 min. Armed with the necessary data and guided by the task sheet, the groups went back into the classroom and made attempts at formulating and solving the problem.

During this stage, it was observed that the teacher had to intervene several times to prompt students on the kind of mathematics they needed to use. On one occasion, everyone stopped work while the teacher taught some basic concepts of trigonometry in a whole-class, "frontal teaching" style for about 15 min. The lesson ended with some students presenting their groups' models and solutions, and all task sheets and feedback sheets were collected.

3.2 *Students' Work*

Figure 11.1 contains two samples of students' work. From the working, it would appear that the students were able to apply simple concepts in trigonometry to the real problem of determining the space required between two cars in adjacent parking spaces. In Fig. 11.1(a), the student made an assumption on the space needed for a person to comfortably get out of a car ("60 cm"). Based on the average length of the car door (driver's side), the "minimum opening angle" required in sub-task 2 above was computed. However, from the diagram drawn, it is evident that this student had made an error, demonstrating inability to relate the assumption to the real life situation. In Fig. 11.1(b), another student had attempted to provide an answer to sub-task 4. Instead of using exact measurements, the student gave a general formula, in terms of angle α, length of car door L, and the width of the car. However, the student did not seem to realise that the required space can be very different if the configuration involves angled parking.

In reality, if angled parking is used, then the actual space occupied by one individual parking space will be as depicted in Fig. 11.2, where L and w are the length and width of the space for one vehicle. These may be estimated from the average dimensions of a vehicle, giving allowance for some space all around the vehicle.

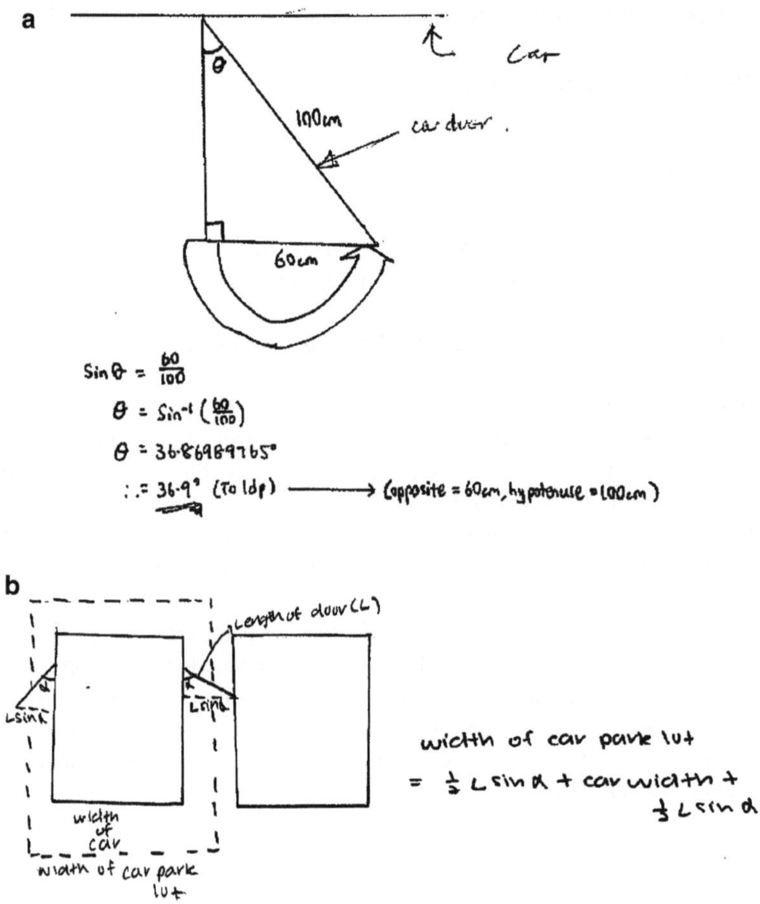

Fig. 11.1 Samples of students' work demonstrating ability to apply trigonometry. (**a**) Student's attempt at finding a comfortable "car door angle". (**b**) Formula for "width of car park lot" (parking space) derived by student

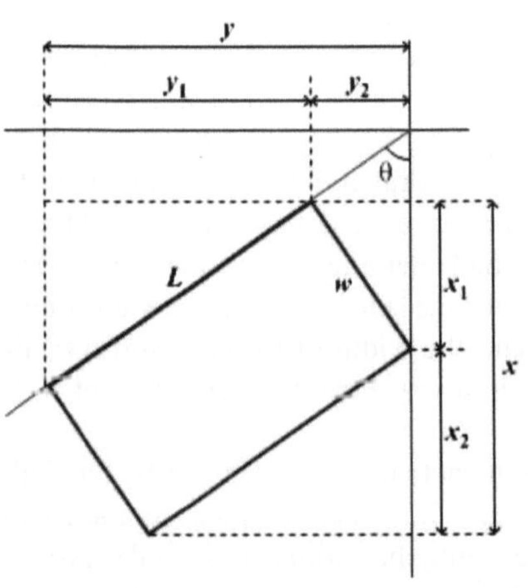

Let θ be the angle between the parking space and the curb.

From Figure 3, we gather that

$x_1 = w\sin\theta$ and $x_2 = L\cos\theta$, giving

$$x = w\sin\theta + L\cos\theta, \qquad \text{... (11.1)}$$

and

$y_1 = L\sin\theta$ and $y_2 = w\cos\theta$, giving

$$y = L\sin\theta + w\cos\theta. \qquad \text{... (11.2)}$$

Fig. 11.2 Actual dimensions of space occupied by one parking space is x by y

Therefore, using Eqs. (11.1) and (11.2), and following the guidelines on the dimensions of parking aisles and access ways given by the local Land Transport Authority (LTA), some layouts of car park arrangements may be constructed. These guidelines are spelt out in the handbook published by the local LTA (2005) which is available in the public domain.

While some students were able to work towards solutions similar to the above, they needed quite a deal of assistance and prompting from the teacher. In addition, they seemed unable to proceed from this point towards a plausible solution to the task. Although one group did manage to provide a layout based on a "45° angle parking", there was no evidence to indicate that they had based their design on earlier calculations. In other words, there was little connection between their final product and the sub-tasks (1)–(4).

3.3 *Student Feedback and Teacher Interview*

From their written feedback, it was observed that most students agreed that they have learnt more about trigonometry and how it can be used in a practical, real life problem-solving situation during the task. In Singapore, it is very common that textbook exercises or items found in assessment books available in the market are used to reinforce concepts taught in the class. Students therefore found it refreshing to be applying school mathematics in a real life problem. One student commented that she "had a great experience" and that her knowledge in trigonometry had "expanded".

Another common positive feeling amongst the students was that they realised how important it was to be able to work as a team and to communicate well when tackling this task as a group. In fact, one student expressed satisfaction (and relief?) that her group managed to "conquer this project with minimal conflicts". She went on to say positively that "overall, I find this project quite interesting". On the other hand, there were also students who did not quite enjoy the task, although they agreed that they did learn more about trigonometry and its uses. A number of students commented that the modelling task was a stressful exercise for them. This could be due to the lack of exposure to such non-routine tasks in mathematics lessons.

Apart from having to solve an unfamiliar problem, students found that they had to do so as a team, and to some, this can be a stressful experience. One student wrote, "I do not like this type of project because I finds (sic) it boring" but quickly added, "But there are fun sides". This student probably does not care much about car park space, and so the problem did not interest her. However, she probably enjoyed the idea of working in groups, collecting data and so on as she had also mentioned the importance of communication and team work in her reflection.

A more negative comment came from a male student, who commented,

> Overall, I feel that it is a waste of time doing this project although I have learned alot (sic) from it. We could actually use the time to a better use like classroom teaching instead of doing this project.

He went on to suggest that the activity could have been carried out after the examinations. To this student, while the task did provide opportunity to learn more about trigonometry and its applications, it did not serve the purpose of preparing him for examinations.

In general, most of the comments were not unexpected and students did not express any strong feelings either way. Nevertheless, the videos of the task implementation, samples of students' work and written feedback had provided much information on the overall experience from which valuable lessons are gained.

The task was designed by the teacher and is adapted from one of the suggested tasks in Ang (2001). The task has been carefully thought through and generally adheres to the principles of design suggested by Galbraith (2006). There is some genuine link with a real world problem that students can identify with, mathematically tractable questions are possible, formulation of solution process is feasible and within reach, and sub-tasks are constructed to provide necessary scaffolding. When asked, the teacher affirmed that he had no prior knowledge of Galbraith's "principles of design", but had simply planned the tasks and worksheets "based on my own understanding and pedagogical sense".

From the videos, it was observed that although all the students have been taught basic trigonometry, they did not seem to have internalised the concepts well enough to be able to apply them when the situation called for it. Much prompting and teaching had to be provided by the teacher who facilitated the activity. Sampled student work indicated that while most students were able to handle simple problems in trigonometry, they had trouble dealing with slightly more complex ones. In retrospect, it may have been better if students were informed of the need to be prepared in the topic before the task was implemented. Alternatively, a separate "pre-task sheet" could be used to help students recall and revisit relevant topics or concepts needed for the task a few days before.

Another observation concerns the facilitation skills of the teacher. The teacher was clearly excited, passionate and enthusiastic about the task. However, in his eagerness to share, he may have spent too much time explaining and presenting his solutions. Although no quantitative data was collected, from the video records, it appears that whenever they were in the classroom, the teacher did a lot more talking than the students thinking or discussing. Even data collection was done under the watchful eyes of the teacher, and every step was guided.

When interviewed, the teacher revealed that he was concerned about students not being able to arrive at the required or expected solution. In hindsight, he agreed that he could have given students more opportunities to discuss and talk among themselves, and attempt to solve the problem. Perhaps instead of providing answers, the teacher could have asked questions that lead students along, and guide them towards a solution. Nonetheless, it is acknowledged such demands on the teacher are not trivial. As pointed out by Leiβ, "a teacher must fulfil a very complex task when combining appropriate teacher interventions and students' self-regulated learning processes" (Leiβ 2005, p. 87) in a modelling task.

Although the modelling task may seem manageable, as this was the first time these students were engaged in such a venture, a more structured approach would probably have worked better. In addition, specific activities and the competence required should have been clearly identified when designing the task (Stillman et al. 2007). The lack of a strong framework probably resulted in the need for the teacher to "over-facilitate" and provide constant intervention in order to help students along with the task.

4 Conclusion

In an environment where examinations often take priority and within a curriculum that is largely traditional, it was indeed a challenge to embark on such a venture and convince students that "modelling is a way of life" (Houston 2001). While students and some teachers may acknowledge that a well designed modelling task can reinforce mathematical concepts and skills, the fact remains that they may not embrace the idea as enthusiastically as they do assessment exercises. Perhaps, as observed by Lingefjärd, mathematical modelling can be used as "a way to summarize and assess the mathematical competencies the students possess" (Lingefjärd 2007, p. 336). Perhaps, one has to find a way to make these assessment processes more explicit within the modelling tasks.

The instructional approach adopted by the teacher may have also played a part in determining the outcome of the experience, as suggested by Legé (2005). Besides thinking about adopting a more "reductionist" or "constructivist" approach when introducing mathematical modelling, the teacher may need to regulate his frequency and level of intervention during the modelling process. That said, it is understandable that in the present case, the teacher was doing all he could to keep students engaged in the activity.

From this experiment with one typical teacher in a modelling class, we gather that the teacher was able to design a reasonable modelling task that is, to some extent, aligned with Galbraith's (2006) principles of design. However, it has also shown that the teacher lacked certain key skills in facilitating modelling activities. Perhaps a more structured and formal professional development programme will be useful in this case.

This experiment has also shown that in general, although they may face difficulties in grasping and applying certain mathematical concepts, students are not opposed to the idea of engaging in mathematical modelling tasks. In this particular case, there was evidence to suggest the students enjoyed the activity, but were at the same time, concerned with its relevance to the curriculum. It would have been better if the activities could be explicitly aligned and linked to the school's mathematics curriculum or scheme of work.

In summary, although the study involved only one typical teacher in one lesson, it provided useful information as well as encouraging signs to indicate that given proper professional training for teachers and clearer alignment of curriculum goals for students, it is possible to promote and further develop the teaching of mathematical modelling in Singapore schools. The next logical step would be for local mathematics educators concerned with mathematical modelling to design and apply a professional development framework that would help teachers move forward in this endeavour.

References

Ang, K. C. (2001). Teaching mathematical modelling in Singapore schools. *The Mathematics Educator, 6*(1), 62–73.

Ang, K. C. (2010). Mathematical modelling in the Singapore curriculum: Opportunities and challenges. In A. Araújo, A. Fernandes, A. Azevedo, & J. F. Rodrigues (Eds.), *Conference proceedings of EIMI 2010: Educational interfaces between mathematics and industry* (pp. 53–61). Lisbon/Bedford: Centro Internacional de Matemática/COMAP.

Galbraith, P. (2006). Real world problems: Developing principles of design. In P. Grootenboer, R. Zevenbergen, & M. Chinnappan (Eds.), *Identities, cultures and learning spaces*. (Proceedings of the 29th annual conference of the Mathematics Education Research Group of Australasia, Canberra, Vol.1, pp. 228–236). Adelaide: MERGA.

Houston, K. (2001). Teaching modelling as a way of life. *Questiones Mathematicae, 24*(Suppl 1), 105–113.

Kaiser, G. (2005). Mathematical modelling in school – Examples and experiences. In H.-W. Henn & G. Kaiser (Eds.), *Mathematikunterricht im Spannungsfeld von Evolution und Evaluation* (pp. 99–108). Hildesheim: Franzbecker.

Land Transport Authority. (2005). *Handbook on vehicle parking provision in development proposals*. Singapore: Land Transport Authority.

Legé, J. (2005). Approaching minimal conditions for the introduction of mathematical modelling. *Teaching Mathematics and Its Applications, 24*(2–3), 90–96.

Leiβ, D. (2005). Teacher interventions versus self-regulated learning? *Teaching Mathematics and Its Applications, 24*(2–3), 75–89.

Lim, S. K. (2002). Mathematics education in Singapore: Looking back and moving on. *The Mathematics Educator, 6*(2), 1–14.

Lingefjärd, T. (2007). Mathematical modelling in teacher education – Necessity or unnecessarily. In W. Blum, P. L. Galbraith, H. Henn, & M. Niss (Eds.), *Modelling and applications in mathematics education* (pp. 333–340). New York: Springer.

Ministry of Education. (2006a). *A guide to teaching and learning of O-level mathematics*. Singapore: Ministry of Education.

Ministry of Education. (2006b). *Mathematics higher 3, (Syllabus 9810)*. Singapore: Ministry of Education.

Stillman, G., Galbraith, P., Brown, J., & Edwards, I. (2007). A framework for success in implementing mathematical modelling in the secondary classroom. In J. Watson & K. Beswick (Eds.), *Mathematics: Essential research, essential practice* (Vol. 2, pp. 688–697). Adelaide: MERGA.

Chapter 12
Students' Mathematical Learning in Modelling Activities

Morten Blomhøj and Tinne Hoff Kjeldsen

Abstract Ten years of experience with analyses of students' learning in a modelling course for first year university students, led us to see modelling as a didactical activity with the dual goal of developing students' modelling competency and enhancing their conceptual learning of mathematical concepts involved. We argue that progress in students' conceptual learning needs to be conceptualised separately from that of progress in their modelling competency. Findings are that modelling activities open a window to the students' images of the mathematical concepts involved; that modelling activities can create and help overcome hidden cognitive conflicts in students' understanding; that reflections within modelling can play an important role for the students' learning of mathematics. These findings are illustrated with a modelling project concerning the world population.

1 Progress in Modelling Competency

In the Danish KOM-project (Niss and Højgaard 2011), mathematical competence is described as a union of eight overlapping competencies of which modelling competency is one. As for all the competencies, progress in modelling competency is spanned in three dimensions: (1) *Degree of coverage*, according to which part of the modelling process the students are working; (2) *Technical level*, according to which kind of mathematics the students are able to use and on how flexible and reflective they use their mathematics; and (3) *Radius of action*, according to the domain of situations in which the students can perform modelling activities and their level of reflection across domains of applications.

M. Blomhøj (✉) • T.H. Kjeldsen
IMFUFA, Department of Science, Systems, and Models,
Roskilde University, Roskilde, Denmark
e-mail: blomhoej@ruc.dk

G.A. Stillman et al. (eds.), *Teaching Mathematical Modelling: Connecting to Research and Practice*, International Perspectives on the Teaching and Learning of Mathematical Modelling, DOI 10.1007/978-94-007-6540-5_12,

In this general framework, progress in the technical level of modelling competency can be seen as embracing the progress in students' learning of mathematical concepts through modelling activities. However, the framework does not really conceptualise progress in students' understanding of mathematical concepts involved in modelling activities. Therefore, in our research we strive to better understand how modelling activities can challenge and support the overcoming of specific learning difficulties related to important mathematical concepts. In this chapter we offer some examples and reflections on how to plan for, notice, and encourage progress in students' conceptual understanding of mathematics involved in modelling activities. The chapter can be seen as a case study contributing to the answering of a more general research question: *How to conceptualise progress in students' mathematical learning in modelling activities?*

2 Students' Conceptual Learning Through Modelling Activities

In general, we find that the potential of modelling activities for enhancing students' learning of mathematical concepts to be connected with principal learning difficulties caused by the epistemology and abstract nature of mathematics. However, in order to be useful for analysing students' modelling activities and for improving the practice of teaching mathematical modelling, these principal difficulties have to be concretised and adhered to particular modelling activities. Hence, in our work we have analysed students' modelling activities in order to pinpoint particular aspects of the mathematical concepts involved, which are causing difficulties. In relation to specific modelling tasks such conceptual difficulties seem to be rather stable across cohorts of students (Blomhøj and Kjeldsen 2010a). To notice, and to design and teach to overcome such particular learning difficulties is a way of enhancing the students' mathematical learning through modelling activities.

The conceptual difficulties that students encounter in modelling activities can, to a large extent be explained as particular instances of principal learning difficulties described in the research literature. In our work we have found it relevant and fruitful to use the concept of students' concept images–signifying a student's total cognitive structure associated with a particular concept–as developed in Tall and Vinner (1981, p. 152). Sfard's (1991) model for formation of mathematical concepts, which emphasises the duality between process and object aspects of mathematical concepts; and the interplay between different representations involved in modelling activities, that is: natural language, types of diagrams (e.g., compartment diagrams), numerical tables, computer programs, symbolic/analytical and graphical representations. We use these theoretical ideas not only to analyse and explain empirical observations, but also as basis for inspiration in our ongoing design of modelling activities, which can contribute to the students' conceptual learning of mathematics.

Based on our analyses with these tools we describe the mathematical understanding of particular mathematical concepts as intended learning outcome for students. This forms the basis for our dialogical interactions with the students during their work and for the redesign and design of new projects. In Blomhøj and Kjeldsen (2010a) we present a case on the integral concept along these lines. In this chapter we present another case from the same modelling course. Here the main mathematical concepts are first order (non-linear) differential equations and their numerical and analytical solutions. Before presenting the case, we provide some background information on the educational context of our work.

3 The Educational Context and the Empirical Basis for the Research

This research is based on experiences and data gathered during more than 10 years of collaboration between the authors on the development and teaching of a course on mathematical modelling of dynamical systems for first (or second) year university students in a general science program at Roskilde University, Denmark. The course is one eighth of a full year program. The general aims of the course are to: (1) develop students' modelling competency; (2) support students' learning of mathematical concepts and methods used for modelling dynamical phenomena, (i.e., differential equations, systems of differential equations and qualitative analyses of such systems); and (3) provide students with experiences of models and modelling in different scientific contexts. However, it should be emphasised, that in this particular study program, the first aim of the course is very much supported also by problem oriented student directed projects that all students at our university conduct each semester. These projects occupy half of a student's workload each semester. Here students can work with the entire modelling process setting up their own model or analysing a model used in science or society (Blomhøj and Kjeldsen 2010b).

The analyses presented below are based on our dialogue with students and their reports on the project concerning the world population (with 10 h over 2 weeks as the total working load for the students). The students have to complete three such projects in the course and for each project the students can choose between three or four modelling problems. In total more than 20 groups have completed this project over the years.

A common element for the projects in the course is that the students must collaborate and document their work with the project in a written report. They are required to write the report as a coherent text that can be read, understood, and validated by any interested person with the same mathematical background as themselves, (e.g., a fellow student not following the course). It is up to the students to structure their report; hence all reports are different. They depend on the particular group of students and on the modelling problem. All projects begin with a set of (quasi) authentic data, and in general all reports must contain the following elements: a description and explanation of which kind of questions the model can provide answers to; a description and explanation in ordinary language of the system they want to model including discussion of the choices and assumptions they made in the modelling process, a presentation of data (if feasible for the mini-project in question), how data were obtained and their role in the modelling process; a description and explanation of a mathematical representation of the system including a presentation of the symbols the students used and the units in which they are measured, a description and explanation of the parameters of the model with considerations about the possibilities of estimating their value; if a computer program is used the students should explain their choice of numerical method, and the numerical treatment of the model must be documented; numerical analyses of the model simulating certain developments including sensibility analyses with documented results often in the form of curves or tables from a computer program; mathematical analysis of the model; interpretation and discussion of the results of the model in relation to the original problem; a final discussion including reflections with regard to the status and applicability of the model(s). These general requirements contribute to the students' awareness of the modelling process in their work and to their modelling competency as well as to their communication competency. In addition we emphasise the importance of collaborative group work in the projects.

Teaching and developing the modelling course together with colleagues for more than 10 years has allowed us to identify particular learning difficulties, which have to be overcome in each of the modelling projects. In fact, many of these learning difficulties are so stable that they can be identified in dialogue with the students essentially every time a group of students are working with that particular modelling project. That is exactly the case with the learning obstacles illustrated in the following.

4 Presentation of the Case: Modelling the Growth of the World Population

The project analysed below is called *Explosive Population Growth.* It belongs to a group of projects from which the students can chose one to work with in the first 3 weeks of the course. In this period, first order differential equations and compartment modelling of dynamical systems are introduced. In compartment models, state variables are represented by compartments (depicted as boxes and measured by their level) and the dynamic of the system is represented by flows between the compartments (depicted by arrows).

The project has been designed deliberately to challenge students to:

- apply transformation of data and linear regression to setup models, and to use graphs and residuals to compare, discuss and reflect upon different models for the world's population;
- setup, solve, and analyse the differential equation for explosive growth (explained to the students as a growth where the rate is proportional to the square of the size of the population);
- estimate parameters for the explosive growth model;
- reflect on what could be relevant criteria for judging the quality of a model;

- experience an interplay between understanding the model mathematically and judging the quality of the model.

Before the students begin the project they have been introduced to linear regression, and they have some experiences with the technique of transformation of data in order to estimate model parameters. Also, they are acquainted with *MatLab* to produce plots of data and residuals. Before the project, the students have worked numerically with *MatLab* and analytically with solving first order ordinary differential equations.

Explosive Population Growth

Describe the growth of the world population in the period 1650–1960 and estimate the population in 2100. Some background for the relevance of knowing how the world population will develop, and a set of data for the world population in year 1650–1960, are provided.

To guide the students' work with this particular modelling project, the general requirements are supplemented with the following guided questions:

- Does the given population data follow an exponential growth?
- For explosive growth, the growth rate is proportional to the square of the population size. Can the given data be described as explosive growth?
- What does the explosive model predict about the world population?
- What is your estimate of the world population in 2100?
- Try to find newer data and discuss the model's prediction in relation to these.

The students worked in groups of two to four. Due to their earlier experiences in the course, they all began the project by creating a plot of data. The students worked with the project in their groups independent of the teacher. Most of the work takes place outside of course hours, but for each mini-project some time is set aside during course hours for the students to work on the project. On such occasions the teacher functions as consultant and the students were encouraged to be well prepared for these sessions so as to take advantage of the possibility to consult with the teacher. As teachers, we support and challenge the students through dialogues and questions such as:

- How do you find the explosive model with the best fit to the data?
- In which period do you think that the explosive model is a good model for describing the world population?

Or more specific questions for the explosive model, such as:

- How many people inhabited the world in the year 0 according to your model?
- According to year model, when did Adam and Eve live?

As teachers we focus the students' attention deliberately on issues, which may cause cognitive conflict in their concept images (Tall and Vinner 1981, p. 152) or foster reflections relevant for their development of modelling competency.

5 Analysis of Students' Work: Learning Mathematics Through Modelling

As mentioned our observations of the students work with this modelling project have been collected over the past 10 years where we have either taken turns teaching the course or in a few cases taught the course together. Our insights into the students' work are based on analyses of dialogue with the groups during their work; the students' group reports; the students' oral presentations of their project in the course, and on the oral individual finals. For each of the projects in the course, we have developed a number of focus points where we are checking and challenging the students' mathematical understanding and their reflections related to the modelling process. These focus points guide our dialogical interactions with the groups, our reading and commenting on their reports, and our questioning at the presentations and at the finals. Often one of us is the examiner while the other is the censor at these finals. We compare and discuss our analyses and experiences in order to redesign the projects, to develop new projects and to search for new interesting focus points.

In our design of the *Explosive Population Growth* project we have tried to balance the development of the students' modelling competency and the enhancing of their conceptual mathematical learning as a dual goal for teaching mathematical modelling. In the following, we will point out some of the difficulties observed in the students' work with this project and discuss the learning potentials in overcoming these difficulties with respect to the dual goal.

Many groups encounter their first problems when they try to write up a differential equation from the verbal representation of the explosive growth model. Reaching the equation $N' = kN^2$ is not trivial for the students and the task creates different kinds of cognitive problems. Firstly, understanding the generality of the verbal representation and then introducing a symbol $N'(t)$ (or similar) to represent the growth rate for all values of the time t causes problems for many students. Secondly, mathematising the expression 'proportional to the square of population size' causes problems. It is difficult for the students to read the expression mathematically. The global meaning of the expression is not easily grasped. Probably, for many students the natural language representation of the derivative of a function is not a well integrated element in their concept image. Students who can read the sense 'for all values of (independent variable for time) the growth rate of the population is given by a constant times the square of the actual size of the population' out of the given expression find it trivial to write the differential equation. Here we see a very close connection between the process of mathematisation as an element in modelling competency and the understanding of the involved mathematical concepts. Thirdly, some groups end up with the equation: $y' = kN^2$ arguing that the growth rate is the derivative hence it must be y', without considering what y' represents. The modelling context of the world population gives us an opportunity to discuss with the students what this particular differential equation actually means. It provides a context, in which the students can read meaning into a differential equation; that the growth rate at any given time depends on (is proportional to) how many individuals there already are; that the population will continue to grow if the constant is positive; that N' will also continue to grow. That this means there is no limit for the size of the population. No equilibrium, is, as we shall see below, not at all obvious to students. When investigating whether the explosive model fits data some groups realise that the differential equation $N' = kN^2$ can be interpreted as a linear relationship between N^2 and N', and that k can be estimated using linear regression on N^2 and N'. However, often these groups are faced with the problem that this can only be done "if we have 'data' for N', and we have not". Even though the students know very well that the differential quotient, if it exists, is the limit of difference quotients it does not automatically occur to them that they can use difference quotients as estimates for N'. Here the modelling context provides opportunities to challenge and strengthen the students' concept images concerning important mathematical concepts. In particular the modelling challenges the students to shift between interpreting N' as a differential quotient in a particular point and as a mathematical object given by a differential equation, which is essential in concept formation according to Sfard (1991).

Some groups use the value found for k to find the explosive model with the best fit to data by means of numerical solution using the standard routine ode45 in *MatLab*. The following dialogue is typical for such groups (T teacher and S students).

T: Are you sure that the solution you have found for the explosive differential equation is the best possible in relation to the given population data?
S1: We have determined the value for k by transformation of data and linear regression and found a good fit, look here. (The students show a plot of their model with data)
T: Okay, but how have you found the solution curve?
S2: We have solved the diff equation with ode45 in MatLab.
T: What initial value did you use?
S2: The first point ….?
T: But why should the first point be the one that will give the best fit to all data?
S1: Hmm right, but what shall we do then?
T: I don't know. Are there other ways to solve the equation?
S: Analytically?
T: You can try, maybe that will give you a way to estimate both the parameter k and the initial value using all given data.

With some help the groups are typically able to develop a solution such as:

$$N'(t) = kN(t)^2 \Rightarrow \frac{1}{N(t)^2}N'(t) = k \Rightarrow \int \frac{1}{N(t)^2}N'(t)\,dt$$
$$= \int k\,dt + c_1 \Rightarrow \int \frac{1}{N(t)^2}\,dN(t) = kt + c_1 \Rightarrow$$
$$\frac{-1}{N(t)} = kt + c_1 \Rightarrow \frac{1}{N(t)} = c - kt \Rightarrow N(t) = \frac{1}{c - kt}$$

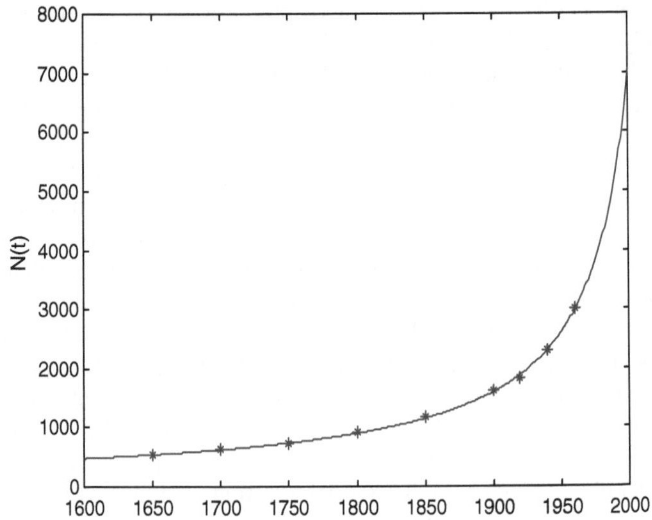

Fig. 12.1 Students' plot of data and their analytical solution (in millions as function of time in years)

c is an arbitrary real number, which can be determined from an initial condition $N_0 = N(t_0)$.

The relation between the integration constant and an initial condition are typically not very clear to students. The same goes for the fact that $N(t)$ is assumed to be free of zeros. In addition the domain of the obtained solution is typically not specified.

Even for groups who solve the equation correctly, typically it is not obvious, and often takes a challenge from the teacher, before the students realise that the second last equation expresses a linear relationship between the reciprocal of the population size and the time, and this can be used to find the two constants in the solution to the given data by means of linear regression. A typical dialogue goes like this:

T: Yes, that is just fine! Can you use all data to estimate values for k and c?
S1: Not really…
T: Look at the last but one equation you have written – what does this equation tell you?
S2: The right hand side is a linear expression.
T: Exactly, and what is on the left hand side?
S1: One over the $N(t)$ – the population size.
T: So how can you use these data to estimate k and c?
S2: We could plot $1/N(t)$ against t and use linear regression.
T: Good idea – try that.

After some time the teacher comes back:

S1: Look, we have drawn this nice solution with the data (see Fig. 12.1).
T: Yes, it looks very nice. Try and see what happens if you plot it up to year 2100.

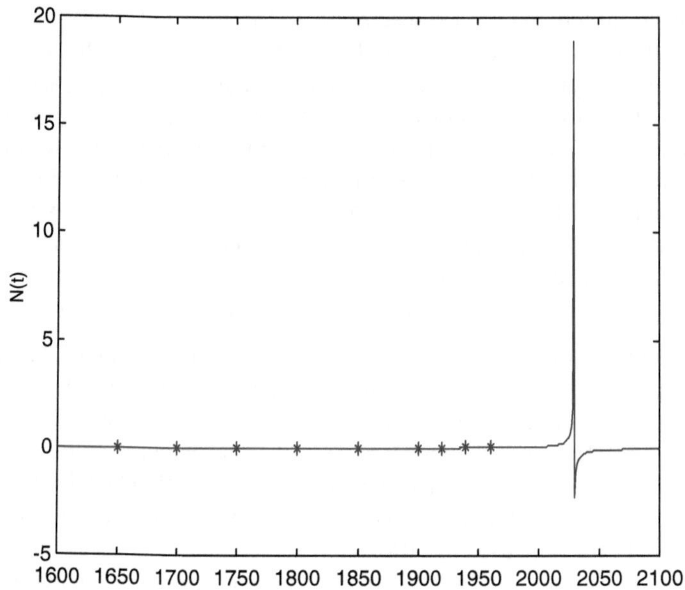

Fig. 12.2 Students' plot of the analytic solution

The teacher leaves the group and enters again after some time:

S2: It grows until year 2025, and then it suddenly becomes negative, but that cannot be right.... we cannot have a negative amount of people on the earth! (see Fig. 12.2).
S1: Is this a numerical mistake in MatLab?
T: No, it is not a numerical solution you have drawn. It is the analytical solution – right? So that is not the explanation!
T: Look at the analytical expression for the solution you found and try to explain the graph.

The teacher leaves the group and enters again after some time

S2: The denominator is zero for $t = 2029.1$. So here the population goes to infinity.
T: What does that mean mathematically for the graph?
S1: What is it called? ... It has a vertical asymptote.
T: Exactly, and for greater values of t, $N(t)$ becomes negative. What do you think of this as a model for the world population?
S1: It must mean that doomsday is near! (The students laugh).

The solution to the differential equation for the explosive growth becomes larger and larger as t approaches the singularity, and for t greater than this the values are negative. In the mathematical domain, this is not a problem. The function is not defined in the singularity; it tends to infinity and minus infinity for t approaching the singularity from left and right respectively; and it tends to zero for t approaching infinity and minus infinity. In the modelling domain however, this is a problem. As the students argued, we cannot have a negative number of people, and we cannot have infinitely many either. In their argumentation and understanding of the modelling situation, the students used the modelling domain to argue about the mathematical domain thus helping them develop their mathematical understanding.

In the final part of the project the students are challenged to find more recent data for the world population and to see if they can find a better model than the explosive model. In the project, most students come to understand and be able to argue as expressed in the following points paraphrased and translated from students' reports:

- The explosive model actually gives a fairly precise description of the given data for the growth of the world's human population from 1650–1980,
- With the same parameter and initial value the model even captures the growth up to year 2000. However, the model predicts that the population will increase still faster and tend to infinity as time approaches year 2029.1, which is unrealistic we hope! Beyond this point of time the population becomes negative. So the explosive model cannot be used in 2100.
- The United Nations' (UN) population model can be used to predict the population until year 2100 based on estimates for the fertility and death rates in different regions. The model predicts ten billion people in 2100. A few groups manage to combine the explosive model and the UN model in one diagram as shown in Fig. 12.3.

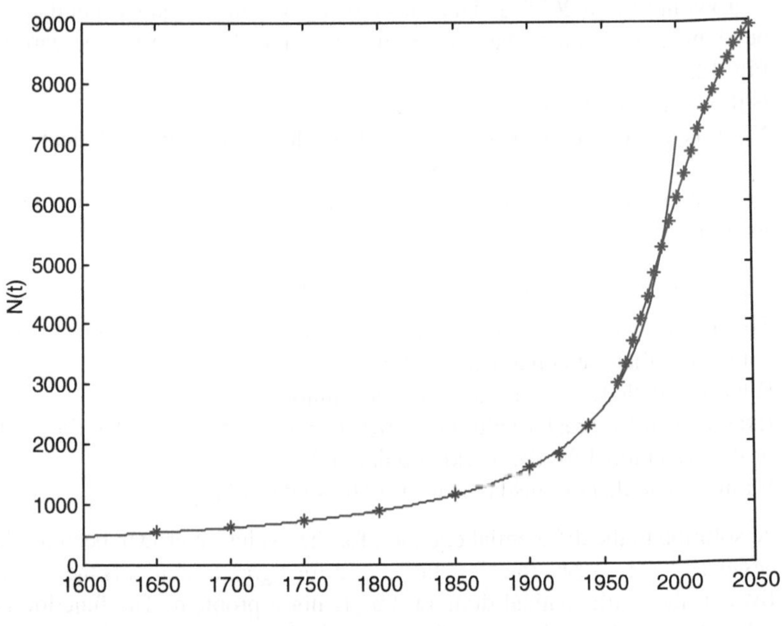

Fig. 12.3 Explosive model (1600–2000) and UN model 1960–2050

6 Discussion

It is our general experience from the course that the modelling context provides a window to students' understanding and their images of the mathematical concepts they work with as well as to their understanding of a mathematical model and modelling. We use dialogue deliberately to open such windows. We have discussed how the project supported the students' formation of the concept of differential equations, how to interpret them, and how to read meaning into the representations mathematically and in the modelling context. The dialogue shows that students do not automatically distinguish between analytical and numerical solutions to differential equations, and they do not have a clear separation of the mathematical domain and the modelling domain in their argumentation. In particular, what cause difficulties for the students is shifting between viewing a differential equation as a relation between the momentary rate of change and the actual size of a certain function (here the size of the population) and viewing it as a relation between a function and its derivative. This phenomenon can be explained by Sfard's (1991) model of concept formation, where the reification of a process into an object is pinpointed as the crucial step in formation of mathematical concepts. The modelling project discussed here is designed deliberately to accentuate potential cognitive conflicts in the students' concept images of differential equations and their solutions. The project and the related dialogue encourage students to reflect on the mathematical meaning of the differential equation as well as on its interpretation and validity as a model for the world population. Therefore we find that the project illustrates very well how the same modelling activity in a course for first year university students can be a didactical vehicle both for developing modelling competency and for enhancing students' conceptual learning of mathematics. At the same time we offer some ideas for the conceptualisation of progress in the students' mathematical learning through modelling activities. We will continue our research on this issue.

References

Blomhøj, M., & Kjeldsen, T. H. (2010a). Learning mathematics through modelling – The case of the integral concept. In B. Sriraman, C. Bergsten, S. Goodchild, G. Pálsdóttir, B. Dahl, & L. Haapasalo (Eds.), *The first sourcebook on Nordic research in mathematics education* (pp. 569–581). Montana: Information Age Publishing.

Blomhøj, M., & Kjeldsen, T. H. (2010b). Mathematical modelling as goal in mathematics education – Developing modelling competency through project work. In B. Sriraman, C. Bergsten, S. Goodchild, G. Pálsdóttir, B. Dahl, & L. Haapasalo (Eds.), *The first sourcebook on Nordic research in mathematics education* (pp. 555–567). Montana: Information Age Publishing.

Niss, M., & Højgaard, T. (Eds.). (2011). *Competencies and mathematical learning – Ideas and inspiration for the development of mathematics teaching and learning in Denmark* (Trans. Niss & Jensen (2002), IMFUFA-text no. 458). Roskilde: IMFUFA, Roskilde University.

Sfard, A. (1991). On the dual nature of mathematical conceptions: Reflections of processes and objects as different sides of the same coin. *Educational Studies in Mathematics, 22*, 1–36.

Tall, D., & Vinner, S. (1981). Concept image and concept definition in mathematics, with particular reference to limits and continuity. *Educational Studies in Mathematics, 12*, 151–169.

Chapter 13
Students' Designing an Ideal Tourism Route as Mathematical Modelling

Chan Chun Ming Eric

Abstract This chapter is part of a larger study investigating Primary 6 (Grade 6) students' engagement in mathematical modelling in a Singapore school and focuses on the mathematical reasoning of one of the modelling tasks attempted by a group of students. The task challenged the students to design an itinerary package for a group of tourists intending to visit a holiday resort island. The design primarily included the need to plan a plausible route to cover places to visit while taking into consideration distance, time, and cost aspects framed within certain task conditions. Through a models-and-modelling perspective, examples of the group's work are presented to show their conceptualisations and mathematical reasoning towards reaching a final considered route.

1 Introduction

Mathematics education reform efforts worldwide are increasingly calling for pedagogies that teach for understanding where students' cognitive reasoning can be deeply engaged (Hiebert et al. 1996). Mathematical modelling has been recognised as able to provide such a platform for students to develop skills and competencies that are required of twenty-first century learners towards enabling them to manage complex systems and solve real-world problems (English and Sriraman 2010). Recent research based on the models-and-modelling view of modelling has suggested that students are capable of developing models and devising systems like sorting, quantifying, weighting, ranking, and organising to give meanings to their conceptual representations (Chan 2008; English and Watters 2005). Engaging in modelling activities thus helps in the promotion of important mathematical reasoning

Chan C.M.E. (✉)
Mathematics and Mathematics Education Academic Group, National Institute of Education, Nanyang Technological University, Singapore
e-mail: eric.chan@nie.edu.sg

G.A. Stillman et al. (eds.), *Teaching Mathematical Modelling: Connecting to Research and Practice*, International Perspectives on the Teaching and Learning of Mathematical Modelling, DOI 10.1007/978-94-007-6540-5_13,

processes such as constructing, explaining, justifying, predicting, conjecturing, and representing (English and Watters 2005).

One key aspect of working on modelling activities lies in the explicit revealing of students' thinking about their solutions through their expressing, testing, and revising of the models they are developing via some representational media (Lesh and Doerr 2003). The refinement of their mathematical ideas helps them to construct models that are useful and meaningful to them and for their clients. This chapter illustrates the conceptualisation and reasoning of a group of Primary 6 students in their design of an ideal tourism route and the revisions of the models towards achieving their ideal route.

2 The Models-and-Modelling View of Modelling

The models-and-modelling view of modelling suggests that students are capable of making meaningful interpretations of situations during mathematical modelling (Sriraman and Lesh 2006). In essence, students construct their own model of a situation that results in sharable and re-usable constructs or conceptual systems. According to Lesh and Doerr (2003), a model consists of elements, relationships, and operations that describe how the elements interact and patterns or rules that apply to the preceding relations or operations. During mathematical modelling, students make significant changes to their underlying ways of thinking about important conceptual systems in mathematics as they express, test, and revise their conceptual and mathematical representations. In this chapter, students make conceptual representations and mathematical translations to describe the system, that is, provide a model of an itinerary package that people can use. To do that, they need to reason and communicate those ideas through behaviours of describing, explaining, justifying and making decisions as they create the representations in the real-world situation. In this regard, students' mental conceptualisations are made visible through some representational media such as expressions of spoken language, written symbols, graphs, diagrams, and experience-based metaphors. In other words, these conceptualisations are continually projected into the external world during mathematical modelling (Lesh and Doerr 2003).

3 Research Design and Method

This chapter stems from a larger study examining students' mathematical reasoning during mathematical modelling. The research question was: What are the characteristics of the mathematical modelling process manifested as mathematical modelling pathways, model development and collaborative discourse for different groups of Primary 6 students? Eighteen groups of students (grouped in fours and fives) in two classes were involved in completing a series of five modelling activities (in different weeks, 1.5 h each). Four groups were video recorded. The four groups were selected by their mathematics teachers who knew them to be vocal and could provide richer displays of problem solving behaviours.

Students' protocols from the transcribed video data were coded and analysed as problem solving behaviours and as episodes using the think-aloud protocol analysis method (Ericsson and Simon 1993). The problem solving behaviour codes were classed under procedural and computational knowledge (K), reasoning (R), meta-cognitive thinking (M) and critical thinking (C). These categories were further subdivided, for example, reasoning included making a conjecture (R3). Modelling episodes were classed under the modelling stages—description, manipulation, prediction, and optimisation—for the purpose of reducing, displaying and interpreting data (Miles and Huberman 1994). Two independent coders and the researcher reviewed the protocols and codes to establish consensus of meanings. In addition to the video recordings, the students' written solutions, and their group journals were collected to triangulate with the transcribed data. In this chapter, only the outcomes related to the model development of one of the modelling activities, *The Itinerary Problem*, is presented.

3.1 The Modelling Task

The Itinerary Problem included a modified version of Singapore's Sentosa Island, an island dedicated as a recreation resort. The scale in the map is represented as 5 km per interval in the grid. The students were expected to design a tourism package to take tourists around an island called Trimbell. In particular, they were to plan a route to visit all the places of interest within a certain duration and at a reasonable cost. Additional information on a separate sheet was provided about accommodation, transportation and places of interest. For example, there were three choices of accommodation, one being the 3-star Prince Hotel offering twin beds at $100 and queen at $120 per night. Supreme Hotel was a 4-star hotel offering twin and queen beds at $120 and $150 respectively per night, and Leisure Hotel was rated 3-stars and offered twin and queen beds at $80 and $120 respectively per night. Other details included a price list for taking a cab with variable costs, a van with a flat rate and a trishaw with the lowest cost to visit the different places of interests as well as the entrance fees for the places of interest.

4 Findings

For the purpose of exemplifying the richness of the students' model development, the work of one group of students has been selected and is used in this chapter for illustrative purposes.

The Itinerary Problem

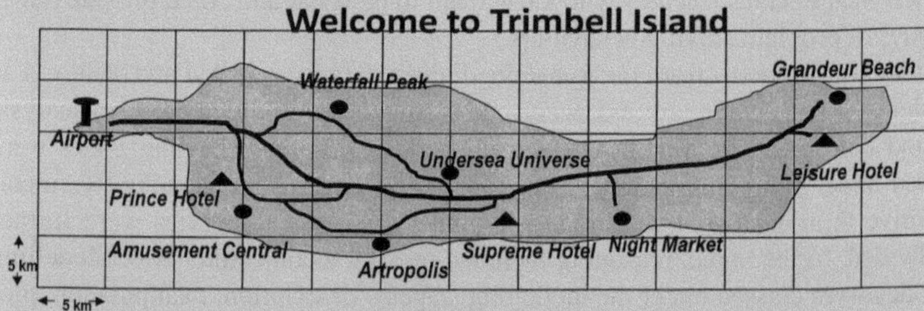

As a school project, your team has been asked to plan a 2-day, 1-night holiday itinerary for tourists who visit Trimbell Island.

Plan an itinerary and budget for a family of 5 (2 parents, and 3 children below 12 years old) showing how they can make good use of their time and money in Trimbell Island. Your itinerary must fit the catch phrase:

"Discover all of Trimbell for as low as $____".

Note
1. Assume that each visit at a place of interest lasts 4 hours.
2. The family would reach the airport at 7.00 am on the first day and would have to be at the airport by 7.00 pm for the flight home on the second day.

Table 13.1 First conceptualisation of route

Conceptualisation of route	Selected mathematical reasoning	Inference and comment
Manipulation stage 1	*I think we should consider factors like we must plan it in a way that it flows so that we don't make rounds and we don't waste money going round and round...*	R3 (Conjecture) – that costs will be lower if overlapping routes can be prevented.
(1st day): AP→AC→AR→NM→LH (2nd day): LH→GB→UU→WP→AP	*Leisure Hotel is cheap and it is easy to go to the places of interest.*	R3 (Conjecture) – appropriate choice of hotel as it was situated along the route in terms of moving forward.

4.1 Models of the Route

An initial route was conceptualised based on visiting all the places spread over two days (Table 13.1). This initial route was very direct with the starting point at one end, the Airport (AP), and proceeded to Grandeur Beach (GB) at the other end before going back to the AP, thus covering all the places of interest. It was a plausible model of a route design based on reasons that avoided 'going round and round'

Table 13.2 Second conceptualisation of route

Conceptualisation of route	Selected mathematical reasoning	Inference and comment
Manipulation stage 1:	*From the airport, we're very tired so we go to Waterfall Peak and after that to Undersea Universe, and at night already, we go to Night Market, and then we go back to Leisure Hotel. Next morning, you'll feel refreshed, so we go to Grandeur Beach. After that we go to Artropolis, and then Amusement Central. ….*	M8 (Impress personal knowledge) – Revising route based on need to have a more relaxed itinerary on the first day.
(1st day): AP → WP → UU → NM → LH (2nd day): LH → GB → AR → AC → AP	*No, you gotta see here, it says how many kilometres down here and then we gotta count. (referring to the scale on the map)*	C2 (Challenge efficacy of idea) – Conceptualisation was challenged. K1 (Delineate mathematics components) – Determining the distance travelled by using the scale in the map.

with respect to the visits so as not to 'waste time and money'. This route represented a plausible initial model but it was yet to be tested for feasibility in terms of the time taken in the travelling.

The second conceptualisation of the route was a revision of the first and this is shown in Table 13.2 where the students' personal preference (from their real-world experience) for a more relaxed itinerary on the first day was taken into consideration in the planning. They preferred the route to the Waterfall Peak (WP) followed by the Undersea Universe (UU) rather than the route where Amusement Central (AC) was located. This resulted in the revision of the route. The students took into consideration the need to find the distance of the route as they used the scales on the map. So far, the students at that juncture had assumed that they could cover all the places over 2 days but they had not factored the time duration as yet.

The third conceptualisation (see Table 13.3) of the route involved working out the time duration for the travelling and the visits. The students found out that they did not have enough time to get back to the airport on the second day if they had visited three places of interest that day. The route was re-conceptualised towards visiting more places on the first day instead as well as altering their bedtime and waking time.

As the students worked out the cost, they decided to exclude the visit to Undersea Universe (UU) because it was expensive. They expressed that they need not visit two places that had to do with water (the other place being Grandeur Beach). This fourth conceptualisation is shown in Table 13.4.

Table 13.3 Third conceptualisation of route

Conceptualisation of route	Selected mathematical reasoning	Inference and comment
Manipulation stage 2:	*I think we should change (the route) because we do not have enough time to cover.*	R4 (Generalise) – Proposing revision of route (after considering previous data).
(1st day): AP → AC → WP → UU → NM → LH	*Next day you have only 12 hours.*	R2 (Make mathematical translations) – Worked out that there was not enough time to visit three places in the second day.
(2nd day): LH → GB → AR → AP	*I think we should cover more places of interest on the first day because second day we have less time.*	R4 (Generalise) – Generalising and justifying for this conceptualisation.
	That's let's say we wake up at 7 o'clock. 8, 9, 10, 11, 12, 1, 2, 3, 4, 5, 6, 7. 7 o'clock exactly (time to reach airport)	R2 (Make mathematical translations) – Calculating time duration to ensure they could reach airport on time based on this new route.

Table 13.4 Fourth conceptualisation of route

Conceptualisation of route	Selected mathematical translation	Inference
Prediction stage 2:	*I still think we either give up this or this because both have to do with the sea. This one (GB) you can get to settle down (relax), this one you only get to see (UU).*	M8 (Impress personal knowledge) – suggest a place of interest to exclude.
(1st day): AP → AC → AR → NM → LH	*Which one is more expensive?*	M3 (Check to understand) – Choice to exclude based on cost.
(2nd day): LH → GB → WP → AP	*This one. Yes, so this one? (UU). So we drop out this.*	M6 (Monitor progress) – Checking to confirm.

It must be noted that the students were not supposed to exclude any place of interest in their visits. Unfortunately this happened because there was a miscommunication during their consultation with the teacher in their attempt to plan an ideal route and the students misinterpreted the teacher's advice. Nevertheless, the revised route was one where the itinerary could fit the time schedules as well as kept within reasonable cost as they compared the expenditure of their package with other groups.

Table 13.5 First conceptualisation of travel cost

Conceptualisation of travel cost	Selected mathematical reasoning	Inference and comment
Manipulation stage 1:	*Because it's $100 per day. But if you use this (cab), it's per km. And imagine how many km will you travel?*	R2 (Make mathematical translations) – Comparing costs based on distance.
The van – flat fare of $100 per day.	*If, 5 km, it'll be $2.50, so I think it's van.* *Then you also have boarding charge, $5 (for cab).* *I don't think this (cab) is cheaper because it's 50 cents per km. How many km are you travelling on that day? So much!*	R3 (Conjecture) – that taking the cab would be more expensive due to the fixed and variable costs as there would be quite a distance to travel.

In summary, the initial conceptualisation began with just a proposed route but along the way, the students delved more deeply into the task by considering variables like distance, time and costs towards getting the final route. Their final route unfortunately was not an ideal route because the students did not manage to cover all the places of interest as they had excluded the Undersea Universe (UU). The fourth conceptualisation of the route was the route the students decided they would recommend to the family visiting Trimbell Island. In this respect, it is considered a workable route but not an optimised route.

4.2 *Travelling Costs*

The conceptualisation of the routes was not a stand-alone endeavour. The problem context demanded that the students also took into account travelling and accommodation costs to determine if the whole travel package would be value for money. An example of how the students reasoned about the travelling cost is exemplified next.

The students' initial thought about the travelling cost was to hire the van as it had a flat rate of $100 (see Table 13.5). They reasoned that taking a cab would cost more although at that point they had not verified that yet. Their reasons were based on their theory that because taking the cab would involve fixed and variable costs, and since they had to travel a reasonable distance, taking a cab would be more expensive. However, when a student worked out the distance to travel to the various places of interest with respect to the cab fares, they found that this was not the case (see Table 13.6). The students worked out how much it would take to travel 50 km by cab, made some projections and found that it cost less than taking the van. They switched their choice to taking the cab. The students' written solution to the modelling problem is shown in Fig. 13.1.

Table 13.6 Second conceptualisation travel cost

Conceptualisation of travel cost	Selected mathematical reasoning	Inference and comment
Manipulation stage 2: The cab – fixed and variable costs	*Hey it's actually cheaper (cab). I've calculated already. 1, 2, 3, 4, 5, 6, 7, 8, 9, 10 (making estimation of distances). 10…10 x 5 so is 50. 50 km … it's cheaper!*	M4 (Realise new information), K2 (Compute), R4 (Generalise) – Working out the cab fare for 50 km and finding it to be cheaper than taking the van.

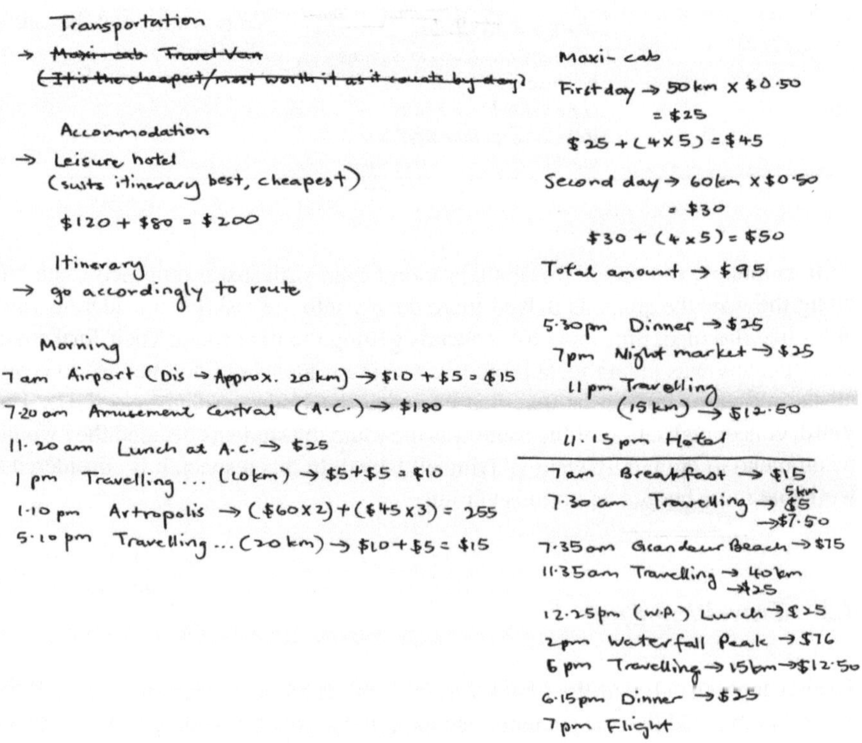
Transportation
→ ~~Maxi-cab Travel Van~~
~~(It is the cheapest/most worth it as it counts by day)~~

Accommodation
→ Leisure hotel
(suits itinerary best, cheapest)

$120 + $80 = $200

Itinerary
→ go accordingly to route

Morning
7 am Airport (Dis → Approx. 20 km) → $10 + $5 = $15
7.20 am Amusement Central (A.C.) → $180
11.20 am Lunch at A.C. → $25
1 pm Travelling … (10 km) → $5+$5 = $10
1.10 pm Artropolis → ($60x2)+($45x3) = 255
5.10 pm Travelling … (20 km) → $10+$5 = $15

Maxi-cab
First day → 50 km x $0.50
= $25
$25 + (4x5) = $45
Second day → 60 km x $0.50
= $30
$30 + (4x5) = $50
Total amount → $95

5.30pm Dinner → $25
7pm Night market → $25
11 pm Travelling
(15 km) → $12.50
11.15 pm Hotel

7 am Breakfast → $15
7.30 am Travelling → 5 km $5
→$7.50
7.35 am Grandeur Beach → $75
11.35am Travelling → 40 km
→$25
12.25pm (W.P.) Lunch → $25
2pm Waterfall Peak → $76
6 pm Travelling → 15 km → $12.50
6.15pm Dinner → $25
7pm Flight

Fig. 13.1 The students' written solution for *The Itinerary Problem*

The solutions clearly distinguish four important aspects taken into account by the students: the itinerary showing the time durations for each place of visit for both days, the entrance cost involved, the cost of transportation related to the distance travelled (their recommended route) for both days, and the amount spent on accommodation. For example, it could be seen that the students have cancelled their initial proposal to use the travel van as they initially thought it was cheaper and better value. The cab fare calculations, shown at the top right of Fig. 13.2, reveals how the group worked out the variable rates for the 2 days. The itinerary shows the timings which were obtained by taking into consideration the distances (shown in brackets) as well as adherence to task conditions of staying in a place of interest for 4 h.

The costs of the visits were evident and the students made their own assumptions by factoring reasonable expenditures for meals. On the whole, each aspect illustrates the students' mathematical reasoning in the overall scheme of conceptualising representations based primarily on establishing distance-time and distance-cost relationships.

5 Discussion and Implications

In this case study, the students' modelling endeavours show how they reasoned towards getting their "ideal" route and the decisions made that had an impact on distance-cost and distance-time relationships. The modelling effort involved expressing, testing, and revising their conceptualisations. Several implications are brought to light with respect to the potential of implementing mathematical modelling in the mathematics curriculum.

5.1 The Pedagogic Platform

Engaging in mathematical modelling activity provides affordances for student-task-teacher interaction and in particular for the modelling task to drive learning. Chan (2008) noted that in such an environment, the students' cognitive engagement was kept high as they made attempts to identify goals and variables, interpret problem situations, interrogate data, make inquiries, monitor their solutions and improve their conceptualisations. This form of instruction differs starkly from learning mathematics in a traditional and didactical manner which is limited in scope in revealing what students are capable of. A modelling platform points to the role that mathematical modelling has in developing students to make sense of complex systems, working within teams, adapting rapidly to a variety of evolving conceptual tools, and developing sharable tools that draw on a variety of disciplines (Sriraman and Lesh 2006) which is more than working with mere computations.

5.2 Modelling and Mathematising

To develop an attractive tour package in light of the given problem was to be able to model a route where all the places of interest could be visited once (without passing by any of the places again except going back to the Airport) with resource efficiency as a key consideration. The students in this study developed their initial model with the goal to plan a route that covered all places of interest but they did not consider other aspects such as the relationships between distance and time or distance and cost. They then revised their initial model into workable ones through greater deliberation on the given data as well as the employment of their own real-life experiences. The testing and revising of their models exemplify the students' sense-making, explaining and decision-making aspects of mathematising towards bringing the real-world situation into the model world and vice-versa. In other words, consistent with what Sriraman and Lesh (2006) have found, these cycles of expressing, testing and revising do not just take the students from pre-mathematised givens to goals when the path is not obvious but also lead to conceptualising givens and goals in productive ways.

6 Concluding Remarks

Engaging students in mathematical modelling paves the way for teachers to redesign ways where students can solve problems in real-world settings. The modelling task such as *The Itinerary Problem* takes on a dimension for students to notice phenomena that are amenable to mathematical thinking, have them clarify the purposes for the modelling situation, brainstorm quantities and variables related to the problem situation, evaluate these quantities and variables towards relevance for working with and developing, manipulating and interpreting models in various ways to obtain results. In the process, the students are forced to manifest their problem solving behaviours and formulate important relationships between variables. It also allows students to assess the appropriateness and limitations of the models and communicate the results. In this respect, mathematical modelling helps students in appreciating mathematics as a critical tool for analysing important issues in their lives, communities, and in society in general (Greer et al. 2007).

References

Chan, C. M. E. (2008). Mathematical modelling experiences for mathematical development in children. *Thinking Classroom, 9*(2), 37–45.

English, L. D., & Sriraman, B. (2010). Problem solving for the 21st century. In L. D. English & B. Sriraman (Eds.), *Theories of mathematics education: Seeking new frontiers* (pp. 263–285). Berlin: Springer.

English, L. D., & Watters, J. J. (2005). Mathematical modelling in third-grade classrooms. *Mathematics Education Research Journal, 16*(3), 59–80.

Ericsson, K. A., & Simon, H. A. (1993). *Protocol analysis: Verbal reports as data* (2nd ed.). Boston: MIT Press.

Greer, B., Verschaffel, L., & Mukhopadhyay, S. (2007). Modelling for life: Mathematics and children's experience. In W. Blum, P. L. Galbraith, H.-W. Henn, & M. Niss (Eds.), *Modelling and applications in mathematics education: The 14th ICMI study* (pp. 89–98). New York: Springer.

Hiebert, J., Carpenter, T. P., Fennema, E., Fuson, K., Human, P., Murray, H., Olivier, A., & Wearne, D. (1996). Problem solving as a basis for reform in curriculum and instruction: The case of mathematics. *Educational Researcher, 25*(4), 12–21.

Lesh, R., & Doerr, H. (2003). Foundations of a models and modeling perspective in mathematics teaching and learning. In R. Lesh & H. M. Doerr (Eds.), *Beyond constructivism: Models and modeling perspectives on mathematics problem solving, learning and teaching* (pp. 3–34). Mahwah: Lawrence Erlbaum.

Miles, M. B., & Huberman, M. A. (1994). *An expanded sourcebook: Qualitative data analysis* (2nd ed.). Thousand Oaks: Sage.

Sriraman, B., & Lesh, R. (2006). Modeling conceptions revisited. *Zentralblatt für Didaktik der Mathematik, 38*(3), 247–254.

Chapter 14
Comparison of Mathematical Modelling Skills of Secondary and Tertiary Students

Juntao Fu and Jinxing Xie

Abstract The primary motivation of this study is to investigate whether there are any differences, and what they are, between the mathematical modelling skills of secondary and tertiary students entering a highly academic university in China. With this aim in mind, we used a multiple-choice question instrument, which has been extensively used in experimental studies in various countries, to assess students' modelling skills. Two tests, each consisting of six questions, were conducted with approximately 200 first-year students from Tsinghua University. The first test was carried out as the students entered the University, and the second test was carried out at the end of the first semester after an introductory unit in mathematical modelling. Statistical analyses were conducted to compare the differences in the results of these two tests.

1 Introduction

During the last two decades, research on mathematical modelling and applications has attracted an increasing interest in the mathematics education field (e.g., Biehler and Leiss 2010; Blum et al. 2002; Frejd and Ärlebäck 2011; Kaiser et al. 2006; Stillman et al. 2010). This increasing trend can be noted from the fact that in recent years there is much research focusing on the teaching and learning of mathematical modelling and applications, ranging across all education levels including primary, secondary, tertiary and teacher education. However, among all the research studies, it seems only a few (e.g., Kaiser 2007) focus explicitly on the comparison of mathematical modelling skills between the students at the secondary and the tertiary levels. This is a little puzzling and the reason might be due to that until now there

J. Fu • J. Xie (✉)
Department of Mathematical Sciences, Tsinghua University, Beijing, China
e-mail: jxie@math.tsinghua.edu.cn

G.A. Stillman et al. (eds.), *Teaching Mathematical Modelling: Connecting to Research and Practice*, International Perspectives on the Teaching and Learning of Mathematical Modelling, DOI 10.1007/978-94-007-6540-5_14,

are actually no roadmaps to sustained implementation of modelling education at all levels. Just as Blum et al. (2002) point out:

> In spite of a variety of existing materials, textbooks, etc., and of many arguments for the inclusion of modelling in mathematics education, why is it that the actual role of applications and mathematical modelling in everyday teaching practice is still rather marginal, for all levels of education? How can this trend be reversed to ensure that applications and mathematical modelling is integrated and preserved at all levels of mathematics education? (p. 165)

This is a big issue and it appears it remains to be resolved. The primary motivation of the current study is to investigate whether there are any differences, and what they are, between the mathematical modelling skills of secondary and tertiary students. To assess students' modelling skills, we adopted a multiple-choice question instrument which has been extensively used in experimental studies in various countries.

As an early application of the multiple-choice question instrument, Haines et al. (2000, 2001) designed two similar tests (Test 1 and Test 2, each consisting of six questions) to examine and measure students' mathematical modelling skills at different stages of a modelling process. As has been clarified in Houston and Neill (2003b), the six questions in each quiz try to identify six types of students' modelling sub-competencies, namely:

> Type 1: Making simplifying assumptions… . Type 2: Clarifying the goal… . Type 3: Formulating the problem… . Type 4: Assigning variables, parameters and constants… . Type 5: Formulating mathematical statements… . Type 6: Selecting a model. (pp. 156–158)

The work by Haines et al. (2000, 2001), including research methods and tests designed, gave an introductory reference for follow-up studies in the United Kingdom (Haines and Crouch 2001, 2005) and spreading to Australia, Ireland, Japan, Germany, Sweden and China (Dan and Xie 2011; Frejd and Ärlebäck 2011; Ikeda et al. 2007; Houston and Neill 2003a, b; Kaiser 2007; Lingefjärd and Holmquist 2005). In particular, Haines and Crouch (2001) paid more attention to the effectiveness of their research tool. They mixed these two tests into one with questions in a random order and asked 42 mathematics undergraduates to finish them within 30 min. According to their analyses, they found that scores with the same question number in these two tests are close, and the comparable manner of the analogue pairs indicates their tests are generally robust. Houston and Neill (2003b) added another question in each test to require students to sketch the graph of a function. They used Test 1 as a pre-test and Test 2 as a post-test in the first year. When it came to the second year, they created a new test comparable to the former ones and took it as a new pre-test. Test 2 was used again as a post-test at the end of the fourth semester. They found that over a 1 year course of study, students generally improved their test scores; however, there was a larger decline on the third test than the second one, indicating that most students did not have their modelling skills consolidated effectively over their summer holiday. The number of test questions was extended to 22 questions by Houston and Neill (2003b), and the issue of assessment validity was addressed in Lingefjärd and Holmquist (2005). Dan and Xie (2011) carried out a study with the students from Logistics Engineering College, an average level institution among Chinese engineering universities. Significant positive correlation between mathematical modelling skills and innovative thinking levels was found. However, correlation between mathematical modelling skills and scores achieved in basic mathematics courses was not so strong.

In the current study, we adopted the two tests (Test 1 and Test 2) originally designed by Haines et al. (2001). We used the data from the tests with students from Tsinghua University, a highly academic university in China, to check whether there are any differences between the mathematical modelling skills of secondary and tertiary students. We also examined whether the students' modelling skills are influenced by other factors such as the scores they achieved in basic mathematics courses.

2 Experiment

The tests were conducted in fall semester of 2010 with 193 first-year students of Tsinghua University. Among these, 157 were boys and the rest were girls. All the students were taking a course called Introduction to Mathematical Modelling, which is a one-semester course aiming to introduce some basic modelling concepts and examples to the first year students. The course consists of fifteen 90-min lectures, once every week. In each lecture, the teacher introduces a modelling instance with close relationship with the students' daily life, including some basic models arising from engineering, public administration, operations management, marketing and economics. Most of the instances are taken from a Chinese textbook by Jiang et al. (2011), which is the most popular one used in Chinese universities for mathematical modelling courses. Here are some examples for the modelling tasks: How do you allocate the number of seats in a committee (i.e., the apportionment problem)? How do you predict the growth of the population size in China and in the world? How much data can red and blue laser compact disks hold? What is the best design for the shape of a Cola can? What's the best time to sell a pig to the market? What is the best policy for inventory management (in particular, the classical economic ordering quantity (EOQ) model and the newsboy model)? Why and how does an airline company use an overbooking strategy? Why are prices usually subject to periodic fluctuations in certain types of markets (in particular, the cobweb model)? How do you calculate the repayment scheme for a fixed rate mortgage? How do you estimate the explosive power for the Trinity detonation with only a video of the detonation (in particular, using the dimensional analysis approach)? What are the advantages of using multi-stage rockets (and why usually are three-stage rockets used)?

During a lecture, the teacher first introduces to the students the modelling context for a specific modelling task, and asks the students to think about and try to solve the problem by themselves for about 15 min. Next the teacher asks some students to report their ideas in front of their classmates, followed by students' discussions guided by the teacher. Finally the teacher presents some well-established mathematical models for the task, pointing out the basic assumptions of the models and analysing the advantages and disadvantages of the models. Although out of the class no homework is assigned to students after each week's lecture, the students are asked to form groups (each group consists of three or four students) to independently pose, discuss and build up mathematical models to solve real world problems they are interested in. At the end of the course, each group is required to submit a course report to the teacher, presenting the out of class achievement of the group. Usually, a very large number of problems and topics from various contexts are covered in the students' reports.

Test 1 was carried out as the first year students entered the University and had just attended the first lecture of the course. Test 2 was carried out at the end of the first semester. All students finished both tests within 20 min. We used the results from the first test as a measure of secondary students' modelling skills, and the results from the second test as a measure of tertiary students' modelling skills. Questions with the same number in the two tests were quite similar and had comparable difficulty, which was proved in Haines and Crouch (2001). Take Question 6 for example.

Question 6 in Test 1. Which one of the following options most closely models the height of a sunflower while it is growing (in terms of time t)?

$$\text{A. } 1-e^{-t} \quad \text{B. } (1-t)^2 \quad \text{C. } t \quad \text{D. } t-t^2 \quad \text{E. } 1/(1+e^{-t})$$

Question 6 in Test 2. Which one of the following options most closely models the speed of a car starting from rest (in terms of time t)?

$$\text{A. } 1-e^{-t} \quad \text{B. } (1-t)^2 \quad \text{C. } t \quad \text{D. } t-t^2 \quad \text{E. } 1/(1+e^{-t})$$

Both of the questions ask the modeller to choose a reasonable model, by reflecting upon the mathematical model to be used in a particular situation and evaluating a solution or comparing the model with the real world problem. So it is reasonable to compare the scores in the two tests to check if students made any progress in modelling skills.

Together with the second test, we also asked the students to report their scores achieved at midterm examinations in Calculus and Linear Algebra, which are the two major basic mathematical courses for the first year students. This information was used to check whether modelling abilities are related to the knowledge of basic mathematics courses.

3 Results

3.1 Comparison of the Two Tests

Each question in the tests has one correct answer worth 2 points and one or several partial correct answers worth 1 point, so full marks for each test is 12. The average scores achieved by the students on the two tests are summarised in Table 14.1.

Table 14.1 Summary of Test 1 and Test 2

Question number	1	2	3	4	5	6	Overall average	Standard deviation
Test 1	1.17	1.35	1.68	1.42	1.91	1.78	9.32	1.86
Test 2	1.69	1.26	1.62	1.47	1.73	1.75	9.53	1.79
Sum	2.87	2.61	3.30	2.89	3.64	3.53	–	–

As can be seen from Table 14.1, the difference of the overall average scores of the two tests is within a gap of only 0.21 points, with almost the same standard deviations. Statistical analyses (in the statistical analyses in this chapter, we use *t*–tests with the significance level of 0.1) show that there is no significant difference between scores achieved by the students on the two tests. We also checked the number of students who performed better in Test 2 and found that 90 students achieved higher scores on Test 2 than Test 1, while 71 students' performance was worse. The difference was statistically insignificant. Therefore, we come to the conclusion that there is no statistically significant difference between the mathematical modelling skills of (upper) secondary and (first year) tertiary students. More precisely, first year students in Tsinghua University did not make any progress in mathematical modelling ability through one semester's study. It disappointed us more or less, as in an analogous study done by Kaiser (2007), the students scored 8.4 in pre-test and 9.6 in post-test (both tests consisted of eight testing questions other than six questions as in our study), which was a significant improvement. Nevertheless, a similar conclusion to our study was also made in Haines et al. (2000) with data from UK students. Some of the possible reasons we think are as follows:

- For quality teaching, "teachers ought to support students' individual modelling routes and encourage multiple solutions" (Blum 2011, p. 24); but we did not do much work along this line in our class or assign such kind of homework for our students. In fact, we put more attention on imposing the well-established models to the students in our class. This motivates us to improve our teaching methods in the future, making our teaching more effective.
- The students entering Tsinghua University are among the best students in China. They already achieved very high scores (with an overall average score of 9.32 out of 12) in the pre-test (Test 1), so it is difficult to improve further. In contrast, German students in Kaiser's study (2007) only scored 8.4 out of 16 on average in the pre-test so they had more of a gap to close.
- One semester is only a short time. It is difficult to improve the students' modelling skills in such a short time.

Since questions with the same number examine the same type of sub-competencies in the process of modelling, if we sum up the scores achieved for two questions with the same number in the two tests, we can identify advantages and disadvantages of our students in the corresponding sub-competency. It is reasonable to add the scores up directly since we have found that our students made little progress through one semester, although the two tests were finished at different times. A similar approach

Table 14.2 Relative scores of Tsinghua and UK students

	Test 1		Test 2		Tests 1 & 2	
Question	UK (%)	Tsinghua (%)	UK (%)	Tsinghua (%)	UK (%)	Tsinghua (%)
1	15.8	12.6	18.0	17.9	17.0	15.2
2	12.0	14.5	7.2	13.2	9.5	13.9
3	14.7	18.1	22.9	17.0	19.1	17.5
4	21.4	15.2	19.7	15.5	20.5	15.3
5	22.6	20.5	20.5	18.1	21.4	19.3
6	13.5	19.1	11.7	18.3	12.5	18.7
Sum	100.0	100.0	100.0	100.0	100.0	100.0

was used by Haines et al. (2000) to check students' understanding at stages within the modelling process. Haines and Crouch (2001) also followed this idea in order to develop a measure of attainment for stages of modelling. Examination of Table 14.1 shows that the students were most successful in solving Questions 3, 5 and 6 (82.6 %, 91.1 % and 88.2 % of the total marks were achieved), whilst Questions 1, 2 and 4 (only 71.6 %, 65.3 % and 72.3 % of the total marks were achieved) seemed to be more challenging for them. This observation suggests, in order to improve the students' modelling skills, we should put more attention to training students' ability in making reasonable assumptions for the real world problems, understanding the goals of the modelling tasks, and identifying required parameters, variables and constants. Among them, the problem of understanding the goals of the modelling tasks was also reported by Houston and Neill (2003a), and Kaiser (2007).

3.2 *Comparison with UK Students*

Now we compare the differences of the achievements of Tsinghua students (in our experiment) with those achieved by the students from two universities in UK (we refer to them as UK students hereafter) as reported by Haines and Crouch (2001). On average, Tsinghua students in our experiment performed better in each question than the UK students. Noticing that the students in Tsinghua are among the best in China, it is unreasonable and not meaningful to compare their scores directly with those of the UK students. Therefore, we decided to analyze the relative scores as shown in Table 14.2. Here the relative score achieved by students in a question means the average score achieved in this question divided by the overall average score for all test questions. For example, Tsinghua students scored 1.63 points in the third question of Test 1 on average and their overall average score is 9.32. Then the relative score in the third question of Test 1 equals 18.1 % (=1.63/9.32). If the relative score achieved by Tsinghua students in a question is much higher than that by UK students, it is reasonable to consider Tsinghua students performed better according to the sub-competency corresponding to the question in the process of modelling.

Table 14.2 shows that UK students performed relatively better in Question 4, and Tsinghua students performed relatively better in Question 6. This indicates that UK students have relatively better understanding than Tsinghua students in assigning parameters, variables and constants, whilst Tsinghua students have relatively better understanding than UK students in assessing and choosing a proper model. The differences in other questions are only marginal. In addition to this, Question 2 seems to be the most difficult one for all the students and Question 5 is the easiest one for them.

3.3 *Relationship with Basic Mathematics Courses*

Serious mathematical modelling needs to make use of much "pure" mathematical knowledge and skills. Questions in our tests also reflect some mathematics thinking, so it is natural to check the relationship between modelling skills and basic mathematics levels of the students. Similarly as in Dan and Xie (2011), firstly, we calculated the correlation coefficients of mathematical modelling skills and both the marks achieved by the students in their midterm examinations for two basic mathematics courses – Calculus and Linear Algebra. We found that the correlation coefficients were both less than 0.2, so no meaningful conclusion can be drawn from the data. However, we achieved a meaningful result after reanalyzing our data from another perspective as below.

We regard those who scored 12 points in Test 1 or Test 2 as skilled modellers, and those who scored no more than 6 points as having poor modelling skills. Similarly, when dealing with the total scores for the two tests, we regard 22–24 points as an indicator for strong modeling skills, and no more than 15 points as an indicator for poor modelling skills. The comparison results are listed in Table 14.3.

Examination of Table 14.3 indicates that the scores achieved by students in Calculus have no significant relationships with modelling skills in our experiment. In contrast, for the Linear Algebra course, all the *p*-values are less than 0.1, which leads to a positive conclusion: Students with strong modelling skills also possess strong skills in Linear Algebra and those with poor modelling skills possess relatively weak skills in Linear Algebra. The reason why this happens is not very clear and needs further investigation and discussion. One possible explanation might be that all the questions in our tests for modelling skills do not need to use knowledge in Calculus, but knowledge in Linear Algebra is relevant to the test questions we used.

4 Discussion and Conclusion

This chapter reports some initial results of comparison of the differences in modelling skills between upper secondary and first year tertiary students, from an experiment conducted at Tsinghua University in China. We recognize the difficulties that

Table 14.3 Relationship between modelling skills and mean scores on basic mathematics courses

		Number of students	Mean scores of calculus	Mean scores of linear algebra
Test 1	12 points	20	89.65	88.13
	No more than 6 points	15	87.60	82.00
	p-value	–	0.26	0.07
Test 2	12 points	18	87.50	85.61
	No more than 6 points	13	83.38	77.88
	p-value	–	0.23	0.08
Tests 1 & 2	22~24 points	22	88.23	87.20
	No more than 15 points	21	88.76	80.48
	p-value	–	0.43	0.07

students met in the process of modelling, which will help us improve our teaching approach. The comparison with UK students (Haines and Crouch 2001) clarifies the distinction of modelling competencies between Chinese and UK students, which to some extent reflects the difference between thinking patterns of the two countries' students. In addition, it needs to be noted that the tests used in our study did not examine some competencies which are necessary for a holistic way of conducting modelling processes (Kaiser 2007). So a research tool that addresses the full range of modelling skills is expected to be developed (Haines and Crouch 2001). For example, it might be reasonable to use an authentic modelling task to test students' modeling skills (e.g., Ludwig and Xu 2010).

Due to the limitations of our experiment (e.g., it was only a very short time between the pre-test and the post-test, and the students were all from a highly academic university in China), findings from the study are not very stimulating. However, the authors think it is a very interesting area to investigate whether there are any differences, and what they are, in modelling skills between students at different levels of education, since the answers to these questions can benefit the curriculum design and qualified teaching for mathematical modelling courses of students at different levels. More studies are expected to be done in the future along this line.

References

Biehler, R., & Leiss, D. (2010). Empirical research on mathematical modelling. *Journal für Mathematik-Didaktik* (Journal for Didactics of Mathematics), *31*(1), 5–8.

Blum, W. (2011). Can modelling be taught and learnt? Some answers from empirical research. In G. Kaiser, W. Blum, R. Borromeo Ferri, & G. Stillman (Eds.), *Trends in teaching and learning of mathematical modelling* (pp. 15–30). New York: Springer.

Blum, W., Alsina, C., Biembengut, M. S., Bouleau, N., Confrey, J., Galbraith, P., Ikeda, T., Lingefjärd, T., Muller, E., Niss, M., Verschaffel, L., Wang, S., Hodgson, B. R., & Henn, H.-W. (2002). ICMI study 14: Applications and modelling in mathematics education – Discussion document. *Educational Studies in Mathematics, 51*, 149–171.

Dan, Q., & Xie, J. (2011). Mathematical modelling skills and creative thinking levels: An experimental study. In G. Kaiser, W. Blum, R. Borromeo Ferri, & G. Stillman (Eds.), *Trends in teaching and learning of mathematical modelling* (pp. 457–466). New York: Springer.

Frejd, P., & Ärlebäck, J. B. (2011). First results from a study investigating Swedish upper secondary students' mathematical modelling competencies. In G. Kaiser, W. Blum, R. Borromeo Ferri, & G. Stillman (Eds.), *Trends in teaching and learning of mathematical modeling* (pp. 407–416). New York: Springer.

Haines, C., & Crouch, R. (2001). Recognizing constructs within mathematical modelling. *Teaching Mathematics and Its Applications, 20*(3), 129–138.

Haines, C., & Crouch, R. (2005). Applying mathematics: Making multiple-choice questions work. *Teaching Mathematics and Its Applications, 24*(2–3), 107–113.

Haines, C., Crouch, R., & Davis, J. (2000). *Mathematical modelling skills: A research instrument* (University of Hertfordshire, Department of Mathematics, Technical Report No. 55). Hatfield: University of Hertfordshire.

Haines, C., Crouch, R., & Davis, J. (2001). Understanding students' modelling skills. In J. F. Matos, W. Blum, S. K. Houston, & S. P. Carreira (Eds.), *Modelling and mathematics education – ICTMA 9: Application in science and technology* (pp. 366–380). Chichester: Horwood.

Houston, K., & Neill, N. (2003a). Assessing modelling skills. In S. J. Lamon, W. A. Parker, & S. K. Houston (Eds.), *Mathematical modelling: A way of life – ICTMA 11* (pp. 155–164). Chichester: Horwood.

Houston, K., & Neill, N. (2003b). Investigating students' modelling skills. In Q. Ye, W. Blum, S. K. Houston, & Q. Jiang (Eds.), *Mathematical modelling in education and culture: ICTMA 10* (pp. 54–66). Chichester: Horwood.

Ikeda, T., Stephens, M., & Matsuzaki, A. (2007). A teaching experiment in mathematical modelling. In C. P. Haines, P. Galbraith, W. Blum, & S. Khan (Eds.), *Mathematical modelling (ICTMA 12): Education, engineering and economics* (pp. 101–109). Chichester: Horwood.

Jiang, Q., Xie, J., & Ye, J. (2011). *Mathematical models* (4th ed.). Beijing: Higher Education Press (in Chinese).

Kaiser, G. (2007). Modelling and modelling competencies in school. In C. P. Haines, P. Galbraith, W. Blum, & S. Khan (Eds.), *Mathematical modelling (ICTMA 12): Education, engineering and economics* (pp. 110–119). Chichester: Horwood.

Kaiser, G., Blomhøj, M., & Sriraman, B. (2006). Towards a didactical theory for mathematical modelling. *ZDM—The International Journal on Mathematics Education, 38*(2), 82–85.

Lingefjärd, T., & Holmquist, M. (2005). To assess students' attitudes, skill and competencies in mathematical modeling. *Teaching Mathematics and Its Applications, 24*(2–3), 123–133.

Ludwig, M., & Xu, B. (2010). A comparative study of modelling competencies among Chinese and German students. *Journal für Mathematik-Didaktik, 31*(1), 77–97.

Stillman, G., Brown, J., & Galbraith, P. (2010). Researching applications and mathematical modelling in mathematics learning and teaching. *Mathematics Education Research Journal, 22*(2), 1–6.

Dan, Q., & Xie, J. (2011). Mathematical modelling skills and creative thinking levels: An experimental study. In G. Kaiser, W. Blum, R. Borromeo Ferri, & G. Stillman (Eds.), *Trends in teaching and learning of mathematical modelling* (pp. 457–466). New York: Springer.
Frejd, P., & Ärlebäck, J. B. (2011). First results from a study investigating Swedish upper secondary students' mathematical modelling competencies. In G. Kaiser, W. Blum, R. Borromeo Ferri, & G. Stillman (Eds.), *Trends in teaching and learning of mathematical modelling* (pp. 407–416). New York: Springer.
Haines, C., & Crouch, R. (2001). Recognizing constructs within mathematical modelling. *Teaching Mathematics and Its Applications, 20*(3), 129–138.
Haines, C., & Crouch, R. (2005). Applying mathematics: Making multiple-choice questions work. *Teaching Mathematics and Its Applications, 24*(2–3), 107–113.
Haines, C., Crouch, R., & Davis, J. (2000). *Mathematical modelling skills: A research instrument* (University of Hertfordshire, Department of Mathematics, Technical Report No. 55). Hatfield: University of Hertfordshire.
Haines, C., Crouch, R., & Davis, J. (2001). Understanding students' modelling skills. In J. F. Matos, W. Blum, S. K. Houston, & S. P. Carreira (Eds.), *Modelling and mathematics education: ICTMA 9: Applications in science and technology* (pp. 366–381). Chichester: Horwood.
Houston, K., & Neill, N. (2003a). Assessing modelling skills. In S. J. Lamon, W. A. Parker, & S. K. Houston (Eds.), *Mathematical modelling: A way of life – ICTMA 11* (pp. 155–164). Chichester: Horwood.
Houston, K., & Neill, N. (2003b). Investigating students' modelling skills. In Q. Ye, W. Blum, S. K. Houston, & Q. Jiang (Eds.), *Mathematical modelling in education and culture: ICTMA 10* (pp. 54–66). Chichester: Horwood.
Ikeda, T., Stephens, M., & Matsuzaki, A. (2007). A teaching experiment in mathematical modelling. In C. P. Haines, P. Galbraith, W. Blum, & S. Khan (Eds.), *Mathematical modelling (ICTMA 12): Education, engineering and economics* (pp. 101–109). Chichester: Horwood.
Jiang, Q., Xie, J., & Ye, J. (2011). *Mathematical models* (4th ed.). Beijing: Higher Education Press (in Chinese).
Kaiser, G. (2007). Modelling and modelling competencies in school. In C. P. Haines, P. Galbraith, W. Blum, & S. Khan (Eds.), *Mathematical modelling (ICTMA 12): Education, engineering and economics* (pp. 110–119). Chichester: Horwood.
Kaiser, G., Blomhøj, M., & Sriraman, B. (2006). Towards a didactical theory for mathematical modelling. *ZDM – The International Journal on Mathematics Education, 38*(2), 82–85.
Lingefjärd, T., & Holmquist, M. (2005). To assess students' attitudes, skills and competencies in mathematical modeling. *Teaching Mathematics and Its Applications, 24*(2–3), 123–133.
Ludwig, M., & Xu, B. (2010). A comparative study of modelling competencies among Chinese and [illegible] students. *Journal für Mathematik-Didaktik, 31*(1), 77–97.
Stillman, G., Brown, J., & Galbraith, P. (2008). Research into the teaching and learning of applications and modelling in Australasia. [illegible]

Chapter 15
Taking Advantage of Incidental School Events to Engage with the Applications of Mathematics: The Case of Surviving the Reconstruction

Vince Geiger, Merrilyn Goos, and Shelley Dole

Abstract This paper reports on one aspect of a 2 year research and development project aimed at enhancing primary and secondary teachers' instructional practices in numeracy. The project made use of Goos' model of numeracy as a basis for assisting teachers to plan for teaching and also to reflect upon the effectiveness of their practice. As part of the project, teachers were challenged to develop learning experiences which were relevant to their own students' lived-in worlds. One teacher took advantage of a major, potentially disruptive, building development within her school to design a sequence of lessons in which students were challenged to adapt to the changes that were associated with the construction. The chapter concludes by discussing the changes to the teacher's disposition towards incorporating events from the students' lived-in worlds into her teaching practice.

1 Introduction

While there are now mature arguments in favour of the inclusion of learning experiences in school classrooms that situate mathematics within authentic, richly contextualised situations (e.g., Hoyles et al. 2002), opportunities for students to use mathematics to investigate issues in the social, political, environmental and economic worlds to which they will eventually contribute cannot yet be considered mainstream

V. Geiger (✉)
Education QLD | Faculty of Education, Australian Catholic University, Brisbane, QLD, Australia
e-mail: vincent.geiger@acu.edu.au

M. Goos • S. Dole
School of Education, The University of Queensland, St Lucia, QLD, Australia

G.A. Stillman et al. (eds.), *Teaching Mathematical Modelling: Connecting to Research and Practice*, International Perspectives on the Teaching and Learning of Mathematical Modelling, DOI 10.1007/978-94-007-6540-5_15,

(Damlamian and Straesser 2009; Stillman 2007). Despite the availability of teaching resources, and in some contexts, the support of educational jurisdictions through curriculum documents (e.g., Queensland Studies Authority 2009), the task of finding applications of mathematics that are genuinely meaningful to students appears to be a difficult challenge for many teachers. Whilst there are substantial bodies of literature devoted to the nature and importance of contextualised mathematics learning and teaching, such approaches to numeracy/mathematical literacy education (e.g., Steen and Turner 2007), and to effective approaches to professional development in mathematics teaching (e.g., Loucks-Horsley et al. 2003), far less is known about how teachers learn about and then appropriate the means to create effective mathematics teaching practices that have a connection with the lived-in world.

This chapter reports on one aspect of a 2 year research and development project aimed at enhancing primary and secondary teachers' planning and teaching practice through attention to context based learning activities. In meeting this aim, teachers were faced with the challenge of developing learning experiences that were relevant to their own students' lived-in worlds. A confounding aspect of this endeavour was how to also take advantage of situations that open the way for the teaching of applications of mathematics as they presented themselves, that is, these are incidental and so cannot be carefully planned. The potentials offered by this challenge are examined in addressing the following research question: Can teachers learn to take advantage of opportunities for the application of mathematics *in situ* within their unique school contexts?

2 Theoretical Framework

Approaches to situating the learning and doing of mathematics within authentic life relevant contexts has been the interest of two distinct but closely aligned branches of mathematics education research: modelling and applications in mathematics and numeracy, which is also known internationally as mathematical literacy. The process of mathematical modelling involves the formulation of a mathematical representation of real world phenomena and then the use of mathematics to derive results from this representation or model. Results are interpreted in terms of the original context and, if necessary, the model is adapted and re-evaluated until an acceptable alignment occurs between model and phenomena.

Numeracy is increasingly seen as a fundamental goal of education as it is a vital element in the formation of informed, reflective and contributing citizens (Steen 2001). This aspect of mathematics education has been recognised internationally through the OECD's Program for International Student Assessment (PISA). According to PISA's definition mathematical literacy is:

> an individual's capacity to identify and understand the role mathematics plays in the world, to make well-founded judgments, and to use and engage with mathematics in ways that meet the needs of that individual's life as a constructive, concerned and reflective citizen. (OECD 2004, p. 15)

Thus, both mathematical modelling and numeracy are concerned with the use of learned-in-school mathematics in the lived-in world away from the school classroom. In order to effectively practice mathematics in the world of work and in the community at large, individuals must also acquire the disposition to think critically and flexibly. The ability to take a critical orientation to the use of mathematics when evaluating or critiquing aspects of students' current cultures or future worlds cannot be assumed to flow from a high level of mathematical knowledge (Jablonka 2003) and so teachers need to design learning activities that provide opportunities to develop these capacities. While theoretical frameworks for both mathematical modelling and for numeracy are apparent in the corpus of literature associated with both fields, few manage to capture the interrelated nature of the elements required for using mathematics in real world practice, especially in relation to critical elements of both enterprises. More recently, however, Goos (2007) proposed a model of numeracy that encompasses four essential elements which are enacted within a perception of mathematics as knowledge-in-action. The model incorporates attention to real-life contexts, the deployment of mathematical knowledge, the use of physical and digital tools, and consideration of students' dispositions towards the use of mathematics. This model incorporates the elements required for using mathematics in real world practice in a tightly integrated way and so was chosen as the theoretical framework for both the design of application tasks and for the analysis of related teaching and learning practice.

Contexts are considered the heart of numeracy and an individual is only considered numerate within contexts in which they can competently apply their mathematical knowledge. This means that it is important for individuals to be numerate in a range of contexts (Steen 2001). The *mathematical knowledge* that underpins numeracy includes concepts, skills, problem solving strategies and the capability to make valid estimations (Zevenbergen 2004). The *use of tools* is a vital element of numeracy. In school and workplace contexts, tools may be representational (symbol systems, graphs, maps, diagrams, drawings, tables, ready reckoners), physical (models, measuring instruments), and digital (computers, software, calculators, internet) (Noss et al. 2000; Zevenbergen 2004). A numerate person has a positive *disposition* to use mathematics when a problem or situation calls for the use of a mathematical approach, that is, they are comfortable with using their mathematical knowledge. This also means that individuals must feel confident in adapting what they know to suit the specifics of a situation and have the flexibility to think about a problem in different ways. The importance of developing positive dispositions to using mathematics has been emphasised in national and international curriculum documents (e.g., National Curriculum Board 2009; OECD 2004).

This model is grounded in a *critical orientation* to numeracy since numerate people not only know and use efficient methods, but also they evaluate the reasonableness of the results obtained and are aware of appropriate and inappropriate uses of mathematical thinking. The critical evaluation of mathematical information is used to make decisions and judgements, add support to arguments, or to challenge an argument or position. Mathematical information and practices can be used to persuade, manipulate, disadvantage or shape opinions about social or political issues (Frankenstein 2001).

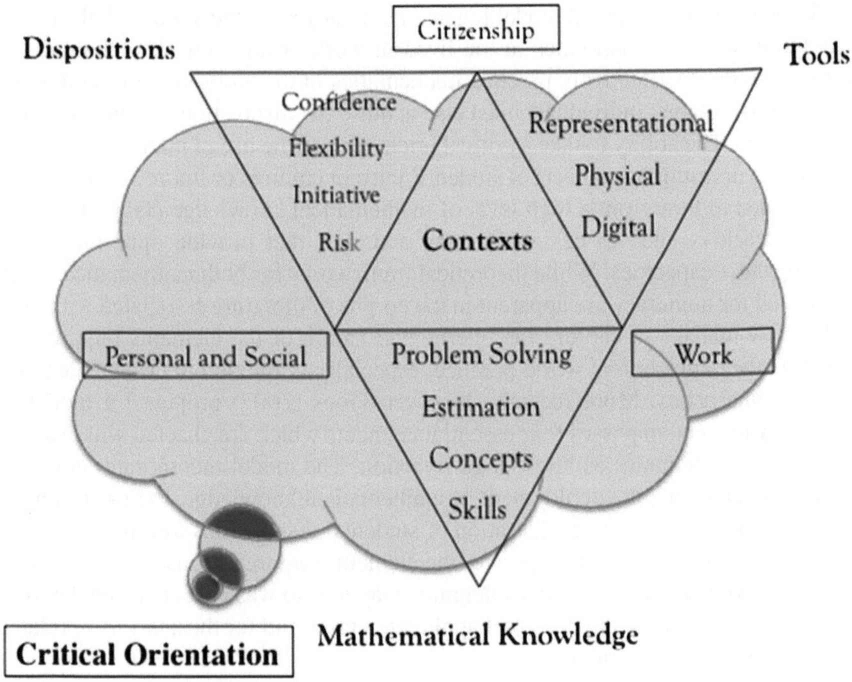

Fig. 15.1 A model for numeracy in the twenty-first century (Goos 2007)

Table 15.1 Descriptions of the elements and critical orientation of the numeracy model

Model component	Descriptions
Mathematical knowledge	Mathematical concepts and skills; problem solving strategies; estimation capacities.
Contexts	Capacity to use mathematical knowledge in a range of contexts, both within schools and beyond school settings
Dispositions	Confidence and willingness to use mathematical approaches to engage with life-related tasks; preparedness to make flexible and adaptive use of mathematical knowledge.
Tools	Use of material (models, measuring instruments), representational (symbol systems, graphs, maps, diagrams, drawings, tables, ready reckoners) and digital (computers, software, calculators, internet) tools to mediate and shape thinking
Critical orientation	Use of mathematical information to: make decisions and judgements; add support to arguments; challenge an argument or position.

The elements of the model can also be represented as the net of a tetrahedron surrounded by and bound together by a critical orientation (Fig. 15.1). Descriptions of the elements of the model and the critical orientation within which these elements interact are summarised in Table 15.1.

This model has been used as a framework to audit mathematics curriculum designs (Goos et al. 2010), for documenting teachers' reflections on their own professional learning in relation to the planning and implementation of numeracy teaching practice (Geiger et al. 2011), and for the analysis of teachers' attempts to design for the teaching of numeracy across the curriculum (Goos et al. 2011).

3 Research Design

The project was conducted over a 2 year period, 2010–2011, and involved two teachers from each of 12 schools – 24 teachers in total. Participating schools were selected by their employing authority after responding to a call for expressions of interest to be involved in a cross-curricular numeracy project. Participants included a mix of primary and secondary teachers. Because of the cross-curricular intent of the project participants included generalist primary teachers who taught across the curriculum and also secondary teachers with specialist subject knowledge (e.g., mathematics, science). Two teachers from each school were included so that they could provide support to each other throughout the project.

An action research approach was chosen for this research-as-development project as this is an appropriate methodology for supporting educational reform through collaborative partnerships between teachers and university researchers (Soomekh and Zeichner 2009). The action research design was consistent with the Loucks-Horsley et al. (2003) framework for professional development that provided direction for the teacher professional learning aspect of the project. In alignment with both of these aspects of the design, project meetings that involved all participants were followed up with school visits. In project meetings, elements of the numeracy model were explored and examples of classroom activities which embodied these elements were demonstrated. Project meetings were also used by teachers to showcase work in progress, to seek feedback on ideas they were preparing for implementation from other project teachers as well as the researchers. Between whole project meetings, members of the research team visited each participating school to collect data and to discuss the achievements and challenges associated with improving numeracy practice. Thus these visits provided on-going support to teachers in their efforts to meet goals they had set for improving their own numeracy practice.

The research component of the project was based on data gathered during whole project meetings and school visits. During whole project meetings teachers were asked to: outline their initial conceptions of numeracy; complete a survey on teachers' confidence with numeracy teaching; and to map their personal progress in numeracy by using the numeracy model as a lens. Researchers' visits to schools involved: recording field notes for lesson observation; pre- and post-lesson teacher interviews and post-lesson interviews with students; and collection of artefacts including teacher planning documents and student work samples.

Data sources used in this report are drawn from a focus group interview with participating teachers in one school and the personal planning documents of an individual teacher who also participated in the same interview. The teacher's planning documents and reports on her own and her students activity when engaging in the applications task were analysed by identifying how plans and action aligned with the elements of the numeracy model.

The school is located in an urban, non-metropolitan setting; the town being the commercial centre of a farming community. The school itself caters for students from a wide range of socio-economic backgrounds. The teacher, who is the focus of this report, was in her third year as a practitioner and had volunteered for the project because of the encouragement she received from the school administration. She also felt that the project provided her with an opportunity to extend her knowledge of mathematics and mathematics pedagogy – areas she admitted to feeling less confident about at the commencement of the project. She was selected as the focus for this chapter because of the unique way in which she took advantage of a potentially disruptive school rebuilding program to develop a task involving the application of mathematics that was relevant to her students lived-in-world.

4 Surviving the Reconstruction

As part of the project, teachers were challenged to develop learning experiences which were relevant to their own students' lived-in-worlds. One teacher, Kym (pseudonym), embraced this challenge by redesigning a unit of her learning plan for her Year 6 class. Within this unit on the topic of Location, Kym chose themes in which she took a contextualised approach to introducing new mathematical knowledge and also to consolidating ideas and concepts students had met earlier. The plan included the following themes:

- *New School Plan* – In response to the injection of significant funds, students were asked to consider what facilities required upgrading and what new facilities were required in redesigning the school. A map needed to be produced to show how these new or up-graded facilities would fit on the current school site.
- *What a Load of Rubbish* – Students were set the task of gathering information about the school's litter patterns and asked to produce a plan for the location of rubbish bins that would optimise students' opportunities of placing litter in the appropriate place.
- *Up the Garden Path* – After being provided with details of a prefabricated path kit, students were asked to show how the pieces would be assembled to form a pathway across a specified square in the school. It was expected that students demonstrated they had considered a variety of possible options and justified their final design.

These tasks were good examples of the type of contextualised approaches to learning and teaching the project aimed to encourage. Each task had the potential to help address each element of the model for numeracy which was introduced to the teachers as a way of structuring their planning and practice. These tasks, however, could have been introduced to students at anytime, or in any other school, as they were not really bound to any particular circumstance that existed within the school at the time of the project. There was another task, however, that drew on happenings within the schools that had immediate and unavoidable consequences for the school community; students, teachers and parents alike.

In the particular instance that is the focus of this chapter, the teacher took advantage of a major, and potentially disruptive, building development within her school to design a sequence of lessons in which students were required to consider how many day to day activities would need to be adapted in order to accommodate the changes that were associated with the construction. Mathematics was seamlessly integrated into the range of adaptations students were asked to investigate because of the way the teacher structured this sequence.

A description of the task, as it appears in Kym's planning documents, appears below:

> The construction at the school has caused many changes. The school map is now out of date and new parents would get confused trying to use it. Your task is to alter the out of date map so that it reflects the changes to the school. Then design a tour of the school for new parents that will show them where everything is.
>
> You need to mark your tour on the map. Write a tour guide to go along with it so that parents can do the tour alone, with instructions like, "On your left you will see the parent entrance to the office. If you look to your right you will see the years 6 and 7 building."
>
> The tour is to be no longer than 10 min. When you have designed your tour, test it out on a group to check for errors and timing.
>
> Before final drafts, compare journeys. Discuss differences in routes, school highlights and times taken for the journey. Discuss any glitches that came up in organising tours. Precision in map reading and instructions is important so people don't get lost.

To begin the task students were given a plan of the school before the construction began and asked to label key landmarks. They were then asked to redesign the plan to include the new buildings that were already being built within the school. As they were required to produce a scaled plan, the task included the particular challenge for students of determining distances and locations without getting too close to the buildings under construction. The task also included a requirement to develop instructions for a tour of the school, that the Year 6 students would take responsibility for, as part of an introduction to the school for prep students' parents. These instructions needed to accommodate the new buildings. Students' plans and instructions were then tested by swapping plans with other students and noting how easily these assisted non-authors to tour the school from the perspective of ease of interpretation and also coverage of important areas in the school. The experiences of using other students' directions were shared in a whole class discussion.

Kym reported a high level of engagement by students with the task. She argued that the inquiry approach, within which the task was conducted, allowed students an element of control over their own learning which contributed to this engagement.

Kym also believed that there were two more contributing factors to students' engagement – the outdoor nature of the activity, and the relevance of the task to students' personal circumstances.

Kym: Students have control of where they want to take it. The students tell me this approach allows them to look into "stuff that we want to learn about".

5 The Construction Through the Lens of the Numeracy Model

This task allowed students to engage with all elements of the numeracy model, although more strongly in some areas than another. There was a clear and engaging *context* based on an obvious and disruptive building development within the school. *Mathematical knowledge* was addressed through engagement with the activity as students were required to construct scale maps and make accurate use of the language of location in order to provide directions for other students in their role as school tour guides. Kym attempted to promote students' *dispositions* towards mathematics and towards the use of mathematics by introducing a context that students' found relevant in an immediate sense because it was affecting their routine activities on a daily basis. In order to accommodate challenges associated with the task, such as establishing the location of the new building on a map without approaching the site too closely, students were required to think flexibly and adaptively and to show initiative and take intellectual risks. Kym also asked students to use representational *tools* such as maps and made use of digital *tools* herself through the use of Google Earth to provide students with a plan view of the school. The whole activity was embedded in a *critical orientation* as students used the elements of the task to make decisions about the best route for a school tour.

While this task provides evidence of Kym's skill as a designer of context driven mathematical tasks, it also provides insight into her own *disposition* as a creator of students' learning experiences in mathematics. Through this episode Kym has shown a capacity to think flexibly and adaptably herself, not in the way she is hoping to promote in her students' use of mathematics, but in the way she is open to, and able to take advantage of, new and unexpected opportunities to design tasks that enhance her own numeracy practice as a teacher. This disposition has also allowed her to connect with other elements of the numeracy model within her practice. She has recognised the potential of a context for engaging her students in their learning and for situating a specific type of *mathematical knowledge* within her students' world of experience. To do this she needed to determine a unit of mathematics, and thus what specific *mathematical knowledge* was best aligned to this context. Kym has also considered what *tools* were needed to support the task. In this case, she provided plan views of the school before the construction and Google Earth images of the school, as well as physical tools such as measuring tapes and rulers. In order for there to be a connection between elements of the numeracy model, Kym had to think of an overarching problem or issue for students to address. She did this in a way that was very personal to students by asking them to develop a set of instructions for a tour of the school that the students themselves would make use of as guides to new parents.

6 Conclusion

While the aim of the project was to assist teachers to enhance their planning and teaching practices through the use of a research based model of numeracy (Goos 2007), it became apparent that the model not only offered a framework for guiding teachers' work when planning students' learning experiences but also a way of thinking about how teachers change their own practice. In this project, one teacher took advantage of a major, potentially disruptive, building development within her school to design a sequence of lessons in which students were challenged to adapt to the changes that were associated with a school rebuilding program. In doing so, she was still able to incorporate all of the elements of a rich applications task as identified by the model of numeracy which guided the development of tasks and teacher action through the project.

Kym's planning documents demonstrated that she had the capacity to develop innovative tasks which engaged students in addition to paying due attention to all the elements of numeracy identified in Goos' model (2007). It was not only her awareness of the elements of this model alone that allowed her to take advantage of the unusual circumstances of her school for the betterment of her students' mathematics education; but also her own disposition towards this unexpected circumstance. In a similar way to how she was attempting to promote flexible and adaptive thinking in her students, in relation to the use of mathematics in context, she demonstrated initiative and took a professional risk in building an element of her mathematics teaching program around what many teachers would consider a disruptive event. It was then, from this position, that she was able to design an activity that wove the elements of the numeracy model into students' learning in a seamless fashion.

How other teachers appropriate the ways of thinking, reasoning and discourse that promote the development of such a disposition is a topic worthy of future research effort. Without an understanding of how such a disposition might be appropriated by teachers, it is difficult to see how the inclusion of authentic and relevant contexts for learning mathematics can ever be an accepted part of mainstream mathematics education at primary or secondary levels.

References

Damlamian, A., & Straesser, R. (2009). ICMI Study 20: Educational interfaces between mathematics and industry. *ZDM – The International Journal on Mathematics Education, 41*(4), 525–533.

Frankenstein, M. (2001). *Reading the world with math: Goals for a critical mathematical literacy curriculum.* Paper presented as the Keynote address delivered at the 18th biennial conference of the Australian Association of Mathematics Teachers, Canberra.

Geiger, V., Goos, M., & Dole, S. (2011). Teacher professional learning in numeracy: Trajectories through a model for numeracy in the 21st century. In J. Clark, B. Kissane, B. Mousley, T. Spencer, & S. Thornton (Eds.), *Shaping the future of mathematics education* (Proceedings of the 33rd annual conference of the Mathematics Education Research Group of Australasia, Alice Springs, pp. 297–305). Adelaide: MERGA.

Goos, M. (2007). *Developing numeracy in the learning areas (middle years).* Paper presented at the South Australian Literacy and Numeracy Expo, Adelaide.

Goos, M., Geiger, V., & Dole, S. (2010). Auditing the numeracy demands of the middle years curriculum. In L. Sparrow, B. Kissane, & C. Hurst (Eds.), *Shaping the future of mathematics education* (Proceedings of the 33rd annual conference of the Mathematics Education Research Group of Australasia, Freemantle, pp. 210–217). Adelaide: MERGA.

Goos, M., Geiger, V., & Dole, S. (2011). Teachers' personal conceptions of numeracy. In B. Ubuz (Ed.), *Proceedings of the 35th conference of the International Group for the Psychology of Mathematics Education* (Vol. 2, pp. 457–464). Ankara: PME.

Hoyles, C., Wolf, A., Molyneux-Hodgson, S., & Kent, P. (2002). Foreword and executive summary. *Mathematical skills in the workplace.* Final Report to the Science, Technology and Mathematics Council. London: Institute of Education, University of London; Science, Technology and Mathematics Council.

Jablonka, E. (2003). Mathematical literacy. In A. Bishop, M. A. Clements, C. Keitel, J. Kilpatrick, & F. Leung (Eds.), *Second international handbook of mathematics education* (Vol. 1, pp. 75–102). Dordrecht: Kluwer.

Loucks-Horsley, S., Love, N., Stiles, K., Mundry, S., & Hewson, P. (2003). *Designing professional development for teachers of science and mathematics* (2nd ed.). Thousand Oaks: Corwin Press.

National Curriculum Board. (2009). *Shape of the Australian curriculum: Mathematics.* Retrieved January 7, 2011, from http://www.acara.edu.au/verve/_resources/Australian_Curriculum_-_Maths.pdf

Noss, R., Hoyles, C., & Pozzi, S. (2000). Working knowledge: Mathematics in use. In A. Bessot & J. Ridgeway (Eds.), *Education for mathematics in the workplace* (pp. 17–35). Dordrecht: Kluwer.

OECD. (2004). *Learning for tomorrow's world: First results from PISA 2003.* Paris: OECD.

Queensland Studies Authority. (2009). *Senior syllabus in mathematics B.* Brisbane: Queensland Studies Authority.

Soomekh, B., & Zeichner, K. (2009). Action research for educational reform: Remodelling action research theories and practices in local contexts. *Educational Action Research, 17*(1), 5–21.

Steen, L. (2001). The case for quantitative literacy. In L. Steen (Ed.), *Mathematics and democracy: The case for quantitative literacy* (pp. 1–22). Princeton: National Council on Education and the Disciplines.

Steen, L., & Turner, R., (with Burkhardt, H.). (2007). Developing mathematical literacy. In W. Blum, P. L. Galbraith, H.-W. Henn, & M. Niss (Eds.), *Modelling and applications in mathematics education* (pp. 285–294). New York: Springer.

Stillman, G. (2007). Implementation case study: Sustaining curriculum change. In W. Blum, P. L. Galbraith, H.-W. Henn, & M. Niss (Eds.), *Modelling and applications in mathematics education* (pp. 497–502). New York: Springer.

Zevenbergen, R. (2004). Technologizing numeracy: Intergenerational differences in working mathematically in new times. *Educational Studies in Mathematics, 56*(1), 97–117.

Chapter 16
The Development of Modelling Competencies by Year 9 Students: Effects of a Modelling Project

Susanne Grünewald

Abstract For several years the competency of mathematical modelling has been prescribed as one of the central competencies in German educational standards in mathematics. This chapter reports results of the study of my master thesis, during which the effects of a short-term modelling project to the development of modelling competencies were evaluated. To evaluate the modelling project I developed a modelling test based on one designed by Houston, Haines, Crouch, Izard and others previously, which focussed on the development of the sub-competencies of mathematical modelling of students. Furthermore, five students were interviewed at the beginning and the end of the project to gain deeper insight into their development of modelling competencies. The results show that it is possible to foster at least students' sub-competencies of mathematical modelling within a modelling project.

1 Introduction

For several years the competency of mathematical modelling has been prescribed as one of the central competencies in German educational standards in mathematics, but still mathematical modelling plays a minor role in German mathematics lessons. As a consequence, mainly two questions have to be answered: In which way can authentic modelling problems be integrated into everyday mathematics lessons to foster the modelling competencies of students and in which way can modelling competencies efficiently be measured? In an attempt to give preliminary answers to these questions a modelling project with approximately 160 Year 9 students participating took place during 4 days in February 2010 in Hamburg.

S. Grünewald (✉)
Faculty of Education, Didactics of Mathematics, University of Hamburg, Hamburg, Germany
e-mail: Susanne.Gruenewald@uni-hamburg.de

G.A. Stillman et al. (eds.), *Teaching Mathematical Modelling: Connecting to Research and Practice*, International Perspectives on the Teaching and Learning of Mathematical Modelling, DOI 10.1007/978-94-007-6540-5_16,

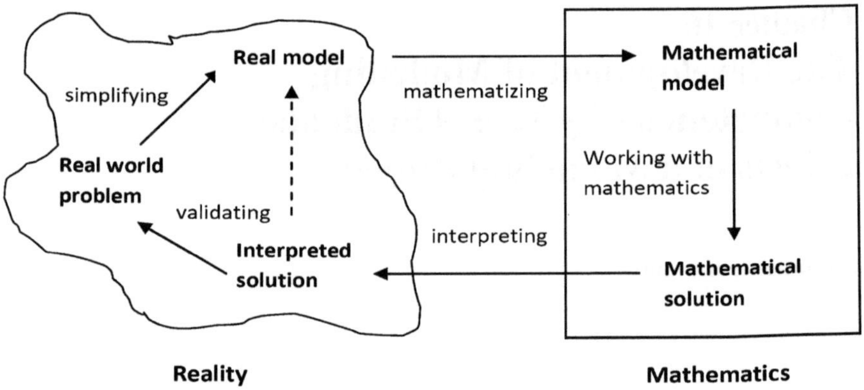

Fig. 16.1 Modelling cycle (Adapted from Maaß 2006)

In the first part of this chapter the theoretical framework will be summarised. Afterwards the structure and course of this "modelling week" will be presented as well as the design of the study and the methods of data collection. Finally, the results of the evaluation will be described and discussed.

2 Theoretical Framework

2.1 Mathematical Modelling

Mathematical modelling is a highly discussed topic of didactics of mathematics internationally. There are different perspectives of mathematical modelling that include different representations of the modelling process as a cycle as well as goals and modelling competencies. An overview is given, for instance, in Kaiser and Sriraman (2006).

Basically, mathematical modelling describes the process of solving real world problems by using mathematical instruments. The ideal-typical process of mathematical modelling is usually illustrated in the form of a cycle. The article at hand refers to a didactical modelling cycle adapted from Maaß (2006) (see Fig. 16.1). The entire process of modelling depends on the assumptions and the choice of mathematical resources (Maaß 2007). In addition, the modelling process is typically portrayed as a cycle with linear sequential steps. In reality, however, such processes are characterised by frequent switching between the various stages of modelling cycles (Maaß 2007; Schwarz et al. 2008).

2.2 Modelling Competencies

The exact definition of modelling competencies is dependent on the particular underlying concept of mathematical modelling (Zöttl 2010). Widely accepted is that modelling competencies include abilities and a willingness to solve real-world problems by using mathematical modelling (Maaß 2004, 2006). Sub-competencies of mathematical modelling as an essential part of modelling competencies are based on the underlying modelling cycle and include the abilities needed to perform the different steps of the cycle (Kaiser and Schwarz 2006). Based on the modelling cycle from Maaß (2006) the following sub-competencies of mathematical modelling are distinguishable:

- Simplifying the real-world problem (I1),
- Clarifying the goal (I2),
- Defining the problem (I3),
- Assigning central variables and their relations (I4),
- Formulating mathematical statements (I5),
- Selecting a model (I6),
- Interpreting the solution in a real-world context (I7),
- Validating the appropriateness of the solution (I8).

According to Maaß (2004, 2006), Zöttl (2010) and Stillman (2011), metacognitive competencies play a significant role for the modelling competencies. Non-existent or very little meta-knowledge about the modelling process as a result may lead to considerable problems while working on modelling tasks, for example, at the transitions between the different phases of the modelling process.

3 Design of the Study

3.1 The Modelling Project

The modelling project took place in February 2010 in Hamburg. During 4 days approximately 160 students of the Year 9 of an academic high school tackled one of the following authentic complex modelling problems in small groups of 4–6 students:

- "Optimal" positioning of three rescue helicopters in a skiing area in South Tirol. For this modelling problem the positions of villages and ski slopes were given as well as the number of skiing accidents for each place. The students had to define "optimal" on their own.
- Definition of more suitable dress sizes, based on measurements carried out in advance.
- Automatic watering of gardens: aim of this modelling problem was to find the "optimal" positioning of lawn sprinklers for a pre-defined garden. Again, the students had to define "optimal" on their own.
- Optimal tariffs for communication networks (phone, mobile phone, internet) for different users. The students had to research tariffs on their own.

During the modelling project each small group of students was mentored by one university student. The university students were instructed to introduce the students to the modelling cycle, but in fact only a few of them presented the modelling cycle.

Table 16.1 Number of girls and boys with different mathematical grades (self-reported)

	Mathematical grade				
Gender	1	2	3	4	5
Girls	6	22	17	17	1
Boys	5	22	31	11	3

Note. For mathematical grade, *1* means "very good" and *5* "fail"

At the end of the modelling project a poster presentation of the different solving processes and solutions of the small groups took place.

One of the main goals of the modelling project was the fostering of the modelling competencies of the students. To evaluate the project the modelling competencies of the students were tested at the beginning and the end of the modelling project. A total of 135 students (63 girls and 72 boys) took part in both tests and are in the following considered as the sample. Students mathematical performance varied from 1 to 5 where 1 means "very good" and 5 "fail" (see Table 16.1).

3.2 Research Questions

The main research question of the evaluation of the modelling project was:

- To what extent can students' modelling competencies be fostered within a modelling project?

  In addition to this, three more questions were investigated by evaluating the data:

- Is there a relationship between the development of the students' modelling competencies and their mathematical abilities?
- Are there gender differences in the development of the students' modelling competencies?
- Do specific sub-competencies develop differently?

3.3 Methods of Data Collection

The study was designed in a pre- and post-format and includes qualitative and quantitative research components. At the beginning and at the end of the modelling project all students filled in a modelling test to assess the sub-competencies of mathematical modelling. The tests were supplemented by semi-open pre- and post-interviews with the same five students each time.

Item 8 : Distance

Lisa leaves the highway A21 at the exit Bad Oldesloe-Süd, when she sees this road sign. “Look”, Lisa tells her little brother Felix, “this sign shows that Lübeck and Hamburg are 73 km away from each other.”

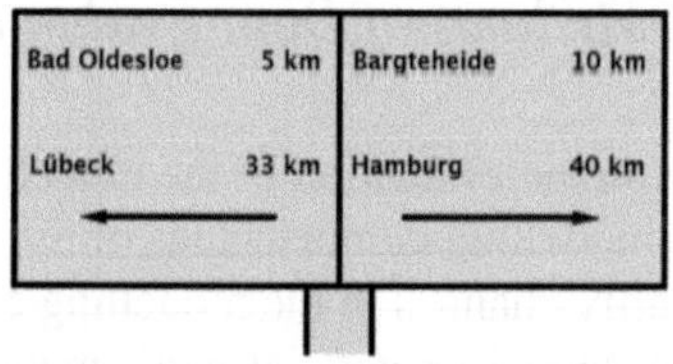

Which of the following comments on the statement of Lisa is **the most appropriate**?

(A)	“Lisa is right, because in order to drive to Lübeck or Hamburg you have to drive on the opposite direction.”
(B)	“Lübeck and Hamburg don’t have to be situated at this street, so Lisa doesn’t have to be right.”
(C)	“Lübeck and Hamburg don’t have to be situated at this street. Lisa would be right if she said that the cities are at most 73 km away from each other.”
(D)	“Lisa is right – it says on the road sign that Lübeck and Hamburg are 73 km away from each other.”

Fig. 16.2 Validating the appropriateness of the solution

3.3.1 Modelling Test

To evaluate the impact of the project, tests in a pre-post-design were developed based on already existing modelling tests designed and reworked by Houston, Haines, Crouch, Izard and others (e.g., Haines and Crouch 2001; Houston and Neill 2003). Eight items in multiple-choice format were developed by the researcher to measure the development of the sub-competencies related to the different phases of the modelling process (see Fig. 16.2 for an example).

Each item formulated a different modelling problem in a specific phase of the modelling process and asked for the next appropriate step. For each item four possible answers are given, one of which is correct, one is partly right and two are wrong.

The tests were divided into part A and part B. Part A contained two questions about gender and previous grades in mathematics classes. Part B contained eight questions about modelling tasks. To maximise the comparability of the pre- and post-test, the items and structure of the modelling tests were most closely corresponded to each other. At the beginning and at the end of the modelling project all students present filled in these tests.

3.3.2 Interviews

In addition to the collection of the modelling tests of all students, five students (two girls and three boys with different mathematical skills) were interviewed after completing the test to additionally measure the development of metacognitive modelling competencies. The pre- and post-interviews were conducted as guided interviews (Flick 2000) to achieve the most comparability of the answers of the students. To study the development of modelling competencies the same five students were interviewed for about 15 min at the beginning and at the end of the modelling project. During both interviews the students were asked about their answers to the tests. In the course of the second interview, in addition the modelling processes during the modelling project and mathematical modelling itself were discussed.

3.4 Methods of Data Evaluation

Based on the evaluation of the modelling tests and the interviews, the development of the modelling competencies of the students was reconstructed. It was of interest to identify changes in the modelling competencies as a whole as well as dependencies of gender or general mathematical skills. To answer the research questions, firstly the chosen answers in the modelling tests were analysed with Excel and afterwards set in relation with the evaluation of the interviews.

For the evaluation of the tests the individual response options of the different items were rated differently, according to a partial credit system. In each case one response option was correct and therefore rated with 2 points. One response option was partially correct and rated with 1 point and two response options were incorrect and evaluated with 0 points. The marking of several response options for one item led to the rating of the item as incorrect. Thus, the highest achievable score on the 8-item individual test is 16 points. Starting from this database, first the average overall score of the students as well as the achieved average scores per item were calculated. In addition, the average scores of the students were calculated differentiated by gender and by mathematical grade. Furthermore, how many students achieved a higher, the same, or a lower score on the post-test in comparison to the pre-test were analysed. Finally, for each item the number of correct, partially correct and incorrect answers given were calculated. The distribution of points within the pre- and post-tests as well as the changes between the results of the two test dates were checked with tests of significance to their reasonableness, namely *t*-tests respectively and one-way ANOVA.

The analysis of the transcripts of the interviews was based on key questions. Subsequently, the analyses were evaluated in terms of how the results and statements of the interviewees are connected to the items of the modelling tests and the sub-competencies of mathematical modelling as well as whether the students were able to develop metacognitive modelling skills.

4 Results

In the following the results of the empirical study are presented, first the outcomes of the evaluation of the modelling test and afterwards the analysis of the interviews. Taken as a whole the evaluation of the modelling tests and interviews showed a significant increase in the sub-competencies of mathematical modelling of the students. There were no significant differences between the improvements differentiated by gender or mathematical grade. The biggest changes in the scores were identified in items which test competencies that are not fostered by usual mathematical tasks.

	Group Statistics		Independent Samples Test						
			Leven's Test for Equality of Variances		*t*-test for Equality of Means				
								95% CI of the Difference	
	N	Mean	*F*	Sig.	*t*	df	*P* (2-tailed)	Lower	Upper
Boys (Pre-Test) Girls (Pre-Test)	72 63	10.92 10.98	5.536	.20	-.183 -.180	133 117.827	.855 .858	-.799 -.811	.664 .676
Boys (Post-Test) Girls (Post-Test)	72 63	12.46 12.70	.366	.546	-.756 -.753	133 128.281	.451 .453	-.869 -.871	.388 .391

Fig. 16.3 Comparison of the overall scores (differentiated by gender)

4.1 *Results of the Modelling Test*

The students achieved an average overall score of 10.95 points in the pre-test and in the post-test 12.57 points. All in all, the average overall score of the students showed a statistically significant increase from the pre- to the post-test, $t(134)=-12.279$, $p=.000$, $d=-1.622$, 95 % CI [−1.884, −1.361]. The girls attained an average overall score of 10.98 points in the pre- and 12.7 points in the post-test. The boys achieved an average overall score of 10.92 points in the pre- and 12.46 points in the post-test. The differences between the results of the girls and boys are not statistically significant (see Fig. 16.3).

Figure 16.4 shows the achieved average overall scores of the students differentiated by mathematical performance grades. Students across the spectrum of grades were able to improve their average overall score. There is an obvious correlation between the grade in mathematics and the achieved score, the better the grade in mathematics, the better the achieved average overall score. However, the differences between the extent of improvement differentiated by mathematical grade are not statistically significant.

Considering the achieved average scores per item, there were also significant improvements in the scores of all items which relate to Maaß's modelling sub-competencies (2006), except item I7 (Interpreting the solution in a real-world context) (see Fig. 16.5). The biggest changes in the scores were identified in items which test sub-competencies that are not fostered by usual mathematical tasks, for example the sub-competence of validating the appropriateness of a solution (I8). There were also no significant differences between the results per item differentiated by gender or mathematical grade.

4.2 *Results of the Interviews*

The analysis of the interviews verified the results of the evaluation of the tests. The students' statements in the interviews resonated with the obvious improvement of the sub-competencies of mathematical modelling in the tests, especially of the

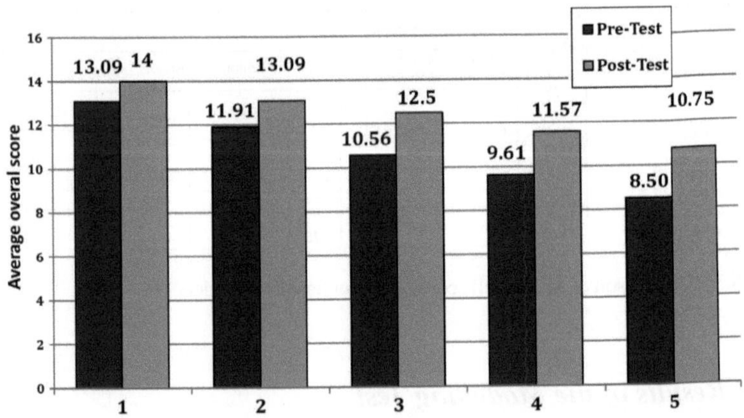

Fig. 16.4 Comparison of the overall scores (differentiated by mathematical performance grades)

Item for Maaß's modelling sub-competencies	Paired Samples Statistics	Paired Samples Test				t	df	p (2-tailed)
	Mean	$\mu_1 - \mu_2$	Std. Deviation	95% CI of the Difference				
				Lower	Upper			
I1 (Pre-Test) I1 (Post-Test)	1.16 1.38	-.215	.737	-.340	-.089	-3.387	134	.001
I2 (Pre-Test) I2 (Post-Test)	1.25 1.56	-.304	.614	-.408	-.199	-5.745	134	.000
I3 (Pre-Test) I3 (Post-Test)	1.43 1.70	-.267	.693	-.385	-.149	-4.469	134	.000
I4 (Pre-Test) I4 (Post-Test)	1.63 1.78	-.148	.580	-.247	-.049	-2.969	134	.004
I5 (Pre-Test) I5 (Post-Test)	1.61 1.83	-.222	.582	-.321	-.123	-4.439	134	.000
I6 (Pre-Test) I6 (Post-Test)	1.84 1.92	-.074	.379	-.139	-.010	-2.270	134	.25
I7 (Pre-Test) I7 (Post-Test)	1.32 1.25	.067	.576	-.031	.165	1.346	134	.181
I8 (Pre-Test) I8 (Post-Test)	0.70 1.16	-.459	.741	-.585	-.333	-7.204	134	.000

Fig. 16.5 Comparison of the results per item

validation sub-competence. The statements of the students at the beginning of the modelling project show that most of the students solved the tasks in the modelling test intuitively. During the second interview the students were more able to explain their chosen answers in the test; nevertheless, they still had problems in generalising their work on one modelling problem. During the modelling project, only one of the interviewed students became familiar with the modelling cycle and only this boy was able to abstract from the single modelling process he had undertaken, so he was the only one of the interviewed students who showed recognisable developments of metacognitive modelling competencies.

5 Discussion

The evaluation of the modelling tests and interviews showed a significant increase in the sub-competencies of mathematical modelling during the modelling project. A correlation between the extent of improvement and gender was not found; however, the mathematical performance grade did indeed have an influence on the achieved average scores. Certainly the results of the study are only able to show tendencies and have to be treated very cautiously. On the one hand, this empirical study measured only short-term effects of a modelling project and considered a comparatively small group of students. On the other hand, the design of the modelling test causes strong limitations of the study, because different items were used in the pre- and post-test and due to time limitations only one item was used to measure each sub-competence of mathematical modelling. However, the results of the study – that it is possible to foster students' modelling competencies by integrating modelling problems into mathematics lessons – correspond to the outcomes of other projects aiming at promoting students' modelling competencies (Maaß 2004; Zöttl 2010). Furthermore, the results of the study are in line with the findings of Frejd and Ärlebäck (2011), who used the same test design (albeit with different and more actual items) to evaluate Swedish upper secondary students' modelling competencies. In their study there was no significant impact on the final average scores of the students regarding gender too, and the mathematical grades of the students also had a significant influence on the achieved overall scores of the students.

References

Flick, U. (2000). *Qualitative forschung*. Reinbek: Rowohlt.

Frejd, P., & Ärlebäck, J. B. (2011). First results from a study investigating Swedish upper secondary students' mathematical modelling competencies. In G. Kaiser, W. Blum, R. Borromeo Ferri, & G. Stillman (Eds.), *Trends in teaching and learning of mathematical modelling* (pp. 407–416). New York: Springer.

Haines, C., & Crouch, R. (2001). Recognizing constructs within mathematical modelling. *Teaching Mathematics and Its Applications, 20*(3), 129–138.

Houston, K., & Neill, N. (2003). Assessing modelling skills. In S. J. Lamon, W. A. Parker, & S. K. Houston (Eds.), *Mathematical modelling: A way of life ICTMA11* (pp. 155–164). Chichester: Horwood.

Kaiser, G., & Schwarz, B. (2006). Modellierungskompetenzen – Entwicklung im Unterricht und ihre Messung. In *Beiträge zum Mathematikunterricht 2006* (pp. 56–58). Hildesheim: Franzbecker.

Kaiser, G., & Sriraman, B. (2006). Theory usage and theoretical trends in Europe – A survey and preliminary analysis of CERME4 research reports. *Zentralblatt für Didaktik der Mathematik, 38*(1), 22–51.

Maaß, K. (2004). *Mathematisches Modellieren im Unterricht. Ergebnisse einer empirischen Studie*. Hildesheim: Franzbecker.

Maaß, K. (2006). What are modeling competencies? *Zentralblatt für Didaktik der Mathematik, 38*(2), 113–142.

Maaß, K. (2007). *Mathematisches Modellieren. Aufgaben für die Sekundarstufe I*. Berlin: Cornelsen.

Schwarz, B., Kaiser, G., & Buchholtz, N. (2008). Vertiefende qualitative Analyse zur professionellen Kompetenz angehender Mathematiklehrkräfte am Beispiel von Modellierung und Realitätsbezügen. In S. Blömeke, G. Kaiser, & R. Lehmann (Eds.), *Professionelle Kompetenz angehender Lehrerinnen und Lehrer. Wissen, Überzeugungen und Lerngelegenheiten deutscher Mathematikstudierender und -referendare. Erste Ergebnisse zur Wirksamkeit der Lehrerausbildung* (pp. 391–424). Münster: Waxmann.

Stillman, G. (2011). Applying metacognitive knowledge and strategies in applications and modelling tasks at secondary school. In G. Kaiser, W. Blum, R. Borromeo Ferri, & G. Stillman (Eds.), *Trends in teaching and learning of mathematical modelling: ICTMA14* (pp. 165–180). Dordrecht: Springer.

Zöttl, L. (2010). *Modellierungskompetenz fördern mit heuristischen Lösungsbeispielen*. Hildesheim: Franzbecker.

Chapter 17
Evidence of a Dual Modelling Cycle: Through a Teaching Practice Example for Pre-service Teachers

Akio Matsuzaki and Akihiko Saeki

Abstract A proposed dual modelling cycle is illustrated with a typical dual modelling cycle example. The purpose is to verify some stages of a *dual modelling cycle*. In a typical case, the dual modelling cycle indicates switching between two modelling cycles, for example between those for an *Oil Tank Task* and a *Toilet Paper Tube Task*. Some modellers might imagine a toilet paper tube from an *Oil Tank Task* because the tank shape and the spiral banister are similar to an intact toilet paper tube and when it is slit, respectively. Verification methods are (1) to conduct experimental classes for pre-service teachers, and (2) to analyse their worksheets for evidence of the dual modelling cycle. We confirm that the intended transition to answer the initial *Oil Tank Task* was enabled by consideration of the other model/situation.

1 Dual Modelling Cycle Framework for Responding to Diversities of Modellers

Each modeller's modelling cycle potentially differs from another's. We distinguish three types of modelling cycle (single, double and dual) based on what we call a *dual modelling cycle* framework as theoretically elaborated and illustrated in Saeki and Matsuzaki (2013). Models and connections between variables used to construct models change during modelling (Borromeo Ferri 2007; Matsuzaki 2007, 2011; Stillman 1996; Stillman and Galbraith 1998), and a modeller might tackle tasks or

A. Matsuzaki (✉)
Faculty of Education, Saitama University, Saitama, Japan
e-mail: makio@mail.saitama-u.ac.jp

A. Saeki
Graduate School of Education, Naruto University of Education, Naruto, Japan
e-mail: asaeki@naruto-u.ac.jp

G.A. Stillman et al. (eds.), *Teaching Mathematical Modelling: Connecting to Research and Practice*, International Perspectives on the Teaching and Learning of Mathematical Modelling, DOI 10.1007/978-94-007-6540-5_17,

problems along the way that are different from the initial modelling task. A dual modelling cycle framework is constructed from the modelling cycle of Blum and Leiß (2007), and is duplicated for progression through a corresponding modelling task to progress the modelling of the initial task. In a typical case, the dual modelling cycle indicates switching between these modelling cycles.

The purpose of this chapter is to verify the existence of some stages of a dual modelling cycle. The first method of verifying was to conduct experimental classes for undergraduate university students who were interested in mathematics education or would like to become a mathematics teacher. The second method was to analyse their worksheets for evidence of a dual modelling cycle. Through these analyses, we would like to develop teaching materials matched to each school level.

2 Modelling Tasks and Dual Modelling Cycle

In this chapter, we use the following two modelling tasks that are distinguished and described in a dual modelling cycle.

2.1 *Modelling Tasks*

The *Oil Tank Task* is an initial modelling task and we provide the modeller with the following second task, the *Toilet Paper Tube Task*, to explore the length of a spiral banister of an oil tank.

Oil Tank Task

We would like to measure the length of a spiral banister of a heavy oil tank. However, it is out of bounds except for the persons concerned with the safety control of the oil tank. Therefore we cannot measure with a tape measure directly.

We found out the sizes of the diameter (9.766 m) at the bottom and the height (10.772 m) of an oil tank by speaking to the persons concerned. Based on these data, we decided to find the length of the spiral banister of the oil tank.

(O1) What kind of methods do you use to answer?

(O2) Let's actually answer.

Toilet Paper Tube Task

(T1) Have you ever opened a toilet paper tube? (Yes or No)
(T2) What kind of shape do you think will form when the toilet paper tube is opened along its slit? Please select from the following shapes and write also about the reason for this selection.

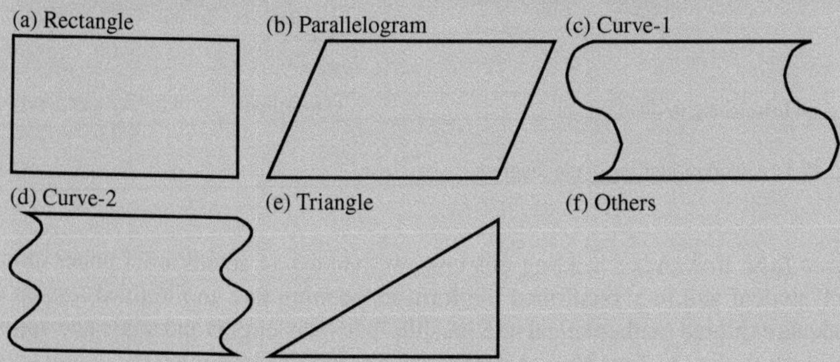

Stillman (2007), commenting on the issue of task authenticity for upper secondary level students, distinguished the following two directions of task authenticity, "Firstly, there is objective authenticity that comes from the real world. Secondly, there is subjective authenticity that comes from the situation being modelled being authentic to the modeller" (p. 464). We prepare the above *Toilet Paper Tube Task* mainly from the perspective of the second direction of task authenticity.

2.2 *Dual Modelling Cycle*

Now we set these two tasks in our dual modelling cycle framework as described in Chap. 7, and we distinguish three stages of modelling progress (see Fig. 17.1). The *Oil Tank Task* is treated in the first modelling cycle and the *Toilet Paper Tube Task* is treated in the second modelling cycle. So the first method of verifying is to develop an imagined model of an oil tank by developing a model of a toilet paper tube. Firstly, we will explain the experimental classes based on these three stages.

We describe the flow of experimental classes and how the tasks were treated. In the first lesson, we presented the *Oil Tank Task* to the students and asked them to solve the problem (O1). The students showed their own solution(s), and we asked them to point out the necessary things to solve their solution(s). In the second lesson, we presented a picture of a toilet paper tube that is one of several similar models of an oil tank with a spiral guard rail or banister (e.g., a sign pole which can be seen in the front of a barber's shop or a screw), and the students tackled the *Toilet*

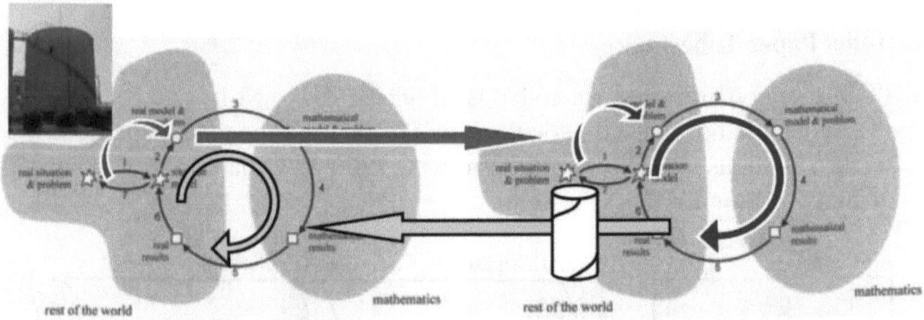

[Modelling cycle of oil tank] [Modelling cycle of toilet paper tube]

Fig. 17.1 A dual modelling cycle diagram

Paper Tube Task. After tackling this task, we handed an actual toilet paper tube to each student and they confirmed the form its opening has. In the third lesson, the students explored mathematical relationships between lengths and angle and applied the result to the *Oil Tank Task*. After these activities, we presented the actual length of the spiral banister, and the students were asked about the difference between the actual length and their mathematical result for a spiral banister.

3 The First Method of Verifying the Purpose of Our Research

We conducted two classes with pre-service teachers ($N=66$) who were undergraduate university students preparing to be mathematics teachers. In the future they will mainly become primary and lower secondary school teachers. They were in the third year of their 4-year course. The first class was conducted in 2nd semester 2010 and the second in 1st semester 2011. The subject studied was *Teaching Secondary School Mathematics* and the teacher of both classes was the first author. The goals of this subject were to understand teaching materials and methods of secondary mathematics. Selected teaching materials in the subject were intended to be applied mathematics or links between mathematics and other subjects. It is necessary for us to understand and analyse responses of the students, because we would like to develop teaching materials matched to each school level and have plans for lesson implementation.

3.1 First Stage: Transition from Modelling Cycle of Oil Tank to Toilet Paper Tube Cycle

Firstly, the university students tackled problem (O1) of the *Oil Tank Task*. Some students might find an answer directly to the *Oil Tank Task* whilst others might set

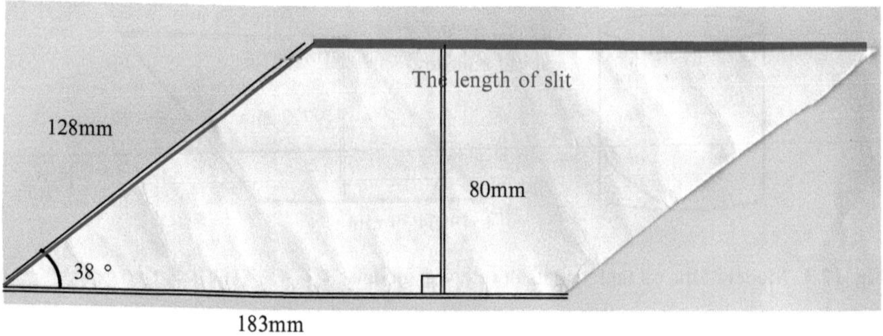

Fig. 17.2 Dimensions of a toilet paper tube

an original task for themselves. In this teaching practice the students were asked to foreshadow their own solution, and the teacher asked, 'Are there similar things to a spiral banister of the oil tank?' The teacher intended that the students would implement a transition from the modelling cycle of an oil tank. Following this the teacher provided an actual toilet paper tube to each student. If modellers do not imagine a toilet paper tube from the *Oil Tank Task*, the teacher can present a toilet paper tube. At this point, the *Toilet Paper Tube Task* could be: 'How long is a spiral of a toilet paper tube?' The shape of an oil tank and a spiral banister are similar to an intact toilet paper tube and when it is slit, respectively.

3.2 Second Stage: Modelling of Toilet Paper Tube

Secondly, the students tackled problems (T1) and (T2) of the *Toilet Paper Tube Task*. Thus, the new situation and the problem in the modelling cycle of a toilet paper tube are based on the *Toilet Paper Tube Task* which differs from the initial real situation and problem located in the modelling cycle of the oil tank of a dual modelling cycle. Following this, the students opened the toilet paper tube along its slit. The shape of the opened toilet paper tube is a parallelogram and it is easy to measure the lengths (slit is 183 mm) and angles (see Fig. 17.2).

3.3 Third Stage: Transition from Modelling Cycle of Toilet Paper Tube to Oil Tank Cycle

Thirdly, the students tackled the problem (O1) again and the problem (O2) of the *Oil Tank Task* based on these data from a toilet paper tube (Fig. 17.2). As shown in Fig. 17.3, necessary data to explore the length of a spiral banister of an oil tank are slope 26° and height from ground to the spiral banister (1.050 m) in addition to the initial data (diameter of tank is 9.766 m at the bottom and the height is 10.772 m) which is given in the task statement.

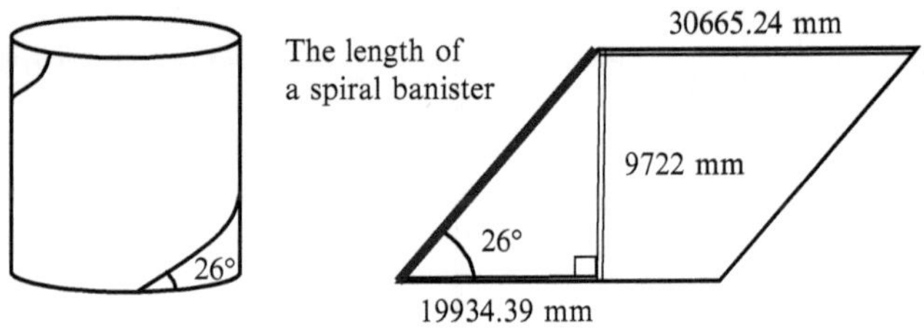

Fig. 17.3 Model of the oil tank and its developing models

The length of the projection on the ground from the beginning to the end of the spiral banister is $\dfrac{10,772-1,050}{\tan 26^{\circ}} = \dfrac{9,722}{0.4877} = 19,934.39$ mm.

Thus, the length of a spiral banister= $\dfrac{10,772-1,050}{\sin 26^{\circ}} = \dfrac{9,722}{0.4384} = 22,176.09$ mm.

Actual length determined by using a tape measure was 19.4 m. (19,400 mm).

4 Analysis of Students' Responses and Evidence of a Dual Modelling Cycle

4.1 *First Stage: Transition from Modelling Cycle of Oil Tank to Toilet Paper Tube Cycle*

The students answered problem (O1) of the *Oil Tank Task* by responding with what they saw as necessary additional data except for the known dimensions of the diameter at the bottom and the height of an oil tank. The following data were what the students wanted to know to solve the task: the beginning point (the height from the ground) and the end point of the banister, the width of a banister or space between body of a tank and the banister, and how many times the rail spiralled around the tank. Some students (4 of 66) drew a diagram of an oil tank viewed from the top to indicate necessary data for leading to an answer indicating times around the spiral steps and width of space between the tank and a banister (as shown in Fig. 17.4).

Almost all students (63 of 66) answered that they would try to draw a development and some of them used Archimedes theorem as the method for finding the length of the spiral banister. Chie, for example, showed the height of a banister from the ground and considered the circular nature of the spiral (see Fig. 17.5). Nao described a model as her method to develop a solution as the following: "make a straight line opened along with a spiral banister" (see Fig. 17.6). Subsequently most

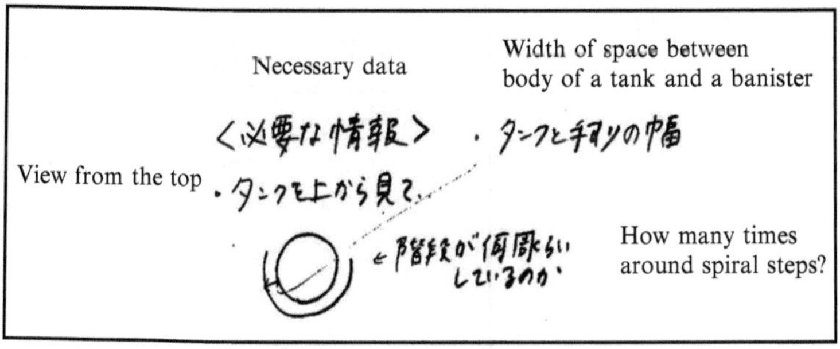

Fig. 17.4 Evidence of necessary data that Miri wants to know leading to her answer

Fig. 17.5 Chie's drawing of a projection figure

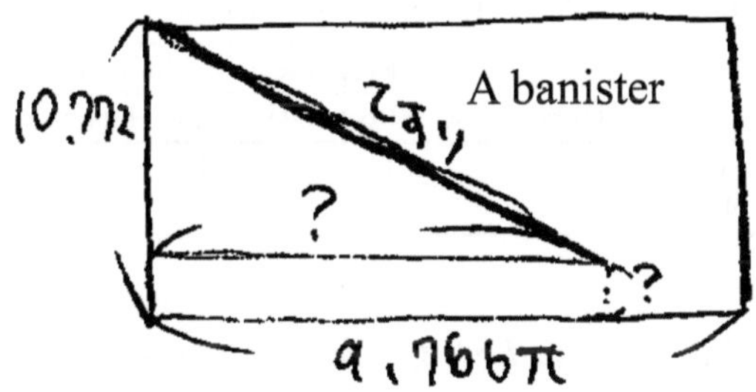

Fig. 17.6 Nao's drawing of a projection figure

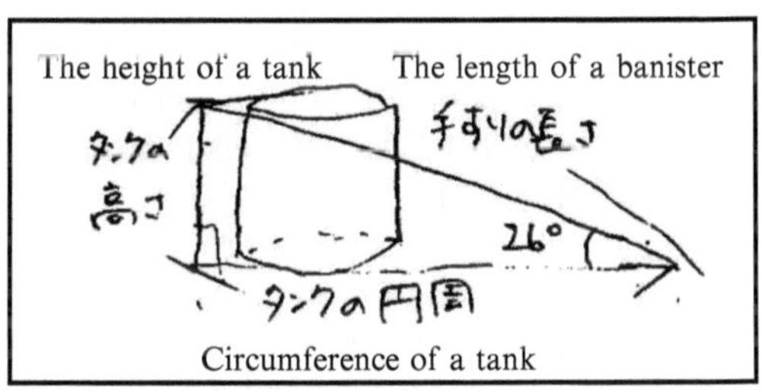

students (56 of 66) drew the development as a rectangle and tried to answer the problem as exploring the shortest distance.

As other solution ideas, two students (2 of 66) raised alternative ideas: The first was to make a scale model of an oil tank. Taku wrote: "I will make a model of an oil tank (1/10 or 1/100 scale) and measure a spiral banister model made by a string, and extend it to an actual size". A second idea was to draw a parallelogram or triangle as a projection figure. Ken drew a parallelogram as a projection figure, but he mistook the position of the banister (see Fig. 17.7). A projection figure for an oil tank is not a triangle. Similarly as for problem (T2) of the *Toilet Paper Tube Task*, this idea is a mistake (see Saeki and Matsuzaki 2013). Next, we present the *Toilet Paper Tube Task* (Sect. 2.1) for providing a transition from the modelling cycle of an oil tank to the modelling cycle of a toilet paper tube.

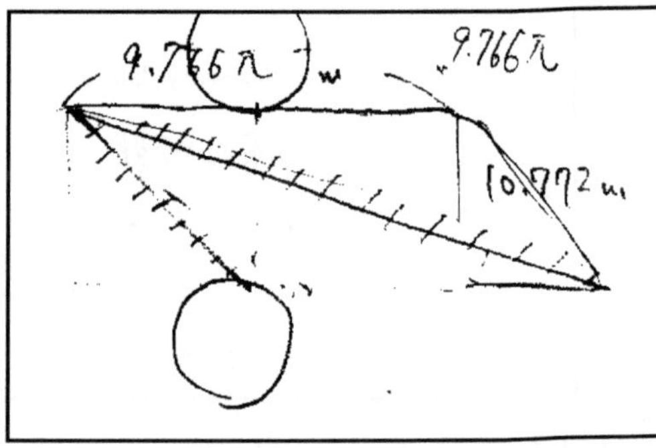

Fig. 17.7 Ken's erroneous projection figure

Table 17.1 Pooled responses for problems (T1) and (T2) of the *Toilet Paper Tube Task*

	Response to T2 (form of open tube)					
T1 (opened tube or not)	Rectangle	Parallelogram	Curve-1	Curve-2	Triangle	Others
Yes	1	31	1	0	2	1
No	0	27	0	1	2	0

4.2 Second Stage: Modelling of Toilet Paper Tube

The teacher provided a toilet paper tube to each student and they tackled the *Toilet Paper Tube Task* with this. Initially, the students responded to the problems without opening a toilet paper tube along its slit. Pooled results from the two classes of responses for each problem of the *Toilet Paper Tube Task* are given in Table 17.1.

Almost all responses (58 of 66) for the shape formed when a toilet paper tube is opened along its slit (T2) were (b) Parallelogram. This was not related to whether they had ever opened a toilet paper tube (T1). Subsequently, all students opened the slit of the toilet paper tube and confirmed or disproved their answers. Almost all students (63 of 66) had answered problem (O1) of the *Oil Tank Task* based on a rectangle, but they answered parallelogram in the *Toilet Paper Tube Task*. From this result, we can say that the intended transition, the first stage, from the modelling cycle of an oil tank to the modelling cycle of a toilet paper tube was successful. On the other hand, there were a few students who were not able to find the height of the toilet paper tube in the parallelogram formed from opening the tube.

4.3 Third Stage: Transition from Modelling Cycle of Toilet Paper Tube to Oil Tank Cycle

The students tried to answer the initial *Oil Tank Task* again based on the results of step 5 of the Blum and Leiß (2007) cycle, that is "working mathematically", in the modelling cycle of a toilet paper tube. Most students (61 of 66) found their answers

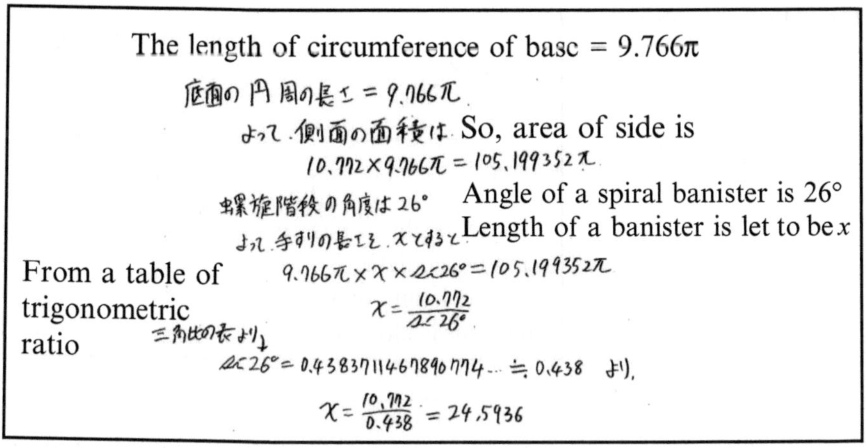

Fig. 17.8 Example of working mathematically by Osa

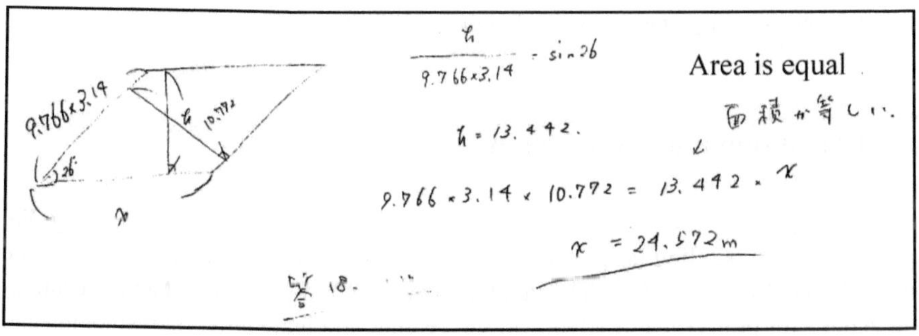

Fig. 17.9 An evidence of working mathematically by Ryo

by using a trigonometric ratio (see Fig. 17.8) and one student found his answer by focusing on equal area of parallelogram (see Fig. 17.9).

In addition, it is noted that a few students (2 of 66) did not find the height of the oil tank by developing a parallelogram model. These students did not apply the relationship between dimensions of a toilet paper tube to the relationship between dimensions of an oil tank. Although all the students were able to explore the length of a spiral banister of the oil tank mathematically, there was a significant error giving a result higher than the actual measurement by using a tape measure. We have to take into consideration many more of the variables that the students wanted to know for finding the answer that were pointed out at the beginning of this section. Thus, we need to refine the model.

Some further suggestions that the students raised for tackling this task included considering the space between the body of the oil tank and the spiral banister (580 mm), measurement mistake of slope angle (26°), or number of rotations of the spiral banister around the tank. The following point regarding lacking data was

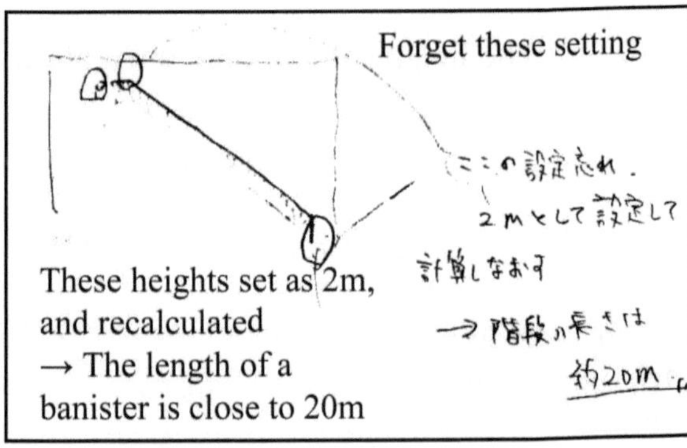

Fig. 17.10 Lacking data identified by Tomi

raised by Tomi. The beginning point of the spiral banister had not been considered (see Fig. 17.10). Tomi pointed out that the end point of a spiral banister is lower than the top of the oil tank and the beginning of the rail must also be higher than the bottom of the oil tank. Tomi assumed that these heights in total were 2 m to refine the calculation. Actually, the measured height is 820 mm.

5 Discussion and Conclusion

In this chapter we verified some stages of a *dual modelling cycle* empirically. We prepared two modelling tasks: the first one is the *Oil Tank Task* and the second one is the *Toilet Paper Tube Task*. The method of verification was to conduct experimental classes for pre-service teachers who were undergraduate university students and to analyse their worksheets for evidence of a dual modelling cycle. For discussing this evidence, we distinguished three stages for the dual modelling cycle.

Before the transition from the first stage, most students drew the development as a rectangle and tried to answer the problem (O1). However, after the transition at the first stage, most students answered that the opened form of toilet paper tube along its slit was a parallelogram (T2). This intended transition was to enable them to answer the *Oil Tank Task* again by another model. At the third stage, all the students were able to explore the length of a spiral banister of the oil tank mathematically. However, the solution differed significantly from the actual measurement. For a more acceptable solution modellers have to take into consideration many more variables. Thus, we would hope that modellers could refine the model through the dual modelling cycle. A dual modelling cycle framework as proposed appears to have potential for drawing modellers' attention to the possibility of solving the task at hand by first solving a related problem as Polya (1988) suggested.

Acknowledgement Mr. Kurokawa Hiroshi (Tokushima office, Nisshinbo Holdings Inc.) allowed us to measure a heavy oil tank and introduced the staff concerned. Mr. Kurebayashi Yutaka (Tokai Seishi Kogyo Co., LTD.) gave us a factory tour and explained the process of making the toilet paper tube. Mr. Kainuma Kenji (Anan office, Asahi Synchrotech Corporation) provided important plan data of the heavy oil tank. Thanks to all concerned.

References

Blum, W., & Leiß, D. (2007). How do students and teachers deal with modelling problems? In C. Haines, P. Galbraith, W. Blum, & S. Khan (Eds.), *Mathematical modelling (ICTMA12): Education, engineering and economics* (pp. 222–231). Chichester: Horwood.

Borromeo Ferri, R. (2007). Modelling problems from a cognitive perspective. In C. Haines, P. Galbraith, W. Blum, & S. Khan (Eds.), *Mathematical modelling (ICTMA12): Education, engineering and economics* (pp. 260–270). Chichester: Horwood.

Matsuzaki, A. (2007). How might we share models through cooperative mathematical modelling? Focus on situations based on individual experiences. In W. Blum, P. Galbraith, H.-W. Henn, & M. Niss (Eds.), *Modelling and applications in mathematics education: The 14th ICMI study* (pp. 357–364). New York: Springer.

Matsuzaki, A. (2011). Using response analysis mapping to display modellers' mathematical modelling progress. In G. Kaiser, W. Blum, R. Borromeo Ferri, & G. Stillman (Eds.), *Trends in teaching and learning of mathematical modelling: ICTMA14* (pp. 499–507). New York: Springer.

Polya, G. (1988). *How to solve it: a new aspect of mathematical method* (1st Princeton Science Library ed.). Princeton: Princeton University Press.

Saeki, A., & Matsuzaki, A. (2013). Dual modelling cycle framework for responding to the diversities of modellers. In G. A. Stillman, G. Kaiser, W. Blum, & J. Brown (Eds.), *Teaching mathematical modelling: Connecting to research and practice* (pp. 89–99). Dordrecht: Springer.

Stillman, G. (1996). Mathematical processing and cognitive demand in problem solving. *Mathematics Education Research Journal, 8*(2), 174–197.

Stillman, G. (2007). Upper secondary perspectives on applications and modelling. In W. Blum, P. Galbraith, H.-W. Henn, & M. Niss (Eds.), *Modelling and applications in mathematics education: The 14th ICMI study* (pp. 463–468). New York: Springer.

Stillman, G., & Galbraith, P. (1998). Applying mathematics with real world connections: Metacognitive characteristics of secondary students. *Educational Studies in Mathematics, 36*(2), 157–195.

Chapter 18
Considering Multiple Solutions for Modelling Problems – Design and First Results from the MultiMa-Project

Stanislaw Schukajlow and André Krug

Abstract This chapter deals with the research project MultiMa that investigates the effects of treating multiple solutions while solving modelling problems on student achievements and motivation in mathematics. First, the starting points of MultiMa are discussed followed by the effects of treating multiple solutions from theoretical and empirical points of view. The MultiMa project goals and methods including important research steps are discussed. Finally, a modelling task is presented and first results of the explorative video study are introduced. The analysis of videos shows that students had only minor problems while solving the task, where two outcomes were demanded. If they could find one solution, the second outcome was developed also. Further, students seldom compared different outcomes and did not discuss why there were differences in outcomes.

1 Introduction: Starting Points of the Research Project MultiMa

The main starting point of the project, *Multiple solutions for Mathematics teaching oriented towards students' self-regulation (MultiMa*[1]*)*, is the disappointing results of German students in large-scale assessment studies. In the TIMSS and PISA

[1]The research project MultiMa, directed by Stanislaw Schukajlow, has been funded by the German Research Foundation [Deutsche Forschungsgemeinschaft] since 2011.

S. Schukajlow (✉)
Department of Mathematics, University of Paderborn,
Warburgerstr. 100, 33098 Paderborn, Germany
e-mail: schustan@math.upb.de

A. Krug
Department of Mathematics, University of Kassel,
Heinrich-Plett-Str. 40, 34132 Kassel, Germany
e-mail: akrug@mathematik.uni-kassel.de

G.A. Stillman et al. (eds.), *Teaching Mathematical Modelling: Connecting to Research and Practice*, International Perspectives on the Teaching and Learning of Mathematical Modelling, DOI 10.1007/978-94-007-6540-5_18,

studies German students were average in mathematics, but significantly worse than Japanese or Finnish students (e.g., OECD 2004). The reasons for this result have been intensively analysed and discussed over the past decade. One possible reason is that students could not transfer their knowledge to the new context, because they have so-called "inert knowledge" (Whitehead 1962). The inert knowledge could not be activated while solving a new problem. With their inert knowledge the students even have difficulties with solving the types of tasks already known to them when the context was slightly changed. The consequence of this is that teaching mathematics should support the flexibility of knowledge in the classroom.

In the TIMSS video study differences in teaching methods in Japan, the USA and Germany have been found. In Japan, for example, where students achieve the best results in mathematics, the lessons were organised in a different way.

> Japanese teachers frequently posed mathematics problems that were new for their students and then asked them to develop a solution method on their own. After allowing time to work on the problem, Japanese teachers engaged students in presenting and discussing alternative solution methods and then teachers summarized the mathematical points of the lesson. Especially revealing was the way in which these features were combined into a pattern or system that characterized a distinctive method of teaching observed in eighth grade Japanese mathematics lessons. These features of practice offer an alternative to those seen in the United States and Germany (Hiebert et al. 2003, p. 16).

Thus, the posing of tasks that have several solutions, the fostering of the development of several solution methods, and the presentation and discussion of these solution methods in the classroom are the characteristic features of Japanese teaching method. To what extent these features are crucial for the performance of Japanese students in mathematics and whether these teaching methods or parts of them can also be used effectively in other countries is still an open question in teaching education.

Further, analysis of large-scale studies and video surveys shows the quality of learning is not determined only by the surface processes that can be observed directly like, for example, different ways of organising students' learning in individual, pair or group work. That is why a statement like "using of a special type of surface structure, (e.g. group work), can guarantee students' high performance" is not true. Such surface structures can only give possibilities for individual development of students' competencies in the particular domain. The consequence is that we need the analysis of the students' cognitive learning processes and not just the surface structures that can be observed directly in the classroom. The indicator of such learning processes may be the *cognitive activation* of students. Some empirical studies show that cognitive activation is a key variable that influences students' achievements (Baumert et al. 2010). Cognitive activation is a construct that is founded on different conceptualisations. In the German – Swiss video-based classroom study directed by Klieme, Pauli and Reusser, the important parts are the challenging tasks, activation of prior knowledge and a content-related discourse in the classroom (Klieme et al. 2006). Thus, in the context of cognitive activation the following two questions are important: What tasks should be used in the classroom? and How should these tasks be treated?

The leading question of the MultiMa project focuses on one specific aspect that is connected to the two questions above: What do we know about the effect of treating multiple solutions on students' achievements?

2 Multiple Solutions: Theoretical and Empirical Background

What is the basis of the effects of treating multiple solutions? Analysis of theories of learning shows that the active development of multiple solutions or representations[2] is an important part of *constructivist* theories of learning in general. According to proponents of these theories, students' engagement with different solutions helps to improve their domain-specific knowledge and competencies. For example, in cognitive flexibility theory Spiro et al. (1988) emphasise the necessity of constructing multiple representations and connecting them with each other. They point out that it is significant to:

> (a) allow an important role for *multiple representations*,
> (b) view learning as a multidirectional and multiperspectival "criss-crossing" of cases and concepts,
> (c) foster the ability to assemble diverse knowledge sources to adaptively fit the needs of a particular knowledge application situation. (p. 383)

Learning new skills occurs by exploring them from various perspectives. Solving problems with different methods stimulates constructing several mental representations and enables multiple mental representations of a new domain. Therefore, the use of different solution methods can improve cognitive flexibility as well as students' academic achievements. Mathematics teachers ought to encourage their students to consider more than one possibility when solving problems (Silver et al. 2005). One of the important points while treating multiple solutions is the theoretical perspective promoting the coherence between different solution methods and concepts. Students' willingness to develop multiple solutions is a crucial prerequisite for the improvement of their mathematical knowledge. There is the first evidence (see Yoshimura and Yanagimoto Chap. 21 this volume) that the majority of students in a small Japanese study indicated willingness to find different solutions.

Other arguments for treating multiple solutions in the classroom are (see also Leikin and Levav-Waynberg 2007):

- a deep understanding of the subject, which has to be learned,
- additional opportunities to optimise a solution,

[2]Multiple solutions can be developed using the same representation or using different representations such as arithmetic or graphical ones. In the recent study we do not distinguish between solutions and representations (but see Große and Renkl 2006).

- stimulation of connections among different mathematical topics or among various representations of the same concept, and
- support of students' self-regulation.

Some empirical studies have investigated teaching practice and teacher beliefs regarding treating multiple solutions. Research about teaching practice and teachers' beliefs shows that teachers are reluctant to solve problems in different ways in the classroom and do so very rarely (see Silver et al. 2005). Some teachers think a presentation of multiple solutions may confuse students (see Leikin and Levav-Waynberg 2007). Silver et al. (2005) assume that teachers' poor content knowledge and pedagogical content knowledge prevent them from using multiple solutions. Some case studies show that capable and confident teachers encourage students' use of multiple solutions, comparing them in the classroom (Ball 1993; Richland et al. 2007).

We have found several studies which investigated the influence of treating multiple solutions on students' achievements in mathematics (see detailed review by Schukajlow and Blum 2011). These studies investigated the question of how problems that could be solved with multiple solution methods should be treated in the classroom. Worked-out examples were used as a form of teaching how to answer the question whether it is better to compare different solution methods and different representations or to reflect on one solution method at a time. The mathematics topics were combinatorics, probability and algebra. One of the important results of these studies is that when the students have sufficient prior knowledge, contrasting different solution methods can improve the students' procedural and conceptual knowledge as well as their procedural flexibility (Große and Renkl 2006; Rittle-Johnson and Star 2009). When the students' prior knowledge is poor, they show better results when reflecting on one solution method at a time (Rittle-Johnson et al. 2009).

The following research question still remains open: What advantages do treating the tasks, which have various solutions, have in comparison with those, which have only one solution? This problem is investigated within the framework of the MultiMa project.

3 The Goals and Method of the MultiMa Project

The main goal of MultiMa is to investigate treating multiple solutions of modelling problems in learning environments oriented towards self-regulation. The main hypothesis of the project is that the treatment of multiple solutions in learning environments oriented towards self-regulation has a more positive effect on students' performance (in solving modelling as well as intra-mathematical problems) and motivation than the treatment of only one solution.

The project consists of three parts. The first step was to carry out an explorative study and preliminary results of this are discussed in the final section of this chapter. The main research question of this explorative study is: How do students deal with the problems that allow multiple outcomes?

The second step was to develop instruments that measure special parts of modelling competency and to investigate:

How do students identify the data they need to solve a modelling problem,
How do students identify the data they lack, and
How do students assume the data they lack?

The third part of the project is the comparison of two types of treatment that differ according to the possibilities and demands to develop multiple solutions (see first results by Schukajlow and Krug 2012).

Before the effect of treating multiple solutions on students' performance and motivation can be explored, what kinds of solutions can be developed while solving modelling problems have to be clarified. Analysis of problem solving shows that different solutions could be developed with multiple solution methods and/or multiple outcomes. Similar principles can be used for analysing the modelling process. Even if different mathematical procedures are used to solve a modelling task, the same result/outcome could be attained. However, if different assumptions are made and different real models are developed, one has different outcomes applying the same mathematical procedures. In the MultiMa project we focus on investigating the effects of treating different solutions resulting from different assumptions and leading to different outcomes. Variation in the solution methods while solving modelling problems will be the focus of further studies.

4 Solving Problems That Demand Multiple Outcomes

The problem *Parachuting MO* that is presented below requires developing two outcomes.

Parachuting MO

When "parachuting", a plane takes jumpers to the altitude of about 4000 metres. From there they jump off the plane. Before a jumper opens his parachute, he free falls about 3000 metres. At an altitude of about 1000 m the parachute opens and the sportsman glides to the landing place. While falling, the jumper is carried off target by the wind. Deviations at different stages are shown in the table below.

Wind speed	Side deviation per thousand metres during free fall	Side deviation per thousand metres while gliding
light	60 metres	540 metres
middle	160 metres	1440 metres
strong	340 metres	3060 metres

What distance does the parachutist cover during the entire jump? Find two possible solutions by estimating missing data.

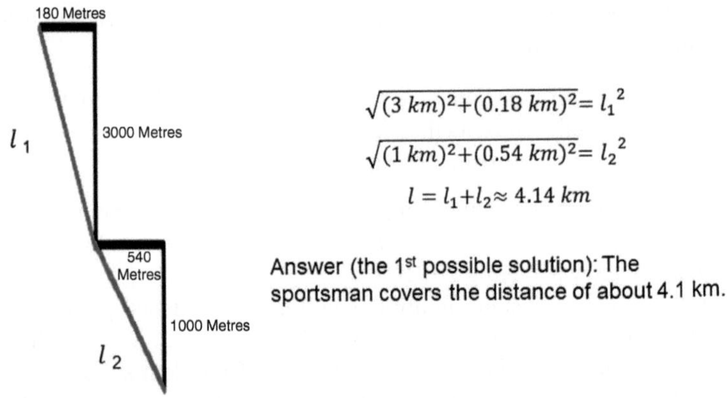

Fig. 18.1 A solution of problem *Parachuting MO*

To solve this problem, students have to make the following assumptions, amongst others:

- the deviation remains constant at the different stages of the jump,
- the wind has the same direction during the entire jump,
- the wind speed is light, middle or strong during the respective stages,
- the parachute opened, for example, at 1,000 metres above the earth.

In the solution presented in Fig. 18.1, the wind is light so the deviation at the falling stage is 180 m and at the gliding stage 540 m. With the help of Pythagoras' theorem you can calculate the hypotenuses in the right-angled triangles and add up the results. The entire jump is approximately 4.14 km. The first possible solution is "The parachutist covers the distance of about 4.1 km". The second solution you can find by analogy when, for example, the assumption is that the wind is strong during both stages. Both solutions were developed using different assumptions but the same solution method but the outcomes are different.

5 Students' Difficulties While Solving Modelling Problems with Multiple Outcomes

5.1 Research Questions and the Method

The explorative study was carried out to clarify the following research questions:

How do students deal with problems that allow multiple outcomes?
How do students react to the demand to develop two outcomes?

The sample for the study consists of six pairs of Year 9 students (ten female and two male) who solved four modelling problems that demand two outcomes. One of these problems was the task, *Parachuting MO*. It was possible to solve these problems using Pythagoras' Theorem as the mathematical procedure. Two months before the study was carried out, intra-mathematical problems and simple modelling problems, the solution of which demands using the Pythagoras' Theorem, had been treated in the classroom. The students were from a German middle track school (Realschule) and were about 15 years old. The students' performance in mathematics varied from very good to poor. Their mathematics teachers reported on the treating of the tasks with connection to reality that demand simple assumptions in the classrooms. The process of solving the problems was filmed from two different locations. The first camera was placed in front of the students. The second camera was pointed to their work sheets. Immediately after solving the problems, one student of each pair was interviewed. The interview was based on the "stimulated recall" method. During this interview, the student watched the video and from time to time was asked to explain his/her way of solving the problem. The interview was also filmed. In the next section we present the preliminary results that are based on the analysis of the processes of solving the modelling problem, *Parachuting MO*.

5.2 *Preliminary Results*

In this section we analyse (a) the general difficulties of students solving modelling problems and (b) their specific difficulties that refer to the demand to develop two outcomes. The collected data were analysed using the modelling cycle by Blum and Leiß (2007) as a structural scheme.

5.2.1 General Difficulties with Modelling Problems

The analysis showed students' general difficulties in (1) *understanding*, (2) *making assumptions*, (3) *idealising and structuring a situation model* as well as in *constructing a mathematical model*. All students applied the same mathematical method for calculating the falling distance. Some students validated their results using their mathematical knowledge.

Firstly, with respect to understanding, some students misunderstood the question of the task and did not construct a correct situation model. For example, they calculated the duration of the parachutist's fall instead of the distance. In addition, students had difficulties with understanding some terms. For example, most students were not sure what the term "side deviation" meant.

While developing a mathematical model, some students did not differentiate between two right-angled triangles, and used only one triangle to calculate the distance covered by the parachutist. The students who identified right-angled triangles used Pythagoras' Theorem for calculating the hypotenuses. Other mathematical methods like drawing the triangles using a suitable scale, measuring the length of hypotenuses in the sketch and calculating their real length were not applied by the students.

There were some students who validated their results. One of them, for instance, said, "If the parachutist falls vertically, he covers 4,000 m". He thought that the distance of less than 4,000 m is impossible and the distance of 4,080 m is far too little (although even the distance of about 4,050 m is possible, if a light wind blows).

5.2.2 Specific Difficulties Related to Developing Two Outcomes

The analysis of the second research question deals with one specific feature of the task *Parachuting MO*, namely the development of two solutions to the same problem. The results of analysis indicate that the demand to develop two outcomes did not impede students' solution process. Some difficulties that are specific to dealing with such tasks were identified. These are now elaborated.

The students had only minor difficulties with the development of the second outcome whilst solving the task *Parachuting MO*. When they found the first solution, they could find the second result as well. However, while solving the task *Parachuting MO* students had difficulties with understanding the demand to develop two outcomes. In this task the demand is formulated as follows: "Find two possible solutions by estimating the missing data". First, some students supposed they had to estimate using their prior knowledge only, instead of using the information given in the task. Students only had to assume how strong the wind blows in each falling stage (light, middle or strong) and at what distance the parachute has to be opened. They, however, did not need to make assumptions of the relevant side deviation using only their experience. The required data could be found in the table.

Students overlooked various possibilities to obtain different outcomes. For example, they varied the wind power but not the altitude at which the parachute opens when solving the *Parachuting MO* problem.

The analysis of one of the DISUM[3] studies shows that students feel bored when solving the same modelling problem with changed numbers. We were unsure whether they would be motivated in looking for a second solution. In this study we noticed no problems with motivating students while developing the second outcome. One possible reason is that the development of the second solution is part of the task here and because of that is interesting to students.

Although students validated the results, they rarely compared different outcomes on their own. The comparison of solution methods is an important part of teaching tasks with multiple solutions and can improve students' performance and cognitive flexibility. Thus, we assume that prompting of comparison of different outcomes and solution methods is crucial for improvement of modelling competency. That is why stimulating discussions by teacher about the following questions is essential: How great are the differences between the outcomes? Why do the outcomes differ?

[3]The research project DISUM (Didactical intervention modes for mathematics teaching oriented towards self-regulation and directed by tasks) led by W. Blum, R. Messner and R. Pekrun.

6 Discussion and Conclusion

In the previous section we analysed students' difficulties while solving the modelling problem *Parachuting MO*, the solution of which demands two outcomes. Firstly, we found that students had some difficulties that refer to the activities described in modelling cycles. Students' process of solving the problem was impeded because of the misconception of the meaning of some terms of the task, general difficulties with understanding the real situation, estimation of missing data and development of the mathematical model. All students used the same mathematical method (Pythagoras' Theorem) for solving the problem. One reason for their choice of this mathematical procedure was that it allows the calculation of the length of one side in the right-angled triangle quickly and exactly in comparison with other mathematical methods. Students' good prior knowledge of Pythagoras' Theorem could also influence their choice of mathematical procedure. The students had practised relatively recently solving tasks that required Pythagoras' Theorem before the study was carried out.

Secondly, students were motivated to develop two solutions of the same problem. This result is in line with findings by Yoshimura and Yanagimoto (2013). As motivation is an important prerequisite for learning, we assume that there are favourable conditions for treating multiple solutions in the classroom. However, only students who found the first solution were capable of finding the second one by varying the assumptions. We assumed that the development of the second outcome guides students to a deeper understanding of the real situation and stimulates students' practising in modelling. This hypothesis should be tested in future research. The observed students varied the wind power but not the distances. Looking for the simplest solution method may be one reason for these students' solving paths. Although some students validated their mathematical results, they did not compare two outcomes with each other and did not reflect on their similarities or differences on their own. Such a reflection has to become a part of teaching multiple solutions. How teachers can successfully do this in the classroom is an open question.

In conclusion, we have to point out the following limitations of this study: only one modelling problem was offered to students and the data we analysed were based on the sample of six student pairs only. Increasing the number of students and varied tasks are important for validation of the results of this study.

References

Ball, D. L. (1993). With an eye on the mathematical horizon: Dilemmas of teaching elementary school mathematics. *The Elementary School Journal, 93*(4), 373–397.

Baumert, J., Kunter, M., Blum, W., Brunner, M., Dubberke, T., Jordan, A., et al. (2010). Teachers' mathematical knowledge, cognitive activation in the classroom, and student progress. *American Educational Research Journal, 47*(1), 133–180.

Blum, W., & Leiß, D. (2007). How do students and teachers deal with mathematical modelling problems? The example sugarloaf and the DISUM project. In C. Haines, P. L. Galbraith,

W. Blum, & S. Khan (Eds.), *Mathematical modelling (ICTMA 12): Education, engineering and economics* (pp. 222–231). Chichester: Horwood.

Große, C. S., & Renkl, A. (2006). Effects of multiple solution methods mathematics learning. *Learning and Instruction, 16*(2), 122–138.

Hiebert, J., Gallimore, R., Garnier, H., Givvin, K. B., Hollingsworth, H., Jacobs, J., et al. (2003). *Teaching mathematics in seven countries. Results from the TIMSS 1999 video study*. Washington, DC: NCES.

Klieme, E., Lipowsky, F., Rakoczy, K., & Ratzka, N. (2006). Qualitatsdimensionen und Wirksamkeit von Mathematikunterricht. Theoretische Grundlagen und ausgewählte Ergebnisse des Projekts "Pythagoras". In M. Prenzel & L. Allolio-Nacke (Eds.), *Untersuchungen zur Bildungsqualitat von Schule. Abschlussbericht des DFG-Schwerpunktprogramms* (p. 127). Munster: Waxmann.

Leikin, R., & Levav-Waynberg, A. (2007). Exploring mathematics teacher knowledge to explain the gap between theory-based recommendations and school practice in the use of connecting tasks. *Educational Studies, 66*(3), 349–371.

OECD. (2004). *Learning for tomorrow's world: First results from PISA 2003*. Paris: OECD.

Richland, L. E., Zur, O., & Holyoak, K. J. (2007). Cognitive supports for analogies in the mathematics classroom. *Science, 316*, 1128–1129.

Rittle-Johnson, B., & Star, J. R. (2009). Compared with what? The effects of different comparisons on conceptual knowledge and procedural flexibility for equation solving. *Journal of Educational Psychology, 101*(3), 529–544.

Rittle-Johnson, B., Star, J. R., & Durkin, K. (2009). The importance of prior knowledge when comparing examples: Influences on conceptual and procedural knowledge of equation solving. *Journal of Educational Psychology, 101*(4), 836–852.

Schukajlow, S., & Blum, W. (2011). Zur Rolle von multiplen Lösungen in einem kompetenzorientierten Mathematikunterricht. In K. Eilerts, A. H. Hilligus, G. Kaiser, & P. Bender (Eds.), *Kompetenzorientierung in Schule und Lehrerbildung – Perspektiven der bildungspolitischen Diskussion, der empirischen Bildungsforschung und der Mathematik-Didaktik. Festschrift für Hans-Dieter Rinkens* (pp. 249–267). Münster: LIT Verlag.

Schukajlow, S., & Krug, A. (2012). Effects of treating multiple solutions on students' self-regulation, self-efficacy and value. In T. Y. Tso (Ed.), *Proceedings of the 36th conference of the International Group for the Psychology of Mathematics Education* (Vol. 4, pp. 59–66). Taipei: PME.

Silver, E. A., Ghousseini, H., Gosen, D., Charalambous, C. Y., & Font Strawhun, B. T. (2005). Moving from rhetoric to praxis: Issues faced by teachers in having students consider multiple solutions for problems in the mathematics classroom. *The Journal of Mathematical Behavior, 24*(3–4), 287–301.

Spiro, R. J., Coulson, R. L., Feltovich, P. J., & Anderson, D. K. (1988). Cognitive flexibility theory: Advanced knowledge aquisition in ill-structured domains. In *The tenth annual conference of the cognitive science society* (pp. 375–383). Hillsdale: Lawrence Erlbaum.

Yoshimura, N., & Yanagimoto, A. (2013). Mathematical modelling of a social problem: Pension tax issues. In G. Stillman, G. Kaiser, W. Blum, & J. Brown (Eds.), *Teaching mathematical modelling: Connecting to research and practice* (pp. 241–251). New York: Springer.

Whitehead, A. N. (1962). *The aims of education*. London: Ernest Benn.

Chapter 19
Challenges in Modelling Challenges: Intents and Purposes

Gloria Ann Stillman, Jill P. Brown, and Peter Galbraith

Abstract Extra curricular modelling events for school students are increasing in popularity. The experience from the perspective of participating Australian and Singaporean Year 10 and 11 students in one such event, the A. B. Paterson College Mathematical Modelling Challenge, is investigated. The aim is to gauge whether their experience meets the intended purposes of mentors that the whole event be considered an inherently valuable learning experience about modelling and application of mathematics to real situations. Findings from a questionnaire and other data included motivation for choice of situation to model, decision making about approach, and reporting of findings were mainly goal oriented rather than performance oriented, despite the Challenge ostensibly being a competition.

1 Introduction

Extra-curricular modelling experiences for school students appear to be on the rise in several countries (see Bracke and Geiger 2011; Brown and Redmond 2007; Kaiser and Schwarz 2010; Kaiser et al. 2011; Kaland et al. 2010; Lee and Ng in press). These take a variety of forms and descriptions: "modelling weeks" (Göttlich

G.A. Stillman (✉)
Education VIC | Faculty of Education,
Australian Catholic University, Ballarat, Australia
e-mail: gloria.stillman@acu.edu.au

J.P. Brown
Education VIC | Faculty of Education,
Australian Catholic University, Melbourne, VIC, Australia
e-mail: jill.brown@acu.edu.au

P. Galbraith
School of Education, The University of Queensland, St Lucia, QLD, Australia
e-mail: p.galbraith@uq.edu.au

G.A. Stillman et al. (eds.), *Teaching Mathematical Modelling: Connecting to Research and Practice*, International Perspectives on the Teaching and Learning of Mathematical Modelling, DOI 10.1007/978-94-007-6540-5_19,

2010; Kaiser et al. 2011), "modelling challenges" (Brown and Redmond 2007), "modelling outreach" as occurred in Singapore in 2010 (Lee and Ng in press) and "modelling days" (Bracke and Geiger 2011). The commonality among all these is that their focus is on mathematical modelling activity, they are events that are external to normal classroom activity, they involve experts external to the school and are usually not at the student's school. It seems an appropriate time to pause and look at the intents, purposes and outcomes of one of these events and whether these are realised in the context of local curricula supporting the teaching of modelling.

2 A Selection of Previous Research Related to Extra-curricular Modelling Events

There has been some previous research in the context of these extra-curricular modelling events and a selection will be briefly overviewed. Kaiser and several colleagues, for example, have researched the feasibility of using "authentic modelling problems, which promote the whole range of modelling competencies" and expand the range of situations where the students can demonstrate their modelling skills (Kaiser and Schwarz 2010, pp. 54–55). To this end, in the main, the problems used are suggested by applied mathematicians motivated by their work in industry, problems which do not necessarily have known solutions. Features peculiar to these problems, and seen as desirable by the researchers, are complexity and openness to the modeller's viewpoint. Evaluation of various modelling events conducted in Hamburg has allowed the researchers to conclude that these tasks are suitable for successful completion in an autonomous manner by upper secondary students (aged 16–19 years) of different mathematical performance levels in the context of schooling in Hamburg (see Kaiser and Schwarz 2006, 2010; Kaland et al. 2010). In early work evaluating extra-curricular modelling activities conducted from 2001 to 2004, Kaiser and Schwarz (2006) observed that "motivation of the students often increased when the problems were part of their own life experiences or if a practical purpose could be identified directly. Also if the students were asked to develop problems autonomously, motivation increased" (p. 205). Before undertaking the modelling activity, the majority of students professed a static view of what constitutes mathematics but after the event a more application oriented view was evident (p. 207). Working in groups allowed them to cope better with the "uncertainty which is characteristic of modelling activities", produce more solution attempts, and learn new ways of working in mathematics (p. 207).

Bracke and Geiger (2011) report on an annual mathematical modelling week for secondary students held since 1993 as well as modelling days for secondary students in Years 7–13 and more recently primary students (Years 2–5). As mathematical modelling has become more prominent in local curricula, two questions of relevance they investigated with Year 9 students (14 years old) are: Was experiencing a program of five modelling tasks in regular lessons advantageous over students without this experience when it came to solving a single modelling task as happens in one of these modelling events? What kind of attitude do these naïve modellers develop towards modelling during a single event? When reports from the students were compared, the experienced modellers were superior in finding a solution and quality of documentation. However, although the experienced modellers finished more quickly, the naïve modellers produced deeper solutions on average. The self-reported attitude of the experienced modellers towards modelling was, not surprisingly, "quite positive" and "better" than that of the naïve modellers (p. 548). The authors conclude that instead of short, one-off modelling events, "one should try to create a long-term modelling experience" throughout the existing curriculum (p. 548). On the other hand, modelling events allow deeper analysis than usually occurs in schools due to timetabling restrictions in the latter.

3 Structure and Context of the A. B. Paterson Modelling Challenge

The 2-day A. B. Paterson Modelling Challenge for Years 4–11 students has been held annually on the Gold Coast in Australia since 2003. Since its inception, one or more of the authors have been involved in mentoring modelling groups in the Years 10–11 level. Students in the Challenge at Year 10–11 level are asked to choose their own situations from which they pose their own questions to answer (see Galbraith et al. 2010). Students come from a variety of schools in Queensland and since 2009 some have also come from Singapore schools. Students are placed in mixed school groups by the organisers including students from both countries.

The Challenge begins for students in the Year 10–11 level with a presentation by the mentors on the nature of modelling during which the modelling cycle is presented as a scaffold and students are given a short modelling task (e.g., optimum location for a hospital) to tackle on the first morning. After morning tea student groups begin the process of choosing a situation of their own to model, posing a problem and generating questions to answer. They have access to the internet in computer laboratories as well as any materials they have brought with them. Subsequently, they begin modelling the situation in earnest. From this point on the mentor's role is mainly supervisory and to intervene as little as practicable. Sometimes there are a few iterations of this process before the group decides on their topic and approach. This is a challenging process but is seen as an essential ingredient of the Challenge by us as it separates modelling as real world problem solving from other educational "modelling" approaches. The final product from the group modelling is a poster report that is displayed publicly for parents and guests to view at the awards ceremony at the end of the second day. Just before taking their posters to be displayed each group gives a 5 min oral presentation of their problem and modelling solution to the mentor and the other teams in their class.

Curricula in Queensland schools actively promote investigation and modelling activity (QSA 2008, 2009) and have done so for many years. The rationale of the Queensland Mathematics B syllabus for Years 11 and 12, for example, requires that the mathematical content be used "to develop mathematical modelling and problem-solving strategies and skills" (QSA 2008, p. 1). Two of the global aims of this syllabus are the development of "the ability to recognise when problems are suitable for mathematical analysis and solution, and be able to attempt such analysis and solve problems with confidence" and the development of "positive attitudes to the learning and practice of mathematics" (p. 3). In Singapore, mathematical modelling is a more recent addition being added to the mathematics framework in the Secondary Mathematics Syllabus in 2006 (MOE 2006) for implementation in 2007 although extended investigations in the form of Interdisciplinary Project Work have been around since 2000 (Ng 2011). Similar aims with respect to developing positive attitudes to learning and using mathematics are expressed in the Singapore Syllabus.

As mentors at the Challenge, our intention is that students engage deeply with modelling a problem derived from a real world situation of their choice, work autonomously in teams to produce models in order to answer questions they pose about the situation and communicate and evaluate this solution via a poster and oral presentation. Many educators believe motivation significantly influences student engagement in learning and learning activities (Kasmer and Kim 2011). By motivation we mean "the inclination to do certain things and avoid doing some others" (Hannula 2006, p. 165). Brahier (2011) elaborates on the three dimensions of motivation according to Ford (1992): goals, emotions and self-efficacy. Motivation deriving from goals refers to an individual engaging "in an activity because some external reward or punishment is associated with it or because the person believes the activity is inherently useful and has a desire to learn from the experience" (p. 5). Goal orientation and performance orientation are different perspectives on why a person might engage in a task (particularly a challenging modelling task), persevere in mathematical exploration and tackle more challenging mathematics (Kasmer and Kim 2011). In the previous statement concerning goal motivation the first alternative is indicative of performance oriented motivation whilst the second indicates goal orientation. Our intention in the Challenge is for students to demonstrate the latter rather than the former. Emotional motivation "refers to how interest and curiosity can drive" the desire to learn about a topic and so engagement in a task (Brahier 2011, p. 5). This is one reason we allow freedom of choice in what to model. Self-efficacy involves the notion that personal agency beliefs allow students to feel capable of succeeding at a task so as to engage in it. As authentic modelling tasks often bring an initial sense of uncertainty and being at risk of failing to progress (see Kaiser and Schwarz 2006) or complete the task, positive personal agency beliefs are essential for students to persevere and enjoy the challenge of what they are doing during the 2 days of the Modelling Challenge.

An open question with regards to modelling is: What motivates students to choose a particular situation to model when the choice is theirs? Julie (2007) has pointed out that "the issues and contexts learners prefer for mathematics investigation is a largely under-researched area" (p. 193). Previously, from categorising the topics modelled by groups in the first years of the Challenge, we found that many of the situations chosen by the students have been from social or ecological contexts such as the spread of HIV/AIDS or water shortages, respectively (Galbraith et al. 2010).

Considering the foregoing, our focus question for our research was: Does the student experience from the student's perspective meet our intended purposes as mentors, namely the whole event be considered inherently valuable as a learning experience about modelling and application of mathematics to real situations?

4 Data Collection and Analysis

At the 2010 Challenge, we mentored 70 students in 19 groups of 3–4 students. Group membership was mixed with respect to schools, countries (from Australia or Singapore), gender and year level (10 or 11) as much as possible. Students completed a 12 item questionnaire over the 2 days of the Challenge. All questions were open and required a written response (see Appendix for examples). Questionnaires were completed as a group. In addition, student work during the challenge was photographed as a digital record, and the final presentations of several groups to the combined group were videorecorded. Final posters from several groups were also made available to the researchers. For the analysis each group's response to the questionnaire was categorised according to emergent themes in the data. In order to identify these themes, the researchers qualitatively analysed the questionnaire responses and used data from the other sources to clarify and confirm their classifying codes. The analysis involved intensive scrutiny of the data from a particular group and across the corpus of data from all groups to develop and refine categories related to these themes (Richards 2005). To gain a window into the students' experience of the Challenge we focused on their motivation in choosing a particular situation to model, their decision making for key points of progress in the event, perceived difficulties, their most interesting learning outcomes and their familiarity from their regular classroom with similar activities. Findings from a selection of the questions on the questionnaire are presented.

5 Findings

5.1 *Motivation for Choice of Modelling Contexts*

Students were asked for reasons for deciding on their choice of real world problematic situation to model. Categorised responses are shown in Table 19.1. Motivation for particular contexts was mainly goal oriented with many comments indicating the task was considered inherently useful and group members were interested in learning from their modelling experience (e.g., "We are in the twenty-first century, and where are we going in the future? Our population is something that needs to be seriously considered. The theory of Australia's population being 'out of control', is it true, is it real, is it the future?"). Interest and natural curiosity indicating emotional motivation did play a role in choice of context for some groups. For example, the Mathematical Mythbusters group had their natural curiosity piqued by Hollywood stunts so they modelled three to determine their feasibility.

Table 19.1 Reasons for choice of situation to model

Response category	Example responses
Pragmatic reasons	
Explored particular context previously	"One of us have done work on topic beforehand."
Strengths of the group members	"Most of us are very good at Physics."
Only viable idea considered	"All our other ideas failed."
Evokes curiosity or interest	"We considered many different models but combined two that were of interest to the group."
Currency of topic	"Currently featured in news → very topical"
Involves complexity or challenge	"Because it is an interesting and appropriately complex situation to model"
Serious societal issue	"Because oil is essential to power the world and produce many of the world's goods."
Personal issue or personally relevant	"Personally faced by us every day."
Significant problem	"There is a significant problem in current systems of timetabling in schools."
Mathematically tractable or interesting	"There's a lot of math behind how energy is made."

5.2 *Decision Making Re Choice of Approach and Content of Poster*

When questioned about their decision making re approach used, most groups wrote about examining the conditions in the real world situation resulting in an approach based on modelling the effect of the most critical variable (e.g., temperature for the Car Baby Safety Group) or taking a comparative approach (Queensland Population Group). Group processes such as brainstorming, discussion and division of labour underpinned decision making in four groups. Others used criteria such as efficiency and data availability or schemas such as "Practicality, Feasibility and Manageability" (Traffic Congestion Group). Attempts to simplify a complex situation determined the approach of two other groups.

As a culminating product from their modelling, student groups produced posters and presented orally. Decisions about content to include were subject to group discussion and consensus. A typical response was: "Group discussion and teamwork lead [sic] to agreeing on the important factors" (Queensland Population Group). The power of communicative devices was highlighted by other groups indicating final selection was based on providing supporting evidence, illustrating points and explaining thinking. Relevance and importance to the solution or predictions were criteria used by several groups. The "points the public need to know" was the overriding selection criterion for another group. Others wrote about imagining themselves in the situation to gain "a first hand feel as to which are the key aspects". All these responses are goal oriented. Only one group appeared performance oriented in their reporting using competitive considerations: "We decided to use the mathematics which best displayed our mathematical abilities…" (World Oil Group).

Table 19.2 Perceived difficulties with whole process

Response category	Example responses
Managing the task	
Developing an overall aim and focus	"We came up with a significant amount of ideas we could potentially have worked with, and it was quite a hard choice"
Keeping the aim in mind	"Keeping in mind the initial aim throughout the process."
Relating small scale investigations with the situation as a whole Changing approach	"Relating small scale investigations with the real world rocket forces and dimensions." "Having to change our approach half way through due to lack of knowledge and time."
Group processes Working in teams; collaborating	"Working together as each of us is different (in personality)" "making sure all group members were on the same page."
Gathering information/data	"Finding all the information"
Suitability/availability of data	"Finding appropriate data for oil consumption"
Consolidating the data	"Consolidating the data"
Formulation	"Formulating our ideas onto the graph"
Estimating quantities	"Finding a reasonable estimate of the world's supply of oil"
Understanding the concept globally	"Understanding the concept – arrow dynamics in archery"
Representing data in tables	"Making sure our tables were set up right – population data"
Model building Developing/understanding a formula for the model Linking factors	"Another difficult part was developing a formula to suit our model" "Understanding some of the formulas used to form the model." "Combining the equations for displacement with the influence of the drag force air resistance"
Model Solution Doing calculations or converting units	"Calculations – projectile motion" "Changing m^2 to km^2"
Interim goals Establishing/interpreting interim goals	"Trying to isolate theta." "Working out why time couldn't go below 0.5."
Verification	"Making sure our table's results correspond with the real data."

5.3 *Perceived Difficulties with Whole Process*

Students were asked about their perceptions of the most difficult aspects of participating in the Challenge. Several groups raised organisational difficulties associated with managing such an open task or managing group processes as they worked in teams (see Table 19.2 for examples). Gathering information or data about their chosen problematic situation was a concern for several groups. In some instances this was indicative of misunderstanding that in modelling not every datum value need be known, rather reasonable estimates of missing quantities can be made. Others experienced difficulties with different stages of the modelling process such as formulation, model building, model solution, establishing and interpreting interim goals and verification. Report writing, presenting to others or evaluation or refinement of models were not mentioned.

5.4 *Most Interesting Learning Outcomes from Involvement in the Event*

The majority of groups were most interested in the outcome of the problem(s) they had formulated in the chosen situation they were modelling or contextual details they had discovered during their investigation of this situation. For example, the group Mathematical Mythbusters discovered from their modelling of a movie stunt from Diehard IV where a car crashes into a hovering helicopter that: "It is possible to get a car to crash into a helicopter although doing it is suicidal."

Meta-aspects of engaging in modelling activity were seen as the most interesting outcome by several student groups. Participation in the Challenge highlighted the importance of domain knowledge for the Archery Group who pointed out "that, in order to complete a modelling and problem solving problem, there needs to be a large investigation into the topic." Others were intrigued by the number of factors contributing to a particular situation needing consideration. The Tennis Serve Group commented: "There were more factors than expected involved in determining time, angle and velocity of a tennis serve". The complexity of modelling real world situations was of most interest to three groups. The Emergency Response Group, for instance, was surprised by "the complexities involved in scheduling" which they were using to model responses of emergency vehicles to an accident. They had great difficulty simplifying a realistic beginning scenario to become tractable to their mathematical expertise. Engaging in the Challenge brought the utility of mathematics to the attention of others who commented on its almost universal applicability and usefulness. The Traffic Congestion Group, for example, noted: "Mathematics can be applied (almost everywhere)"; whilst the Australian Population Group marvelled how "maths could unlock complicated future scenarios such as population".

Students in three groups noted as their most interesting learning outcome the new mathematics they had learnt from other group members with such knowledge, from internet sources or with assistance of mentors. These included how to calculate velocity, projectile motion, Leslie matrices, solving equations simultaneously to find points of intersection and the relationship between velocity and air resistance, particularly drag coefficients and the velocity squared relationship. Perhaps, in response to the difficulties they had with the openness of choice of task, members of the Energy Efficiency Group commented that they had learnt the focusing nature of a task, commenting: "It's easier to work when you have a task" as finding an idea to investigate in the first place was noted as their most difficult challenge.

5.5 *Frequency of Open Modelling Activities in School*

Even though modelling has been a part of upper secondary school curricula in Queensland for some time, this does not mean all students have necessarily experienced extended modelling tasks in the classroom particularly where the choice of what to model is theirs. The students in the Challenge were no exception with some members of six groups indicating they had never done this and of another eight groups, rarely. As the group modelling the effects of alcohol on stopping distance when driving a car pointed out: "Most of the time we're given problems to solve instead of us having to find a problem and solve it." A member of the Cinema Profit Group noted that, "In Singapore, little to no such lessons are conducted due to the rigorous and demanding curriculum which the students are undertaking." However, the Singaporean Year 10 student from the Tennis Serve Group wrote that these activities were done "very often" in her school and that she had done such activities over the "whole of this year. It is called project work in Singapore." The Singaporean members of the Car Baby Safety Group had done such activities "not very often" adding, "This maths is much more practical to the real world."

6 Discussion and Conclusions

The format of the Year 10–11 part of the Challenge corresponds with elements found by other researchers (e.g., Kaiser and Schwarz 2006, 2010) in similar extra curricular modelling events as desirable for increased student motivation such as complexity, being able to develop problems autonomously and working in groups. Similarly, it had the potential to raise awareness of an application-oriented view of mathematics by the nature of the tasks students chose and worked on. Having opportunities to recognise problems in real situations suitable for mathematical analysis and being confident enough to attempt such analysis with the support of group members was in alignment with the local curriculum documents in the educational systems from which students came.

Student choice of situation to model was motivated, in the main, by a goal oriented perspective (Kasmer and Kim 2011). Indicators of emotions being involved in task engagement were also present in the form of natural curiosity and interest driving choice of task and perseverance in quite challenging and complex tasks. Decision making about approach to adopt in tackling modelling of their chosen situation and reporting results was also mainly goal oriented although one group was clearly performance oriented in reporting results (encouraged by the advertised competitive nature of the Challenge). Students perceived as most difficult about the Challenge managing such an open task or managing group processes as they worked in teams. It became evident in responses to other questions that students who had experienced modelling tasks in the classroom might not have tackled such open problems. Others had not experienced similar activities at all and certainly had not worked in teams in mathematics previously. Gathering information or data about their chosen situation was a concern for several groups; whilst others experienced various difficulties with different stages of the modelling process.

Students were most interested in the outcomes of their modelling indicating they found the whole experience intrinsically interesting. Meta-aspects of engaging in modelling activity were nominated as their most interesting learning outcome by several groups with complexity of modelling real situations and the importance of domain knowledge and factors to be considered coming to the fore. There was evidence that students became more acutely aware of an application-oriented view of mathematics. As in previous events (see Kaiser and Schwarz 2006), students reported learning new mathematics mainly facilitated by working with students from different school contexts. Even though the vast majority reported having not participated in similar school activities, all groups were able to choose a situation to model and engage in attempts to model it over the 2 days to an extent that enabled them to report increased understanding of the complexity of the situation and to describe and critique their attempts at mathematical analysis of self initiated questions about the situation. In terms of student experience, our intended purposes as mentors were met in that the whole event was considered by most as inherently valuable as a learning experience about modelling and application of mathematics to real situations.

Appendix – Relevant Questionnaire Questions

Q 1: Why did you decide on this particular situation to model?
Q 3: How did you come to choose the approach you adopted?
Q 7: How did you decide on what was most important to include in your report?
Q10: What were the most interesting things you learned by being involved in the Modelling Challenge?
Q11: What were the most difficult parts of the whole process?
Q12: How often would you have done similar types of activities in your mathematics classroom at school?

References

Bracke, M., & Geiger, A. (2011). Real-world modelling in regular lessons: A long term experiment. In G. Kaiser, W. Blum, R. Borromeo Ferri, & G. Stillman (Eds.), *Trends in teaching and learning of mathematical modelling* (pp. 529–549). New York: Springer.

Brahier, D. J. (2011). Motivation and disposition: The hidden curriculum. In D. J. Brahier & W. R. Speer (Eds.), *Motivation and disposition: Pathways to learning mathematics* (pp. 1–8). Reston, VA: NCTM.

Brown, R., & Redmond, T. (2007). Collective argumentation and modelling mathematics practices outside the classroom. In J. Watson & K. Beswick (Eds.), *Mathematics: Essential research, essential practice* (Proceedings of MERGA30, Hobart, Vol. 1, pp. 163–171). Adelaide: MERGA.

Ford, M. E. (1992). *Motivating humans: Goals, emotions and personal agency beliefs*. Newbury Park, CA: Sage.

Galbraith, P., Stillman, G., & Brown, J. (2010). Turning ideas into modeling problems. In R. Lesh, P. L. Galbraith, C. R. Haines, & A. Hurford (Eds.), *Modeling students' mathematical modelling competencies* (pp. 133–144). New York: Springer.

Göttlich, S. (2010). Modelling with students – A practical approach. In A. Araújo, A. Fernandes, A. Azevedo, & J. F. Rodrigues (Eds.), *Conference proceedings of EIMI 2010* (pp. 235–242). Lisbon/Bedford: Centro Internacional de Matemática/COMAP.

Hannula, M. S. (2006). Motivation in mathematics: Goals reflected in emotions. *Educational Studies in Mathematics, 63*(2), 165–178.

Julie, C. (2007). Learners' context preferences and mathematical literacy. In C. Haines, P. Galbraith, W. Blum, & S. Khan (Eds.), *Mathematical modeling: Education, engineering and economics* (pp. 195–202). Chichester: Horwood.

Kaiser, G., & Schwarz, B. (2006). Mathematical modelling as bridge between school and university. *ZDM, 38*(2), 196–208.

Kaiser, G., & Schwarz, B. (2010). Authentic modelling problems in mathematics education – Examples and experiences. *Journal für Mathematik-Didaktik, 31*(1), 51–76.

Kaiser, G., Schwarz, B., & Buchholtz, N. (2011). Authentic modelling problems in mathematics education. In G. Kaiser, W. Blum, R. Borromeo Ferri, & G. Stillman (Eds.), *Trends in teaching and learning of mathematical modelling* (pp. 591–601). New York: Springer.

Kaland, C., Kaiser, G., Ortlieb, C. P., & Struckmeier, J. (2010). Authentic modelling problems in mathematics education. In A. Araújo, A. Fernandes, A. Azevedo, & J. F. Rodrigues (Eds.), *Conference proceedings of EIMI 2010* (pp. 321–332). Lisbon/Bedford: Centro Internacional de Matemática/COMAP.

Kasmer, L., & Kim, O.-K. (2011). Using prediction to motivate personal investment in problem solving. In D. J. Brahier & W. R. Speer (Eds.), *Motivation and disposition: Pathways to learning mathematics* (pp. 157–168). Reston, VA: NCTM.

Lee, N. H., & Ng, K. E. D. (Eds.) (in press). *Mathematical modelling – From theory to practice.* Singapore: World Scientific.

Ministry of Education (MOE). (2006). *Secondary mathematics syllabus*. Singapore: Ministry of Education.

Ng, K. E. D. (2011). Mathematical knowledge application and student difficulties in a design-based interdisciplinary project. In G. Kaiser, W. Blum, R. Borromeo Ferri, & G. Stillman (Eds.), *Trends in teaching and learning of mathematical modelling* (pp. 107–116). New York: Springer.

Queensland Schools Authority (QSA). (2008). *Mathematics B senior syllabus*. Brisbane: Author.

Queensland Schools Authority (QSA). (2009). *Year 10 guidelines: Mathematics*. Brisbane: Author.

Richards, L. (2005). *Handling qualitative data: A practical guide*. London: Sage.

Chapter 20
Mathematical Modelling of a Real-World Problem: The Decreasing Number of Bluefin Tuna

Akira Yanagimoto and Noboru Yoshimura

Abstract Nature and the constructed world in which students live have provided suitable sources of tasks for junior high school students in the past. The decreasing number of bluefin tuna was considered another appropriate context. Students were asked to provide a mathematical solution to this problem, utilising data on the number of bluefin and the catch of bluefin in the past. Three different mathematical approaches, appropriate to the educational level of the students involved, were taken to considering the decline in bluefin tuna. Year 7 students used a proportional model, Year 8 students used linear function models, and in Year 9 students used recurring formula models. Each set of models was appropriate to the mathematics in the curricula at the level. The aim of this mathematical modelling activity was to develop people with the skills to lead the way in our society.

1 Introduction

Problems in the natural world (Matsumiya et al. 1989; Yanagimoto 2003) and built environment (Yanagimoto and Yoshimura 2001) have been the stimuli for tasks for junior high school students in the past. The decreasing number of bluefin is another such problem and was the subject matter for teaching material for mathematical modelling. An educational experiment for Japanese junior high school students was conducted using this context. The consumption of tuna has increased internationally, and

A. Yanagimoto (✉)
Faculty of Education, Kyoto University of Education, Kyoto, Japan
e-mail: ayanagi@kyokyo-u.ac.jp

N. Yoshimura
Faculty of Education, Tennoji J.H.S. attached to Osaka Kyoiku University, Osaka, Japan
e-mail: noboru@cc.osaka-kyoiku.ac.jp

G.A. Stillman et al. (eds.), *Teaching Mathematical Modelling: Connecting to Research and Practice*, International Perspectives on the Teaching and Learning of Mathematical Modelling, DOI 10.1007/978-94-007-6540-5_20,

the number of existing bluefin has decreased drastically, resulting in an international problem. According to an estimate by the science committee of the International Commission for the Conservation of Atlantic Tunas (ICCAT, see www.iccat.int), the volume of bluefin inhabiting the Mediterranean Sea and the Atlantic Ocean has decreased from a peak figure of about 300,000 tons in 1974 to the present level of a little under 80,000 tons. Each ICCAT participant country agreed to reduce the bluefin tuna fishery frames for the Mediterranean Sea and the Atlantic Ocean in 2010 to 13,500 tons, approximately a 40 % decrease in comparison with 2009, and took various other measures. However, according to the estimation of the science committee, the bluefin tuna of the area are subject to indiscriminate fishing. For example, in 2007, there were fish catches of 61,000 tons, even though the fishery frames of 2007 allowed for only 29,500 tons. Environmentalist groups may put strong pressure on the government of every country to work on bluefin tuna protection, and Monaco proposed an embargo on the catching of bluefin tuna. The Monaco proposal prohibits international transactions by the Washington Convention. This proposal was later voted on at a Washington Convention international conference held in Doha, Qatar, in March 2010 (Black 2010).

Bluefin tuna consumption in Japan, at approximately 43,000 tons (2008), accounts for approximately 80 % of the world consumption and almost 40 %, approximately 17,000 tons, are Atlantic bluefin tunas that are imported along with other bluefin tunas for consumption in Japan. Therefore the Monaco proposal was a very serious matter for Japan. Early in 2010 this news was frequently reported on TV and in newspapers in Japan. Many Japanese were interested in this news.

"Essential characterisations of modelling and applications involve posing and solving problems located in the real world, which…includes…general contexts of living as they impact on individuals, groups and communities" (Niss et al. 2007, p. 17). To this end, using the context just described, the authors created teaching materials focused on mathematical modelling that aimed to allow students to predict the future survival quantity of bluefin tuna. This was based on the quantity of past bluefin tuna survival and the data of fish catches. We decided to implement this as lessons for junior high school students across three different year levels, 7, 8, and 9, at one school. The aim of this mathematical modelling activity was to develop students with the skills to lead the way in our society in keeping with one of the goals of modelling, namely "to provide experiences…that contribute to education for life after school…enhancing the quality of life" (Niss et al. 2007, p. 19), in this instance "ensuring…environmental well being" (p. 18).

2 Teaching Practices

2.1 The Proportion Model for Year 7 Students

In Year 7, the task was implemented with 152 students, in four classes, from 'T' national junior high school. Two 1-h sessions were allocated to the task on different days in March 2010. The teacher was the usual teacher, Kenichi Taketoshi. The aim

Table 20.1 Catches of bluefin tuna and the survival quantity in Atlantic area (1974–2009)

Year	Fish catches (10,000 tons)	Quantity of survival (10,000 tons)	Year	Fish catches (10,000 tons)	Quantity of survival (10,000 tons)
1973			1992	3.5	
1974	2.4	30	1993	3.7	
1975	2.7		1994	4.9	
1976	2.8		1995	5.0	
1977	2.4		1996	5.3	
1978	2.0		1997	4.9	
1979	1.9		1998	3.9	
1980	2.0		1999	3.5	
1981	2.0		2000	3.7	
1982	2.5		2001	3.7	
1983	2.4		2002	3.7	
1984	2.7		2003	3.4	
1985	2.4		2004	3.3	
1986	2.1		2005	3.8	
1987	2.0		2006	3.2	
1988	2.7		2007	3.4	
1989	2.4		2008	–	
1990	2.6		2009	–	8
1991	3.0		2010		

of the implementation was (a) training students in mathematical modelling, and (b) making students aware of the problem of the decreasing number of bluefin tuna.

The first period: Students were directed to study the issue of bluefin tuna beforehand. Subsequently, several students were asked to present what they had learned. They looked at the species of the tuna, world consumption levels, fishing grounds, the problem of the extinction crisis, and general information about bluefin tuna including the news of the international conference. The teacher distributed to students an article from a recent newspaper and documents explaining the various kinds of tuna.

Table 20.1, which shows fish catches of bluefin tuna and the number of bluefin tuna in existence, was distributed to the students. It was confirmed that the number of bluefin tuna inhabiting the Atlantic decreased from approximately 300,000 tons in 1974 to approximately 80,000 tons in 2009, according to the International Commission for the Conservation of Atlantic Tunas (ICCAT). Data in Table 20.1 were taken from a graph (Organization for the Promotion of Responsible Tuna Fisheries 2006) (http://www.oprt.or.jp/c27.htm).

Next, students drew a rough graph of bluefin tuna fish catches in the Atlantic using the data from Table 20.1. The aim was to help students grasp the numerical change in the list. Finally, students worked on posing various mathematical problems from the table and the graph about the number of bluefin tuna that have survived, and tried to answer one of the problems by themselves. This was intended to be an incentive to study the next learning content, and also to examine their problem-posing abilities.

Table 20.2 Catches of bluefin tuna and survival quantity in Atlantic area (1974–2020)

Year	Fish catches (10,000 tons)	Quantity of survival (10,000 tons)	Year	Fish catches (10,000 tons)	Quantity of survival (10,000 tons)
1973			2008	–	
1974		30	2009	–	8
•	•	•	2010		
•	•	•	2011		
•	•	•	2012		
2000	3.7	?	2013		
2001	3.7		2014		
2002	3.7		2015		
2003	3.4		2016		
2004	3.3		2017		
2005	3.8		2018		
2006	3.2		2019		
2007	3.4		2020		

The second period: Table 20.2 showing fish catches of bluefin tuna and the number of bluefin tuna in existence and a learning handout were distributed to the students. The following problems were posed: When is the extinction time? Can we recover the number of bluefin tuna? Students attempted to solve the problems by using proportion under the teacher's guidance as follows.

Firstly, students estimated the quantity of bluefin tuna from 2000 surviving in 2009. The data showed that over 35 years 1974–2009, the quantity of fish reduced by 22 ten thousand tons. To estimate the proportion surviving in 2000, assume the decrease is constant and hence:

$30-8=22$ (10,000 tons), $2009-1974=35$, $2000-1974=26$ (years)
$22 \div 35 \approx 0.6286 \approx 0.63$: yearly average decrease over the 35 years 1974–2009
$0.6286 \times 26 \approx 16.3$ (10,000 tons): the decrement from 1974 to 2000
$30-16.3=13.7$ (10,000 tons): the quantity of bluefin tuna of 2000

Secondly, the students defined variables x and y such that x represented the number of years since 2000 and y represented the decrease of bluefin tuna in 10,000 tons. They found $y=-0.63$ using a negative constant.

Thirdly, students estimated the survival quantity and predicted when extinction would occur. They undertook calculations using the proportional expression. They filled in the decrement of the bluefin tuna and the estimated quantity of surviving bluefin tuna from 2000 through 2020. They used electronic calculators, Sharp EL-N422-x, for the calculation. The number of bluefin tuna decreased steadily in this calculation. Consequently, it was predicted that they would become extinct in 2022.

Fourthly, students considered the yearly average fish catches. They were able to come up with a yearly average of 35,300 tons, based on the fish catches in Table 20.2 from 2000 through 2007. Fifthly, students considered fish catches to maintain the current bluefin tuna population. They determined that the annual fish catches must be $3.53-0.63=2.9$ (10,000 tons) if the current bluefin tuna population is to be maintained.

Fig. 20.1 Estimation from Table 20.2

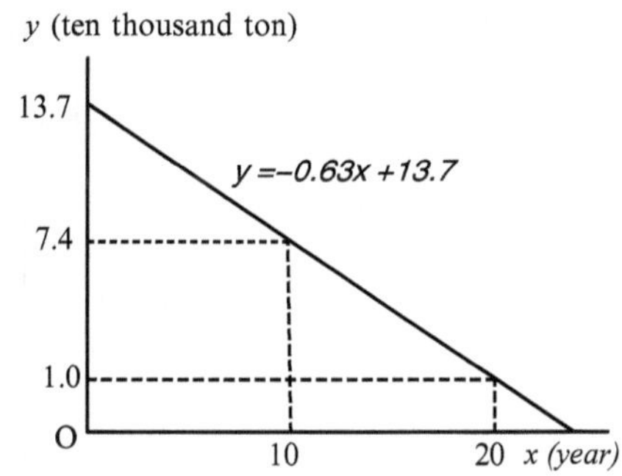

Lastly, students thought about the proportional expression of the increment supposing fish catches to be zero. If we suppose the fish catches are 0 tons from 2010, it is $y = 2.9x$, where y is expressed in an expression of x as 10,000 y tons with the increment of bluefin tuna after x years from 2010. In this case, when does the number of bluefin tunas recover to 30 tons of 1974? They performed the calculations using the proportional expression $y = 2.9x$, and filled in the increment of bluefin tuna and the quantity of estimated survival from 2010 through 2020 in the table. It was predicted that the number of bluefin tunas increased steadily and the amount of the increase from 2010 to 2017 was $8 + 2.9 \times 8 = 31.2$ (10,000 tons) that is more than 300,000 tons. Finally, they looked back on the solving process and the overall problem-solving task; but there was not enough time to complete it.

2.2 *The Linear Function Model for Year 8 Students*

In Year 8, the task was implemented with 151 students, in four classes, from 'T' national junior high school. Two hours were allocated to the task on different days in March 2010. The teacher was the usual teacher, Noboru Yoshimura. The aim of the implementation was the same as in Year 7. There was nothing the students had to do beforehand, but the rest of the first period of the lesson was similar to Year 7. Table 20.2 and a learning handout were distributed to the students in the second period. Students did the calculation using the linear function as follows.

Firstly, students estimated the quantity of bluefin tuna of 2000 as in Year 7. Then they thought about the linear function of the quantity of surviving bluefin tuna. Students expressed y as a function of x in 10,000 y tons for the quantity of surviving bluefin tuna after x years from 2000. The relationship the students determined in class was $y = -0.63x + 13.7$, where 13.7 is the quantity of bluefin tuna in 2000.

Secondly, students undertook calculations using the linear function formula and completed the quantity of estimated surviving bluefin tuna from 2000 through 2020. They used the same calculators for the calculation. The results were the same as for Year 7. In addition, they drew a linear function graph and calculated the value x when $x = 0$ (see Fig. 20.1). Yearly average fish catches and the fish catches to maintain the current survival rate were the same as for Year 7 too.

Fig. 20.2 Model where no catch from 2010

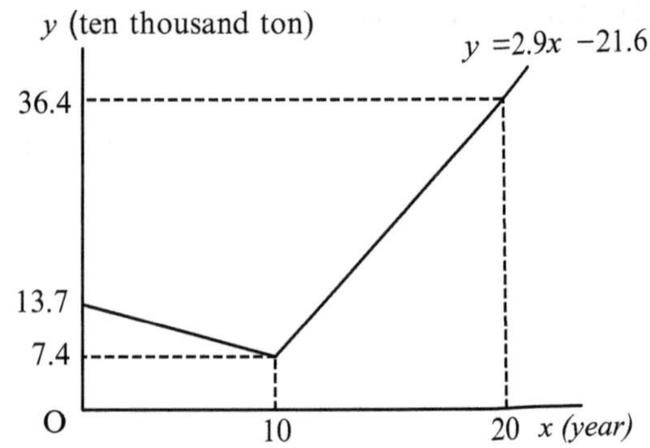

Thirdly, students thought about the linear function of the quantity of survival of bluefin tuna supposing fish catches to be zero. If we suppose that fish catches are 0 tons from 2010, what is the linear function? It becomes $y = 2.9x - 21.6$, where y is expressed in terms of x as 10,000 y tons for the quantity of surviving bluefin tuna after x years from 2000.

Fourthly, students calculated using the linear function formula $y = 2.9x - 21.6$, to complete the quantities of estimated survival of bluefin tuna from 2010 through 2020 of the table. In this case, it was predicted that the number of bluefin tuna increased steadily and the quantity of surviving bluefin tuna was more than 300,000 tons in 2018. In addition they drew a linear function graph and confirmed the state of the increase and decrease of this function (see Fig. 20.2).

Finally, validating and interpreting the results was discussed. It was confirmed that there was an average decrease of 6,300 tons a year, and a natural increase of 29,000 tons a year, as the basis of the said calculation. However, we did not have time to discuss whether this prediction was valid or not.

2.3 *The Recurring Formula Model for Year 9 Students*

In Year 9, the task was implemented with 157 students, in four classes, from 'T' national junior high school. Two hours were allocated to the task on different days in March 2010. The teacher was the usual teacher, Noboru Yoshimura, who was the same teacher as for Year 8. The aim of the implementation was the same as for Year 7. Again, there was nothing the students had to do beforehand, and the rest of the content was similar to the Year 7 implementation. Table 20.2 and a learning handout were distributed to the students in the second period. Students did the modelling calculation by using the recurring formula as follows:

First, similar to the Year 7 students, the Year 9 students found that the quantity of the decrease in bluefin tuna is 6,300 tons a year on average, and that estimated quantity of surviving bluefin tuna in 2000 is calculated to be approximately 136,000 tons, and that the average catch is 35,300 tons a year, calculating from fish catches in Table 20.2 from 2000 through 2007.

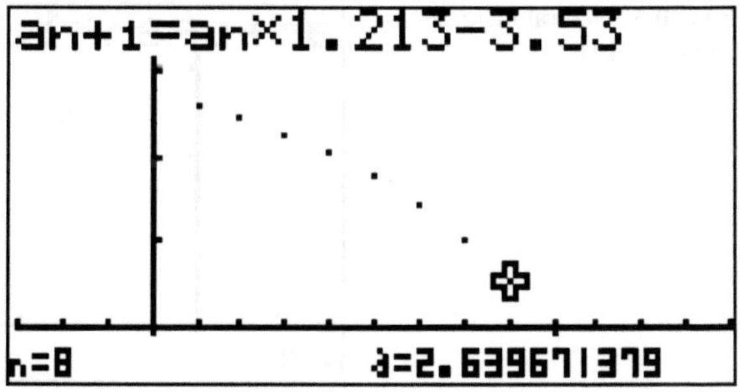

Fig. 20.3 Natural increase rate of 21.3 %

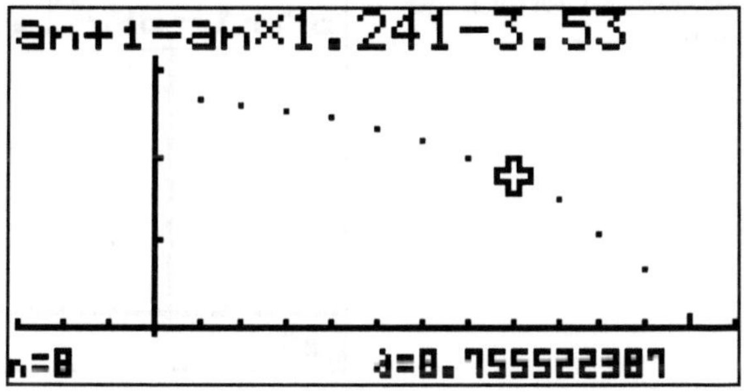

Fig. 20.4 Natural increase rate of 24.1 %

Secondly, students calculated the natural average yearly increase to be 29,000 tons, using the figures 6,300 tons, the yearly average decrease, and 35,300 tons, the yearly average fish catch. They divided this by 136,000 tons, the estimated surviving bluefin tuna in 2000. Thus, they came up with an approximate figure of 21.3 % as the natural increase rate of bluefin tuna.

Thirdly, students calculated the quantity of surviving bluefin tuna in 2001, using the survival quantity of 136,000 tons in 2000, natural increase rate of 21.3 %, and the average fish catch of 35,300 tons. That is, $13.6 \times 1.213 - 3.53 \cong 13.0$ (10,000 tons), which is 130,000 tons. Similarly, $13.0 \times 1.213 - 3.53 \cong 12.2$ (10,000 tons). So the quantity of surviving bluefin tuna in 2002 was 122,000 tons.

Fourthly, students continued doing this recurring formula calculation. For this calculation they used COMP menu of a CASIO fx-9700GE graphic calculators and found that the number of bluefin tuna decreased steadily in this calculation. It was predicted that the quantity of bluefin tuna would drop by 26,396 tons in 2008, to –0.328 tons in 2009, which would mean extinction. Here they used the iteration calculation of COMP menu, not the recursion mode of TABLE menu. We can easily see the change in the graph in the recursion mode of the graphic calculator menu (see Fig. 20.3).

Fifthly, suppose the quantity of surviving bluefin tuna in 2000 is 136,000 tons, and the annual fish catch 35,300 tons. Then how large a natural increase rate in percent would be needed for the quantity of surviving bluefin tuna in 2009 to become approximately 80,000 tons? Students successfully determined the increase rate by trial and error with a graphic calculator. It was approximately 24.1 %. Here they used the recursion mode of the graphic calculator (see Figs 20.4 and 20.5).

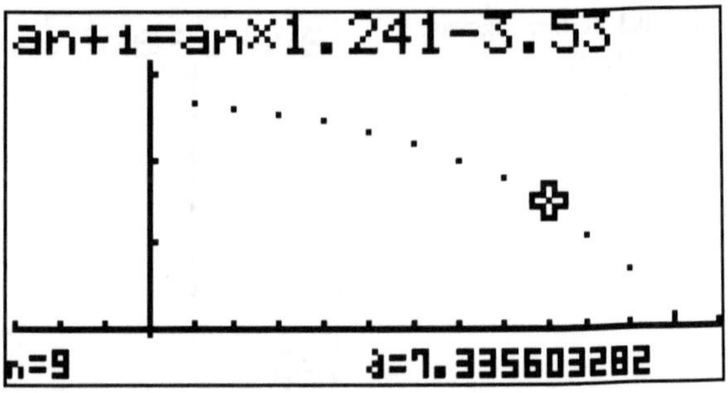

Fig. 20.5 Natural increase rate of 24.1 %

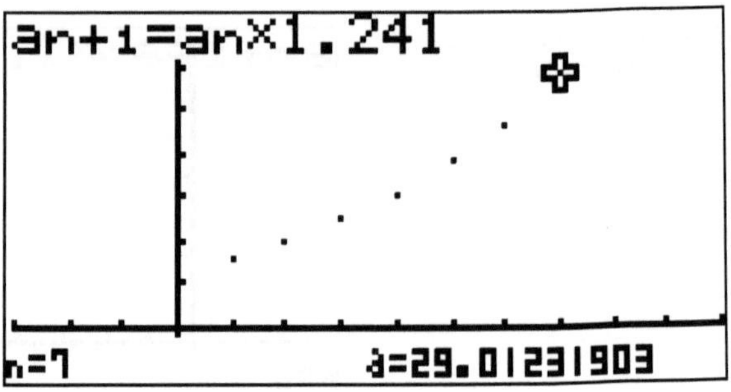

Fig. 20.6 No fish catch after 2010

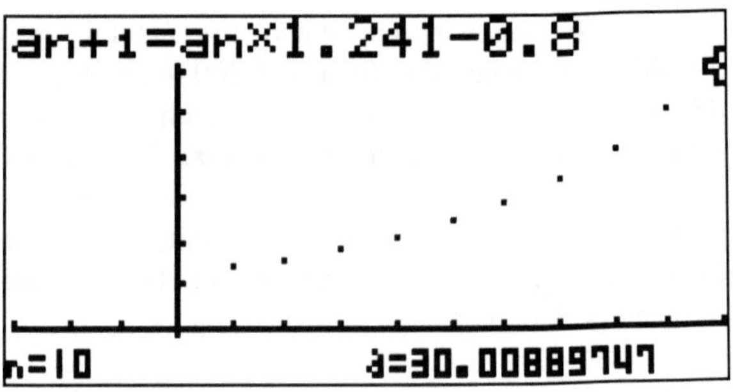

Fig. 20.7 0.8 tons fish catch after 2010

Sixthly, students calculated the quantity of surviving bluefin tuna again, using the natural increase rate of 24.1 %, and wrote in the quantity of estimated survival bluefin tuna from 2000 through 2015 on their worksheets. The bluefin tuna would become extinct in 2013 in this case (see Fig. 20.5). In addition, they copied the rough graph by hand onto their worksheets from the screen of the graphic calculator.

Seventhly, supposing that there will be no fish catch after 2010, it was calculated that in 2017 the quantity of surviving bluefin tuna would recover to the level of 1974: 300,000 tons ($a_0 = 6.4$ in 2010, $a_{n+1} = a_n \times 1.241$, then, $a_7 = 29.0$, see Fig. 20.6).

Eighthly, it was assumed that the quantity of surviving bluefin tuna would recover to 300,000 tons in 2020. The question posed was: What would the fish catch be in this case? Students tried to figure out the quantity of the fish catch by trial and error with a graphic calculator, and succeeded. It was approximately 8,000 tons (see Fig. 20.7).

Finally, we come to validating and interpreting the results. Though there was not enough time for students to think about it thoroughly, students were able to see that the calculations undertaken were based on the present quantity of surviving bluefin tuna, the quantity of fish catches, the natural increase rate, etcetera.

3 Evaluation of Student Performance

As schooling year level increases, many Japanese students are not interested in mathematics. In addition, they feel that mathematics is not useful for them. These are serious situations in mathematics education in Japan. Although students of 'T' national junior high school are higher level, and they are more interested in mathematics than general students, they show the same tendency. The implementation of mathematical modelling and applications, as discussed in this chapter, are effective in changing this situation as is illustrated below.

3.1 *Mathematics in Junior High School*

In this teaching sequence Japanese junior high school students were taught about the decreasing number of bluefin tuna using mathematical modelling materials. An assessment of students' attitude towards mathematics was carried out by administering a pre-test before teaching and a post-test after teaching. The tests included the following question: When you think about real world problems, do you think that mathematics is helpful? The answer to the question was selected from: Very much; A little; Not so much; Not at all. The results are shown in Table 20.3.

Across all year levels students were able to recognize the usefulness of mathematics through these lesson sequences. The results reveal that the number of students who answered very much or a little increased from pre- to post-test for students for all year levels (from 86 to 91 % for Year 7 students, from 74 to 89 % for Year 8 students, and from 66 to 84 % for Year 9 students). This confirms more students were able to recognise the usefulness of mathematics after this lesson sequence.

Table 20.3 Student recognition of the usefulness of mathematics (%)

			Response (%)			
Year level	N	Test	Very much	A little	Not so much	Not at all
7		Pre	26	60	12	2
	152	Post	44	47	9	0
8		Pre	23	51	22	4
	151	Post	38	51	9	2
9		Pre	20	46	30	4
	157	Post	29	55	14	2

Table 20.4 Student interest in context (%)

	Response (%)			
Year level	Very much	A little	Not so much	Not at all
7	23	62	11	4
8	22	54	17	7
9	26	53	14	7

3.2 Students' Interest in This Lesson

The problem of the decreasing bluefin tuna was the stimulus for this lesson sequence, because it is an important international environmental problem in our world. The assessment of student interest in this topic was done by administering a post-test after the classes were conducted. The students were asked the following question: Were you interested in this learning content of calculating the survival quantity of bluefin tuna? The answer to the question was chosen from the following selections: Very much; A little; Not so much; Not at all. The results are shown in Table 20.4.

About 80 % of the students in each year level chose answers very much or a little, showing a interest in this learning context. This result is higher than the results from lessons researched previously (e.g., Yanagimoto et al. 1993; Yanagimoto and Yoshimura 2001). However, these lesson sequences were implemented when the topic used was timely in the world.

4 Conclusion

Firstly, the experimental lessons have shown it is possible to use mathematical modelling teaching materials for "posing and solving problems located in the real-world" (Niss et al. 2007, p. 17) in this case dealing with the declining bluefin tuna population for instructing junior high school students. For Years 7 and 8, it is appropriately placed in the mathematics curriculum in Japan. Students learn the proportion function in Year 7, and linear functions in Year 8, so this content is easy to treat as applied mathematics teaching materials at the end of the topic. For Year 9, the recurring formula calculation is part of the content of the Year 10 mathematics curriculum in Japan, but Year 9 students can treat it appropriately by using graphic calculators including the iteration and recursion calculations. It is suggested that these teaching materials are effective in large classes as a teaching material for the introduction period of mathematical modelling. However, strategies such as group learning, "typical for learning environments where modelling is required" (Niss et al. 2007), might be necessary for the more advanced stages of mathematical modelling. Generally there is no placement of these kinds of tasks in the Japanese mathematics curriculum. In this implementation, students had to solve the problem with a mathematical model given by the teacher, as they had little previous experience of mathematical modelling activities.

Importantly, the students' perception of mathematics changed for the better as a result of this educational experiment. When asked, students who answered that *mathematics is useful for solving real-life problems* accounted for 91 % of the Year 7, 89 % of the Year 8, and 84 % of the Year 9 students. Students who thought *this teaching material was interesting* accounted for 85 % of Year 7, 76 % of Year 8, and 79 % of the Year 9 students. Our future challenge is to promote the development of mathematical modelling teaching materials from the viewpoint of nurturing future citizens in Japanese society. The aim of this mathematical modelling activity is to develop leaders in our society similar to one of the goals of using applications and modelling in school curricula (Niss et al. 2007).

References

Black, R. (2010, March 18). Bluefin tuna ban proposal meets rejection. *BBC News*. Retrieved from http://news.bbc.co.uk

Matsumiya, T., Yanagimoto, A., & Mori, Y. (1989). Mathematics of a lake – Problem solving in the real world. In W. Blum, M. Niss, & I. Huntley (Eds.), *Modelling, applications and applied problem solving* (pp. 87–97). Chichester: Ellis Horwood.

Niss, M., Blum, W., & Galbraith, P. (2007). Introduction. In W. Blum, P. L. Galbraith, H.-W. Henn, & M. Niss (Eds.), *Modelling and applications in mathematics education: The 14th ICMI study* (pp. 3–32). New York: Springer.

Organization for the Promotion of Responsible Tuna Fisheries. (2006). *A graph of fish catches of Bluefin tuna*. Retrieved from http://www.oprt.or.jp/c27.htm

Yanagimoto, A. (2003). Environmental problems and mathematical modelling. In S. J. Lamon, W. A. Parker, & S. K. Houston (Eds.), *Mathematical modelling: A way of life, ICTMA11* (pp. 53–60). Chichester: Ellis Horwood.

Yanagimoto, A., & Yoshimura, N. (2001). Mathematics of traffic safety – Composite real mathematics approach. In J. F. Matos, W. Blum, S. K. Houston, & S. P. Carreira (Eds.), *Modelling and mathematics education – ICTMA9* (pp. 401–410). Chichester: Ellis Horwood.

Yanagimoto, A., Mori, Y., & Matsumiya, T. (1993). Mathematics of track and field events – Composite real mathematics approach. In T. Breiteig, I. Huntley, & G. Kaiser-Messmer (Eds.), *Teaching and learning mathematics in context* (pp. 115–126). Chichester: Ellis Horwood.

References

Chapter 21
Mathematical Modelling of a Social Problem: Pension Tax Issues

Noboru Yoshimura and Akira Yanagimoto

Abstract Problems in students' social world, the natural world and built environment have proved to be fruitful sources of tasks for junior high school students in the past. Pension tax issues were taken up as a teaching material for mathematical modelling, and an educational experiment for junior high school students was conducted. Students were asked to think about the mathematical solution of the prefecture tax rate on the basis of the data about Osaka Prefecture. A proportion model for the 7th grade students and a linear function model for the 9th grade students were used in this experiment. As a result of the teaching experiment, the percentage of students who felt that mathematics was useful for solving real-world problems increased. Students showed high interest in the task context but only 50–60 % indicated a desire to find other solutions than the first one.

1 Introduction

Problems in students' social world (Yanagimoto et al. 1993) the natural world (Matsumiya et al. 1989; Yanagimoto 2003) and built environment (Yanagimoto and Yoshimura 2001) have proved to be fruitful sources of tasks for junior high school students in the past. Pension tax issues were considered to be a prime candidate for mathematical modelling. It is important to have a pension so that at the age of retirement you will have enough money coming in each month to cover

N. Yoshimura (✉)
Faculty of Education, Tennoji J.H.S. attached to Osaka Kyoiku University, Osaka, Japan
e-mail: noboru@cc.osaka-kyoiku.ac.jp

A. Yanagimoto
Faculty of Education, Kyoto University of Education, Kyoto, Japan
e-mail: ayanagi@kyokyo-u.ac.jp

G.A. Stillman et al. (eds.), *Teaching Mathematical Modelling: Connecting to Research and Practice*, International Perspectives on the Teaching and Learning of Mathematical Modelling, DOI 10.1007/978-94-007-6540-5_21,

the cost of your bills and allow you to live a comfortable life. Currently, young people's lack of payment into the national pension funds is becoming a serious social problem. For years the Japanese nation has spent beyond its means. As a result, it has become difficult for the Japanese government to maintain the levels of social security such as healthcare and old-age pensions. Japan is at a fiscal tipping point. There is a need for the Japanese government to make drastic changes to the pension system. This issue is so perplexing that it is impossible to find an answer immediately.

In this situation, the original pension reform was deemed suitable as teaching material for mathematical modelling for junior high school students in years 7 and 9. This is intended to be a mathematical modelling activity that will develop people to lead the way in our society in keeping with one of the goals of applications and modelling and their curricular embedding (Niss et al. 2007).

2 Teaching Practices

2.1 The Proportion Model for 7th Grade Students

In the 7th grade, the task was implemented with 152 students in four classes from 'T' national junior high school. Three hours were allocated to the task on different days in December 2010. The teacher was the usual teacher, Noboru Yoshimura. The aims of the implementation were (a) training students in mathematical modelling, and (b) making students aware of pension reform plans. The teacher led all students in a class as is the usual traditional Japanese style lesson.

The first period: Students discussed an extract of the explanation document and a recent newspaper article about pension issues to understand the context of the pension issues. The original system that prefecture tax money is used to cover the basic portion was introduced. Students were asked to think about the mathematical solution of the prefecture tax rate problem on the basis of the data for Osaka prefecture.

These data are as follows:

(1) Osaka prefecture has a population of 8.8 million. The population distribution by age is equal to each other and the average life expectancy in Osaka pref. is just 80 years old.
(2) The average income of wage-earners, 70 % of whom are 20–60 years old, is 4 million yen annually.
(3) People over 61 years old are eligible for a y thousand yen pension when pension tax rate for the income of wage-earners is x%.

Students calculated using proportion. After deliberations, students calculated as follows:

(a) The estimated population over 61 years of age is

$$(8.8 \div 80) \times (80 - 60) = 2.2 \,(\text{million people})$$

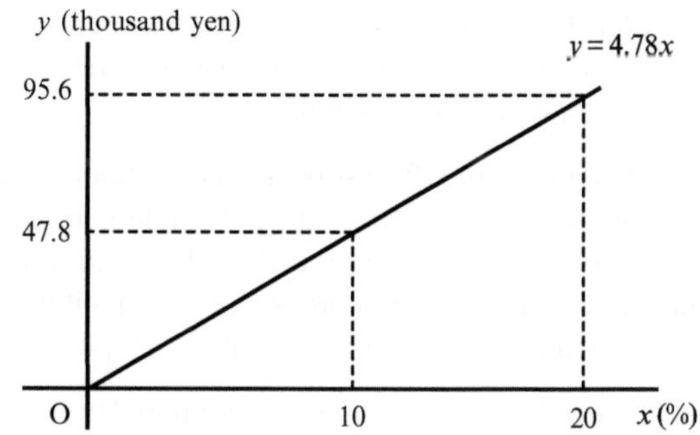

Fig. 21.1 The proportion of pension payments

(b) Proportional expression of pension payments

$$y=(61-19)\times(8.8\div 80)\times 0.7\times 4{,}000\times x\div 100\div 2.2\div 12$$
$$\therefore y=4.78x$$

(c) Drawing the graph (see Fig. 21.1)

Students calculated values of y, when $x=10$ and $x=20$. In consideration of welfare payments (from 62 to 81 thousand yen) based on an aged single person household, students reached the appropriate decision that the pension tax rate of 13–17 % is better.

The second period: Supposing that the mandatory retirement age is extended from the current 60 years old to 65, those over 66 years old are eligible for a y thousand yen pension, when the pension tax rate for the income of wage-earners is x%. The content was approximately similar to the first period. Students were left to solve the issue by themselves. They expressed y in an expression of x as y thousand yen pension a month charging a pension tax of x% on income of wage-earners. The expression became $y=7.16x$.

Charging a pension tax of 9 %, pension payments are 64.4 thousand yen a month. Charging a pension tax of 11 %, pension payments are 78.8 thousand yen a month. In consideration to welfare payments (from 62 to 81 thousand yen) based on an aged single-person household, students reached the appropriate decision that the pension tax rate of 9–11 % is better.

They looked back on their solving process and the setup conditions of this problem. They were then given an assignment in which they had to search for another solution.

The third period: Students discussed their assignment. Students were asked to think about the mathematical solution for the prefecture tax rate on the basis of the other conditions of Osaka prefecture.

These modified conditions are as follows:

The people over 66 years old are eligible for a pension with the following conditions:

1. Aged single-person households are a third of the population of those over 66 years old and are eligible for a y thousand yen pension.
2. Aged couple households are a third of the population of those over 66 years old and are eligible for 70 % of aged single-person households' payments.

3. Elderly households living with children or grandchildren are a third of the population of those over 66 years old and are eligible for 30 % of aged single-person households' payments.

Students were left to solve the issue. Some students were told that when there is a y thousand yen pension for aged single-person households, pension payments for aged couple households are $0.7y$ thousand yen and pension payments for aged households living with children or grandchildren are $0.3y$ thousand yen. Students calculated by a way of thinking of the proportion as follows.

(a) A third of the over 66 years old population

$$(8.8 \div 80) \times (80 - 65) \div 3 = 0.55\ (\text{million people})$$

(b) The quantity of estimated total amount of pension payments

$$(0.55y + 0.55 \times 0.7y + 0.55 \times 0.3y) \times 12 = 0.55 \times 2y \times 12\ (\text{billion yen})$$

(c) Quantity of estimated total amount of pension taxes

$$4{,}000 \times (x \div 100) \times (65 - 19) \times 0.11 \times 0.7\ (\text{billion yen})$$

(d) Proportional expression of pension payments

$$0.55 \times 2y \times 12 = 4{,}000 \times (x \div 100) \times (65 - 19) \times 0.11 \times 0.7 \quad \therefore y = 1.07x$$

(e) Drawing the graph

Students calculated the values of y, when $x = 5$ and $x = 10$. When the pension tax rate is 6 %, aged single-person households are eligible for about a 64 thousand yen pension, aged couple households are eligible for about a monthly 90 (45 × 2) thousand yen pension payment and aged households living with children or grandchildren are eligible for about a 19 thousand yen pension. In consideration of welfare payments (from 62 to 81 thousand yen) based on aged single person household, students reached the appropriate decision that the pension tax rate of 6 % is better. The amount of pension taxes is reduced from 600 to 240 thousand yen a year. The students also reached the appropriate decision that the pension tax rate of 6 % is better. They reviewed their solving process and the conditions of the problem. They were given a report assignment where they had to obtain the original solution.

2.2 *The Linear Function Model for the 9th-Grade Students*

In a second teaching experiment conducted with the 9th grade students, the linear function model was used. The task was implemented with 157 students on different days in May 2011. The intended aims were as for Year 7.

The first period: Students discussed the extract of the explanation documents and the recent newspaper article about pension issues. The teacher explained Japan's pension system. It is complex. The national pension program is a single-tier one that provides a basic pension for the entire populace. The welfare pension program and mutual aid pension program are a two-tier system. The lower tier is the basic pension of the national pension program. The upper tier is a pension geared to each individual's income. In other words, the insured person pays premiums and receives benefits in proportion to his or her income. The future of the pension insurance system is threatened by a number of issues, such as rising pension payments in accordance with the aging of the population, a diminishing number of new participants caused by a low birth-rate, and a larger number of people shunning the system. Reforms for keeping the system stable and sustainable are therefore being implemented from the long-term perspective. Students understood the context around these pension issues.

The original system that prefecture tax money is used to cover the basic portion was introduced. Students were asked to think about the mathematical solution to the prefecture tax rate on the basis of the data for Osaka prefecture.

These data are as follows:

(1), (2) are the same as for the first period for 7th grade.
(3) People over 66 years old are eligible.
(4) The contributions from the government to those over 66 years old per person are 30 thousand yen per month per person.

Students calculated using a linear function as follows:

(a) The quantity of estimated total amount of pension payments without any contributions from the government

$$(y-3)\times(8.8\div 80)\times(80-65)\times 12=(y-3)\times 0.11\times(80-65)\times 12\ (\text{billion yen})$$

(b) The quantity of estimated total amount of pension taxes

$$4{,}000\times(x\div 100)\times(60-19)\times(8.8\div 80)\times 0.7=28\times x\times(60-19)\times 0.11\ (\text{billion yen})$$

(c) Linear function of pension payments

$$(y-3)\times 0.11\times(80-65)\times 12=28\times x\times(60-19)\times 0.11\ \therefore\ y=6.4x+30$$

(d) Drawing the graph (see Fig. 21.2)

In consideration of welfare payments (from 62 to 81 thousand yen) based on aged single-person households, students reached the appropriate decision that the pension tax rate of 6 % is better. When the pension tax rate is 6 %, aged single-person households are eligible for a monthly pension of about 68 thousand yen. The amount of pension taxes is reduced to 240 thousand yen a year. They reflected on their process and the setup conditions of this problem.

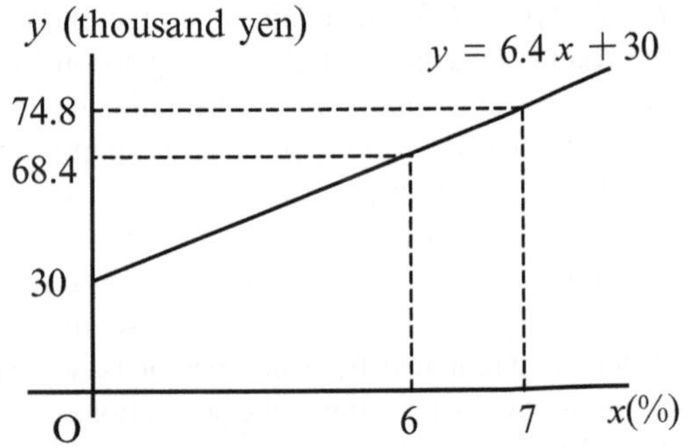

Fig. 21.2 The linear function of pension payments

Students discussed this result and they searched for another solution. They were asked to think about the mathematical solution to the prefecture tax rate based on additional conditions. The additional conditions (i), (ii), (iii) are the same as those for the third period for the 7th grade students.

They expressed y in an expression of x. The equation became $y=9.6x+30$. In consideration of welfare payments (from 62 to 81 thousand yen) based on an aged single-person household, students reached the appropriate decision that the pension tax rate of 4 % is better. When pension tax rate is 4 %, aged single-person households are eligible for a monthly pension payment of about 68 thousand yen. The amount of pension taxes is reduced to 160 thousand yen a year. Again, they reviewed their process and the setup conditions of the problem. They were given an assignment in which they had to search for another solution.

The second period: Students discussed their assignment. They considered dividing the income earners into three groups without changing the overall average income. Students were asked to think about the mathematical solution of the prefecture tax rate.

The three groups are as follows.

(i) Pension tax rate for the group of 80 % of wage-earners is x%. The average income of those who earn less than 4 million yen is 3 million yen.
(ii) Pension tax rate for the group of 10 % of wage-earners is $1.5x$%. The average income of those who earn more than 4 and less than 8 million yen is 6 million yen.
(iii) Pension tax rate for the group of 10 % of wage-earners is $2x$%. The average income of those who earn more than 8 million yen is 10 million yen.

Students calculated by using a linear function as follows.

(a) The quantity of estimated total amount of pension payments except the contributions from the government

$$(y-30)\times(8.8\div 80)\times(80-65)\times 12=(y-30)\times(80-65)\times 0.11\times 12\ (\text{billion yen})$$

(b) The quantity of estimated total amount of pension taxes

$$
\begin{aligned}
&3{,}000\times(x\div 100)\times(60-19)\times 0.11\times 0.7\times 0.8+6{,}000\times(1.5x\div 100)\\
&\quad\times(60-19)\times 0.11\times 0.7\times 0.1+10{,}000\times(2x\div 100)\times(60-19)\times 0.11\\
&\quad\times 0.7\times 0.1=(24x+9x+20x)\times(60-19)\times(8.8\div 80)\times 0.7=1521.1x\times 0.11
\end{aligned}
$$

(c) Linear function of pension payments

$$(y-30)\times(80-65)\times 0.11\times 12=1521.1x\times 0.11 \quad \therefore y=8.5x+30$$

(d) Drawing the graph

In consideration of welfare payments (from 62 to 81 thousand yen) based on an aged single-person household, students reached the appropriate decision that the pension tax rate of 4 % is better. In that case pension payments are 64 thousand yen a month.

Pension tax rate for the group of 80 % of wage-earners is 4 %, pension taxes are 120 thousand yen pension a year. Pension tax rate for the group of 10 % of wage-earners is 6 %, pension taxes are 360 thousand yen pension a year. Pension tax rate for the group of 10 % of wage-earners is 8 %, pension taxes are 800 thousand yen pension a year.

3 Evaluation of Student Performance

Students who are not interested in mathematics as their grade in school advances are a growing trend in Japan. They also feel that mathematics will not be of help. These are serious situations in mathematical education in Japan. We found the same tendency in a survey of students of 'T' national junior high school students who are at a higher level. A survey of 'T' national junior high school students' to evaluate if the teaching material for mathematical modelling discussed in this chapter changes some aspects of this situation was conducted.

3.1 Mathematics in the Junior High School

The assessment of students' evaluation about the utility of mathematics in real problems was done by carrying out a pre-test before teaching and a post-test after teaching. Students were asked:

When you think about real problems, do you think that the mathematics is helpful?

The answer to the question above was chosen from the following choices: (A) Very helpful, (B) Agreeably helpful, (C) Not so helpful, (D) Not helpful at all. The results were as shown in Table 21.1.

The result reveals that the number of students who answered (A) or (B) were 87 % in the pre-test, 94 % in the post-test for 7th grade students, and 82 % in the pre-test, 95 % in post-test for 9th grade students. Students commented that "Pension

Table 21.1 Students' evaluation of the usefulness of mathematics (%)

Year			Response (%)			
Level	N	Test	A	B	C	D
7	152	Pre	40	47	11	2
		Post	53	41	6	0
9	157	Pre	29	53	17	1
		Post	56	39	4	1

tax is roughly proportional to pension payment", "I was able to understand pension tax issues", "I could understand balanced profits and losses", "I could understand the flow of money using the mathematics" and so on, as the reasons for changing their answers between the two tests. Thus more students were able to appreciate the helpfulness of mathematics through experiencing this single teaching practice.

3.2 *Students' Interest in This Lesson*

We used the pension tax issue in this teaching practice. This is a serious modern social problem in Japan. The assessment of students' interest in the topic was carried out using a post-test after teaching. Students were asked:

Were you interested in this learning content of pension tax issue?

The answer to the question above was chosen from the following choices: (A) Very much, (B) Rather, (C) Not so much, (D) Never. The results are shown in Table 21.2 together with results from previous cohorts using different contexts for comparison.

This table shows that the students who participated in this study responded with almost the same tendency as those in Yanagimoto et al. (1993) when sport was used and Yanagimoto and Yoshimura (2001) when traffic safety was used. About 30–60 % students chose the answers (A) or (B) in the various aspects of the track and field events task, and about 55–75 % students chose the same answers in the traffic safety context. Interest depends on how much students are related to the learning context. In this learning context about 70–80 % of students chose the answers (A) or (B), and showed more interest than in the other two studies. This learning context included a socially serious problem, a topic discussed frequently in Japan. In general, students' interests are influenced by their own involvement and the currency of topics.

The reasons given by students for answering (A) or (B) included: "I have never gone through pension tax issues with a critical eye", "We will face the same problem in future", "I could learn a bit about how society works." This practice aroused many students' curiosity. However, the reasons for other students choosing (C) or (D) included "It has nothing to do with me", "It happens in a distant world unrelated to me", "The problem is so complicated." This suggests that some lacked a feeling

Table 21.2 Student interest (%) in different contexts

Teaching context	Year Level	N	Response (%) A	B	C	D
Pension tax	7	152	16	56	24	4
	9	157	20	62	15	3
Track and field	9	157				
1. Designing an athletics track			6	25	53	16
2. Recording running broad jump			13	39	34	14
3. Location of mark point in passing a baton			18	43	29	10
Traffic safety	9	160				
1. Pedestrian crossing time			12	50	32	6
2. Stopping distance of cars			28	46	22	4
3. Time to yellow signal			13	44	38	5

Table 21.3 Students' willingness (%) to seek other solutions

Year Level	N	Response (%) A	B	C	D
7	152	12	37	44	7
9	157	10	54	32	4

of motivation for solving the task and it might be necessary to improve the specific teaching guidelines to overcome this.

3.3 *Students' Willingness for Seeking Other Solutions*

All students found one of the solutions to the pension tax issue. An assessment of the students' willingness to find the other solutions was included in the post-test after teaching. We asked:

Do you think you want to seek the other solution to pension tax issue after this learning?

The answer to the question above was chosen from the following choices: (A) Very much, (B) A little, (C) Not so much, (D) Never. The results are in Table 21.3.

The result reveals that the number of students who answered (A) or (B) were 49 % for 7th grade students and 64 % for 9th grade students. This implies many students showed willingness to seek the other solutions. From our observations, the mathematical modelling teaching materials opened up society for students more and boosted awareness for them as members of society; but students who did not have any patience for the task, were relying on others and had a tendency to avoid challenging tasks.

4 Conclusion

Firstly, it is possible to use mathematical modelling teaching materials which deal with the pension tax issues for instructing junior high school students. Students learn the proportion function in 7th grade, and linear function in 8th grade, so this content is easy to treat as applied mathematics teaching materials. It is thought that these teaching materials are effective in a large class as a teaching material for the introduction period of mathematical modelling. There has been no coursework at these levels of schooling like this performed practice in Japan previously and students are hardly used to mathematical modelling activities. It is best that a teacher gives students the problem with a mathematical model modified by teachers rather than the students develop the model themselves. However, more open teaching practices such as using group learning (Antonius et al. 2007) are necessary for the more advanced stages.

Secondly, one of the goals of dealing with teaching material for mathematical modelling would be to promote excellence in problem solving and also to make students aware of the usefulness of mathematics. The students' evaluation of mathematics changed for the better as a result of this educational experiment. When asked, the percentage of students who answered that mathematics is useful for solving real-world problems jumped from 87 to 94 % of the 7th grade students and from 82 to 95 % of the 9th grade students. The mathematical modelling teaching materials aroused students' curiosity, encouraged students to open up society more and boosted awareness as members of society. Students who thought this teaching material was interesting accounted for 72 % of the 7th grade students and 82 % of the 9th grade students. Students who indicated willingness to seek other solutions accounted for 49 % of the 7th grade students and 64 % of the 9th grade students. These are general surveys, however. We need to gain deeper insights into the level of performance of the students on the tasks by further analysis.

Maaß (2010) has developed a scheme for classifying modelling tasks so other potential users can be guided in terms of task features, objectives and target audience. This teaching material has a general focus in terms of the whole process as a modelling activity in the classification scheme for modelling tasks (Maaß 2010), though the teacher led students. There has been consideration about the contexts which interest students in other literature (e.g., Julie 2007). Some of this has been related to the modelling of social issues in schooling contexts (e.g. Julie and Mudaly 2007). Indeed, Galbraith et al. (2010) and Caron and Bélair (2007) suggest that social contexts have several benefits as topics for modelling. This teaching material appears to have interested many students because it is a social context and related to their future lives, in spite of the topic being chosen by teachers.

Our future challenge is to promote the development of mathematical modelling teaching materials from the viewpoint of nurturing pioneers who will be trailblazers in our society. The aim of this mathematical modelling activity is to develop people who can lead the way in our society in keeping with one of the goals of applications and modelling and their curricular embedding (Niss et al. 2007).

References

Antonius, S., Haines, C., Jensen, T. H., & Niss, M. (with Burkhardt, H.). (2007). Classroom activities and the teacher. In W. Blum, P. L. Galbraith, H.-W. Henn, & M. Niss (Eds.), *Modelling and applications in mathematics education* (pp. 295–308). New York: Springer.

Caron, F., & Bélair, J. (2007). Exploring university students' competencies in modelling. In C. Haines, P. Galbraith, W. Blum, & S. Khan (Eds.), *Mathematical modelling ICTMA 12: Education, engineering and economics* (pp. 120–129). Chichester: Horwood.

Galbraith, P., Stillman, G., & Brown, J. (2010). Turning ideas into modelling problems. In R. Lesh, P. Galbraith, C. R. Haines, & A. Hurford (Eds.), *Modelling students' mathematical competencies* (pp. 133–144). New York: Springer.

Julie, C. (2007). Learner's context preferences and mathematical literacy. In C. Haines, P. Galbraith, W. Blum, & S. Khan (Eds.), *Mathematical modelling: Education, engineering and economics* (pp. 194–202). Chichester: Horwood.

Julie, C., & Mudaly, V. (2007). Mathematical modelling of social issues in school mathematics in South Africa. In W. Blum, P. L. Galbraith, H.-W. Henn, & M. Niss (Eds.), *Modelling and applications in mathematics education: The 14th ICMI study* (pp. 503–510). New York: Springer.

Maaß, K. (2010). Classification scheme for modelling tasks. *Journal für Mathematik-Didaktik* (Journal for Didactics of Mathematics), *31*(2), 285–311.

Matsumiya, T., Yanagimoto, A., & Mori, Y. (1989). Mathematics of a lake – Problem solving in the real world. In W. Blum, M. Niss, & I. Huntley (Eds.), *Modelling, applications and applied problem solving* (pp. 87–97). Chichester: Ellis Horwood.

Niss, M., Blum, W., & Galbraith, P. (2007). Introduction. In W. Blum, P. L. Galbraith, H.-W. Henn, & M. Niss (Eds.), *Modelling and applications in mathematics education: The 14th ICMI study* (pp. 3–32). New York: Springer.

Yanagimoto, A. (2003). Environmental problems and mathematical modelling. In S. J. Lamon, W. A. Parker, & S. K. Houston (Eds.), *Mathematical modelling: A way of life, ICTMA11* (pp. 53–60). Chichester: Ellis Horwood.

Yanagimoto, A., & Yoshimura, N. (2001). Mathematics of traffic safety – Composite real mathematics approach. In J. F. Matos, W. Blum, K. Houston, & S. P. Carreira (Eds.), *Modelling and mathematics education* (pp. 401–410). Chichester: Ellis Horwood.

Yanagimoto, A., Mori, Y., & Matsumiya, T. (1993). Mathematics of track and field events – Composite real mathematics approach. In T. Breiteig, I. Huntley, & G. Kaiser-Messmer (Eds.), *Teaching and learning mathematics in context* (pp. 115–126). Chichester: Ellis Horwood.

Part III
Pedagogical Issues for Teaching and Learning

Chapter 22
Pedagogical Reflections on the Role of Modelling in Mathematics Instruction

Toshikazu Ikeda

Abstract From a broad perspective, there are two categories of pedagogical aims of modelling. The first is where modelling itself is treated as an objective. The second is mathematical knowledge construction, where modelling is treated as a means to an end. For the first aim three key questions are considered: (1) How can the teacher set an appropriate situation so that students realise the necessity of solving a real-world problem? (2) How can the teacher assist students' abstraction processes? (3) How can the teacher show students the necessity of controlling various assumptions? For the second aim, three principles are suggested: (1) expanding and clarifying real-world situations satisfying a developed original model, (2) expanding and integrating mathematical knowledge by setting up a concrete situation so that students can consider it, and (3) refining and clarifying the developed mathematical methods by treating instances of the same contexts repeatedly.

1 Background of This Study

1.1 *Pedagogical Aims of Modelling*

Regarded from a wide perspective, the pedagogical aims of modelling can be regarded as two. The first is modelling itself, with modelling treated as the objective. Of course, it is divisible into several more detailed aims (Blomhøj 2009; Blum and Niss 1991). Regarding this point, the author suggests the issue of a teacher's role: 'How does a teacher cultivate students' thinking about modelling?' The second aim

T. Ikeda (✉)
College of Education and Human Sciences, Department of School Education, Yokohama National University, Yokohama, Japan
e-mail: toshi@ynu.ac.jp

G.A. Stillman et al. (eds.), *Teaching Mathematical Modelling: Connecting to Research and Practice*, International Perspectives on the Teaching and Learning of Mathematical Modelling, DOI 10.1007/978-94-007-6540-5_22,

is mathematics knowledge construction, with modelling treated as a method to achieve a goal. The issue of the relation between modelling and mathematics knowledge construction is selected to answer questions such as 'How is modelling related to mathematics knowledge construction in the teaching of mathematics?'

1.2 *Pedagogical Reflections on the Role of Modelling*

When considering mathematics instruction, we must distinguish school mathematics from mathematics and modelling outside of school. By considering the content or activities related to modelling and mathematics outside of school, it is possible to design a school mathematics curriculum, teaching plan, and so on that are expected to be more relevant to students' interests. Nevertheless, these products should be examined after due reflection upon students' activities and understanding. It is important to contrast the activities of scientists and mathematicians with students' activities and thereby derive practical issues or suggestions that might be adapted to the future teaching of mathematics.

1.2.1 Pedagogical Reflection as Objectives

When examining modelling activities outside of school mathematics from the perspective of cultivating students' thinking about modelling, three characteristics can be observed from contrasting students' activities against those of scientists and others who use mathematics as a tool.

First, the problems and contexts that scientists and others encounter are familiar to scientists. They are those very situations in which they are interested. Therefore, they have clear reasons and incentive and motivation to solve a problem. Nevertheless, problems and contexts that students encounter are not always familiar to them. Students do not always know the reason why there is a need to solve a given real-world problem. This difference creates an issue for a teacher of how to present a real-world problem to students.

Second, scientists and others who use mathematics as a tool observe a situation or a phenomenon over a long duration. In other words, they have thought extensively about the structure of a problem, particularly which variables are important and which are not. In contrast, students have little or no experience at observing a situation or phenomenon before encountering a problem presented by the teacher. Moreover, it is difficult for students to observe a realistic set of circumstances in a classroom. This difference creates an issue for a teacher of how to assist students' process of abstraction from a real world into mathematics.

Third, scientists who use mathematics as a tool can illustrate modelling processes and modelling skills very well. They have experienced modelling processes numerous times. However, students who have not experienced modelling have no idea how to mathematise a real-world situation into mathematical terms and contexts. Real-world problems have many variables. Students typically have no idea about how to treat numerous variables to solve a problem. This difference presents an issue to the teacher of how to support students' thinking so that they can acquire modelling competency.

The three issues described above are discussed in the first parts of this paper.

1.2.2 Pedagogical Reflection as Methods

Next, we can consider mathematical knowledge construction by which modelling is treated as a means to an end. By examining the activities of scientists and others who use mathematics as a tool and those of mathematicians who construct more abstract mathematics, the following are addressed by contrasting them with students' activities.

First is a suggestion that mathematics education be conducted by analysing the activities of scientists who use mathematics as a tool. Most scientists and others build up a mathematical model based on mathematical principles and processes that were developed by someone many years ago. After producing an original model, they begin to gather additional concrete examples which satisfy the original model. In the philosophy of science, Hesse (1966) discerned analogies of three types in scientific models: positive analogies, negative analogies, and neutral analogies. Positive analogies are those features which are known, or thought, to be shared by two systems. Negative analogies are those features which are known or thought to be present in one system, but which are absent in the other. Neutral analogies are those features for which status as positive or negative analogies is uncertain. Of those three types, neutral analogies are by far the most interesting because they suggest ways to test the limits of our models, lighting a path to scientific advancement. Using the idea of Hesse in modelling, expanding and clarifying real-world situations that satisfy a developed original model is crucial. The original model is applicable over a wide range if more extra concrete examples are applied in the model. Building up the original model and extending the range in which the original model is applied is an extremely important activity for students.

Second is the difference between students' activities and scientists' activities. This analysis shifts our emphasis to mathematicians who construct more abstract mathematics. Unlike scientists and others, students have not acquired the necessary mathematics skills or knowledge to solve the problem, and mathematics is unfamiliar to students. In other words, for scientists who use mathematics as a tool, mathematics is viewed as a developed system, but for students, mathematics should best be viewed as a developing system. In some cases, students can solve a problem by application of acquired mathematical knowledge. However in some cases, they cannot solve a problem even through application of all mathematical knowledge acquired previously. However, students' perceptions about the limitations of their mathematical knowledge draw on a breakthrough in acquiring new mathematical knowledge. It is a meaningful motivation for students to understand the necessity of acquiring new mathematical knowledge.

When we specifically examine these aspects, recognizing that 'mathematics has not been developed yet,' and that 'mathematics is not always familiar for them,' mathematicians' activity is brought into relief. Mathematicians construct more abstract mathematics based on their concrete mathematics. More abstract mathematical methods, which are now being developed, are not familiar even to many mathematicians. We can see similar situations with students from this perspective. Students construct mathematics in their internal world, and mathematics is not always familiar. Therefore, a mathematician's typical approach such as 'consideration by producing a concrete model' is emphasised (Iyanaga 2008, pp. 71–95). Even for mathematicians, constructing new abstract mathematics is not easy, so they produce concrete models and consider the mathematical principles using the model.

What principles can we elicit to teach mathematical knowledge through modelling from these analyses? These issues are discussed in the second part of this chapter.

2 Teacher's Role Derived from Pedagogical Reflections as Objectives

2.1 *Why Do Students Solve Problems?*

We first specifically examine the first characteristic by which the immediate problem situation is familiar to scientists and others who use mathematics as tools: they have clear motivations to solve problems. In this regard, students' situations differ greatly. The following questions should be addressed: 'Why do students solve problems?' and 'How can a teacher select a material, and establish a learning situation?'

Regarding the question of 'What is an appropriate modelling task?', Galbraith (2007, pp. 181–184) pointed out two issues: (T1) consistency with an avowed purpose, and (T2) introducing real-world modelling tasks. Regarding (T2), two additional issues are pointed out: (a) the importance of using models based on experience, and (b) motivation. Regarding (a), additional questions can be derived such as 'Does the problem situation concern the surroundings of students at present, in the past, or in the future?' and 'Is it concerned with most students or only a few students?' If the first question is related to the students' future, then an additional question is suggested: 'Is it related to a situation confronted as a citizen, as individuals, or in their profession or vocation?' The PISA category related to contexts is extremely useful in relation to this point (PISA 2009). Regarding '(b) motivation,' the teacher's role is to clarify the reason why someone must solve the problem, then to set the appropriate situation so that students can accept the problem posed by someone as their own problem.

For instance, students are very familiar with relay races. To win a relay is important for a typical class. Therefore, a teacher might ask students how to win in a relay race in a school sports competitions (Osawa 2004). Several issues can then be inferred by students, such as the runner order, how to pass the baton, and so on. It is important that the teacher clarifies types of issues to be considered and thereafter

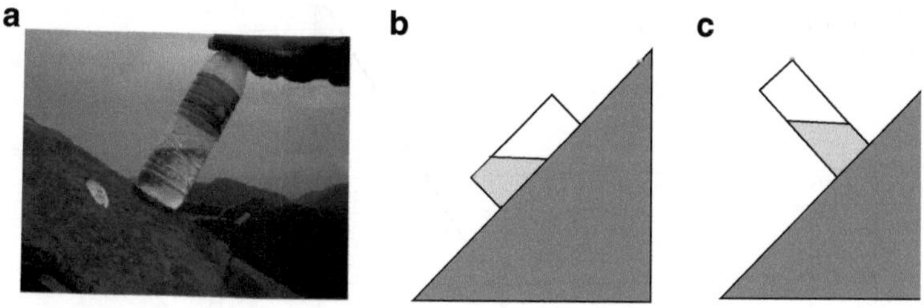

Fig. 22.1 Great Wall problem (**a**) using a bottle, (**b**) Method 1, (**c**) Method 2

specifically examine the baton pass, but not specifically examine the baton pass without clarifying issues to be considered. Students' awareness that the baton pass is one issue related to winning a relay race is crucial.

A problem derived from students' school life will be readily accepted as their own problem. How can we treat a problem derived from someone else (not students)? In such cases, the teacher should at least explain the background related to why the problem arose. We can select two illustrative examples. Figure 22.1 is the Great Wall of China. On the top of a mountain, we can feel that it is apparently very steep. The question 'How steep is it?' is derived if we were actually climbing the Great Wall. How many degrees of this inclination angle is the steepest slope? When treating such a problem, it is important for the teacher to let students surmise the answer. For example, the teacher asks students to select the following alternatives of this inclination angle: (1) 25°, (2) 35°, or (3) 45°. By selecting an alternative, students might want to clarify whether their own answer is correct or not. It is one method for the teacher to let students build desire to solve the problem. After selecting alternatives, the teacher explains the situation as shown in Fig. 22.1.

'Let's presume we were there in summer. Then we have a plastic bottle of water and pen. Is there any method to measure the inclination angle?' The answer is depicted in Fig. 22.1a. It is one way to present the geometrical figure exactly as shown in Fig. 22.1b, c if students have no idea of how to measure the inclination angle. Students can understand that this mathematical problem can arise in a real-world situation. By asking 'Which angle is the inclination angle?' students have a chance to solve the problem. For students, the setting in Fig. 22.1b supports their discovery of the inclination angle. Another setting, shown in Fig. 22.1c, is a little bit more difficult to work with. In ordinary mathematics teaching in Japan, the alternative angle and corresponding angle are taught in a purely geometrical situation. Finding relations between angles in a purely geometrical situation is not useful for their life and is also not surprising for students. However, in this situation, ascertaining the relation between angles is extremely useful: we can measure an angle that is physically impossible to measure by measuring an angle that can be measured easily. Ascertaining relations is crucial. The angle is 35°, which is not so steep. Why do we feel that it is so steep? Additional questions are derived after solving the problem. The reason to solve the suggested problem is clear for students.

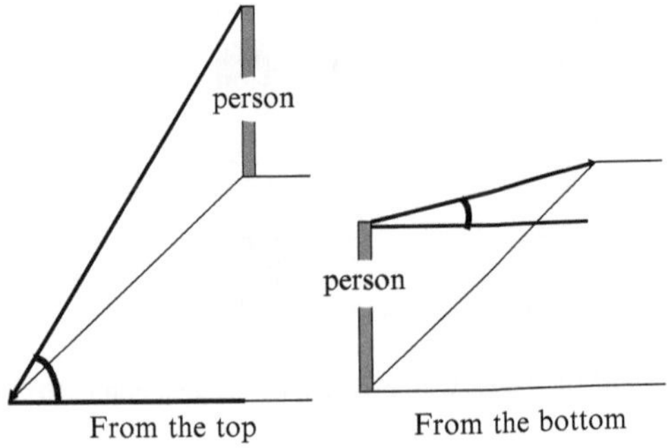

Fig. 22.2 Explanation by geometrisation

Fig. 22.3 Three sizes of crucifixes

Psychological aspects will affect human feelings, but this feeling is explainable by drawing figures as shown in Fig. 22.2.

From the top, the inclination angle is apparently large, but from the bottom it is not so big. We can consider one more example. There is the very big and very old Berliner Dom in Germany. After looking at the inside of this dome, there is a souvenir shop just at the exit. Three sizes of crucifixes are sold, as presented in Fig. 22.3, which suggests a question. Are these three similar mathematically? It is also one method for the teacher to entice students to form their conjectures of whether these three crucifixes are similar or not. By supposition, this problem might be accepted as a students' own problem: 'How can we judge that in a souvenir shop? I have no measure and no protractor. Is there any good method?' It is necessary to formulate the problem. Here, particularly addressing the right triangle is effective (Fig. 22.4). The formulated question is 'Is the relation of two right triangles similar?'

Several methods exist. Direct comparison of two inclination angles is one (Fig. 22.5). Another is to use three crucifixes to judge the similarity of two of them (Fig. 22.6). The largest one and the middle one are similar if the point is touched. In this case, it is not touched, so these two triangles are not similar. Moreover, we must devote attention to the implicit assumptions of two methods. We must check two assumptions: that the crucifix is linearly symmetric, and another small right triangle has not been considered yet.

Fig. 22.4 Right triangles

Fig. 22.5 First method

Fig. 22.6 Second method

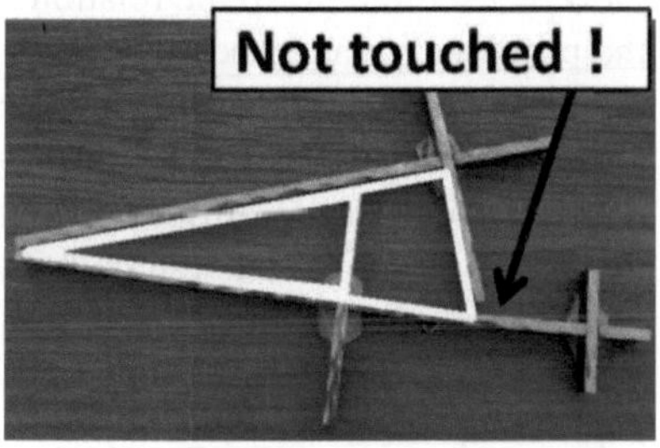

When the teacher selects the materials, it is suggested that the teacher identify the purposes of using mathematics. Niss (2008) categorised the following three purposes to use mathematics: understand, action, and design. Teaching processes as well as modelling processes depend on the purpose or area of other disciplines. For example, particularly addressing validation, the process differs according to the aims—understand and action and design. In understanding, the model is validated by contrasting it with real-world phenomena. However, in action and design, plural models are compared by discussing the merits and shortcomings of each.

2.2 *Students' Abstraction Processes*

Next is the second question of the first issue. For scientists and others who use mathematics as tools, the situation and related phenomena of the confronted problem have been observed for a long time. However, students confront the problem situation in a classroom immediately. Therefore, if students cannot see or experience a real problem situation, then it is necessary for the teacher to prepare concrete models so that students can observe or manipulate them.

Abstraction is an important phase of mathematisation. Abstraction means that when 'the meanings of same terms and things are grasped only by first calling to mind more familiar things and tracing out connections between them and what we do not understand', they are designated as "abstract" (Dewey 2007, p. 66). The essential structure of a problem situation and related phenomena, which are abstract for students, can be grasped only by observing and manipulating a concrete situation and phenomena. Discerning the essential structure is difficult for students who have little experience with concrete real-world situations. Using a concrete model is crucial for students.

I explain two examples. First is a sunset. 'Here is Cabo da Roca. It is at the extreme western part of Portugal. What a wonderful sunset! However, because I became careless, I missed seeing the sunset from the beginning. Unfortunately, the sun has almost set. How long does it take from the beginning of sunset to the end of sunset? Which do you think is correct among the following alternatives: 2, 4, 8, or 16 minutes?' It is difficult to imagine the relation between the sun and earth. A globe is useful to grasp the relation between them. We can manipulate and observe the positional relation between them. The teacher can ask 'What is the situation in the concrete model regarding the beginning and end of sunset? When viewed from above, we can ascertain the positional relation between them (Fig. 22.7). The location of the end of the sunset is easy to ascertain. By asking why that is so, the line tangent to two circles is elicited. Then, the teacher can ask about the location of the beginning of the sunset. Similarly, another line, which is tangent to two circles, is elicited. To calculate the time of the sunset, it is necessary to ascertain the central angle comprising two tangent points. Therefore, by regarding the distance between the sun and earth as OB in Fig. 22.8, it becomes possible to solve the right triangle OAB.

The time of sunset is calculated as 24 (hours) $\times$ 60 (min) $\times$ $2\theta/360°$. The radius of the sun is 700,000 km and distance between the sun and earth is 150,000,000 km. The ratio between the radius of the sun (OA) and the distance between them (AB) is 7:1,500. Using the reduction method—by constructing a right triangle of which two lengths are 1.75 and 37.5 cm—produces a very thin right triangle. It is difficult to measure such an angle that depends on the strictness of construction (Fig. 22.13). Someone will measure it as around 1°, others as around 0.5°. The time of sunset is 4 min in the case of 0.5° and 8 min in the case of 1°. Students will be able to understand that the method to construct the figure is limited.

Another example is the situation derived from a journey to northern Europe in summer, to Bergen in Norway. 'The north latitude of Bergen is 60°. Past 9 o'clock, past 10 o'clock, it is still not dark. Before 11 o'clock, it is just getting dark (Fig. 22.9). Why is the night time (sunset–sunrise) so short?' The purpose to solve this problem is 'to understand'. Manipulating and observing the globe is effective to find out

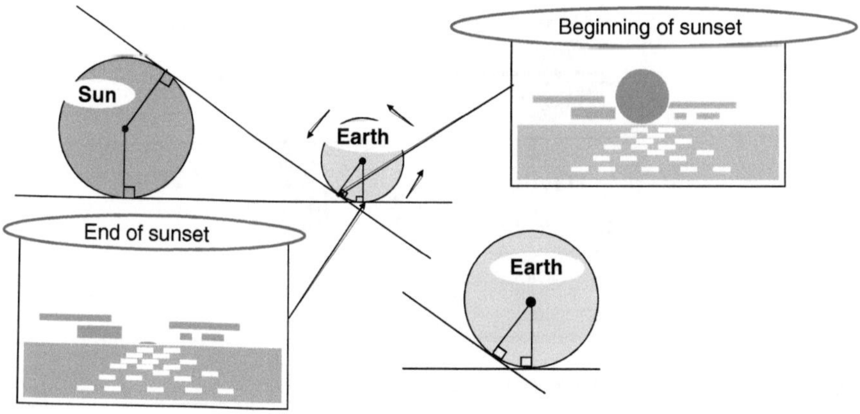

Fig. 22.7 Positional relation between sun and earth

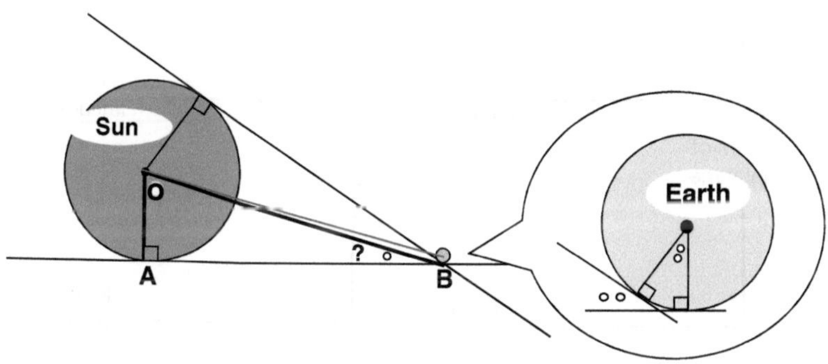

Fig. 22.8 Assumptions to calculate the time of sunset

Fig. 22.9 Sunset in Bergen

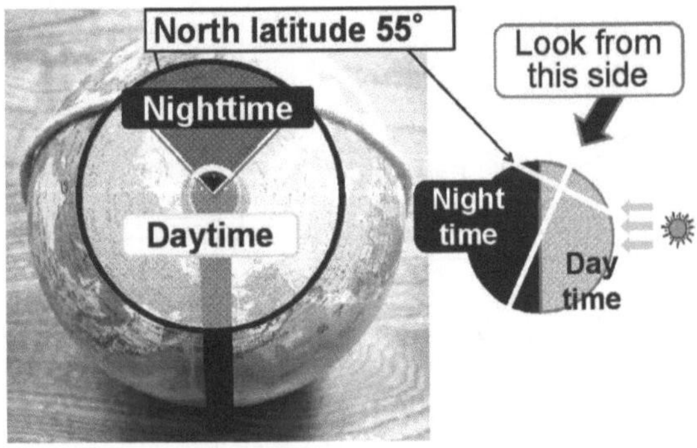

Fig. 22.10 Observation using a concrete model

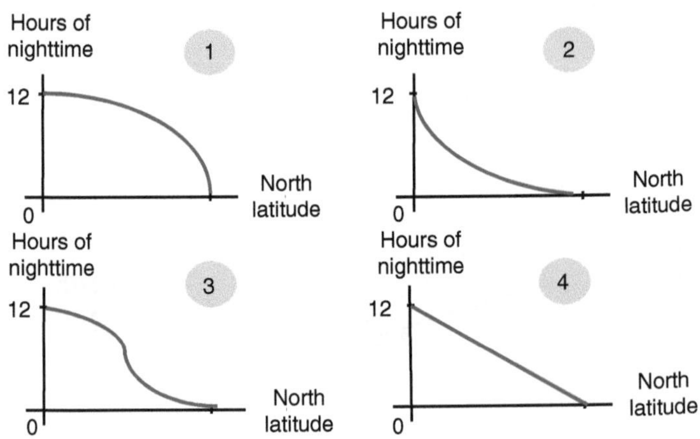

Fig. 22.11 Four graphs

which part in a globe is night and which quantity corresponds to night time. The necessary structure is elicited from manipulation or observation using the concrete model. Considering special cases, such as countries near the equator, is effective for estimation. In countries near the equator, we can realize by observing the globe that daytime and night time are equal: 12 h. What about in high-latitude countries? Using some additional tools, we can realize that night time at high latitudes is shorter than daytime (Fig. 22.10). Night time decreases as latitude increases. How does night time decrease with latitude?

Can we draw a graph for which the *x* axis is north latitude and the *y* axis is hours of night time? Of the following four graphs (Fig. 22.11), which graph is correct? It is also possible to obtain data about 'angle of night' using a globe and draw the graph (Fig. 22.12). We can judge that (1) is correct.

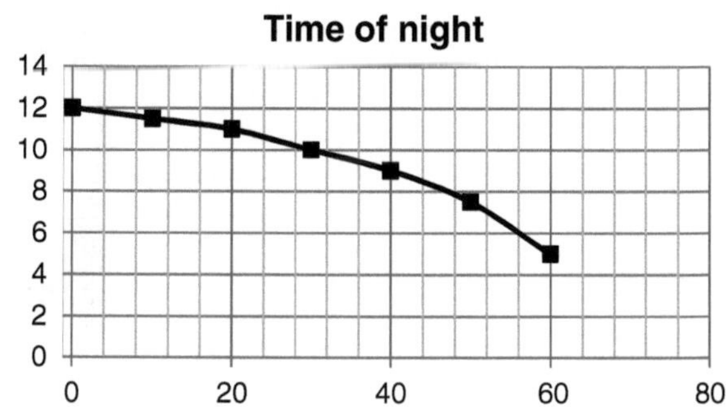

Fig. 22.12 Data and a graph in a concrete model

2.3 Fostering Students' Modelling Skills

The following discussion relates to the third question. For scientists and others who use mathematics as a tool, modelling processes and modelling skills are familiar. However, in school, students do not know it well. Control of assumptions is a challenge for students.

In a previous study (Ikeda and Stephens 1998), we asked undergraduate students to solve an oil pipeline problem. Our analyses of their reports revealed that even students who are good at mathematics set too many variables to solve it. In contrast, another student formulated a problem which ignored the essence of the problem. Responses of many students did not even make sense. How can the teacher help students to realize how to control many variables to solve a real-world problem? One method is to inform them of why they need to control the number of variables. They can generate related variables; subsequently they need to check whether or not the generated variables affect problem solving. Additionally, they must consider 'Is it possible to solve this using my acquired mathematics knowledge?' It was a good occasion for students to understand why they need to control many variables when encountering a conflicting situation. For example, they will appreciate key ideas to set assumptions as simple as possible at the beginning, modifying them gradually later into more general situations.

As one example, the mirror problem is illustrative. This problem is an extremely popular one in Japan, and many Japanese teachers have treated this problem (Matsumoto 2000; Shimada 1990, pp. 41–54). 'What size of mirror do we need to see at least your entire face? The following is only a small part of classroom teaching for Japanese junior high school students at grade 9 (Ikeda and Stephens 2010). After students' experiments, the teacher asked students, Is the following sentence true or false: At least a half-size mirror is needed to see my whole face? Students responded, it might be true because it is apparently half in this drawn figure. Another student said, it might be false because, if the mirror is far from my face, it is sufficient to use a small mirror. Then the teacher said to students, Let's draw a figure to check the answer. By picking up one student's drawing (Fig. 22.13), various variables were

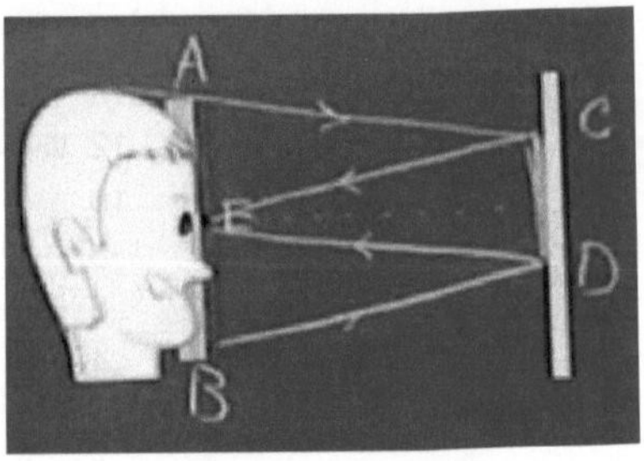

Fig. 22.13 Students' drawing

elicited from students: How about the face width? Are three points, namely the point of the eye, the point of head and the point of chin, on the same line? Is the relation of two lines, namely the face and mirror, parallel or not? and Is the eye located at the midpoint between the point of head and the point of the chin? After writing all variables on the blackboard, the teacher asked students, How can we treat these variables in this problem situation? Setting assumptions is necessary for students to proceed. For example, I pick up one variable, Is the relation of two lines of the face and the mirror parallel or not? Two students said, It is apparently easy to solve the problem if the relation of the two lines is parallel, and If the angle is not a right angle, then it is too difficult to solve the problem. Hearing two opinions, another student stated it from a different perspective, However, the relation of the two lines is not always parallel in a real situation. This situation is very conflicting. Students can appreciate the key idea that setting up an assumption so that the relation of two lines can be parallel at first, and in case they are not parallel, this can be considered later.

After solving the longitudinal (vertical) direction, it is important to consider the side width (horizontal direction) by changing the assumption. In this case, the difference from the previous analysis is the number of eyes. When we see one ear with two eyes, the mirror is larger than the longitudinal (vertical) direction (Fig. 22.14). However, when we see one ear with one eye, the mirror is smaller than the longitudinal length. That is to say, the mirror size is the width between the left eye and left ear (Fig. 22.15). After solving the problem, checking the answer is important. 'Is it okay in any situation? Is there any mistake?' If the distance between two eyes is great, the centre is invisible, as it would be for a horse. We should add the assumption that the 'width between two eyes is shorter than double of width between left eye and left ear'.

It is also possible to consider the general case: The mirror size when the face length is 1; y, the declination angle of the mirror; θ, location of eye in face; $\alpha:\beta$, $\alpha+\beta=1$; distance between face and mirror; d (Fig. 22.16). By applying the similar relation of triangle, the following general formula is elicited: $y=d\cdot\beta\cdot\sin\theta/(2d-\beta\cdot\cos\theta)+d\cdot\alpha\cdot\sin\theta/(2d+\alpha\cdot\cos\theta)$. In the special case such as $\theta=90°$, we can check that y is 1/2.

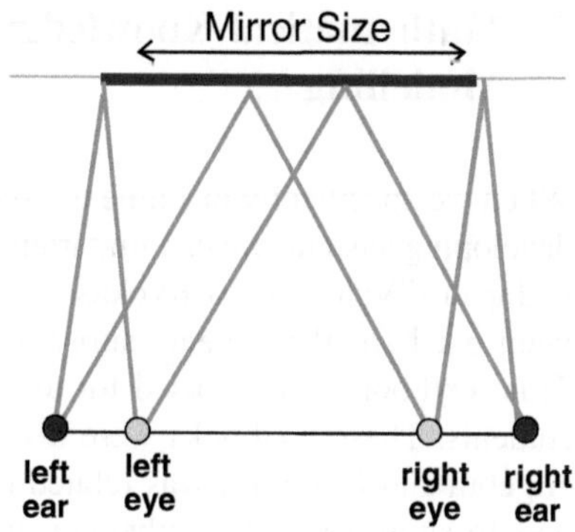

Fig. 22.14 Seeing one ear with two eyes

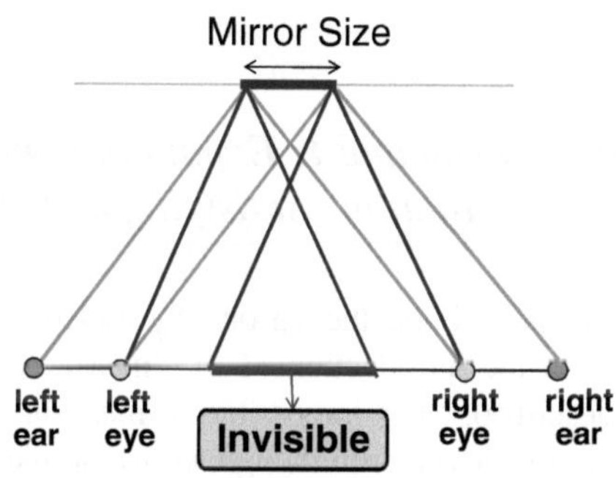

Fig. 22.15 Seeing one ear with one eye

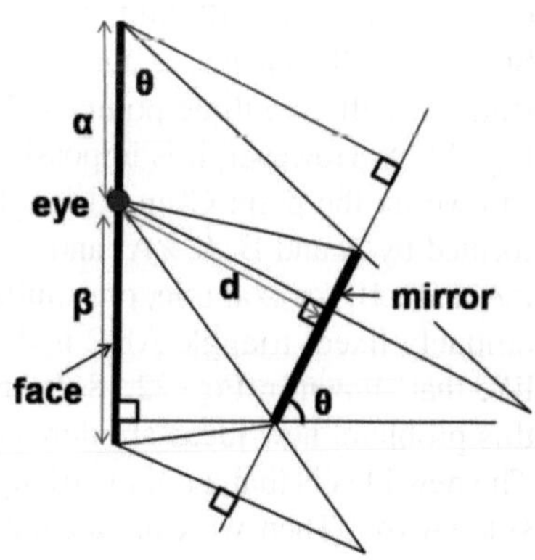

Fig. 22.16 General case

3 Mathematical Knowledge Construction Through Modelling

When we specifically examine the standpoint by which mathematics is viewed as a developing system, we can find traditional textbooks published before World War 2 in Japan (Tyuto Gakko Kyokasho Kabushiki Kaisya 1943). These textbooks had been used for the 5 years subsequent to the 6 years of elementary education. This textbook was not used for teaching everyone, but was used instead for elite students. These textbooks were divided into two parts: Dai Ichirui was related to Algebra and Dai Nirui was related to Geometry. The objective of these textbooks was to show students mathematical concepts and methods abstracted from real-world situations. By contrasting scientists' and mathematicians' activities with students' activities related to mathematics knowledge construction through modelling as described in 1.2.2, we elicit three principles by reflecting on historical textbooks, even though these are not described explicitly in a teacher's textbooks.

3.1 Principle 1: Expanding and Clarifying Real-World Situations Satisfying a Developed Original Model

In this textbook, the reason why students must solve a problem is emphasised. At grade 7, in the introduction of the 'measurement' unit, a problem with an illustration is described. 'It is impossible to measure the distance directly from here to an island across the sea even though we would like to know it. Of course it is possible to judge the distance with the eye. However, we would like to know it precisely, and it is necessary to devise some way of doing it.' The explanation is intended to mathematise a problem from a real-world situation into an abstract representation. 'Is there any method to measure the distance AC?' (Fig. 22.17) A previously developed system is recalled. Students will think, If we have a scale figure, then it might be possible .... Students will translate a concrete situation into an abstract representation. Then students will take three points and draw a scale figure, namely triangle A′B′C′ in Fig. 22.18. However, it is impossible to measure A′C′ because C′ is not fixed. How can we fix the point C′ in a triangle A′B′C′? One method is to measure the angle formed by A and B. If $\angle A'$ and $\angle B'$ is fixed, then point C′ is fixed. By measuring $\angle A$ and $\angle B$, for example, presuming that $\angle A$ is 110° and $\angle B$ is 28°, then point C is uniquely fixed: triangle ABC is determined. Students can draw a scale figure just like that shown in Fig. 22.18. It becomes possible to measure A′C′. After solving this problem, new ideas are derived by reflecting their thinking process in a class. The new idea is that a unique triangle is determined by two angles and the included side (ASA). Then we can calculate the real distance using the ratio between two similar triangles. It was called the 'reduction method'.

It is important to clarify the situations in which this idea is likely to be useful. By expanding real-world situations satisfying the developed ideas, this kind of idea will

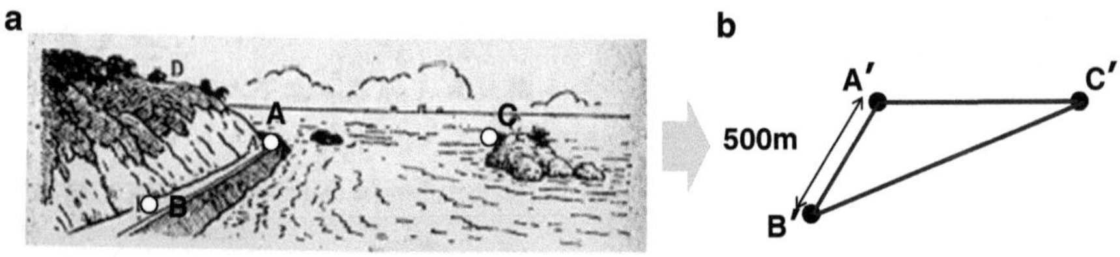

Fig. 22.17 (**a**) Island problem (**b**) Mathematisation

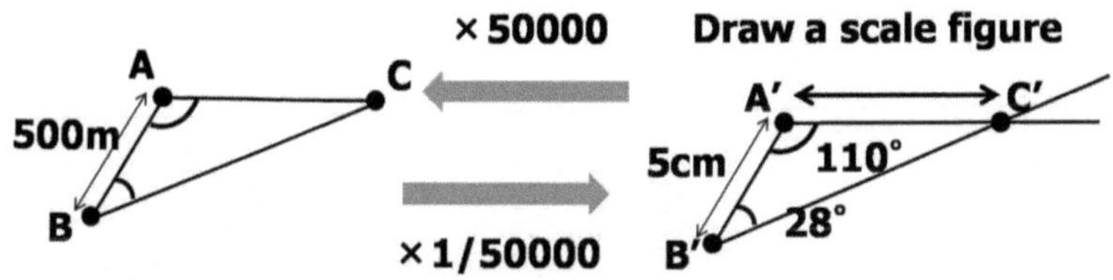

Fig. 22.18 Drawing a scale figure

be clarified. This activity is suggested by scientists' activity (see Sect. 1.2.2). Up to now, we have examined how to measure the distance between one point and another point for which it was impossible to measure. Therefore, the question is posed: Are there any other situations satisfying these new ideas? Students will encounter new situations, namely how to measure distances between two distant points. As one example, Fig. 22.19 is the distance between two trees across the river. It is impossible to apply the previously developed method. It is therefore necessary for students to develop new ideas.

Some additional activity to expand real-world situations satisfying previously developed methods is sought. Fig. 22.20 shows an illustration of a monument. The problem presented here is 'let us consider how to measure the height of an object that is far from here'. A real-world situation differs from previous situations; however, from mathematical points of view, we can identify it as a similar case. We can apply the previously developed method.

3.2 *Principle 2: Expanding and Integrating Mathematical Knowledge by Setting Up a Concrete Situation That Students Can Consider*

When considering the balance between modelling and constructing mathematical knowledge, it is suggested for us to examine the dual meaning of a model. The first role is to build up the model to mathematise and solve real-world problems. In this role, the activity is abstraction from the real world to a mathematical model. The

Fig. 22.19 Distance between trees across the river

Fig. 22.20 Height of some distant edifice

second role is to build up the model to test the validity of mathematical concepts. The activity is translation of mathematical concepts to a real-world model. This activity is suggested by the mathematician's activity, as noted in Sect. 1.2.2. In the teaching of mathematics, clarifying which world the problem is derived from is important for the teacher.

In this textbook, we can regard the second role of a model. Mathematical questions are posed for students at first, such as 'Let's consider how to determine the code (plus or minus) if you multiply a positive or a negative number by a negative number.' It is difficult for students to consider this problem without a concrete situation. Therefore, a concretized model is developed: an upward slope from west to east. In this concretised model, if we go east x m horizontally, then height increases 0.2 times x m. The mathematical question is translated into a concretised situation as, 'If x is a negative number, then how can we define the rule?' By considering the concretised model (Fig. 22.21), mathematical rules such as 'a positive number times a negative number is a negative number' are determined.

One more question is posed: a negative number times a negative number. We should develop a different kind of concretised model, with a downward slope from west to east. The mathematical problem is translated to a concretized model as 'If we go east x m in the horizontal direction, then the height is increased -0.2 times x m. If x is a negative number, then how can we define the rule?' By considering the concretised model (Fig. 22.22), mathematical rules are determined so that a previous rule applied in a positive number is also included in a new developed rule.

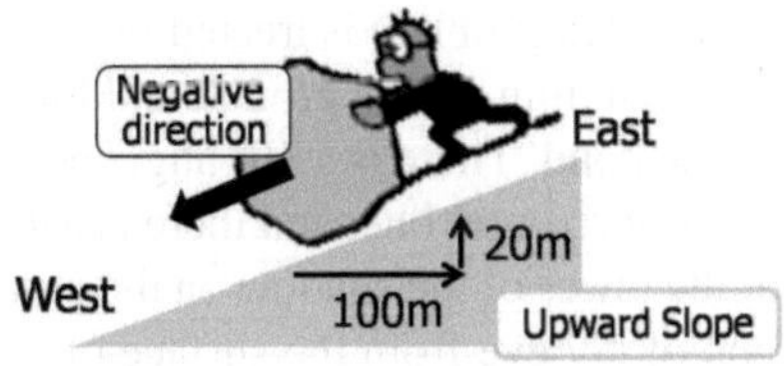

Fig. 22.21 Concretized model of 'a positive number times a negative number'

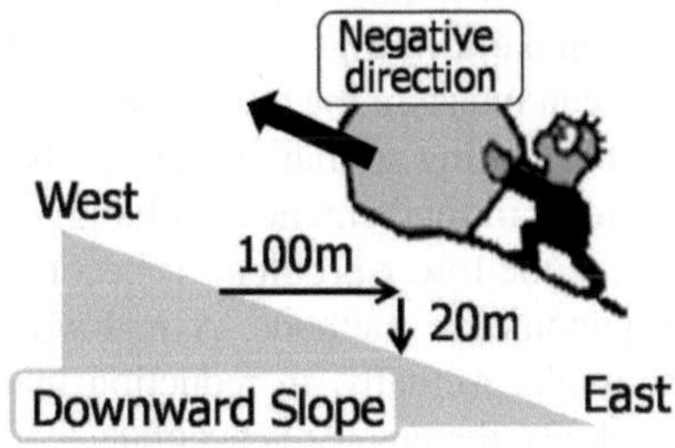

Fig. 22.22 Concretized model of 'a negative number times a negative number'

At the junior high school level, many problems are posed which are non-realistic situations. When students think about the real world, they cannot make sense of it. For example, a simultaneous equation problem might be posed such as 'He bought a cake for 230 yen and a loaf of bread for 80 yen, and the total cost is just 2,000 yen. How many cakes and loaves of bread has he bought?' When considering a real-world situation, students wonder 'Why did he not know the number? He bought them.' Some teachers want to develop more realistic situations so that simultaneous equations become effective in a real-world situation. However, we should consider alternative methods. We regard this problem as derived from the mathematical world, and seek to develop a concretised model. No problem arises even though the problem is non-realistic, if we truly think about it.

3.3 *Principle 3: Refining and Clarifying Developed Mathematical Methods by Treating Instances of Similar Contexts Repeatedly*

In these historical textbooks, the same context is treated repeatedly. For example, at grade 7, an open-ended problem is treated like this. 'Let's devise methods for how to measure the height of a mountain using the angle between a line from my eye to the top of the mountain and a horizontal line, and the angle between a line from my eye to the top of the mountain of the shadow on the pond and a horizontal line.' Then during grade 9, before and during learning of trigonometric methods, the formulated problem is treated twice. The formulated problem is cast in these terms: I am standing on the ground and looking at the top of the tower and a shadow of it on the pond. There is a 30° angle between a line from my eye to the top of the tower and a horizontal line, and there is a 40° angle between a line from my eye to the top of the tower of the shadow on the pond and a horizontal line. The height of the location of my eye from the surface of water is 3 m. Find the height of the tower from the surface of water by constructing the geometric figure (Fig. 22.23a). This problem is solved by construction (Fig. 22.23b). To construct the figure, 3 m is reduced to 3 cm: 1:100 scale reduction. Then take 30° and make a line. Next, take 40°, however it is impossible to do it. Therefore, taking 3 cm from the opposite side and make a line just like Fig. 22.23b. The intersection is the top of the tower. We can measure the tower's scale height then make the length 100 times.

By treating similar contexts, it becomes possible to appreciate the different subsystems in mathematics, including an awareness of their strengths and limitations. Using the tower problem, it becomes possible to contrast reduction methods with trigonometric methods. A real-world problem situation is treated as a concrete example. Actually, in reduction methods, the error is large, but in trigonometric methods, the answer is accurate.

It is possible to select two more problems that were treated before. The first is 'How long does it take from the beginning of sunset to the end of sunset?' We can see that reduction methods engender large errors. That is the important limitation of reduction methods. However, if students study trigonometric methods, then they can calculate the time accurately. Sin θ equals 7/1,500, or 0.00466. Using a trigonometric table (or a calculator), students can apply a double-angle formula, triple-angle formula, and fourth-angle formula. Students can restrict the range of θ, and also the range of time. Consequently, the time of the sunset is between 2 and 2.7 min. Using a calculator, the answer can be found as about 2.1 min. We can check the time by contrasting the results of real measurement.

Another example is that of night time. A further question comes to mind: 'Is it possible to develop a function for north latitude and night time?' By formulating a mathematical problem, trigonometric methods become applicable (Fig. 22.24). The location of city with latitude θ is P. The assumption is that the midsummer angle α is 23.4°. Line M is perpendicular to line L passing through point P. Viewed from above (Right figure in Fig. 22.24), PR is diameter, β is the pursuing angle, and the time from sunset to sunrise is described as '$24 \times 2\beta/360$'.

By solving the problem mathematically, we can obtain the following function, in which x is the north latitude (degree) and y is night time (hours):

$$y = (2/15) \times \cos^{-1}\{\tan(23.4) \times \tan x\} \qquad \text{formula (22.1)}$$

It is possible for students to draw a graph using trigonometric functions with technology (Fig. 22.25). What can students find that out from the graph? In middle-latitude countries such as Japan, the difference of night time is short, but in high-latitude countries such as Norway, the difference of night time is huge, even

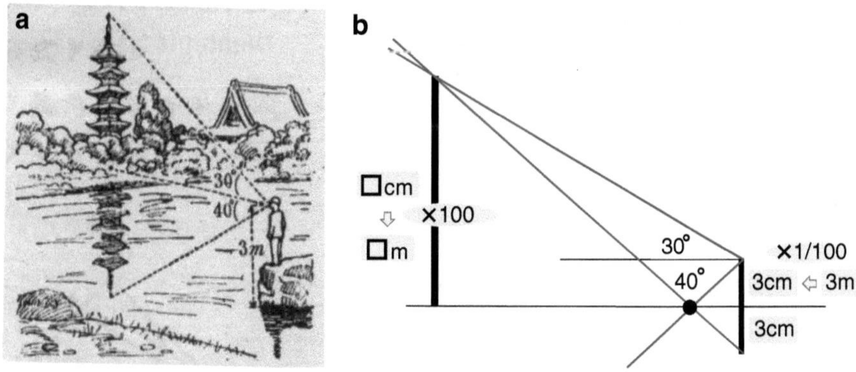

Fig. 22.23 (**a**) Tower problem, (**b**) Reduction method

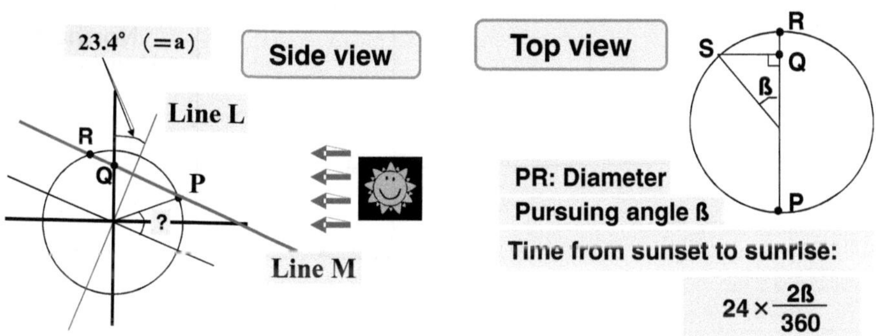

Fig. 22.24 Formulated problem of night time

though the latitude interval is shorter than that of Japan. At latitudes greater than 66.6°, it is 'night of the midnight sun'. Another important issue is that the graph presents the values obtained when x is a negative number (Fig. 22.26). How can students interpret this result? Actually, it shows the function for the southern hemisphere. The mathematics presented more information than the students considered.

After developing the function for midsummer, we can generate the night time function after t days from the day of the vernal equinox. The 'tan 23.4' in formula (22.1) is changed into 'tan θ', in which θ means the declination of axis of earth. In fact, tan θ is represented as follows from the vernal equinox to the autumnal equinox.

$$\tan\theta = \tan(23.4)\times\left|\tan\alpha\right| / \sqrt{\tan^2\theta + 1}, \quad \alpha : 360t/360$$

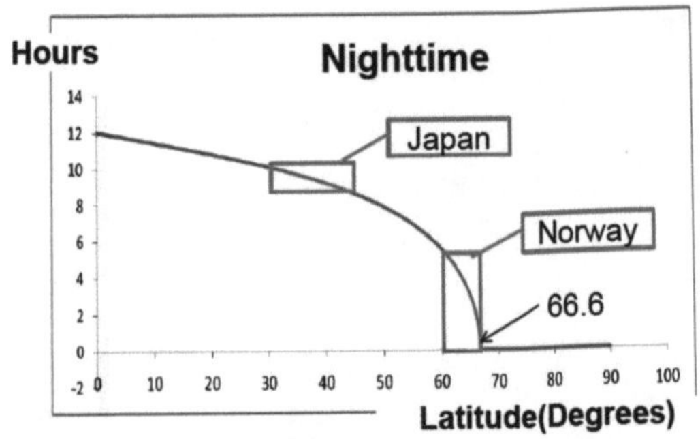

Fig. 22.25 Graph of night time

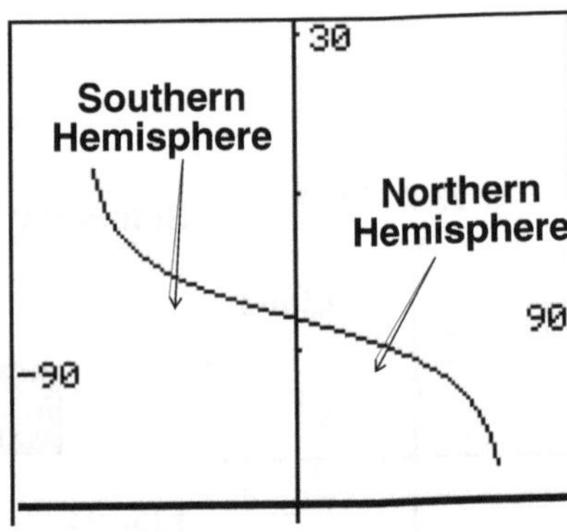

Fig. 22.26 Graph presents x as a negative number

Then, the night time function is produced as follows. In the formula, 'SIGN' is the function used in Excel (Microsoft Corp.) which presents '1' when the number is positive and presents '−1' when the number is negative.

$$y = (2/15) \times \cos^{-1}\left\{\tan x \times \tan(23.4) \times SIGN(180-\alpha) \times |\tan\alpha| / \sqrt{\tan^2\alpha + 1}\right\}$$

By devoting attention to the refraction of air, the definition of sunset and sunrise and so on, the following formula is modified: night time is shorter than daytime as 1/6 h at vernal equinox.

$$\mathrm{y} = \left(\frac{2}{15}\right) \times \cos^{-1}\left\{\tan x \times \tan(23.4) \times SIGN(180-\alpha) \times \frac{|\tan\alpha|}{\sqrt{\tan^2\alpha + 1}}\right\} - 1/6$$

The different sub-systems in mathematics can be appreciated, along with an awareness of their strengths and limitations. It is also possible to contrast deductive methods with inductive methods by referencing a real-world situation as a concrete example.

References

Blomhøj, M. (2009). Different perspectives in research on the teaching and learning of mathematical modelling – Categorising the TSG21 papers. In M. Blomhøj & S. Carreira (Eds.), *Mathematical applications and modelling in the teaching and learning of mathematics* (IMFUFA tekst nr.461, pp. 1–18). Roskilde: Roskilde University, Department of Science, Systems and Models.

Blum, W., & Niss, M. (1991). Applied mathematical problem solving, modelling, applications, and links to other subjects: State, trends and issues in mathematics instruction. *Educational Studies in Mathematics, 22*(1), 37–68.

Dewey, J. (2007). *How we think.* Stilwell: Digireads.com Publishing.

Galbraith, P. (2007). Authenticity and goals – Overview. In W. Blum, P. L. Galbraith, H.-W. Henn, & M. Niss (Eds.), *Modelling and applications in mathematics education. The 14th ICMI study* (pp. 181–184). New York: Springer.

Hesse, M. B. (1966). *Models and analogies in science*. Notre Dame: University of Notre Dame Press.

Ikeda, T., & Stephens, M. (1998). The influence of problem format on students' approaches to mathematical modelling. In P. Galbraith, W. Blum, G. Booker, & I. D. Huntley (Eds.), *Mathematical modelling – Teaching and assessment in a technology-rich world* (pp. 223–232). Chichester: Horwood Publishing.

Ikeda, T., & Stephens, M. (2010). Three teaching principles for fostering students' thinking about modelling: An experimental teaching program for ninth grade students in Japan. *Journal of Mathematical Modelling and Applications, 2*(1), 49–59.

Iyanaga, S. (2008). *How to learn mathematics*. Tokyo: Chikumasyobou Pubilisher (in Japanese).

Matsumoto, S. (2000). Activities that expand mathematical world through making mathematical models – Considering length of mirrors that whole body can be reflected. *Journal of Japan Society of Mathematical Education, Mathematical Education, 82*(1), 10–17 (in Japanese).

Niss, M. (2008). Perspectives on the balance between applications and modelling and 'pure' mathematics in the teaching and learning of mathematics. In M. Menghini, F. Furinghetti, L. Giancardi, & F. Arzarello (Eds.), *The first century of the International Commission on Mathematical Instruction (1908–2008)* (pp. 69–84). Roma: Instituto DELLA, Enciclopedia Italiana, Fondata da Giovanni Treccani.

Osawa, H. (2004). Development of applications and modelling by action research in Japanese secondary school. In H.-W. Henn & W. Blum (Eds.), *ICMI Study 14: Applications and modelling in mathematics education. Pre-conference volume* (pp. 199–204). Dortmund: University of Dortmund.

Programme for International Student Assessment – PISA. (2009). *PISA 2009 assessment framework – Key competencies in reading, mathematics and science*. Paris: OECD.

Shimada, S. (1990). *Collected problems for teachers*. Tokyo: Kyouritsu Publisher (In Japanese).

Tyuto Gakko Kyokasho Kabushiki Kaisya. (1943). *Mathematics 1–3 (Textbooks at grade 7–9) Dai Ichirui and Nirui*. Tokyo: Sanseido (in Japanese).

Chapter 23
Complex Modelling Problems in Co-operative, Self-Directed Learning Environments

Gabriele Kaiser and Peter Stender

Abstract After a short description of the development of the current discussion on teaching and learning mathematical modelling special modelling activities are described. These so-called modelling weeks or days deal with complex modelling problems, which are focusing on students in an autonomous learning environment. Evaluation results of these modelling activities are presented pointing out, that most of the participating students from upper secondary level appraised these positively. Finally scaffolding as an approach to support students' independent modelling processes is discussed and first results of an empirical study on interventions in modelling activities with complex modelling problems are presented.

1 Theoretical Framework for Modelling in School

Currently, responsible educational institutions like politicians, curriculum designers and educational committees agree worldwide on the aim for mathematics education to include mathematical modelling into mathematics education departing from the central goal of mathematical literacy put forward by the large-scale international study PISA. Under this perspective mathematical education shall offer students the opportunity to become a responsible citizen, who is able to solve mathematics-related problems using mathematical means in a competent and reflective way. In order to achieve this goal real world examples tackled independently by the students need to be incorporated into mathematics education.

G. Kaiser (✉)
Faculty of Education, University of Hamburg, Hamburg, Germany
e-mail: gabriele.kaiser@uni-hamburg.de

P. Stender
Faculty of Education, Psychology, Human Movement, University of Hamburg, Hamburg, Germany

G.A. Stillman et al. (eds.), *Teaching Mathematical Modelling: Connecting to Research and Practice*, International Perspectives on the Teaching and Learning of Mathematical Modelling, DOI 10.1007/978-94-007-6540-5_23,

However, many different goals and intentions combined with the teaching of mathematical modelling have been developed in the last decades leading to different approaches to the teaching and learning of mathematical modelling (for an overview see Blum et al. 2007). In principle two different directions can be identified: (a) under the headline 'mathematics for applications, models and modelling', the usage of mathematics of extra-mathematical contexts for extra-mathematical purposes is emphasised as an important activity in itself; in contrast to (b) the headline 'applications, models and modelling for the learning of mathematics', in which the utilisation of mathematics in extra-mathematical contexts for extra-mathematical purposes, which can support the learning of mathematics in various ways, is in the foreground. This contrast has shaped the debate on the teaching and learning of mathematical modelling for decades since its beginning and has significant influence on the way of integration of modelling examples into mathematics education and the kind of modelling examples used.

Various descriptions of goals for the inclusion of mathematical modelling into classrooms have been developed in the last decades ranging from pedagogical goals, which emphasise within modelling activities that students acquire abilities to understand central aspects of our world in a better way to subject-related goals, which intend to structure learning processes or introduce new mathematical concepts and methods including modelling as illustrations (see Blum et al. 2007). Departing from these descriptions Kaiser and Sriraman (2006) developed a classification of the debate on mathematical modelling in school distinguishing several perspectives on mathematical modelling according to the aims pursued with mathematical modelling. The projects described in this chapter are embedded in the perspective of the so-called realistic or applied modelling fostering pragmatic-utilitarian goals and continuing traditions of pragmatically oriented approaches developed within approaches coming from applied mathematics. A key characteristic of these various perspectives is the way how the mathematical modelling process is understood, and how the relation between mathematics and the rest of the world is described. Analyses show, that the modelling processes are differently specified by the various perspectives and streams within the modelling debate. The realistic or applied perspective as the theoretical basis of the projects described in this chapter put the solution of the original problem in the foreground in contrast to theory development as other perspectives do. Within this realistic perspective on modelling and closely connected perspectives, various modelling cycles have been developed, but they all share as a common core the description of the idealised process of mathematical modelling as a cyclic process to solve real problems by using mathematics. The following phases are central components of the modelling cycle: The given real problem is simplified in order to build a real model of the situation, amongst other things many assumptions have to be made, and central influencing factors have to be identified. To create a mathematical model the real model has to be translated into mathematics. Inside the mathematical model mathematical results are worked out by using mathematics. Interpreting the mathematical results leads to real results, which have to be validated within the real situation and/or within the real model. It might be necessary to carry out single parts of the modelling process or the whole process once more or even several times. The version of the modelling cycle in Fig. 23.1 is used in the empirical studies, which will be

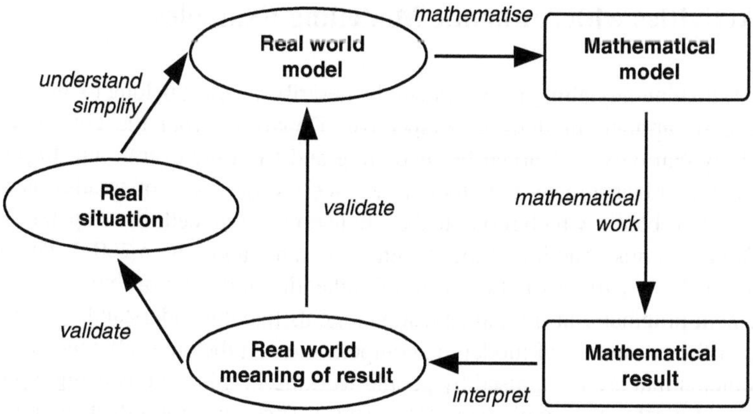

Fig. 23.1 Modelling cycle

described in the following sections and is a slight variation of a description proposed by Maaß (2005) and a further development of earlier proposals.

Although the modelling process is described as cyclical, it is nowadays a consensus that modelling processes consist of moving forward and backwards from the real world to mathematics in order to solve the real world problem with several mini-loops within the process (Borromeo Ferri 2011).

The competencies needed for these kinds of modelling processes are still the topic of current controversial debate. The following consensus has been reached in the last decade (for an overview see Blomhøj 2011): It seems to be inevitable that students develop partial competencies for conducting the single phases of a modelling process as well as an overall competency to carry out the whole modelling process on their own. In addition metacognitive modelling competencies are necessary, that means the ability and the willingness to monitor and reflect about one's own modelling process based on metacognitive knowledge. Furthermore social competencies such as the ability to work in a group and to communicate about and via mathematics are emphasised. Especially metacognitive and social competencies seem to be necessary for the complex modelling examples proposed by the realistic or applied perspective.

Summarising the current state-of-the-art, one can state, that despite the prolific research on modelling there is still a lack of strong empirical evidence for, how to integrate modelling examples into mathematical teaching, although several research projects have been launched in the last years evaluating the effects of various ways of including modelling examples into school practice (for an overview on the more recent debate see Kaiser et al. 2011 and the 14th ICMI Study, Blum et al. 2007). It is especially an open question, how complex authentic modelling problems put forward by the realistic or applied perspective on modelling can be integrated into mathematics education, what kind of learning environment is necessary, whether a change in the role of the teacher to a coach or mentor of the students is needed. In the following we will discuss research activities around this distinctive approach to including modelling into mathematics education, namely the usage of complex modelling examples into co-operative, self-directed learning environments in school.

2 Activities with Authentic Modelling Examples

In the following modelling activities will be described, which belong to the so-called realistic or applied modelling perspective. Departing from the consensus that modelling can only be learned by modelling and from the overall aim to promote responsible citizenship by the students we see the necessity of treating authentic modelling problems, which promote the whole range of modelling competencies and broaden the radius of action of the students (Blomhøj and Jensen 2003). The central feature of these activities is the usage of authentic modelling problems in order to implement pragmatic-utilitarian educational goals like the understanding of the real world or the promotion of modelling competencies. Authentic problems are defined as problems that are recognised by people working in this field as being a problem they might meet in their daily work (Niss 1992). These examples then should articulate the relevance of mathematics in daily life, the environment and the sciences, and impart competencies to apply mathematics in daily life, the environment and the sciences. The conducting of project-oriented modelling activities embedded in co-operative, self-directed learning environments is considered to be a powerful and effective way to promote these goals. In the following we will describe only briefly these activities and refer the reader for details to Kaiser and Schwarz (2010).

The project 'Mathematical Modelling in School' was established in 2000 by the Department of Mathematics (Jens Struckmeier and Claus Peter Ortlieb) in co-operation with the working group on Didactics of Mathematics at the Department of Education (Gabriele Kaiser) at the University of Hamburg and aimed to establish a bridge between university and school. It has undergone several changes in the last years, but the project has kept the two components, namely complex modelling examples dealt with by students in school tutored by future teachers, who carry out these activities in the frame of their university study, as part of their mathematical and their didactical courses.

Since its establishment, one aim of the project has been that the participating students will acquire competencies to enable them to carry out modelling examples self-directed, that is they should be able to extract mathematical questions from the given problem fields and develop on their own the solutions of real-world problems. The purpose of this project is not to provide a comprehensive overview about relevant fields of the application of mathematics. An overarching goal is that students' experiences with mathematics and their mathematical world views or mathematical beliefs are broadened. This kind of approach can be described as a holistic approach, using the terminology of Blomhøj and Jensen (2003), that is a whole mathematical modelling process is carried out covering all modelling competencies described above. Adapting a phrase by Schoenfeld on problem solving one can say that modelling is not a 'spectator sport', but can only be learnt through own activities. The teaching-and-learning-process is characterised as co-operative self-directed learning, that is, a process in which the students decide upon their ways of tackling the problem and no fast intervention by the tutors takes place. The future teachers working as tutors are expected to offer no more than just assistance, if mathematical means are needed or if the students are heading into a cul-de-sac. With this kind of teaching approach the students experience long phases of helplessness and insecurity, which is an important aspect of modelling and a necessary phase within the modelling process. Overall, no high technical level of mathematical knowledge is expected, the complexity of the problems comes from the context, and from making assumptions, simplifications, and interpretations. The challenge does not come from the algorithms used, but lies in the independent realisation of the modelling process.

Within this project modelling activities in various organisational forms were carried out. In the last years we have concentrated on the following two forms:

- modelling activities using the form of a modelling week within the school year, which take place in the university with students from upper secondary level, twice a year with about 200 students from schools from Hamburg and its surroundings; and
- modelling activities in the form of so-called modelling days lasting 3 days and which are carried out in schools with students at the end of the lower secondary level covering the whole age cohort of this school and which are offered by us once a year with several hundred students participating.

The idea of these project oriented-modelling weeks has been developed at the University of Kaiserslautern by the working group of Helmut Neunzert (for descriptions and examples see for example Kaiser et al. 2013), who has already been running modelling weeks for more than a decade. In contrast to other forms of modelling weeks, we aim at working with average students who are not especially interested in mathematics.

In the following we describe one example in detail, the optimal design of a bus stop, which was used in several modelling activities in the last years.

The Bus Stop Problem refers to the following situation: A bus drives through in area, where people live on both sides of the road. The question is: At which distance should the bus company place the bus stops? The underlying problem is that a very small distance between two bus-stops leads to an increase in the travel-time of the bus; on the other hand, big distances between the bus stops lead to long walking distances from the home of the people to the bus stop. These two contradicting issues bring up the question of whether there is an optimal bus-stop-distance.

The following assumptions can be made in order to develop a real world model:

- The bus route is a straight line on a major street.
- The overall travel time includes the walking time to the bus stop, the driving time between the bus stops and the time the bus needs to accelerate and the people to get on the bus.
- The best distance between two bus-stops is found, if the overall travel time for a randomly picked traveller is minimal.

In order to develop a mathematical model and, based on this, a mathematical solution, the unknown distance between two bus stops, the length of the total bus line and the velocity of the bus need to be taken into account. To calculate the walking time to the bus stops, additional assumptions are necessary with respect to the total walking distance and the walking velocity of the person. As a third influential group of factors, the time for decelerating, alighting from the bus, boarding and accelerating, needs to be estimated. This leads to a function describing the travel time as a function of the walking time, the travel time between the bus stops and the time for boarding and alighting. The minimum of the travel time in relation to the total travel distance can be calculated in different ways according to the level of mathematical knowledge, either by evaluating a graph or by determining the extreme value.

Before we present the results of the evaluation, we describe the modelling weeks from the perspective of the students using the written words of two female students, who wrote this report for their school journal (school name and the names of teachers and students are pseudonyms).

How much acid sulphur can the Rhine withstand?

Mathematical modelling week 27/01/12 – 03/02/2012

Math is not just mindless opening the book, use numbers and write down results. The question raised often in class: 'And for what do I need it?' was heard by the UHH (University of Hamburg, authors), who designed for students the mathematical modelling week. The motto was 'Use maths in real life.' 22 schools participated in the 7th modelling week, including our school August-Bebel comprehensive school. The maths courses from Ms. Miller and Mr. Smith of S2 (second semester of year 12, authors) came together for a week and worked intensively on the topic, 'pollutant dispersion in the Rhine'.

The week was launched with a plenary meeting of all participating schools. The various proposed themes were presented and the leading professors presented the weekly schedule. Then we were assigned to two maths students. The first finding: maths students can also look good!

With an easy entry task we warmed ourselves up. The following 3 days were characterised by much confusion, but also new insights and many aha moments. Despite some obstacles, we mastered our group task with flying colours. Some of us went on longer in poster design, others in formulas and graphs. The enquiry was at the forefront too, because nothing runs without background knowledge! On the last day there was an exhibition of all the posters, as well as a Powerpoint presentation, which was presented by the computer scientists amongst us, well, compact and consistent. To conclude, our poster was chosen as second best and as prize there was chocolate for everybody, a lanyard and a Rubik's Cube. So the week was over quickly, many new experiences were made and the question, where maths is hidden in everyday life has been definitely clarified for us. The modelling week is a great project of the UNI Hamburg. Thanks for the chance to get to know maths differently!

Sarah Lueck, Laura Moeller, S2

August-Bebel Comprehensive School

3 Empirical Results of the Evaluation

In the following we report about results of the evaluation of the modelling activities described in the last section limited to the evaluation of four modelling weeks from 2009 to 2011. The modelling week in September 2009 has not been evaluated due to organizational problems. We evaluated the beliefs and the appraisal of the

Table 23.1 Number of participating students in the evaluation

Date	Number of students participating in the evaluation
March 2009	281
March 2010	170
September 2010	162
March 2011	182

participating students of their tackled modelling example. A description of the development of modelling competencies can be found amongst others in Kaiser (2007). For this evaluation we used a questionnaire with seven half-standardised questions on the beliefs of students about mathematics and mathematics teaching and their interest in mathematics teaching. In addition three open questions on the appraisal of modelling examples tackled in the modelling week and four closed questions to be answered on a 5-point-Likert-scale were asked. The questionnaire was filled in at the end of the modelling weeks by most of the participating students. The open questions were evaluated based on methods of Grounded Theory (see Strauss and Corbin 1990). We evaluated the open questionnaires using in-vivo-codes, that is, codes extracted out of the text written by the students as verbatim quotation and grouped them to related quotations under a theoretical perspective. In the following we describe the appraisal of the modelling examples due to space limitations and refer for results of the open questions restricted to the first modelling week to Kaiser and Schwarz (2010).

Table 23.1 shows the number of students who returned the questionnaire, the return rate was approximately 75 % of the overall participants.

3.1 *General Results of the Evaluation*

One central aspect is the question, whether the students want to have modelling examples included in their usual mathematics lessons or whether they want to keep these kinds of ambitious examples out of their usual lessons. This aspect is dealt with in the next question, where the students are asked the following question: "Should these examples be increasingly dealt with as part of regular maths classes or would you reject this?"

As shown in Fig. 23.2, the majority of students gave strong approval to the question whether these examples shall be dealt with in usual mathematics lessons, but with varying degrees ranging from 56 to 75 %. A significant number of students ranging from 21 to 34 % of the participating group rejected the inclusion of these examples in their usual mathematics teaching.

The variety of the answers is unexpectedly high, however as further analysis showed this was dependent on other factors, mainly the kind of modelling examples chosen and the learning success experienced by the students. An analysis of the negative answers shows as the main reason for this answer time constraints, the students are expecting. Other reasons given by the students such as "not interesting," "not

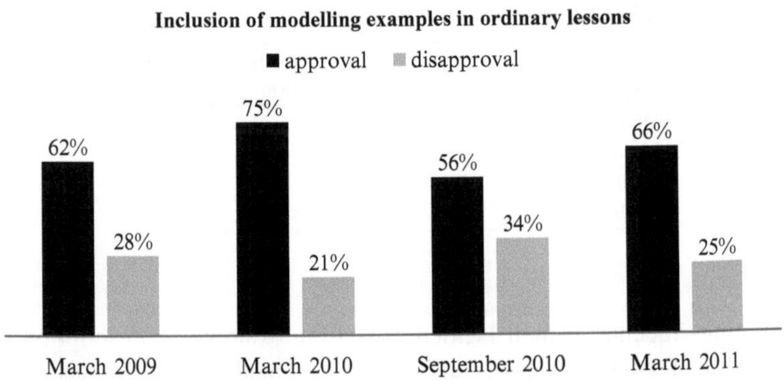

Fig. 23.2 Inclusion of modelling examples in ordinary lessons

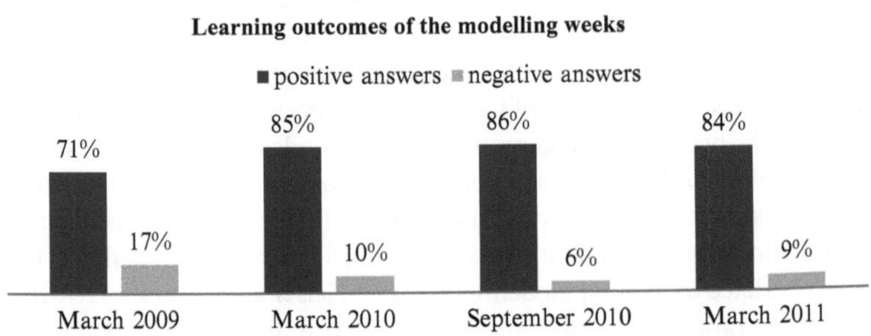

Fig. 23.3 Learning outcomes of the modelling weeks

relevant", "too difficult" or "imprecise" are less important compared to the time constraints and given by similar numbers of students. The results clearly show the time pressure students are experiencing in upper secondary level. Furthermore it shows that many students were simply not acquainted with independent work on open problems.

Another important part of the questionnaire dealt with the appreciation of the learning outcomes of the modelling week. The students answered the question "In your opinion, what did you learn when dealing with the modelling example?" Answers to the question were prevailingly positive ranging from 71 to 86 %, with only a small group of students not seeing any learning outcomes from the modelling week (see Fig. 23.3).

From Fig. 23.4 it can be seen, in which areas the students identified by themselves their learning results. In 2009, other codes were used, which yielded incomparable results and are therefore omitted.

These answers point out that the highest outcome was experienced in the area modelling competencies, however other aspects were important too such as working techniques, mathematical competencies and social competencies.

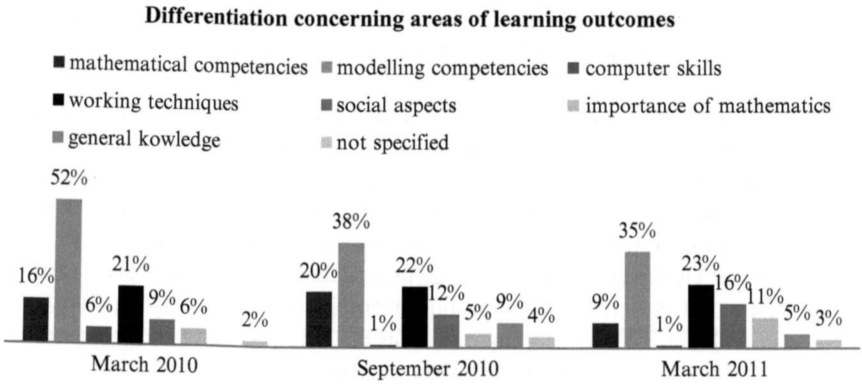

Fig. 23.4 Differentiation concerning areas of learning outcomes

Table 23.2 Number of students dealing with the most often chosen examples

Modelling example	Number of students
Optimal design of windpark	113
Design of ski jump inrun	63
Development of venereal disease of ladybugs	78
Optimal position of rescue helicopter	124
Optimal planning of water irrigation systems	199
Optimal design of bus stops	91

3.2 *Results of the Appraisal of the Modelling Examples*

In the following we concentrate on the evaluation of those modelling examples, which were dealt with most often within the modelling weeks in order to avoid a bias by examples only tackled rarely as chosen examples. Table 23.2 shows the frequency with which particular examples have been chosen.

The students were asked to rate the modelling example dealt with during the modelling week with respect to their interest in the content, the difficulty, the strength of the given structure, and the strength of the relation to the real world.

The answers to the question "How interesting was the problem, that you dealt with?" show an overall positive assessment. All examples except the ski jumping inrun were evaluated as interesting or even very interesting by about 60 % of the students. Only the ski jumping inrun was evaluated more negatively with only 42 % describing it as interesting or very interesting.

The question "How did you rate the difficulty of the modelling problem?" showed an overall judgement of the examples as being of adequate difficulty varying from 60 to 80 % of the students. An exception is the example venereal diseases of ladybugs, which is based on systems of differential equations and indeed very ambitious. The bus stop example was rated by nearly 70 % as being of adequate difficulty.

The question "How highly was the example structured?" showed, that apart from the rescue helicopter example about 40 % of the students evaluated the examples as reasonably structured. Only the rescue helicopter example was rated by less than 10 % of the students as unstructured. These results point to the fact, that a majority of the students were satisfied with the given structure of the problem description.

The answers to the question "How realistic was the modelling example?" showed some variety, but as a global tendency most examples were seen as being realistic or even very realistic. The ski jumping inrun was the only exception and was assessed by 40 % of the students as being unrealistic or even very unrealistic.

To summarise, the students' evaluation of the modelling weeks yielded positive results. According to the students' opinion the modelling tasks were adequately structured, mostly realistic, adequately difficult and interesting. Furthermore, most students described their impression of learning success in several areas, not only those limited to modelling competencies. However, not all students wished to include these kinds of modelling activities in ordinary teaching, mainly due to time constraints.

4 Interventions in Co-operative, Self-Directed Modelling Activities – First Results of an Empirical Study

In the following we will describe first results of a project focusing on the role of the teacher in co-operative, self-directed modelling activities with complex modelling examples. As it is the main goal of these kinds of activities, described in Sect. 2, to support the students in their self-directed modelling activities, the question arises: How can teachers support the students without destroying the independency of these modelling activities? The theoretical approach of scaffolding as a comprehensive, long-term approach combined with interventions as direct and immediate adaptive actions by the teachers form the theoretical framework of this project.

4.1 Theoretical Framework – Scaffolding and Adaptive Interventions

Teaching and learning modelling is a complex interactional process influenced by many factors ranging from the learning environment to teachers' knowledge. In the last decade within the framework of constructivist teaching-and-learning approaches emphasising own knowledge construction by the students and based on approaches for adaptive teaching, the pedagogical approach of scaffolding has been developed. Scaffolding aims at tailored and temporary support that teachers can offer students during autonomous teaching-and-learning-processes. The central goal of scaffolding approaches, as has been pointed out by Van de Pol et al. (2010) in their survey of the state-of-the-art, is to enable students to solve a problem on their own. Therefore adequate measures are provided in order to support the students in case they are not able to solve the given problem or when they are stuck. The support focuses on the means at the cognitive level (such as required strategies and concepts) and measures at the metacognitive level (such as instructing self-regulated learning). The main principle is a consequent orientation on the individual learning process. Van de Pol et al. (2010) name this 'contingency' as one of three central attributes of scaffolding. When students work on complex modelling tasks and can choose mathematical algorithms on their own, the teacher must be able to understand the thinking processes of the students and decide in a short time, if the student's way is goal-oriented or not. Depending on how self-regulated students are in this process, the teacher tries to reduce the support given to them, which is called 'fading' by Van de Pol et al. (2010), because the teachers are 'transferring the responsibility' to their students. Hammond and Gibbons (2005) define the interaction in classroom between teacher and students as micro scaffolding or interactional scaffolding, that is the assistance teachers give during teaching in order to enable students to reach a solution, they otherwise would not have achieved. Although, these interventions cannot be planned in advance in detail, the question arises, whether it is possible to develop guidelines for these kinds of interventions beforehand. Referring to the teaching and learning of modelling, the question arises, whether these interventions carried out by the teachers during the modelling process by the students are adequate. The adequacy of these interventions may depend on many variables amongst others on the circumstances when they are given, the modelling competency of the students and so on. Therefore, adequate interventions have to be a consequence of teachers' diagnosis of students' difficulties while solving mathematical problems or by other activities happening in the classroom. Hammond and Gibbons (2005) describe interventions as special kinds of micro scaffolding and distinguish it from macro scaffolding as a way by which teachers can provide or foster different didactical settings.

In the didactical research on scaffolding and interventions several theoretical approaches for the support of learners by teachers are discussed. A distinction between different kinds of interventions – well-known in the German debate – is offered by the taxonomy of assistance developed by Zech (1998), which refers to the principle of minimal help developed by Aebli (1983). This taxonomy developed by Zech (1998) for problem solving activities departs from the norm, that the intensity and strength of the intervention shall increase step by step in relation to the lack of success of the students and that these interventions shall support the students to develop a solution on their own, if possible. This taxonomy differentiates motivational, feedback, general-strategic, content-oriented strategic and content-oriented assistance. The intensity of the intervention increases gradually from the motivational assistance to the content-oriented assistance. This classification has been used in modelling activities already for a long time, but only at a practical level without empirical evaluation on its efficiency. Based on this categorisation, Leiß (2007) evaluated in a laboratory study within the DISUM project (cf. Blum 2011) the usage of various kinds of support given by teachers in modelling processes. The main results of Leiß's study were, amongst others, that strategic interventions are included in the intervention-repertoire of the observed teachers only very marginally and that the teachers often choose indirect advice where students have to find only one step by themselves in order to overcome the difficulty. Further studies from Link (2011) and Beutel and Krosanke (2012) achieved partly different results. Link's study departed from the taxonomy developed by Zech and identified in laboratory studies a high amount of general-strategic interventions. Both studies found that, in particular, strategic interventions lead to metacognitive activities of learners (for details see Link 2011 and for results by Beutel and Krosanke see next section).

Beside studies on this principle of minimal help, the role of metacognition within mathematical modelling for a basis of possible interventions was studied by Stillman et al. (2007). These studies identify mental or cognitive blockages, which prevent students from successful modelling. They emphasise the necessity of metacognitive activities of the students, that is students should observe their own modelling process as "looking over one's own shoulder". Stillman (2011) claims the necessity for teachers' reflections about students' metacognitive activity within the specific situation and with respect to the teacher's role in the modelling process and calls that a meta-metacognitive process. It is consensus within these studies that teacher interventions are necessary in order to facilitate reflective learning, for example, as teachers actions 'on the fly'.

4.2 Methodical Approach and First Results of an Own Study

In the following we will describe first results of a study, which aims to develop possible methods of micro scaffolding, or to phrase it differently, measures of adaptive interventions within self-directed complex modelling processes. In addition, the effects of these methods of micro scaffolding shall be evaluated. The study is embedded in ongoing more comprehensive activities at the University of Hamburg, which aim at implementing complex modelling activities into schools, amongst others modelling days and modelling weeks. In addition it is intended to develop a teacher training course for pre-service and in-service teachers where (future) teachers will acquire competencies to introduce and support mathematical modelling in school.

The design of the empirical study is shaped by the modelling days, carried out in February 2011 and 2012 at a higher track school of Hamburg. The whole school age cohort of year 9 (age 14–15 years) participated in these modelling days, about 160 students, with about 50 % female students. The groups of students have been supervised by future teachers, who were prepared for tutoring mathematical modelling in the frame of their master studies. The students were offered a choice between several modelling examples, amongst others the modelling task, *The Bus Stop*. In 2011 two groups of students, supervised by a female tutor, were videotaped for the whole period, two and a half days, the last half day was reserved for presentation. The tutor intended and practised an explicit usage of various kinds of scaffolding and intervention methods, amongst others the reflective usage of the modelling cycle described in Sect. 1 as a means for metacognitive aspects. After certain interventions, stimulated recall with selected students was carried out by using the respective video-taped sequence (for details of this method see Busse and Borromeo Ferri 2003). The stimulated recall was done on an individual basis, selected video-taped sequences were shown to the students and the students were asked, whether the experienced and just seen special kind of intervention was helpful or disturbing, adequate or not adequate for them. Likewise, stimulated recall was carried out with the tutor, and she was asked about her reasons for the intervention and the adequacy of her intervention. Both stimulated recall interviews were executed by other future teachers, who were closer to students than a mature researcher would have been. In 2012 ten groups of students with ten future teachers as tutors were video-taped. One group was analysed in detail in the frame of a master thesis (Beutel and Krosanke 2012). In total 160 h of video-taped lessons, 41 interviews as stimulated recall lasting six hours and 36 min were sampled. After a screening of the material, about 15 % of the video-taped lessons and 100 % of the interview material was transcribed verbatim following transcription rules.

The analysis of the large data set follows various methods, amongst others, as an overall framework the heuristics by Polya (1945), and the classification of the interventions developed by Zech (1998) with the increasing degree of intensity of offered help. In addition, interventions are connected to the various phases of the modelling process (see Table 23.3).

The ongoing evaluation of the data yielded the following preliminary results: A variety of interventions is within the scope of novice teachers, at least, if they are trained adequately. We will exemplify this result by a short example from the modelling weeks in 2011, in which three students, two females and one male worked on the *Bus Stop Problem*, supported by a female future teacher, who was at the end of her university study.

The three students Ute, Klaus and Sarah were trying to figure out how to calculate the acceleration distance in order to find out the whole distance between the bus stops. In the middle of their discussion the tutor enters the table and intervenes within a longer discussion of about 4 min three times.

The first intervention was a content-oriented strategic help which can be read from the following dialogue:

T: So you want to pass on from the time to the distance in order to see ….

The second intervention was a motivational help:

T: Have you thought about how you can calculate that?

The third intervention was a motivational and feedback help:

T: Think again by yourselves, you are on the right way.

From the first results as reconstructed, no uniform evaluation pattern of the students to the same intervention occurs, that is, students react to intervention of a teacher in very different ways. For example, Ute assessed the effects of the above

Table 23.3 Connections of modelling phase and intervention

	Motivational help	Feedback help	General-strategic help	Content-oriented strategic help	Content help
In all modelling phases	You can solve the problem	You are on the right way.	Which part of the modelling cycle are you currently working on?		
Formulate real world model	The problem is not really difficult for you.	Think once more about the assumptions made.	Have you simplified the task sufficiently?	Try to represent the relations between the influential variables in a diagram.	Develop a diagram, in which the first column represents the independent variable and the second the dependent variable.
Translate real world model in mathematical model			Which mathematical methods might be appropriate?		
Work within mathematical model		Check your calculations.		How about trying to find a formula (if possible)?	Have you checked this step?
Interpret the mathematical results, validate the results in real world situations		You are nearly finished, but an important step is missing.	Think about which assumptions you could change to adjust your model to reality.	What happens in extreme cases?	

described interventions positively when questioned by the reviewer as to whether they had the feeling that they needed help, as follows:

Ute: Oh God, yes- Thus, that moment we didn't know how to calculate it… it was something that was dealing with acceleration; thus, I think she came at the right moment.

Ute: (…) at that moment we didn't know how to continue, we also didn't know whether we would do it correctly, yes. Thus I thought it was good, that she had come.

Sarah reacted more indifferently to these interventions and answered the question, whether the help of the teacher was helpful as follows:

Sarah: Yes, exactly that moment, it did not really help us that she has said that we should continue alone.

However, she saw positive effects of the motivational help:

Sarah. Eh, yes, I think so. Therefore we knew that we were on the right track and not totally wrong and therefore we were able to continue well.

The third student, Klaus, reacted negatively and saw neither a reason nor a use in the intervention by the tutor:

Klaus: No, no, not really. We were just a bit thinking about, so that, not really needed it, in that sense, no not.

Similarly to the other student, he felt motivated by the motivational intervention, but devaluated it quite strongly:

Klaus: Eh, yes, that motivates a little bit, but, yes, but to get the result it didn't really help, but a little bit as a kind of motivation or so, but…

In the interview the tutor emphasised the motivational aspects of her intervention:

Tutor: I think that it had the influence, … they continued to calculate at that point where they were thinking to calculate… Possibly because I said: 'Yes, I think it is already quite good!'

She identified the necessity to encourage students to think by themselves as the main reason for her feedback help, and the students' own thinking process and its encouragement as the underlying focus for her interventions:

Tutor: Basically, a concrete question came up, a concrete question, which they could have solved by means of their mathematical knowledge and they should have got the idea by themselves, to have a look into a book or so and therefore I do not believe, that I should have done more there.

In addition, she stressed the support of metacognitive activities, namely the strengthening of self-monitoring activities as another reason for her kind of chosen support based on earlier experiences she had made:

T: Exactly. Rather, I think, that I in a way felt that (student's work, authors) as being a bit disorderly. (…) Something could develop, where I (would) possibly need to say something. If I sit down, then most often something occurs on one's own. 'We are here, we are going to make that …' And then sometimes they already settle themselves on their own, without one needs to have said anything.

To summarise, the effectiveness and appropriateness of teacher interventions seem to strongly depend on how these measures are accepted and adopted by the students, probably dependent on their intellectual capacity, beliefs on mathematics and mathematics education, self-confidence and so on. Due to this variety of influencing factors one might expect that no right or adequate intervention for all types of students can be developed. On the teacher's side a careful analysis of the actual situation is necessary before intervening, the teacher needs to have clear goals for the intervention. Strategic and content-related strategic interventions are possible and can even be done by novice teachers, probably necessitating careful training in advance. But which kind of strategic and content-related strategy has to be chosen strongly depends on the given modelling problem, so there are no easy recipes to follow.

However, one powerful and effective intervention could be identified so far, namely a general strategic intervention, the request to the students given by the teacher to present the state of their work to him/her, when entering the table. This intervention is on the one hand a prerequisite for an adequate scaffold by the teacher, because scaffolding has to be based on a careful diagnosis of students' work, if it shall be efficient and successful. On the other hand as the synthesis of meta-studies by Hattie (2009) has pointed out this kind of intervention is a central part of an effective feedback and is closely related to the kind of feedback questions Hattie (2009) identifies as being effective, namely the "feed up" question 'Where am I going?' and the "feed back" question 'How am I going?' and supports the "feed forward" question 'Where to next?' The reflections of the novice teacher clearly points out that she had these kinds of elements of an effective feedback in mind. The ongoing evaluation of the data sampled will probably lead to a better understanding, how these kinds of feedback questions and further interventions can be connected to the teaching and learning of mathematical modelling, referring amongst others to the explicit usage of the modelling cycle as a meta-cognitive measure.

References

Aebli, H. (1983). *Zwölf Grundformen des Lehrens*. Stuttgart: Klett-Cotta.

Beutel, M., & Krosanke, N. (2012). Rekonstruktion von Handlungsabläufen in komplexen Modellierungsprozessen – Schülerprobleme und Lehrerverhalten. Unpublished master thesis, University of Hamburg, Hamburg.

Blomhøj, M. (2011). Modelling competency: Teaching, learning and assessing competencies – Overview. In G. Kaiser, W. Blum, R. Borromeo Ferri, & G. Stillman (Eds.), *Trends in teaching and learning of mathematical modelling* (pp. 343–349). New York: Springer.

Blomhøj, M., & Jensen, T. H. (2003). Developing mathematical modelling competence: Conceptual clarification and educational planning. *Teaching Mathematics and Its Applications, 22*(3), 123–139.

Blum, W. (2011). Can modelling be taught and learnt? Some answers from empirical research. In G. Kaiser, W. Blum, R. Borromeo Ferri, & G. Stillman (Eds.), *Trends in teaching and learning of mathematical modelling* (pp. 15–30). New York: Springer.

Blum, W., Galbraith, P. L., Henn, H.-W., & Niss, M. (Eds.). (2007). *Modeling and applications in mathematics education. The 14th ICMI study*. New York: Springer.

Borromeo Ferri, R. (2011). *Wege zur Innenwelt des mathematischen Modellierens*. Wiesbaden: Vieweg-Teubner.

Busse, A., & Borromeo Ferri, R. (2003). Methodological reflections on a three-step-design combining observation, stimulated recall and interview. *Zentralblatt für Didaktik der Mathematik, 35*(6), 257–264.

Hammond, J., & Gibbons, P. (2005). Putting scaffolding to work: The contribution of scaffolding in articulating ESL education. *Prospect, 20*(1), 6–30.

Hattie, J. (2009). *Visible learning. A synthesis of over 800 meta-analyses relating to achievement*. London: Routledge.

Kaiser, G. (2007). Modelling and modelling competencies in school. In C. Haines, P. Galbraith, W. Blum, & S. Khan (Eds.), *Mathematical modelling (ICTMA 12): Education, engineering and economics* (pp. 110–119). Chichester: Horwood.

Kaiser, G., & Schwarz, B. (2010). Authentic modelling problems in mathematics education – Examples and experiences. *Journal für Mathematik-Didaktik, 31*(1), 51–76.

Kaiser, G., & Sriraman, B. (2006). A global survey of international perspectives on modeling in mathematics education. *ZDM – The International Journal on Mathematics Education, 38*(3), 302–310.

Kaiser, G., Blum, W., Borromeo Ferri, R., & Stillman, G. (Eds.). (2011). *Trends in teaching and learning of mathematical modelling. ICTMA 14*. Dordrecht: Springer.

Kaiser, G., Bracke, M., Göttlich, S., & Kaland, C. (2013). Realistic complex modelling problems in mathematics education. In R. Strässer & A. Damlamian (Eds.), *Educational interfaces between mathematics and industry (20th ICMI study)*. New York: Springer.

Leiß, D. (2007). *Hilf mir es selbst zu tun*. Hildesheim: Franzbecker.

Link, F. (2011). *Problemlöseprozesse selbständigkeitsorientiert begleiten: Kontexte und Bedeutungen strategischer Lehrerinterventionen in der Sekundarstufe I*. Wiesbaden: Vieweg+Teubner/Springer.

Maaß, K. (2005). Modellieren im Mathematikunterricht der Sekundarstufe I. *Journal für Mathematik-Didaktik, 26*(2), 114–142.

Niss, M. (1992). *Applications and modelling in school mathematics – Directions for future development*. Roskilde: IMFUFA Roskilde Universitetscenter.

Polya, G. (1945). *How to solve it*. Princeton: Princeton University Press.

Stillman, G. (2011). Applying metacognitive knowledge and strategies in applications and modelling tasks at secondary school. In G. Kaiser, W. Blum, R. Borromeo Ferri, & G. Stillman (Eds.), *Trends in teaching and learning of mathematical modelling. ICTMA14* (pp. 165–180). Dordrecht: Springer.

Stillman, G., Galbraith, P., Brown, J., & Edwards, I. (2007). A framework for success in implementing mathematical modelling in the secondary classroom. In J. Watson & K. Beswick (Eds.), *Mathematics: Essential research, essential practice* (Vol. 2, pp. 688–697). Adelaide: MERGA.

Strauss, A., & Corbin, J. (1990). *Basics of qualitative research*. Thousand Oaks: Sage.

Van de Pol, J., Volman, M., & Beishuizen, J. (2010). Scaffolding in teacher-student interaction: A decade of research. *Educational Psychology Review, 22*, 271–293.

Zech, F. (1998). *Grundkurs Mathematikdidaktik*. Weinheim: Beltz Verlag.

[illegible]

Chapter 24
Inducting Year 6 Students into "A Culture of Mathematising as a Practice"

Jill P. Brown

Abstract This chapter describes a classroom teaching experiment conducted in three Year 6 classes where the purpose was to induct students into the mathematising of real world tasks. The focus is on whether applications and mathematical modelling are viewed as being outside of mathematics, as an add-on, or as an integral part of mathematics, requiring students and teachers to operate in a "culture of mathematising as a practice" (Bauersfeld 1993 cited in Yackel and Cobb J Res Math Educ 27(4):459, 1996). Two tasks were used. In one task student mathematising was quite superficial. The second task, introduced with an explicit focus on modelling, resulted in more in-depth mathematising but students still stalled in the modelling process.

1 A Culture of Mathematising as a Practice

In discussing classroom cultures, Bauersfeld describes the extremes as "flexible interpretation versus ritualised reading" (1992, p. 467). He notes that, as teachers strive to embed in their students clear images of what mathematics is they may [inadvertently] provide an image of mathematics as having "an unequivocal relation between object and activity" (p. 467). For example, a picture of three people climbing out of a swimming pool leaving six swimmers in the water playing is perceived as only being able to portray the image of $9-3=6$ with no other interpretation possible (Voigt 1998, p. 197). Bauersfeld argues that this situation, and many others, can and should be interpreted in multiple ways, as he ponders when he asks:

> how could one generate flexibility in mathematising any subject matter — the flexibility that is required later for problem solving — when the student's ascription of meaning is forcibly

J.P. Brown (✉)
Education VIC | Faculty of Education,
Australian Catholic University, Melbourne, VIC, Australia
e-mail: jill.brown@acu.edu.au

G.A. Stillman et al. (eds.), *Teaching Mathematical Modelling: Connecting to Research and Practice*, International Perspectives on the Teaching and Learning of Mathematical Modelling, DOI 10.1007/978-94-007-6540-5_24,

> narrowed to ritualised interpretation, by social conventions rather than by mathematical necessities? It would seem more promising to develop the art of mathematical interpreting rather than the usual funnel-like limiting to prepared, unequivocal reactions. (1992, p. 468)

Suggestions by Bauersfeld to develop this flexibility included tasks requiring students to reflect on their own mathematical activities. In these the teacher provides opportunities for students to "make explicit *their* ways of constructing, analysing, and reasoning" (p. 468)—a task he notes which is much more complex than considering the thinking of others.

What learners do is impacted by what teachers do. Individual teachers' views of mathematics impact on their classroom practice (see e.g., Kaiser and Maaβ 2007). Is mathematics "taken as an objective truth, … or as a practice of shared mathematizing, guided by rules and conventions emerging from this practice" (Bauersfeld 1994, p. 140)? Bauersfeld argues teachers who perceive mathematics as a practice of shared mathematising will develop classroom practices where learners experience "optimal chances for discussions based on intensive experiences and aiming at the negotiation of meanings" (p. 140). These ideas were elaborated by Yackel, Cobb and Wood (e.g., Wood 1993; Yackel and Cobb 1996). Wood (1993) describes such a classroom as one where:

> The teacher acted to initiate and guide students' learning by posing questions and highlighting children's expectations. As students engaged in this discourse, their personal meanings were negotiated until an agreement was reached. The establishment of taken-as-shared meanings between the participants enabled mathematical ideas to be established by members of the class. (p. 20)

The view held by Bauersfeld and others is that teaching and learning mathematics involves "a model of participating in a culture rather than a model of transmitting knowledge. Participating in the process of a mathematics classroom is participating in a culture of using mathematics, or better, a culture of mathematising as practice" (Bauersfeld 1993, p. 4, cited in Yackel and Cobb 1996). The study reported here will attempt to ascertain if the students who are the focus approach their mathematics genuinely believing they have a role to play in interpreting the problem situation and relate the mathematics to themselves and/or the real world in general.

2 Methods

This chapter reports an empirical study of the extent of student engagement with real world tasks. This study is situated within a larger project (Contemporary Teaching and Learning of Mathematics Project, CTLM) with a research and professional development focus. CTLM aims to improve the teaching and learning of mathematics by the provision of professional development to extend teachers' pedagogical content knowledge and hence classroom practice, and thereby improve student learning. Mathematical modelling formed a small part of the project as one approach to the development of mathematical thinking and reasoning. A qualitative inquiry approach following Stake (2005) is taken. The unit of analysis is students working in small groups of 2–4. It is an intrinsic case study as the intention is to gain insight into the phenomenon of upper primary students' mathematising of real world tasks.

Three Year 6 classes, one each from three schools (referred to as School A, B, and C) involved in CTLM, participated. Each school was in the second year of participation in the project. Participation by classes from School A and B was as part of the standard practice of the school visits in the CTLM Project whereas the teacher of the class from School C volunteered to have his students engage in modelling activity. Data collection occurred in the latter part of the year (term 4 of 4). Due to time constraints it was only possible for each class of students to attempt one task. The students' responses to two tasks form the basis of the data collection. The first task was undertaken in a single lesson (approximately 60 min) at Schools A and B. The second task formed part of a two lesson focus on real world tasks at School C. Both tasks were researcher designed. The second task was developed after implementation of the first task, with the intention of having "task context and the mathematics entwined" (Stillman 2002, p. 77) (i.e., context as tapestry).

Data collected included student scripts, photographs of student scripts taken at various stages during task solution, audio recording of elements of student engagement with the tasks, field notes, post lesson discussion with observing teachers, and for the second task, a post task questionnaire completed by the students.

2.1 The Tasks

The first task, *Letters,* saw students first familiarised with examples of wooden letters. These letters are often used on people's bedroom doors. For example, the letters shown belong to Jill (J), Mel (M), and in contrast the other letters belong to Ada.

Letters

A toy store owner asks for help in determining how many of each letter she should include in an order for 400 letters. In addition, students were asked to explain their solution in the form of a letter with enough details such that the toy store owner could apply their solution strategy to a different number of letters.

The second task, *Brass Numerals*, focused on numerals used for house and apartment numbers, for example. In this task, rather than being given a set number to order, after being familiarised with the context, students were asked how many of each numeral should be in stock in a large hardware store, where all stock is on the floor.

Brass Numerals

Recently my friend shifted into a duplex in Ballarat. She noticed that she often received next door's mail in her mailbox so she decided she needed to put some numerals on the fence to tell the postie whose mailbox it was. She lives at No.2 302 in her street. She went to her local Bunnings hardware store but found that there were no 0's or 2's. On several visits to Melbourne she also found that the two Bunnings stores she visited also did not have 2's. After about 6 months she visited the Ballarat Bunnings store again and found she could buy one brass 3 and one brass 2. She asked the local store if they could get in some more 2's and 0's and finally they did. Can you help Bunnings decide how many of each brass numeral they should have in stock?

2.2 *Implementation*

The first task was implemented at Schools A and B with Year 6 students early in term 4. In both schools, the lesson was observed by the classroom teacher as well as several other classroom teachers. In both classes students worked in self-selected groups. The class at School A contained 29 students (15 male, 14 female) aged 11–12 years old. The students worked in nine groups of three and one pair. All but two groups were single gender groups. The School B class included 26 students (13 male, 13 female), working in eight groups of three and one pair. Six groups were of mixed gender. Both lessons were also observed by another researcher.

The second task was implemented in School C with 25 students (15 male, 10 female) who worked in 8 self selected groups over the 2 days. Five students attended only one of the 2 days. Group size ranged from two to four over the 2 days. All groups except one were single gender groups. Prior to presenting the *Numerals Task,* the lesson focused on mathematical modelling, with the intention of inducting students into the importance of context throughout the solving of real world problems. Elements of this included: the nature of mathematical modelling, important aspects of mathematical modelling and the modelling process as exemplified by Galbraith et al. (2007). These ideas were explored further through the context of the following messy real world situation: Dreamworld's thrill rides are very popular and sometimes patrons can be waiting 30–60 min or more for just one ride. Student groups then considered the following: What is the real world mathematical problem? What variables or factors are involved? What assumptions can I make? Ideas were discussed in small groups and then as a whole class. These ideas were linked to the various stages of the modelling cycle as shown in Fig. 24.1 and each stage related back to the real world.

Both tasks presented a somewhat familiar context to students and asked them to solve a specific problem. In the first task, a specific total number of letters were required to be ordered, however it was clear the task involved looking beyond this particular case. The use of communication artefacts, in the *Letters Task*, this was the

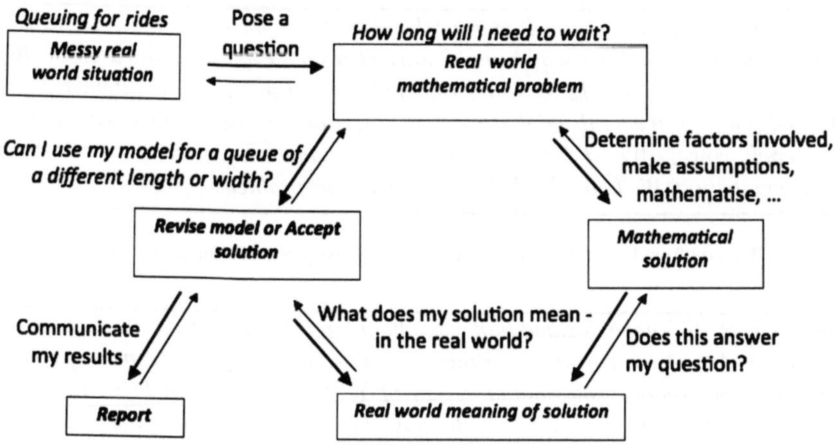

Fig. 24.1 Elements of the Dreamworld situation mapped to the modelling cycle

requirement to write a letter to the toy store owner, has been found previously to be an effective tool in allowing students to demonstrate their understanding to teachers and researchers by making "understanding transparent" (Brown and Edwards 2011, p. 196). The *Brass Numerals Task* also required students to communicate their thinking, albeit in a less specific form, as they were clearly asked to explain how Bunnings should decide how many of each numeral to stock. No time was available for students to present their solutions to each other in the class.

Both tasks involved several layers of complexity. In the first task, letters could be used for example, as a single initial or a full name. If both cases were considered then the task solvers also needed to consider in what proportion this might occur. In contrast, simplifying assumptions could have been made and task solvers assume, for example, that all buyers of the wooden letters were only buying a single initial (i.e., J for Jill). In the *Brass Numerals Task,* the placement of the brass letters was a possible consideration (some locations may have the letterbox and building numbered). The proportion of flats or apartments (and if numbers or letters were used) was a key factor as was the varying length of streets in the local area or Ballarat (although, students, from suburban Melbourne, were more likely to use local knowledge where this was considered).

3 Findings

3.1 The Letters Task

Upon analysis of the data, four non-mutually exclusive approaches were undertaken by the student groups. These approaches can be described as (1) a data approach, (2) an 'equal proportions' approach, (3) an unequal proportions approach, and (4) an

Dear Chris. We have solved your problem and our answer was ... 17 vowels and 15 consonants. This was how we worked it out. First we divided 26 by 400 [sic]. The answer was 15.38 so we rounded it down to 15. We multiplied 26 by 15 and got 390. We realised that we had 10 left over so we put two extra letters for each vowel. We did this because most names have more vowels than consonants in it. We hope this helps you solve your problem.

Fig. 24.2 Excerpt from letter by Group B7 showing 'equal proportions' with adjustments approach

Dear Chris, We have decided that you should get 20 of A, J, L, R, E, K, C, I, S, H, O, M, D, and N because they are most likely to be the most popular letters to be bought. You should get 14 of Y, T, P, G, B, and F because they may be bought and they may not be bought. You should get six U, Z, X, Y, Q, and V because they are unlikely to be bought.

Fig. 24.3 Excerpt from letter to Chris by Group A8

altered problem situation. These will be described first and then the analysis presented. More than one of these approaches could be used by a group.

A *data* approach occurred when students used data in solving the problem. The data may have been carefully collected (e.g., Group A8: *We have taken a telly [sic] in the class and each name in the class has at least one vowel*) or simply a rough estimate (see discussion of Group A9 later in this chapter).

An '*equal proportions*' approach occurred when students decided that Chris should order equal quantities of each letter. As the number of letters in the alphabet (26) is not a factor of 400, this required some re-adjustment. This may have involved simply allocating the remaining ten letters as shown in Fig. 24.2 or having approximately 15 per letter (e.g., group A4 had 17 of A and E, 16 of IJKLMS and 15 of the remaining letters). Group B7, and others, simply allocated all remaining letters to be vowels.

An *unequal proportions approach* occurred when students grouped letters of the alphabet and determined different numbers of wooden letters to be ordered for each group. The number of groups ranged from two to eight. Rationale for determining allocation to groups may have been according to the nature of the letter (i.e., vowel versus consonant), perceived frequency (e.g., popular v unpopular, see Fig. 24.3), or a combination of these.

Finally, the *altered problem* approach saw student groups change the problem task. Examples included, situations where the students avoided difficulties by increasing the number of letters to be ordered. For example, Group A6 proposed, 'Four hundred letters is nowhere near enough letters, and if you work it out, you only get 15.348615 letters. Next time you may have to order around 2,600. Then you will have 100 of each letter.' Group A10 who considered the ratio of buyers who bought their initial versus first name (average length of 5) also argued that more than 400 letters were required. 'We think you should get 1,400 letters. 70 per vowel. 50 per consonant. We worked this out by getting 7 letters per 3 people and we got that by figuring every third person would put their whole name and 2 others would get first initial.'

Table 24.1 Approaches taken to the Letters Task

Approach	School A	School B	Total
Data	A1*, A6[2]*, A9*, A10*	B3*, B6*, B7*	7
Equal proportions	A4, A6[2]*, A7	B1, B2, B3*, B4, B5, B7*, B8	10
Unequal proportions	A1*, A3, A5, A8	B6*	5
Altered problem	A2, A6[1], A9*, A10*	B9	5

Note: *Indicates two approaches undertaken in the same solution. [1]First distinct solution of the group. [2]Second distinct solution of the group

Table 24.1 presents the initial analysis of the classification of the approach(es) taken by the student groups from Schools A and B. Overall, this shows that the equal proportions approach was most common. However, distinct differences can be seen across the two schools. At school A the approaches were almost equally likely to be undertaken, whereas at School B the equal proportions approach dominated student strategies.

3.1.1 Task Complexity

For School A groups A1–A8, it is impossible to ascertain, if the factor related to buying one's whole name versus one's initial was considered. However, both groups A9 and A10 clearly considered this factor thus increasing the complexity of their solution strategy. Both of these groups considered the class of 30 as being representative of a set of customers at the toy shop. Group A9 did a rough tally of letters in the first name of the class members and thus increased the data to cover the possibility of some student deciding to buy letters for their family name in addition to their first name, as illustrated in the following discussion. (All names are pseudonyms.).

Jasmine: Yes. But not all people have an A in their name. We are just like rounding it off to a comfortable number.

Dearne: Just in case they want their last name.

Researcher: So when you say you are rounding, what do you mean by rounding?

Jasmine: We are just adding a few more, so like Dearne said, because some people might want their surname. …

Dearne: Well now, like if all 30 of us go into the shop and we decide we want to have our last name, or our first name with the letters. But we will know that we will have enough letters.

In contrast, group A10 decided to estimate what proportion of customers would buy initials as compared to their first name, "we got that by figuring every third person would put there [sic] whole name and 2 others would get first initial."

As a data approach necessarily required student groups to also use a second approach, it was decided to see if there was any correlation between the use of a data approach and the other three approaches. In addition, those taking a data approach were more likely to be engaging with the real world, in contrast to those not using a

Table 24.2 A closer look at the use of a data approach

Approach	Data	No data considered
Equal proportions	A6[2], B3, B7	A4, A7, B1, B2, B4, B5, B8
Unequal proportions	A1, B6	A3, A5, A8
Altered problem	A9, A10	A2, A6[1], B9

Note: [1]First distinct solution of the group, [2]Second distinct solution of the group

data approach, who tended to ignore real world aspects of the task. Table 24.2 shows the approach taken and whether or not a data approach was considered. It can be seen that student solutions were more likely not to use a data approach than to do so (7 solutions used data, 13 did not). Of the solutions taking data into consideration, there was little difference in approach chosen (equal proportions 3, unequal proportions 2, and altered problem approach 2). In contrast, the solutions where no data were considered tended to be an equal proportions approach (7 of the 13). It must be noted that the number of groups is small and no inferences can be made.

3.2 *The Brass Numerals Task*

In considering the responses to the second task, it quickly became apparent that the analysis was more complex. However, the same four approaches were taken. Details of these are presented in Table 24.3. Table 24.3 includes brief descriptions of details of student approaches. In addition, where student groups took multiple approaches in the course of their solution this is indicted by an arrow.

Four groups altered the problem situation. For example, C1 suggested that at the store, Bunnings should be checking "daily to see if more numerals are needed. [Moreover,] they should have inspectors who drive around and note where numerals are placed, the amount of numerals included in each house number. Inspectors could also check on the computers/ receipts to see the popular numbers bought, and how many customers buy numbers".

Of particular interest was, what did students do in terms of the transitions in the first two phases of the modelling cycle. Namely, firstly from the *messy real world problem* to the *real world problem statement* and secondly from the *real world problem statement* to the *mathematical model*. In the first transition, the focus is on *understanding, structuring, simplifying, and interpreting context.* In the second transition, the focus turns to *making assumptions, formulating, mathematising.* Figure 24.4 maps the student progress with respect to the modelling process.

As can be seen in Fig. 24.4, student progress varied considerably. Two student groups (C4 and C7) were unable to seriously engage with the messy real world situation, preferring to offload the problem to the store owners (or suggest a different store not realising the same situation may occur) rather than investigate proposing a solution themselves. A third group, C1, was unable to formulate a problem from the situation although they were able to undertake some mathematisation of the problem situation as they noted the need for data. Group C2 progressed further, as they

Table 24.3 Analysis of approaches taken at School C to the Letters Task

Approach	No Data considered	Data
Equal proportions	C3[1]: → C5[1]: 40–50 letters, at least 20 C6: 50 of each	C3[2]: A new street with 20 houses, numbered 220 – 240 (sic), more zeros, less ones then from 1–100 it would even out
Unequal proportions	C5[2]: more ones than twos is 20 houses in the street	C4[2]: Our 3 streets – a lot of 1's and 2's. 0–7, 1–26, 2–13, 3–8, 4–9, 5–7, 6–5, 7–6, 8–8, and 9–8. C8: say an average street of 40 houses, need 14 of 1, 2, 3; four 5's, and four of 5–9 and 0. So more digits less than 4, excluding zero
Altered problem	C1:, C2:, C4[1]:, C7	

Note: [1]First distinct solution of the group, [2]Second distinct solution of the group

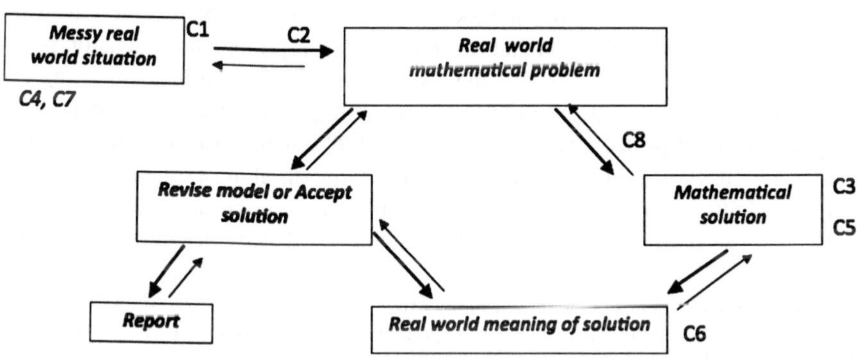

Fig. 24.4 Student group progress in the modelling process

considered types of houses (duplexes, flats, etc.) however they were unable to set a mathematical problem to solve. Group C8 moved beyond the real world problem but were unable to arrive at a mathematical solution. Two groups, C3 and C5, arrived at a mathematical solution for a simple mathematical problem posed by them (C3: How many 2's needed for a street numbered 220–240? C5: What numerals are required for a street of 20 houses?) but were unable to progress further. Group C6 not only posed and solved a problem, but also they were able to consider the real world meaning of their solution (i.e., would the solution fit on the warehouse racks).

4 Discussion and Conclusions

Although students clearly engaged with the *Letters Task*, many student groups perceived the task as a division problem, albeit without an appropriate exact answer as desired. Few engaged with the context or gave any indication that the task required Bauersfeld's "flexible interpretation" (1992 p. 467). The majority of solutions (13 of the 20) failed to make any use of data although it should have been apparent that data were easily accessible.

Subsequently, the revised task, the *Brass Numerals Task* certainly saw the task interpreted beyond a "ritualized reading" (Bauersfeld 1992). Students did engage with the context and several sustained this engagement. They were more likely in this task to see that their role was to make sense of the situation and to mathematise it themselves. In addition, the use of the Dreamworld scenario and modelling cycle diagram provided prior experience. However, successfully undertaking this modelling process remained a difficult task. Reasons for this could lie in the task itself, the facilitation of the task, the length of time available, the need to make the goals of the task more explicit, and/ or previous experiences. There is no doubt that across these three classes, students were not experienced in interpreting mathematical problem situations nor believing this was a normal part of school mathematics. This lack of experience of shared mathematising and negotiation of understanding and meanings contributed to the lack of student belief that they had personal experience and knowledge to bring to the solving of the tasks. In the *Letters Task* this resulted in many students ignoring the real world and perceiving only a (problematic) division problem and in the *Brass Numerals Task* the result was difficulties in mathematising the task or making substantial progress in the modelling task. Finally, it appears that using tasks of the type described by Bauersfeld (1992), in this case tasks that required students to reflect on their mathematics and make their thinking explicit can contribute to Year 6 students perceiving themselves as playing an important role in interpreting the real world problem situation and to relating to mathematics of the real world.

Acknowledgements Thanks to Ann Downton (ACU) and Michael Di Nuzzo (CEOM) for data collection support and to the teacher and student participants without whom this research could not have been undertaken.

References

Bauersfeld, H. (1992). Classroom cultures from a social constructivist's perspective. *Educational Studies in Mathematics, 23*, 467–481.

Bauersfeld, H. (1994). Theoretical perspectives on interaction in the mathematics classroom. In R. Biehler, R. W. Scholz, R. Sträßer, & B. Winkelmann (Eds.), *Didactics of mathematics as a scientific discipline* (pp. 133–146). New York: Kluwer.

Brown, J. P., & Edwards, I. (2011). Modelling tasks: Insight into mathematical understanding. In G. Kaiser, W. Blum, R. Borromeo Ferri, & G. Stillman (Eds.), *Trends in teaching and learning of mathematical modelling, ICTMA14* (pp. 187–197). New York: Springer.

Galbraith, P., Stillman, G., Brown, J., & Edwards, I. (2007). Facilitating middle secondary modelling competencies. In C. Haines, P. Galbraith, W. Blum, & S. Khan (Eds.), *Mathematical modelling (ICTMA12): Education, engineering and economics* (pp. 130–140). Chichester: Horwood.

Kaiser, G., & Maaβ, K. (2007). Modelling in lower secondary mathematics classroom – Problems and opportunities. In W. Blum, P. L. Galbraith, H.-W. Henn, & M. Niss (Eds.), *Modelling and applications in mathematics education, ICMI 14 study volume* (pp. 99–108). New York: Springer.

Stake, R. (2005). Qualitative case studies. In N. Denzin & Y. Lincoln (Eds.), *The Sage handbook of qualitative research* (3rd ed., pp. 443–466). Thousand Oaks: Sage.

Stillman, G. (2002). *Assessing higher order mathematical thinking through applications.* Unpublished Doctor of Philosophy thesis, University of Queensland, Brisbane.

Voigt, J. (1998). The culture of the mathematics classroom: Negotiating the mathematical meaning of empirical phenomena. In J. Voigt, F. Seeger, & U. Waschescio (Eds.), *The culture of the mathematics classroom* (pp. 191–220). Cambridge: Cambridge University Press.

Wood. T. (1993). Creating an environment for learning mathematics: Social interaction perspective. In *Rethinking elementary school mathematics: Insights and issues*, (Journal for research in mathematics education. Monograph, Vol. 6, pp. 15–20, 115–122). Reston: NCTM.

Yackel, E., & Cobb, P. (1996). Sociomathematical norms, argumentation, and autonomy in mathematics. *Journal for Research in Mathematics Education, 27*(4), 458–477.

Chapter 25
A Whole Week of Modelling – Examples and Experiences of Modelling for Students in Mathematics Education

Nils Buchholtz and Sarah Mesrogli

Abstract This chapter describes the arrangement of a project about mathematical modelling at the University of Hamburg. In the first part of the chapter the background and the structure of a modelling week are presented in detail. Subsequently in the second part of the chapter the experiences with one modelling problem are described including problems the students had to face when dealing with the problem. Finally, experiences and selected results of the evaluation of the modelling weeks are taken into focus, in which it became apparent, that the students have different attitudes towards mathematics as a subject with different reasons for their attitude.

1 Introduction

Even though the teaching of mathematical modelling has found its place within the German national mathematics curricula (see KMK 2004), students often do not have the opportunity to experience complex modelling problems, either due to restrictions of time or probably due to the teachers' lack of knowledge of how to treat complex modelling problems adequately (cf. Blum 2011). Since 2009, the project of the 'modelling weeks' offer a possibility for students from upper secondary school to solve complex modelling problems at the University of Hamburg. The continuous work on modelling problems for a few days, away from the school environment, enhances the students' modelling competencies

N. Buchholtz (✉) • S. Mesrogli
Faculty of Education, Psychology and Human Movement,
University of Hamburg, Hamburg, Germany
e-mail: Nils.Buchholtz@uni-hamburg.de

G.A. Stillman et al. (eds.), *Teaching Mathematical Modelling: Connecting to Research and Practice*, International Perspectives on the Teaching and Learning of Mathematical Modelling, DOI 10.1007/978-94-007-6540-5_25,

and motivates them to solve mathematical problems with the help of their own mathematical skills. However, the systematic evaluation of the impact of the project is now beginning. The project is accompanied by regular evaluation. The central questions of our research in this chapter are whether the rather complex modelling problems tackled in the modelling weeks are feasible for the students, what attitudes the students have towards mathematics and whether the visit of the modelling weeks has an influence on their mathematical interest. In the following, we will report about the project and present findings from the evaluation of the project from 2009–2011.

1.1 Framework for the Mathematical Modelling in Hamburg

The approach of teaching students mathematical modelling at the University of Hamburg, which is described in this chapter, belongs to the comprehensive didactical perspective of the *realistic or applied modelling* (Kaiser and Sriraman 2006; Kaiser et al. 2007). A central feature of this approach is the use of *authentic modelling problems* in order to implement different educational goals of mathematical modelling (see Kaiser et al. 2010).

First, pragmatic-utilitarian educational goals such as understanding the real world, the ability to apply mathematics to solve practical problems in general or the promotion of (selected) modelling competencies are the focus. The authentic examples that are used in the teaching of mathematical modelling should therefore explain the relevance of mathematics in daily life, the environment and the sciences, and impart competencies to apply mathematics. Authentic problems are hereby defined as problems that are only a little simplified and, in accordance with the definition of authentic problems by Niss (1992) recognised by people working in this field as being a problem they might meet in their daily work. However, the problems should not be authentic only for experts. According to Vos (2011), our understanding of authenticity of the problems used also includes that this authenticity is obvious to the students and that the problems have an out-of-school purpose.

Secondly, psychological goals play a major role in the teaching of mathematical modelling at the University of Hamburg. The approach takes as essential starting point that – in order to promote modelling competencies – the students need their own experiences with authentic modelling problems. The authenticity of the problems therefore should foster the motivation and attitude of the students towards mathematics in general and hopefully arouse their interest in applied mathematics.

Based on empirical research (see e.g., Kaiser-Meßmer 1986) we see the necessity of using authentic modelling problems in teaching, which promotes the whole range of modelling competencies and broadens the radius of action of the students. Similar proposals have been developed by Haines and Crouch (2006) or at the beginning of the modelling debate by Pollak (1969).

1.2 Structure of the Modelling Week

Modelling weeks are a further development of previous activities between the departments of mathematics and education at the University to include mathematical modelling in school and teacher education. These activities – like seminars about mathematical modelling for future teachers – have been conducted since 2000. They usually offer the possibility for future teachers to make their own experiences with mathematical modelling in schools (see e.g., Kaiser and Schwarz 2006). Since 2009, modelling weeks – conducted twice a year – are also a great opportunity for students from upper secondary level of schools in and around Hamburg to get in touch with the Department of Mathematics and especially with authentic mathematical problems (see Kaiser et al. 2011). Approximately 500 secondary students (about 10 % of the age-cohort of 16–18 year old students in Hamburg) took part in the 2 modelling weeks in 2010. The students were released from school and worked in small groups on complex and authentic modelling problems continuously for 1 week at the Department of Mathematics.

The project was carried out as a joint activity between the Department of Education (Mathematics Didactics) and the Department of Mathematics (Applied Mathematics) and was directed by professors from both departments. In addition, on the part of the university parallel to the modelling week, a seminar was carried out for future teachers who worked as tutors for the students. By this means the students were intensively supervised during their work on the modelling problems whilst on the other hand the future teachers had the opportunity both to gain practical experiences during the project and to reflect on the supervision of the students in the respective seminar. Through the students' independent work on modelling examples, precisely in developing mathematical questions from given problems by themselves and developing solutions for real world problems the promotion of modelling competencies can be achieved. In addition, the students' motivation can be enhanced because at the end of the week, the groups have to present their solution to the other participants.

According to the framework of the realistic modelling approach, the modelling examples on which the students work have to conform to necessary 'authentic' requirements. For example, they have to be suggested by applied mathematicians working in industry, they are often taken from the area of industrial or other technical processes and are still embedded into reality. This leads to the difficulty, that neither the students nor the questioner know an adequate solution for the modelling examples. Additionally, a problematic situation is described in the example and the students have to determine or develop a question, which they can solve, themselves. Various problem definitions and solutions are possible; they all depend on the approach of the modellers.

In Kaiser et al. (2010), we find an overview about the problems used so far and give references to already published descriptions of their implementation:

- Stock price forecast,
- Premium in private health insurance,
- Prediction of fishing quotas,

- Optimal position of rescue helicopters (Kaiser 2005),
- Radio-therapy planning for cancer patients,
- Identification of fingerprints,
- Pricing for internet booking of flights (Kaiser and Schwarz 2006),
- Price calculation of an internet café (Kaiser and Schwarz 2006),
- Traffic flow during the Soccer World Championship in 2006 in Hamburg,
- Construction of an optimised timetable of school,
- Irrigation of a garden,
- Population of ladybugs (Kaiser et al. 2011),
- Chlorination of a swimming pool (Kaiser et al. 2010),
- Optimal planning of bus lines and placement of bus stops, (see Chap. 23, this volume),
- New rules in ski jumping: the influence of a variable inrun.

2 Experiences with a Specific Modelling Example

One of the problems tackled in the modelling week in 2009 investigated the chlorination of a swimming pool. The company *Grundfos*, that markets pumping systems, is searching for a model describing the mixture of chemicals in a swimming pool. The concentration of urea and the number of bacteria in a swimming pool is to be adjusted by a systematic supply of chlorine. A mathematical model is needed, which describes the regulation of pool water quality via the chlorine concentration in swimming pools. We will briefly describe one approach to the solution of this modelling problem and refer for further description to Kaiser et al. (2010). The approach is from a group of students, but shows quite typically the solutions of complex modelling problems of the students in the context of the modelling week in Hamburg.

First, the group created a *real* model, where different factors were listed that influence the quality of the bathing water. The group agreed on examining the concentration of bacteria, urea and chlorine since different sources of information referred to varying factors. These are exposed to the following influences:

Concentration of bacteria

1. New bacteria are added by the bathers.
2. These bacteria proliferate.
3. Chlorine kills the bacteria.

Concentration of urea

1. The concentration of urea is increased by the bathers.
2. Chlorine destroys the urea.

Concentration of chlorine

1. The concentration of chlorine decreases by the killing of bacteria.
2. By neutralising the urea, the concentration of chlorine decreases.
3. Since chlorine is gaseous at room temperature, it also evaporates from the water.
4. Chlorine is added via the pumping system.

It became clear that specific modelling assumptions had to be made. In particular, only *concentrations* were considered. The amount of water in a swimming pool changes continuously in reality. However, the model does not take this into account because the considerations of a varying amount of water complicates the model to such an extent that it is mathematically difficult to handle. The volume of water was therefore considered to stay *constant.* The proliferation rate of bacteria in bathing water also depends on its temperature; therefore, the *water temperature* was considered to stay *constant* as well.

The first attempt to create a *mathematical model* resulted in a rather complicated model, in which several factors of influence were considered. Thus, the model limited itself by the different units of the concentrations of the main influence factors: bacteria concentration *B*, the urea concentration *H*, and the chlorine concentration *C*. The urea and the chlorine concentration are given in [g m^{-3}] whereas the concentration of bacteria is measured in [KbE m^{-3}] as colony forming units. More precisely, the mathematical model examined the *modifications* of concentration beforehand. This was achieved by *functions of change* f_B, f_H and f_C which the students developed (for details see Kaiser et al. 2010):

$$f_B = -v_B CB + z_B V^{-1} A + \lambda B \quad \left[KbE{\cdot}m^{-3}\,\min^{-1}\right]$$

$$f_H = -v_H CH + z_H V^{-1} A \quad \left[g{\cdot}m^{-3}{\cdot}\min^{-1}\right]$$

$$f_C = -n_B BC - n_H HC + Z_C - g{\cdot}C \quad \left[g{\cdot}m^{-3}{\cdot}\min^{-1}\right]$$

For the group, it was also necessary to take into account defined limits per DIN-norm:

$$0.3\frac{g}{m^3} \leq C \leq 0.6\frac{g}{m^3}$$

$$H \leq 1\frac{g}{m^3}$$

$$B \leq 100\frac{KbE}{ml} = 10^8\frac{KbE}{m^3}$$

Already the attempt to compare the different factors made two problems clear, which are typical for a situation also observed by Peretz (2005) and which can be transferred to our experiences with the modelling weeks. After mathematising the *real* situation, the resulting mathematical model is still too complex and needs 'further' simplifying. The functions of change have varying units. Therefore, the values are not comparable and the magnitudes of the single summands are also enormous as the values of the constants and the allowed bacteria concentration clearly show.

The model had to be scaled respectively to make its values comparable and to eliminate these problems. Obviously, the students reached a dead-end situation, because they wanted to take into account so many different factors, but the model developed was far too complicated. We will not describe the whole modelling process of the group, but after the intervention of the future teachers and further elaboration of the model by the students (simplification, scaling), the group in the end was able to create a model, with which different concentrations of chlorine could be calculated concerning the volume of water and the average number of bathers of several swimming pools in Hamburg.

3 Evaluation of the Modelling Week

We contribute our evaluation of the modelling weeks to previous and ongoing research. Kaiser and Schwarz (2010) report on the evaluation of the modelling week in March 2009, they describe the high learning outcomes of the modelling weeks in the area of working techniques, strategies in problem solving, and social aspects. They also identify a strong plea of the students for the inclusion of modelling in usual mathematics lessons. However, in order to analyse the effectiveness of the modelling week at the level of the personal attitudes of the students, we want to identify relevant aspects of the interest of the students from their statements.

For the evaluation of the modelling weeks a questionnaire with mainly open questions was completed at the end of each modelling week. The questions focused on the beliefs of the students about mathematics teaching and on the appreciation of the modelling examples tackled in the modelling week. Not all participating students filled out the questionnaire, but 289 questionnaires in March 2009 and 170 and 162 questionnaires in March and September 2010 respectively were evaluated. In March 2011, the questionnaire was completed by 182 of the 254 participating students. Based on qualitative methods of grounded theory (Strauss and Corbin 1998) we used in-vivo codes for the open questions, that is, codes extracted from the students' answers, written as verbatim quotations, and grouped them to similar quotations under a theoretical perspective. In order to analyse the students' answers in the open questions we transformed the grouped in-vivo-codes into theoretical codes. We used methods of consensual coding for reasons of quality assurance, which means that a coding team consisted of two coders, who conducted all steps described above together.

The question, on which we will concentrate in this chapter, is: *Are you interested in mathematics?*

Figure 25.1 shows the results of the recent modelling weeks. Most of the students who took part in the modelling week are interested in mathematics as a subject.

In the next section, we focus on the students from the March 2011 modelling week. The students gave different and sometimes multiple reasons for their interest or disinterest, but these reasons may not be directly connected to the experiences in the modelling week. The most frequent reasons the students are interested in mathematics as a subject were subordinated under the perspective *formalism* (e.g., mathematics is logical; exact; abstract), *process* (e.g., I like to develop new ideas; think on problems; find solutions) and *use* (e.g., I use mathematics in daily-life; reference to reality; mathematics makes sense). Additionally, reasons were classified

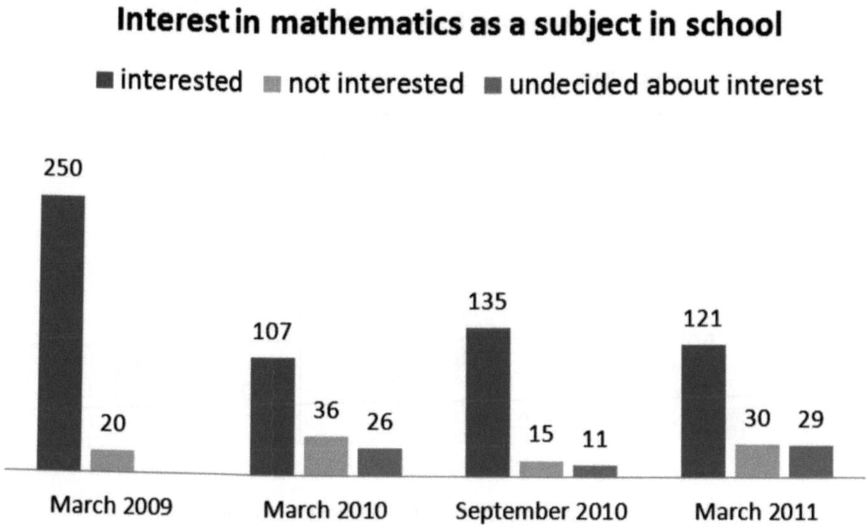

Fig. 25.1 Distribution of interest in mathematics

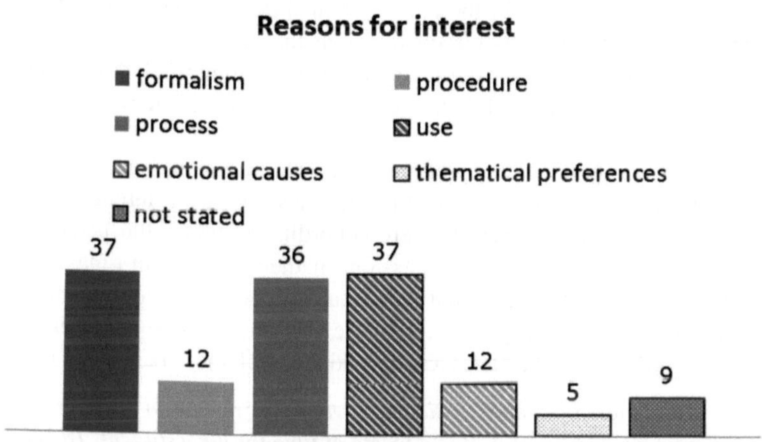

Fig. 25.2 Reasons for interest in mathematics ($n = 121$, multiple coding possible)

under the perspective of *procedure* (e.g., I like to calculate; solve equations) and *emotional causes* (e.g., I feel challenged; fun; interested) as shown in Fig. 25.2. Examples of reasons given by the students:

Yes, because it is such a concrete subject. It is never vague, but always right or wrong. (student, 17 years)

Topics that are related to daily life concern me. Uninteresting and useless things discourage me. (student, 17 years)

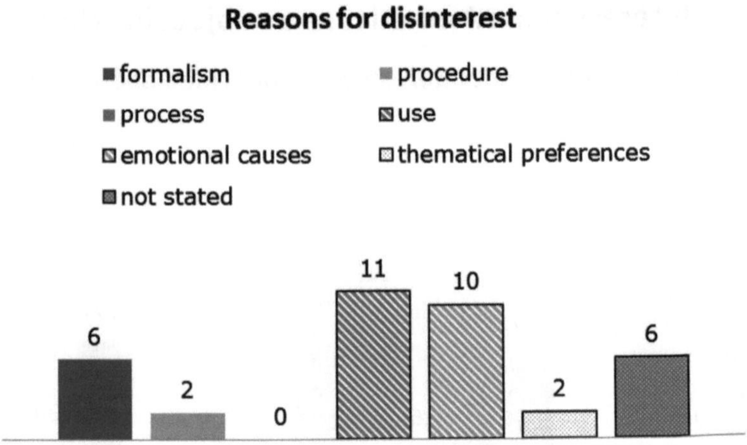

Fig. 25.3 Reasons for disinterest in mathematics (n=30, multiple coding possible)

In addition to reasons that highlight the challenging formal character of mathematics, many reasons for the interest in mathematics were about its utility in everyday life and its process-like character in the search for solutions to given problems. We assume that the experiences of the modelling week have a big impact here. Thirty students indicated disinterest in mathematics. The reasons for disinterest are shown in Fig. 25.3.

Explanations suggest that many of these students are not interested in mathematics as a subject, because their usual mathematics lessons lack reference to the daily-life of the students. These reasons were subordinated under the perspective *use* (e.g., lack of relevance for future profession; mathematics doesn't make sense; no reference to reality). We assume that these reasons could also be influenced by the experiences of the modelling week, because the students also differentiate here between their ordinary mathematics lessons and the week they spent at the university.

It is difficult to understand, unrealistic and I will never need it in my life to the same extent as I now have to learn it. It is a waste of time for me. (student, 18 years)

Also emotional causes play a major role in the disinterest. The students mentioned that they were not motivated or did not like their teacher. In addition, the students, who are disinterested in mathematics as a subject, often give other emotional reasons, for example perceiving themselves as having a lack of understanding:

I would say, mathematics is not 'my' subject, because I do not understand very much and I do not find mathematics very interesting. (student, 16 years)

Actually, I am interested in mathematics, because I would like to understand physical and technical processes. Nevertheless I am often not very good in finding logical coherence and therefore I am often one of the weakest students in mathematics courses. Therefore I do not have much fun. (student, 17 years)

4 Conclusions

When planning an event such as a modelling week, the necessary authenticity of the problems to be used needs to be considered. In addition, whether the used examples are compatible with the students' *mathematical knowledge*, which might not be very high, for example only covering the beginning of calculus, needs to be considered. The modelling competencies of the students differ on a large scale and the authenticity of the given real-world problems often creates difficulties for the students. Nevertheless, teacher intervention should focus only on supporting students in the case of lacking mathematical means or when the students are in a dead-end situation such as the case of the group struggling with the swimming-pool task. After all, experiencing helplessness and insecurity is a central aspect and a necessary phase when dealing with mathematical modelling.

As several comments of the students show, the modelling weeks may not only have outcomes on the level of mathematical skills but also on the level of attitude towards mathematics. So far, we assume that the experiences with the complex and authentic modelling problems actually have an influence on the students' interest in mathematics, although the question about the interest in mathematics unfortunately does not suggest a direct relation to these experiences. For further research, we will concentrate on this relation. First, we conclude that either the experience of helplessness or the surprise of how well a student can actually use known mathematics for solving problems that arise from reality may influence the students' motivation in either a challenging or enhancing way. In particular, students who struggle with comprehension problems or the lack of reference to the real world in their mathematics lessons, can experience the utility of mathematics when using it to answer questions that arise from the reality.

Acknowledgement We thank Christoph Lederich for the evaluation of the data.

References

Blum, W. (2011). Can modelling be taught and learnt? Some answers from empirical research. In G. Kaiser, W. Blum, R. Borromeo Ferri, & G. Stillman (Eds.), *Trends in teaching and learning of mathematical modelling: ICTMA 14* (pp. 585–594). New York: Springer.

Haines, C. R., & Crouch, R. M. (2006). Getting to grips with real world contexts: Developing research in mathematical modelling. In M. Bosch (Ed.), *Proceedings of the fourth congress of the European Society for Research in Mathematics Education CERME4* (pp. 1655–1665). Barcelona: Fundemi IQS-Universitat.

Kaiser, G. (2005). Mathematical modelling in school – Examples and experiences. In H.-W. Henn & G. Kaiser (Eds.), *Mathematikunterricht im Spannungsfeld von evolution und evaluation. Festschrift für Werner Blum* (pp. 99–108). Hildesheim: Franzbecker.

Kaiser, G., & Schwarz, B. (2006). Mathematical modelling as bridge between school and university. *ZDM – The International Journal on Mathematics Education, 38*(2), 196–208.

Kaiser, G., & Schwarz, B. (2010). Authentic modelling problems in mathematics education – Examples and experiences. *Journal für Mathematik-Didaktik, 31*(1), 51–76.

Kaiser, G., & Sriraman, B. (2006). A global survey of international perspectives on modelling in mathematics education. *ZDM – The International Journal on Mathematics Education, 38*(3), 302–310.

Kaiser, G., Sriraman, B., Blomhøj, M., & Garcia, F. J. (2007). Report from the working group modelling and applications – Differentiating perspectives and delineating commonalities. In D. Pitta-Pantazi & G. Philippou (Eds.), *Proceedings of the fifth congress of the European Society for Research in Mathematics Education* (pp. 2035–2041). Larnaca: ERME.

Kaiser, G., Lederich, C., & Rau, V. (2010). Theoretical approaches and examples for modelling in mathematics education. In B. Kaur & J. Dindyal (Eds.), *Mathematical applications and modelling. Yearbook 2010, Association of Mathematics Educators* (pp. 219–246). Singapore: World Scientific.

Kaiser, G., Schwarz, B., & Buchholtz, N. (2011). Authentic modelling problems in mathematics education. In G. Kaiser, W. Blum, R. Borromeo Ferri, & G. Stillman (Eds.), *Trends in teaching and learning of mathematical modelling: ICTMA 14* (pp. 585–594). New York: Springer.

Kaiser-Meßmer, G. (1986). *Anwendungen im mathematikunterricht. Band 1: Theoretische konzeption. Band 2: Empirische Untersuchungen*. Bad Salzdetfurth: Franzbecker.

KMK = The Standing Conference of the Ministers of Education and Cultural Affairs of the Länder in the Federal Republic of Germany. (2004). *Bildungsstandards im Fach Mathematik für den Mittleren Schulabschluss (Jahrgangsstufe 10). Beschluss vom 04.12.2003*. Munich: Wolters Kluwer.

Niss, M. (1992). *Applications and modelling in school mathematics – Directions for future development*. Roskilde: IMFUFA Roskilde Universitetscenter.

Peretz, D. (2005). Inverse mathematical model: Yet another aspect of applications and modeling in undergraduate mathematics for prospective teachers? *Teaching Mathematics and Its Applications, 24*(2–3), 134–142.

Pollak, H. (1969). How can we teach applications of mathematics? *Educational Studies in Mathematics, 2*(1), 393–404.

Strauss, A., & Corbin, J. (1998). *Basics of qualitative research: Techniques and procedures for developing grounded theory* (2nd ed.). Thousand Oaks, CA: Sage.

Vos, P. (2011). What is 'authentic' in the teaching and learning of mathematical modelling? In G. Kaiser, W. Blum, R. Borromeo Ferri, & G. Stillman (Eds.), *Trends in teaching and learning of mathematical modelling: ICTMA 14* (pp. 585–594). New York: Springer.

Chapter 26
Teachers' Self-Perceptions of Their Pedagogical Content Knowledge Related to Modelling – An Empirical Study with Austrian Teachers

Sebastian Kuntze, Hans-Stefan Siller, and Christiane Vogl

Abstract Empirical research into teachers' views and pedagogical content knowledge related to modelling in the mathematics classroom is still relatively scarce, such as in the domain of self-perceptions of modelling-specific pedagogical content knowledge. Consequently, this study concentrates on such views of an Austrian sample of pre-service and in-service mathematics teachers. Both quantitative and qualitative methods were used to explore how teachers perceive their knowledge about possibilities of providing students with specific help in the modelling process and how they see their professional development at university with respect to modelling. The findings show that the mean self-perceptions in both of these areas were not positive, indicating a need for intensified professional development support.

1 Introduction

There is a consensus that modelling is an integral part of mathematics and that it should play an important role in mathematics education. For this reason modelling is agreed to be a big idea for mathematics as a scientific discipline with high relevance for mathematical literacy. Consequently, teachers should be aware of this big idea, know how modelling relates to a variety of curricular content, and

S. Kuntze (✉)
Institute for Mathematics and Informatics,
Ludwigsburg University of Education, Ludwigsburg, Germany
e-mail: kuntze@ph-ludwigsburg.de

H.-S. Siller
Institute for Mathematics, University of Koblenz, Koblenz, Germany

C. Vogl
WIKU Salzburg, Salzburg, Austria

G.A. Stillman et al. (eds.), *Teaching Mathematical Modelling: Connecting to Research and Practice*, International Perspectives on the Teaching and Learning of Mathematical Modelling, DOI 10.1007/978-94-007-6540-5_26,

master instructional strategies linked to modelling. However, empirical research into in-service teachers' knowledge and views associated with modelling is still relatively scarce, despite its importance.

The project ABCmaths ("Awareness of Big Ideas in Mathematics Classrooms", www.abcmaths.net) therefore concentrates on pre-service and in-service teachers' views related to modelling, including self-perceptions held by the teachers of their modelling-specific professional knowledge. In this study, we analyse quantitative and qualitative data from a questionnaire survey on modelling in the mathematics classroom. The results give insight into how teachers think about their own professional development related to mathematical modelling at the university level, as well as into their views about their own abilities of supporting students in modelling processes. The results indicate differences in the teachers' perceptions, which point to a need for intensifying both learning opportunities at the university level and ongoing professional development focusing on modelling.

2 Theoretical Framework

2.1 Modelling – A Big Idea for the Mathematics Classroom

The idea of modelling (Blum 1996; Blum et al. 2007) is an aspect of mathematics that is not only important within mathematics (Siller et al. 2011), but also is crucial for the competencies of mathematically literate citizens (OECD 2003). For instance, national standards in many countries (AECC 2008; KMK 2004; NCTM 2000) and assessment studies in the domain of mathematical literacy (OECD 2003) emphasise the significance of modelling as an aspect of mathematical competency.

Process aspects are in the focus of the idea of modelling, such as the descriptions of the modelling cycle (Blomhøj and Jensen 2003; Blum and Leiß 2005). However, empirical observations have shown that the modelling process is individual and does often not follow the cyclic model (Borromeo Ferri 2006). In the process of modelling, translation processes between the real world context and mathematics (in both directions) play a central role (Blum 2007), an aspect that tends to increase the complexity of modelling problems and to cause many difficulties of learners. Consequently, support for learners often focuses on meta-knowledge about modelling and an increased awareness of the modelling process (Blum 2007; Breitenbücher and Kuntze 2010; Maaß 2006). Further, an appropriate task culture (Neubrand 2002) can enhance instructional quality by a focus on tasks with modelling requirements, on a level of difficulty adapted to the specific groups of learners. Siller (2010) suggests learning opportunities for student-centred classroom settings that aim at fostering modelling competencies. An increasing body of publications with suggestions of "good classroom practice" associated with modelling (Blum et al. 2007) points to a need for developing corresponding professional knowledge of mathematics teachers. For the domain of professional knowledge, we will outline the theoretical background of this study.

2.2 Professional Knowledge and Instruction-Related Views of Mathematics Teachers

Creating cognitively activating learning opportunities related to modelling and helping students when dealing with modelling requirements presupposes professional knowledge linked to modelling in the classroom. Such specific professional knowledge includes knowledge about modelling in mathematics, meta-knowledge about the modelling process (Blomhøj and Jensen 2003; Blum and Leiß 2005; for empirical results see Maaß and Gurlitt 2009; Siller et al. 2011) and implications for teaching, knowledge about specific settings for modelling in the mathematics classroom, knowledge about technology use in the modelling process (Siller and Greefrath 2010), about possibilities of support of students in the modelling process, and views related to tasks (Kuntze 2011; Kuntze and Zöttl 2008). Hence, specific professional knowledge of mathematics teachers should cover a range of aspects – including those mentioned in the previous section. These aspects are necessary for an awareness of modelling as a big idea with relevance for the mathematics classroom.

As an over-all framework for professional knowledge, we refer to Shulman's (1986) categories of "pedagogical content knowledge" (PCK), "content knowledge" and "pedagogical knowledge". For designing learning opportunities in the mathematics classroom, PCK is expected to play a key role. In a more detailed model of professional knowledge (Kuntze 2011), we integrate the spectrum between knowledge and epistemological beliefs/convictions (Törner 2002) as well as the spectrum of so-called levels of globality (Törner 2002) as additional dimensions. For example, the teachers' views about the significance of modelling for the mathematics classroom (Siller et al. 2011) can be described according to this model: they can either be global (general significance of modelling) or rather content-specific (e.g., views about the significance of modelling for certain content areas such as functions). Among such views are also self-perceptions of the teachers' PCK related to modelling, a component of professional knowledge that merits attention, as empirical evidence in this area is scarce.

2.3 Teachers' Self-Perceptions of Pedagogical Content Knowledge Related to Modelling

Self-perceptions of PCK, that is, views of teachers concerning their own PCK, are likely to play a mediating role for the development of professional knowledge. Moreover, they may be considered as indicators of the teachers' PCK, in an analogous way to the predictive role of self-efficacy for competency (Dweck 1986). Further, such self-perceptions afford insight into the teachers' views about their personal needs of professional development. In this sense, these self-perceptions can be considered as "local" individual views, which might frame somewhat "larger" epistemological beliefs, as understood by Törner (2002).

In a first approach to self-perceptions of PCK related to modelling, empirical indicators can use the perspective of a direct reference, that is self-perceptions of PCK when relevant for teacher-student interactions related to modelling in the classroom, or the perspective of a more indirect reference, that is, perceptions of the professional development at university against the background of the requirements of instructional practice. For both of these complementary perspectives, qualitative explanations for the teachers' answers can support the validity of the data. Hence, open questions focusing on corresponding PCK and on views about teacher professional development at the university can help to frame empirical findings related to self-perceptions of PCK in the domain of modelling.

3 Research Questions

Consequently, in order to find out about teachers' self-perceptions of PCK in the domain of modelling, the following two research questions are at the centre of this study:

(a) What self-perceptions of their PCK related to modelling do Austrian pre-service and in-service teachers hold?
(b) What explanations for these self-perceptions can be found in the teachers' answers to open questions related to their modelling-specific PCK?

4 Sample and Methods

In this study, 38 Austrian pre-service and 48 Austrian in-service teachers were asked to answer a pencil-and-paper-questionnaire. All pre-service teachers and 20 in-service teachers answered the questionnaire. The 20 in-service teachers (14 female, 6 male) had a mean age of 32.6 years ($SD=9.97$ years) and had been teaching for a mean time of 6.7 years ($SD=9.51$ years) at academic-track secondary schools. The 38 pre-service teachers (30 female, 8 male) had a mean age of 23.5 years ($SD=3.55$ years) and were preparing to teach in academic-track secondary schools.

The questionnaire consists of several sub-sections. For the research questions above, a multiple-choice section about self-perceptions complemented with open questions about providing help to students in the modelling process, hence focusing on specific PCK, and views related to possibilities of improving modelling-specific professional development at the university. The multiple-choice part of the questionnaire was designed to consist of four indicator-like scales; the data were analysed with quantitative methods based on the answers given in the format of four point Likert scales. The answers to the open questions were analysed with qualitative methods in a bottom-up approach. The scope of this qualitative analysis was to explore possible backgrounds to the quantitative results and to check their validity, according to research question (b).

5 Results

5.1 *Self-Perceptions of PCK Related to Modelling*

Research question (a) focuses on the teachers' self-perceptions of their PCK related to modelling. As mentioned at the end of Sect. 1, these self-perceptions were explored from the perspectives of views about PCK relevant for modelling-specific teacher-student interactions ('direct reference') and views about the teachers' satisfaction with what they had learned at university concerning ways of fostering modelling abilities ('indirect reference'). For both of these perspectives, two scales had been conceived for the questionnaire, covering more than only one aspect per perspective. The 'direct reference' scales focused on self-perceptions of diagnostical knowledge related to the modelling process and on providing modelling-specific help to students. Sample items are given in Table 26.1. The 'indirect reference' items focused on self-perceptions about modelling-specific PCK learned at university and about PCK relevant for technology use in the modelling process. The teachers' answers were coded with values from 1 ("I do not agree at all") to 4 ("I fully agree"). The reliability values for all scales were good, given the low number of items per scale.

Figure 26.1 displays mean values and standard errors for the two sub-samples. The mean values are close to the centre of the scale or in the spectrum of negative answers. The difference between the mean answers of pre-service and in-service teachers for the self-perceptions of PCK about technology use is significant. The data in Fig. 26.1 show that the indicators for self-perceptions of PCK related to modelling do not show high values, neither for pre-service nor for in-service teachers.

Table 26.1 Scales about self-perceptions of PCK and reliability values

Scales Positive self-perception concerning PCK ...	Sample item	Number of items	α (Cronbach)
in the domain of diagnostical knowledge related to the modelling process	I know how to connect the origins of difficulties of learners with (particular) phases of the solution process of modelling tasks.	2	.86
relevant for giving specific help to students in the area of modelling	I can give hints to students which phases of the modelling cycle have to be gone through again in order to improve the quality of the final result.	2	.79
about modelling learned at university	During my professional development at the university I have been able to build up a good knowledge base for teaching modelling tasks.	4	.85
about technology use in modelling processes learned at university	As far as the use of technological tools for solving modelling tasks in the classrooms is concerned, I see my preparation at the university as satisfactory.	2	.76

Table 26.2 Correlations between the scales displayed in Table 26.1

Positive self-perception concerning PCK …	(2)	(3)	(4)
in the domain of diagnostical knowledge related to the modelling process (1)	.80***	.64***	.06
relevant for giving specific help to students in the area of modelling (2)		.65***	.10
about modelling learned at university (3)			.45***
about technology use in modelling processes learned at university (4)			

*** p<0.001

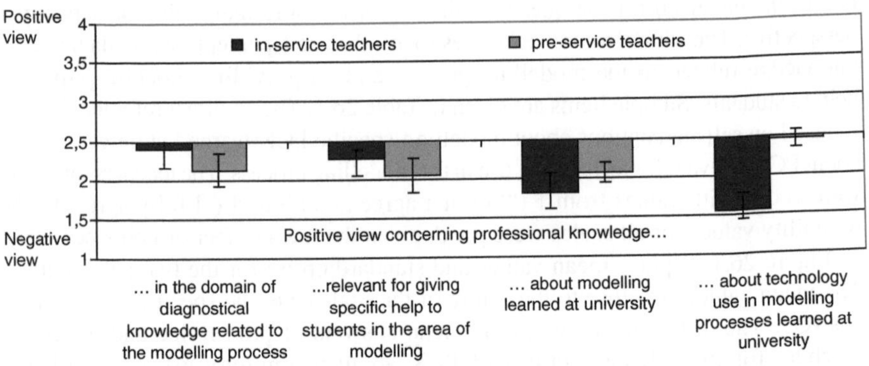

Fig. 26.1 Teachers' self-perceptions of their PCK related to modelling

Especially the in-service teachers saw their PCK development at the university as rather negative.

Table 26.2 shows correlations between the scales considered in this study. The values indicate a relatively high interdependence between the 'direct reference' scales, suggesting that the constructs may not be empirically different. These scales correlate with the self-perceptions of modelling-specific PCK learned at university. This expected correlation appears to underpin the significance of professional development at the university for modelling-specific PCK. Another expected correlation is the interdependence between the 'indirect reference' scales referring to the perceived PCK learning outcomes of professional development at university.

5.2 *Explanations of Self-Perceptions of PCK Related to Modelling*

In order to evaluate whether the findings in Fig. 26.1 correspond to a low modelling-specific PCK, research question (b) focuses on an analysis of the open questions. Relatively closely linked to the first two scales, the teachers were asked which help they would provide students with who are facing problems in the modelling process.

Table 26.3 Examples of answers to the open question on modelling-specific PCK

Teacher	Question: Which possible teacher reactions to potential difficulties of students when working on modelling tasks can you think of?
Claire's answers (pre-service teacher):	Repeating contents again, with which problems have occurred Addressing particular weaknesses of students Using tasks with easily accessible solution path (→ i.e. using also solutions that "make sense")
Fred's answers (in-service teacher):	Re-considering/refining situational model Help when interpreting the mathematical solution Finding further possibilities of solving the problem mathematically

We observed that those in-service and pre-service teachers who were able to give an answer to this open question (about 74 % of all pre-service and 45 % of all in-service teachers) frequently focused on rather general suggestions instead of specific help linked to different phases of the modelling process. Examples of typical answers are given in Table 26.3. Claire's suggestions, for instance, appear to stay on a relatively superficial level, with an emphasis on the teacher's role and little awareness of specific aspects of the modelling process. In contrast, Fred's answers contain aspects specific to modelling and hence suggest that he has corresponding PCK.

Practical experience might play a role: There is a tendency that in-service teachers noted more diverse strategies when asked to do so in the questionnaire, in line with the findings related to research question (a). For example, in-service teachers referred to strategies like using different models, working with analogies, asking well-directed questions or helping students interpreting the results. In contrast, pre-service teachers predominantly listed strategies like giving hints (which were mostly not explained further) or just showing them how they could translate a real-life problem into a mathematical one. In-service teachers also answered that they would help students by giving them help with respect to the content of the problem. None of the pre-service teachers suggested this form of intervention, perhaps a result of a lack of experience with students' work on modelling problems in the classroom.

Qualitative evidence linked to the 'indirect reference' scales can be seen in the sample answers to the question displayed in Table 26.4. For instance, in several answers by in-service and pre-service teachers, there appears to be a demand for more emphasis on practical experience like active work with students in class on modelling tasks. The work on modelling tasks is seen as a valuable preparation activity for building up PCK about modelling by several teachers. This might reflect the process focus and the task culture aspect of modelling-specific PCK. Moreover, meta-knowledge about modelling ("what modelling is") plays a role.

6 Discussion and Conclusions

The teachers' self-perceptions of their PCK related to modelling suggest that there is a need for professional development not only as far as PCK related to modelling is concerned, but also related to the aspect of a pedagogical modelling-specific

Table 26.4 Examples of answers to the open question on modelling-specific PCK

	Question: How can professional development at university be improved concerning the work with modelling tasks in the mathematics classroom?
Answers by pre-service teachers	Practice is lacking not only in the area of modelling in the classroom
	Practice-orientation instead of knowledge orientation
	Putting in place special university courses that put the focus on modelling
	Teacher education: dealing with modelling (for more than one hour)
Answers by in-service teachers	Presenting what modelling is, producing examples and discussing them
	Solving tasks, that can be used also in the school context
	Work on such tasks (university level), for sharing the situation and developing solution strategies

self-efficacy of teachers. This self-efficacy may be supported by positive experiences of the teachers with modelling tasks in the classroom. In particular, professional development support for teachers related to the use of technology in the modelling process might be helpful according to the in-service teachers' self-perceptions. Even if the results have to be interpreted with care, given the limited sample size, the results related to research question (a) indicate that there might be differences between pre-service and in-service teachers concerning the perception of their education at university. These differences might stem from recent developments in the professional development program at the university that include the idea of modelling in relation with mathematical contents and mathematical education goals. Moreover, the pre-service teachers' comparatively high self-efficacy concerning the application of technology when dealing with modelling tasks might not be fully justified – it could be seen as a result of their lack of practice in the mathematics classroom. The pre-service teachers' assessment concerning their relatively good knowledge towards the use of technology might change when they need to use technology in the setting of modelling tasks with all the possible impeding conditions.

However, intensified professional development in the domain of modelling-specific PCK appears to be needed, both in initial teacher education at the university level and at the ongoing professional development level for in-service teachers, even in the case of existing modelling courses at universities. The relatively non-optimistic self-perceptions of the teachers' PCK point to a need even seen from the teachers' own perspective. These findings are in line with prior empirical results about professional knowledge related to modelling (Kuntze 2011; Maaß and Gurlitt 2009; Siller et al. 2011).

The results related to research question (b) raise the issue of the role of instructional practice and classroom experience. Classroom experience might shift the perspective not only of learning at the university, but also of the teachers' PCK.

Hence, the development of self-perceptions of PCK against the experience of pre-service and in-service teachers merits deepened attention in further research, for example, in deepening qualitative studies.

There appear to be differences in the answering patterns of the pre-service and the in-service teaches. Even if the data in Table 26.3 suggest that there are in-service teachers who have developed modelling-specific PCK, the relatively low answering rate of the in-service teachers indicates that there are other in-service teachers who might not have been able to describe any modelling-specific help to learners in the classroom.

As far as the pre-service teachers are concerned, the self-perceptions of modelling-specific PCK tended to be lower, in line with less specific answers to the open question in Table 26.3. However, the pre-service teachers showed a less negative view of their PCK development related to the use of technology for solving modelling tasks experienced at university. These findings appear to draw a picture of professional knowledge in a phase of development. With intensified and specific opportunities of professional learning and support when linking this knowledge to classroom experience, there is the hope that these teachers will continue to build up modelling-specific professional knowledge. However, pre-service learning in the area of modelling should be combined with continuous professional development activities, as the findings related to providing specific help in classroom situations suggest.

Generally, the teachers' answers to the open questions appear to support the validity of the quantitative indicator-like scales. For example, the answers that stay on a relatively superficial level, with an emphasis on the teacher's role and little awareness of specific aspects of the modelling process can be interpreted as non-optimal modelling-specific PCK, in line with the non-optimistic self-perceptions reported by the teachers.

As a conclusion for practice, not only pre-service teachers' professional knowledge but also in-service teachers' professional knowledge concerning modelling should be developed further. In order to meet this demand, one field of activity in the project ABCmaths (www.abcmaths.net) aims at providing teachers with specific learning opportunities connected to modelling. The findings also indicate a need for further research into the structure of professional teacher knowledge concerning modelling. Such research could provide an empirical base for the conception of sustainable professional development activities for teachers about modelling.

Acknowledgements The project ABCmaths ("Awareness of Big Ideas in Mathematics Classrooms", www.abcmaths.net) was funded with support from the European Commission (503215-LLP-1-2009-1-DE-COMENIUS-CMP). This publication reflects the views only of the authors, and the Commission cannot be held responsible for any use which may be made of the information contained therein. We acknowledge the cooperation of Elke Kurz-Milcke, Anika Dreher, Anke Wagner, Claudia Wörn (Ludwigsburg University of Education), Karl-Josef Fuchs, Michael Schneider (University of Salzburg), Stephen Lerman, Peter Winbourne (London South Bank University) and Bernard Murphy (Mathematics in Education and Industry, UK) who collaborated as team members in the project ABCmaths.

References

AECC. (2008). Bildungsstandards im Fach Mathematik. *Österr. Kompetenzzentrum für Mathematikdidaktik*. Klagenfurt: IDM.

Blomhøj, M., & Jensen, T. H. (2003). Developing mathematical modelling competence: Conceptual clarification and educational planning. *Teaching Mathematics and Its Applications, 22*(3), 123–139.

Blum, W. (1996). Anwendungsbezüge im Mathematikunterricht – Trends und Perspektiven. *Schriftenreihe Didaktik der Mathematik, 23*, 15–38.

Blum, W. (2007). Mathematisches Modellieren – zu schwer für Schüler und Lehrer? In GDM (Ed.), *Beiträge zum Mathematikunterricht 2007* (pp. 3–12). Hildesheim: Franzbecker.

Blum, W., & Leiß, D. (2005). Modellieren im Unterricht mit der "Tanken"-Aufgabe. *Mathematik Lehren, 128*, 18–21.

Blum, W., Galbraith, P., Niss, M., & Henn, H.-W. (Eds.). (2007). *Modelling and applications in mathematics education, The 14th ICMI study*. New York: Springer.

Borromeo Ferri, R. (2006). Theoretical and empirical differentiations of phases in the modelling process. *Zentralblatt für Didaktik der Mathematik, 38*(2), 86–93.

Breitenbücher, F., & Kuntze, S. (2010). Eiswaffeln und Kriminalfälle – Ideen zum Lernpotential eines mehrfachen Durchlaufens des Modellierungskreislaufs im Mathematikunterricht. *Der Mathematikunterricht (MU), 56*(4), 33–47.

Dweck, C. (1986). Motivational processes affecting learning. *American Psychologist, 41*(10), 1040–1048.

Kultusministerkonferenz (KMK). (2004). *Bildungsstandards im Fach Mathematik für den mittleren Schulabschluss*. München: Wolters Kluwer.

Kuntze, S. (2011). In-service and prospective teachers' views about modelling tasks in the mathematics classroom – Results of a quantitative empirical study. In G. Kaiser et al. (Eds.), *Trends in teaching and learning of mathematical modelling* (pp. 279–288). New York: Springer.

Kuntze, S., & Zöttl, L. (2008). Überzeugungen von Lehramtsstudierenden zum Lernpotential von Aufgaben mit Modellierungsgehalt. *Mathematica Didactica, 31*, 46–71.

Maaß, K. (2006). What are modelling competencies? *ZDM, 38*(2), 115–118.

Maaß, K., & Gurlitt, J. (2009). Designing a teacher questionnaire to evaluate professional development in modelling. In V. Durand-Guerrier, S. Soury-Vavergne, & F. Arzarello (Eds.), *Proceedings of the sixth Congress of the European Society for Research in Mathematics Education (WGII)*, (pp. 2056–2065). http://ife.ens-lyon.fr/publications/editions-electroniques/cerme6/wg11-02-maa6.pdf

National Council of Teachers of Mathematics (NCTM) (Ed.). (2000). *Principles and standards for school mathematics*. Reston: NCTM.

Neubrand, J. (2002). *Eine Klassifikation mathematischer Aufgaben zur Analyse von Unterrichtssituationen*. Hildesheim: Franzbecker.

OECD. (2003). *The PISA 2003 assessment framework – Mathematics, reading, science and problem solving knowledge and skills*. Paris: OECD.

Shulman, L. (1986). Those who understand: Knowledge growth in teaching. *Educational Researcher, 15*(2), 4–14.

Siller, H.-S. (2010). Modellbilden – ein Thema für fächerübergreifenden Unterricht. *Der Mathematikunterricht, 4*, 28–32.

Siller, H.-S., & Greefrath, G. (2010). Mathematical modelling in class regarding to technology. In V. Durand-Guerrier, S. Soury-Lavergne, & F. Arzarello (Eds.), *Proceedings of the sixth congress of the European Society for Research in Mathematics Education* (pp. 2136–2145). Lyon: INRP.

Siller, H.-S., Kuntze, S., Lerman, S., & Vogl, C. (2011). Modelling as a big idea in mathematics with significance for classroom instruction – How do pre-service teachers see it? In G. Kaiser (Ed.), *Proceedings of the Seventh Congress of the European Society for Research in Mathematics Education (WG 6)*. http://www.cerme7.univ.rzeszow.pl/index.php?id=wg6. Retrieved 6 Sept 2011.

Törner, G. (2002). Mathematical beliefs. In G. Leder, E. Pehkonen, & G. Törner (Eds.), *Beliefs: A hidden variable in mathematics education* (pp. 73–94). Dordrecht: Kluwer.

Chapter 27
A Cross-Sectional Study About Modelling Competency in Secondary School

Matthias Ludwig and Xenia-Rosemarie Reit

Abstract The meaning of mathematical modelling and modelling competencies has been discussed frequently and, in parts also controversially, recently. There is still a debate about how to measure modelling competency and a comprehensive agreement is not yet in sight. In this chapter we applied six stages of mathematical modelling competency to classify student solutions according to their modelling performance and investigated 234 student solution approaches from grades 6 to 11 for the modelling task, *Restringing a Tennis Racket*. We identified four main solution approaches. In the study we analysed the correlation of modelling competency stage and students' grade as well as gender. The study shows that mathematical modelling competency is independent of gender issues in this specific case, and gives insights into the modelling behaviour of students of different grades.

1 Introduction

In the last 10 years there has been a busy and controversial discussion about how mathematical modelling competencies can be evaluated and, especially, how they can be improved (e.g., Blomhøj and Jensen 2003; Kaiser et al. 2011). For this reason a variety of treatments have been established which show, among others, that the modelling competency, as well as incorporated sub-competencies (see Fig. 27.1), can definitely be improved by many different forms of treatments (e.g., DISUM[1]-project). Modelling competency is a quite difficult construct, involving many

[1]Didaktische Interventionsformen für einen selbstständigkeitsorientierten aufgabengesteuerten Unterricht am Beispiel der Mathematik, W. Blum, R. Messner and R. Pekrun.

M. Ludwig (✉) • X.-R. Reit
Institute for Mathematics Education, Goethe-University Frankfurt, Frankfurt, Germany
e-mail: ludwig@math.uni-frankfurt.de

G.A. Stillman et al. (eds.), *Teaching Mathematical Modelling: Connecting to Research and Practice*, International Perspectives on the Teaching and Learning of Mathematical Modelling, DOI 10.1007/978-94-007-6540-5_27,

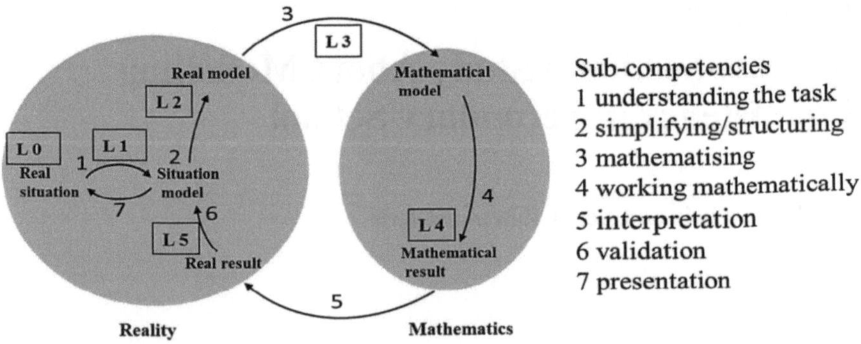

Fig. 27.1 Modelling cycle according to Blum and Leiß (2007) completed with modelling competency stages

factors, reaching from cognitive psychological aspects, like the influence of different thinking styles on the transition process from reality to mathematics (Borromeo Ferri 2010), through to problems concerning the analysis of empirical results, due to its limited measurability. In the context of measuring modelling competency there is still no consistent opinion. Isolated efforts have been made to operationalise modelling competency to have an assessable concept (e.g., Maaß 2007).

The study of Ludwig and Xu (2008) investigated solutions, documented by Chinese and German students of different grades, concerning the mathematical background of peeling a pineapple in a presented, characteristic way. Based on the possibility that there are differences among the two populations, students from grade 9 to 11 from Germany and Shanghai were asked to mathematically explain, that is, to set up a mathematical model of the spiral-shaped appearance of a peeled pineapple, as it is customary in China. Besides the result that the solution quality is correlated to the grade (9–11) of the students, the authors argued that the ability to solve the problem successfully increases with the grade. The research question for the present chapter results from this observation and addresses the assumption that the required mathematical knowledge needs to be cognitively consolidated before it can be used in a targeted manner. This phenomenon forms the basic question of the present chapter. How will students from a wider range of grades solve a modelling task? Can we expect multiple solution strategies? Is there any difference according to gender? To answer these questions we first have to analyse and evaluate the student solutions.

2 Theoretical Framework

This study is based on an evaluation scheme already applied in the study of Ludwig and Xu (2008) which assigns stages to the documented solutions of students, which represent their modelling progress in terms of the modelling cycle of Blum and Leiß (2007). The implementation of the study as a cross section study of students from different grades allows for an investigation of the relationship between modelling competency and grade. The same modelling task is given to all of the students, so that a consistent comparison of performance can be archieved. In addition, by extracting different solution approaches, we are better able to understand the modelling process of students.

2.1 *The Stages*

When elaborating on a mathematical modelling task, one passes through different phases of the modelling cycle, according to Blum and Leiß (2007) (see Fig. 27.1). The transition from one modelling phase to another, that is a progression in terms of the modelling cycle, requires a successful overcoming of cognitive obstacles (Blum 2007). Although we cannot assume a linear cycling through the modelling cycle by the students, as it was verified by Borromeo Ferri (2006) that students follow individual modelling routes when being confronted with modelling tasks, most solutions indicate a more or less detailed solution process. Since it was not part of the study to reveal the actual modelling progress of the students, we concentrated on how far the students have gone through the modelling cycle reflected by their solutions. We assigned so called *modelling competency stages* to each student solution, based on their solution progress within the modelling cycle. We applied a scale of six consecutive stages to categorise the student solutions (Ludwig and Xu 2008), which can be easily integrated in the modelling cycle of Blum and Leiß (2007) (see Fig. 27.1). However we cannot preclude that the students would not have been able to reach a higher stage when having had more time to work on the task. This fact is represented by the terminology "stage" since "stage" reflects a part of a progress which implies a potential achievement of a higher stage.

The definition of *modelling competency stage*, as we use it in this chapter in relation to modelling competency, is based on the competency concept of Weinert (2001), where he points out that competency is an ability which is subject to assessment and used by a person explicitly. In a broader sense this understanding of modelling competency coheres with the definition of Blomhøj and Jensen where they pointed out that modelling competency is "headed for action" (Blomhøj and Jensen 2003).

In the following we will explain the stages as they are applied in this study. For a comparative explanation see Ludwig and Xu (2008). We have to note, that the stages are not distributed equidistantly in general. The reason for this lies in the fact, that the underlying scale is not metric.

- Stage 0 indicates that the student does not understand the task or is not willing to solve it. There are no sketches or notes on the worksheet.
- Stage 1 shows that the student understands the given real situation, but the student is neither able to structure nor to simplify it. The student is also not able to find a connection of real situation and mathematical ideas. There are some reasonable sketches on the worksheet but no simplifications or mathematical formations are identifiable.

- Stage 2 encompasses the development of a real model by simplifying and structuring the real situation. However, the student is not able to transfer this model into a mathematical model. There are sketches of the situation and assumptions to simplify it.
- Stage 3 contains the transfer of the real model into a mathematical model. The student is able to work to a limited extent with this model and produces a concrete result. However, the student is not able to generalise the solution process. The real model is set up, which implies a completion of any sketch with mathematical notations. In addition, reasonable formula approaches are obvious and a solution in terms of a numeric value is identifiable.
- Stage 4 includes a generation of a mathematical question from the real situation. The student is able to work within the mathematical context and to establish a general formula; but this formula is not analysed or validated yet.
- Stage 5 indicates that the student analysed or validated the solution to better adjust the formula to the given situation. Thus, the student gives suggestions for improvement.

3 The Study

3.1 The Real Situation

The string of a tennis racket can only be replaced as a whole, in contrast to the strings of a guitar, which can be easily exchanged individually. In addition, the string needed for the racket has to be cut off from a big string coil, such that the total length must be known in advance. Of course the string should not be cut off too short, since then it is unusable. On the other hand, string should not be wasted.

Based on these considerations the question arises, how the minimal string length of a tennis racket can be determined using the dimension data of the racket. To be able to repair different sorts of rackets, as for example also badminton rackets, a general formula has to be established which is independent of the actual dimension data given on the worksheet. This problem has also been considered by Ludwig (2008); however we focus here on the student solutions, their performance and correlation with gender aspects. An extended version has been published in Ludwig and Reit (2013b).

3.2 Sample and Study Implementation

The study was performed in grammar schools in Bavaria and Baden Wuerttemberg in grades 6, 8, 9, 10 and 11 with students from the age of 12–17. We involved a total of 234 students, all not having had special training in mathematical modelling before.

The study was to be integrated into the normal school routine, so the task had to be able to be processed in a 45-min class, as is usual in Germany. We developed a modelling task which is, according to the German curricula, equally well suited for

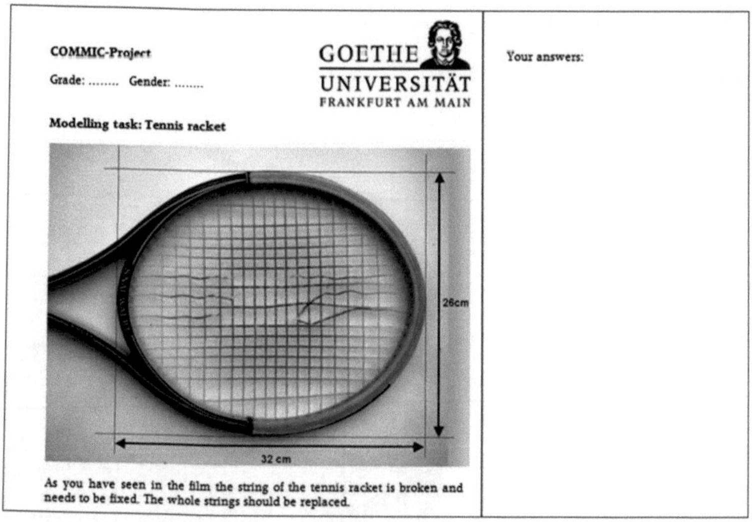

Fig. 27.2 Worksheet with the modelling task

all participants. The underlying mathematical knowledge, for example performing an approximation of the tennis racket by a rectangle, should already be available to 5th graders, according to German curricula.

At the beginning of the lesson a 90 s clip was shown, illustrating the situation of the modelling task. The second phase of the lesson contained working on the actual modelling task in a seatwork setting. Therefore, the students received a worksheet in A4 landscape format (see Fig. 27.2); on the left side there is a picture of the tennis racket with the broken string. This picture is equipped with original specifications, indicating the horizontal and vertical length of the racket area. Additionally, the following two mathematical questions are below the picture (not shown in Fig. 27.2).

(a) It is now up to you to estimate the total length of the string you need for this racket in a mathematical way. Perhaps the dimensions in the picture will help you.
(b) Can you specify a simple formula that an employee in a sports shop can use to calculate the total length of the string of different rackets? The formula can use racket data, which are easy to determine.

4 Solution Approaches

4.1 Approach by Direct Measurement

With a direct measurement approach the student directly measures the string length illustrated on the worksheet with a ruler and translates these values subsequently into real values by calculating the scale using the dimensions given in the picture.

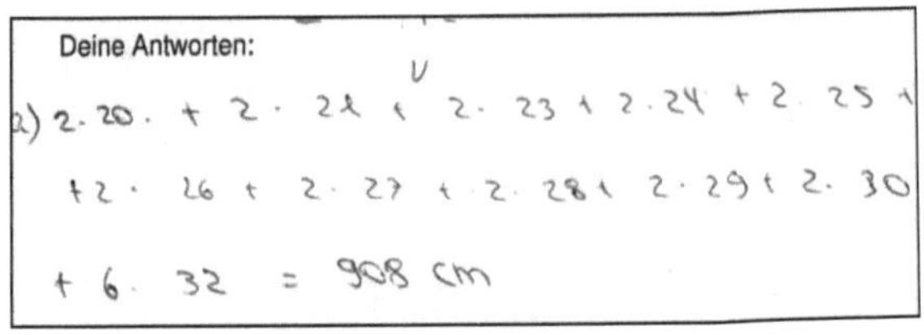

Fig. 27.3 A student solution using direct measurement (rated stage 3)

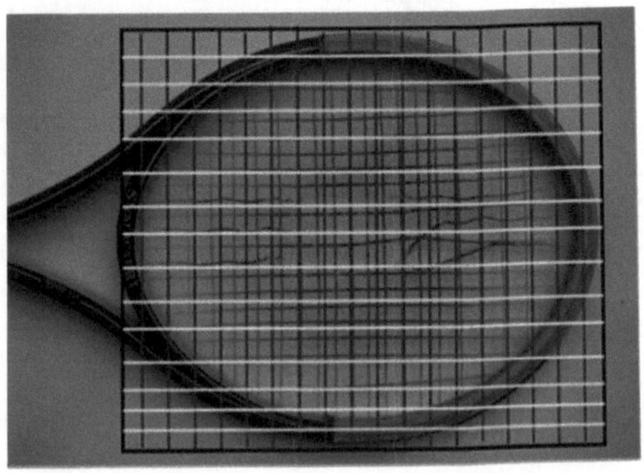

Fig. 27.4 The *rectangle* net overlaying the elliptical racket

This process can also be observed vice versa, that is, calculating the scale first and then summing up the real string lengths to obtain the total length (see Fig. 27.3). The main problem of the direct measurement approach is that it is not possible to set up a general formula based on this approach so it must be classified as stage 3. Most students using this approach stopped after subtask (a), having stated a numeric value for the total string length. Subtask (b) requiring aspects of generalisation, had often not been elaborated to some extent.

4.2 *Rectangle Model*

The rectangle model includes the approximation of the tennis racket area by a rectangle (see Fig. 27.4). Doing this, the respective students lengthen the strings beyond the frame of the racket. They count the vertical and horizontal strings (in our case 18 horizontal and vertical strings) and multiply these values by the given side length (here 32 cm) and width (here 26 cm) of the rectangle, or tennis racket respectively (see Fig. 27.5).

The calculated result of 1,044 cm total string length in Fig. 27.5 qualifies the solution approach for stage 3. Furthermore, the student argues, that the rectangle model is sufficiently exact, because the resulting overestimation intuitively includes the extra string needed to fasten the string on the frame of the racket. This improves the student solution to stage 4. The establishment of variables for the number of horizontal and vertical strings and length and width of the racket leads to a general formula, which is valid for different rackets. This gives rise to rating the student solution at stage 5.

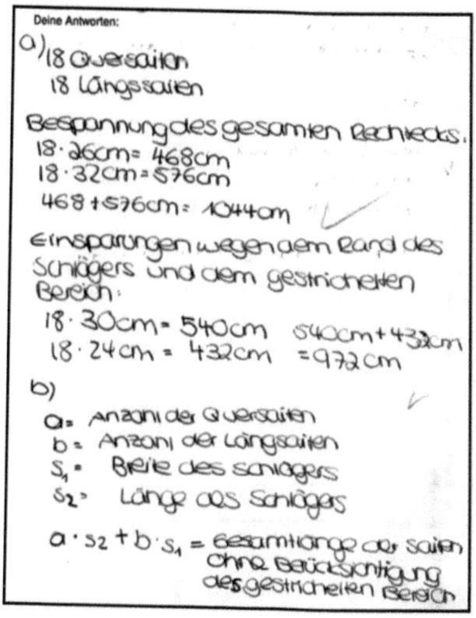

Translation:
Your answers:
a) *18 vertical strings*
18 horizontal strings
Covering of the total rectangle:
18·26cm = 468cm, 18·32cm = 576cm
468 + 576cm = 1044cm
Conservations resulting from the edge of the racket and the dashed area:
18·30cm = 540cm
18·24cm = 972cm 540cm + 432cm = 972cm
b)
a = number of crosses
b = number of mains
s_1 *= width of the racket*
s_2 *= length of the racket*
$a \cdot s_1 + b \cdot s_2$ *= total length of the string regardless of the dashed area*

Fig. 27.5 A rectangle solution of a student in grade 10 (rated stage 5)

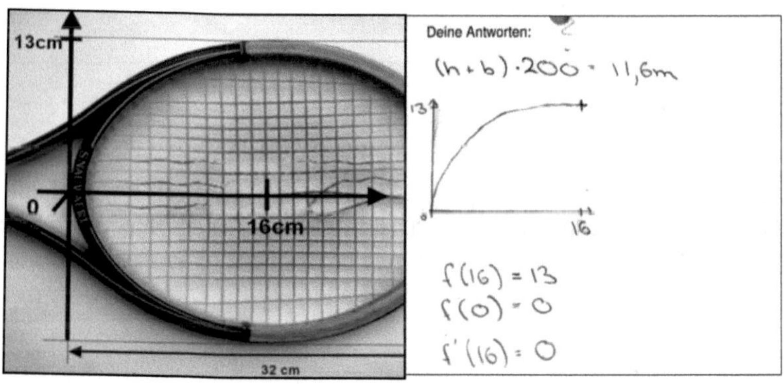

Fig. 27.6 A solution using the function approach (rated stage 2)

4.3 Functional Model

By applying the functional model some students tried to approximate the shape of the elliptical tennis racket with a graph of a function (see Fig. 27.6). To do so, the student implicitly inserted a coordinate system into the picture on the worksheet such that three constraints ($f(16)=13, f(0)=0, f'(16)=0$) can be set up. However, as long as the function type is unknown, the three equations are more or less unusable, even if they are indeed comprehensible. An explicit calculation could not be performed by the student which leads to a rating with stage 2, but the subsequent strategy was outlined perfectly. The student argued that the inverse function of the sketched function is necessary and that the outputs belonging to integer x-values provide the single string lengths.

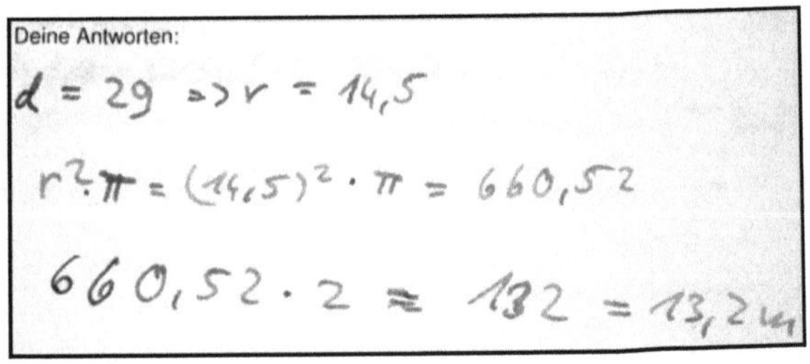

Fig. 27.7 A student solution using a circular model and ignoring units (rated stage 2)

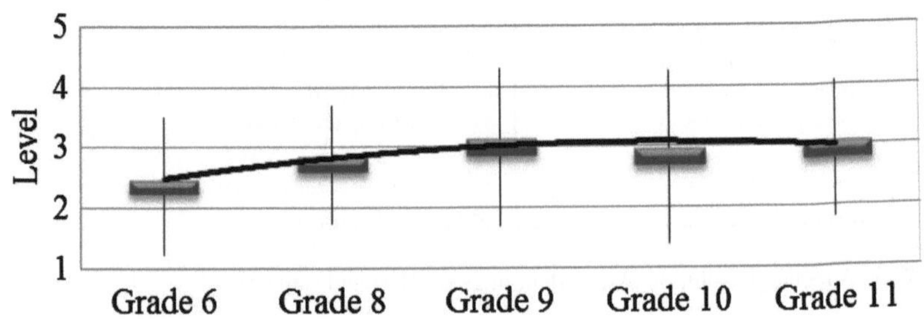

Fig. 27.8 Distribution of stages per grade

4.4 Area Model

The area model, as shown in Fig. 27.7, was used rather infrequently. It became clear, that the student had difficulties in transforming the area of an object with the string length (see Fig. 27.7). One explanation may be the difficult accessibility of a circle or other geometric object by a functional term.

5 Results

The 234 solutions have been analysed by two independent raters classifying each solution approach according to our stages. They found the four main solution approaches and classified them using the six stages. With Cohen's reliability coefficient, $\kappa = 0.789$ and Pearson's correlation coefficient, $r = 0.84$, there was a good inter-rater reliability.

In Fig. 27.8, it is obvious that the students improve their modelling skills between grade 6 and grade 9, at least related to the given task. From grade 9 this development passes over to saturation. A possible assumption is that so long as the required mathematical knowledge is not cognitively consolidated to this point, we have a continuous process of improvement concerning modelling skills. This increase in modelling competency then reaches its maximum and stays at this stage.

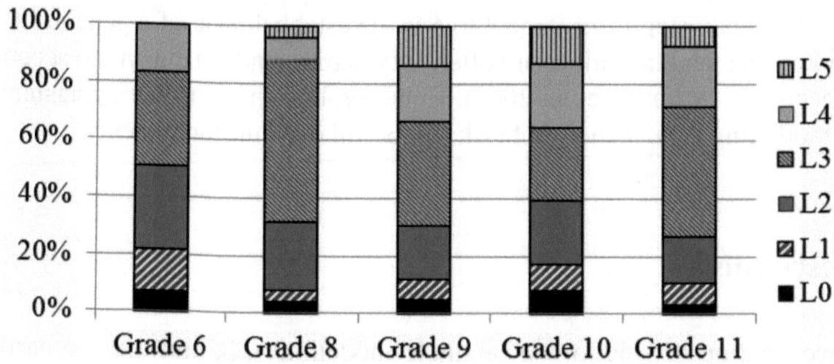

Fig. 27.9 Percentage of students by grade reaching various stages

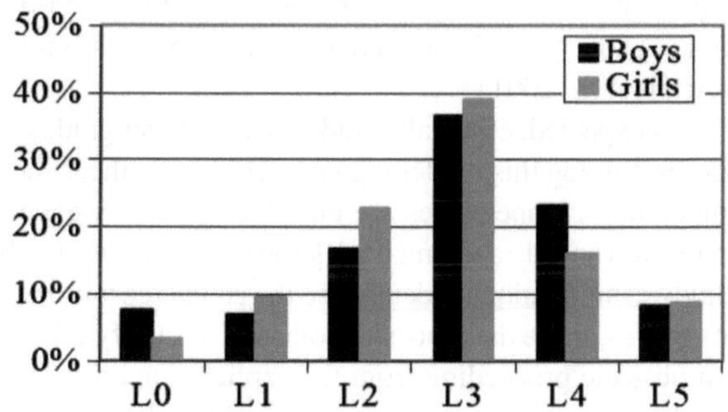

Fig. 27.10 Percentage of boys and girls stopping at various stages

5.1 *Stages Within Grades*

When looking at the distribution of stages within the grades, it is apparent that no student of grade 6 reached stage 5 and more than 80 % of the students of grade 6 and 8 did not reach stage 4 (see Fig. 27.9). Grade 6 and 8 students appear to have had more problems with subtask (b), that is generalising their solution of subtask (a), than the others, since the elaboration of subtask (b) was prerequisite to reach stage 4. The distribution of stages from grade 9 to 11 is not conspicuously different, consistent with the findings of Fig. 27.8.

5.2 *Performance by Gender*

A Mann-Whitney-*U*-Test to detect differences in the modelling competency among boys and girls was not statistically significant ($z=-0.89$, $p=0.374$) (see also Fig. 27.10).

Both, girls and boys have a maximum at stage 3, suggesting that stage 3 is a kind of barrier for the students. Figure 27.9 also clearly shows that most students reached stage 3. The main step from stage 3 to 4 is the establishment of a general formula. It is known that generalisations are often very hard to understand and to accomplish by students (c.f. Stillman et al. 2007). However, this shows that the transfer to the actual modelling task, namely subtask (b), is still difficult for students.

6 Discussion

The investigation of solutions of a single modelling task and the comparison to performance has not been done before explicitly in this manner and gives interesting insights into modelling behaviour of students. A first quite pleasing result of this study was that nearly all students were able to document a more or less useful solution to this modelling task in contrast to PISA 2000 where German students were not yet able to work on mathematical problem solving tasks in a satisfactory manner (Artelt et al. 2001).

As expected, especially students in younger grades seem to have had more difficulties in solving this modelling task. This strengthens the assumption that mathematical modelling competencies are enhanced by experience and practice, even though the more advanced mathematical knowledge might also be an issue. Since the present study cannot uniquely determine the actual reasons for the differences in performance of grades, more data must be collected to clarify this. However, for all grades there is an obstacle proceeding from the explicit numeric result to its generalisation and this could be commensurate with similar difficulties in purely mathematical contexts.

In contrast to the widespread stereotype of mathematically science oriented boys and linguistically gifted girls we did not find statistical evidence of differences in the performance of boys and girls in elaborating on a modelling task. This coincides well with the findings of Ludwig and Reit (2013a) where tests support a general comparability of boys' and girls' modelling competency with reference to the tennis racket task. To explain this in more detail, some more task oriented research is needed. The correlation between solution model and performance has not been investigated in this study. Moreover, with the evaluation of solution approaches it is possible to draw conclusions about the coherence of subject matter and approach used.

References

Artelt, C., Baumert, J., Klieme, E., Neubrand, M., Prenzel, M., Schiefele, U., et al. (2001). *PISA 2000 Zusammenfassung zentraler Befunde*. Berlin: Max-Planck-Institut für Bildungsforschung.

Blomhøj, M., & Jensen, T. H. (2003). Developing mathematical modelling competence: conceptual clarification and educational planning. *Teaching Mathematics and Its Applications, 22*(3), 123–138.

Blum, W. (2007). Modellierungsaufgaben im Mathematikunterricht – Herausforderungen für Schüler und Lehrer. In H. Humenberger et al. (Hrsg.), *Festschrift für Hans-Wolfgang Henn* (pp. 8–22). Hildesheim: Franzbecker.

Blum, W., & Leiß, D. (2007). How do students and teachers deal with mathematical modelling problems? The example "Sugarloaf". In C. Haines, P. Galbraith, W. Blum, & S. Khan (Eds.), *Mathematical modelling ICTMA 12: Education, engineering and economics* (pp. 222–231). Chichester, UK: Horwood.

Borromeo Ferri, R. (2006). Theoretical and empirical differentiations of phases in the modelling process. *ZDM – The International Journal on Mathematics Education, 38*(2), 86–95.

Borromeo Ferri, R. (2010). On the influence of mathematical thinking styles on learners' modelling behaviour. *Journal für Mathematik-Didaktik* (Journal for Didactics of Mathematics), *31*(1), 99–118.

Kaiser, G., Blum, W., Borromeo Ferri, R., & Stillman, G. (2011). *Trends in teaching and learning of mathematical modelling: ICTMA 14*. New York: Springer.

Ludwig, M. (2008). *Mathematik + Sport: Olympische Disziplinen im mathematischen Blick*. Wiesbaden: Vieweg + Teubner Verlag.

Ludwig, M., & Reit, X.-R. (2013a). Comparative study about gender differences in mathematical modelling. In G. Nagarjuna, A. Jamakhadia, & E.M. Sam (Eds.), *Proceedings of EPISTEME 5* (pp. 48–54). Mumbai: Cinnamon Teal.

Ludwig, M., & Reit, X.-R. (2013b). Eine empirische studie zum mathematischen modellieren im sport. In R. Borromeo Ferri, G. Greefrath, & G. Kaiser (Eds.), *Mathematisches modellieren für schule und Hochschule – Theoretische und didaktische hintergründe*. Berlin: Springer Vieweg.

Ludwig, M., & Xu, B. (2008). A comparative study of modelling competencies among Chinese and German students. In M. Blomhøj & S. Carreira (Eds.), *Mathematical applications and modelling in the teaching and learning of mathematics: Proceedings from TSG 21 at ICME 11* (pp. 197–206). Roskilde: IMFUFA, Roskilde University.

Maaß, K. (2007). *Mathematisches modellieren – Aufgaben für die Sekundarstufe*. Berlin: Cornelsen Verlag Scriptor GmbH & Co. KG.

Stillman, G., Galbraith, P., Brown, J., & Edwards, I. (2007). A framework for success in implementing mathematical modelling in the secondary classroom. In J. Watson & K. Beswick (Eds.), *Proceedings of the 30th annual conference of the Mathematics Education Research Group of Australasia* (pp. 688–697). Adelaide: MERGA.

Weinert, F. E. (2001). Vergleichende Leistungsmessung in Schulen – eine umstrittene Selbstverständlichkeit. In F. E. Weinert (Ed.), *Leistungsmessungen in Schulen*. Weinheim: Beltz.

Chapter 28
Teacher Readiness in Mathematical Modelling: Are There Differences Between Pre-service and In-Service Teachers?

Kit Ee Dawn Ng

Abstract Since its introduction in the Singapore mathematics curriculum in 2003, there have been limited efforts at incorporating modelling tasks in schools. One key factor for this relates to the unpreparedness of Singapore teachers for modelling tasks. It is believed that the majority of the teachers have yet to experience the role of a modeller and hence have difficulty acknowledging the potentials of the use of modelling tasks in their classrooms. This chapter presents findings on the initial modelling experiences of pre-service teachers in primary mathematics education in Singapore. The findings will be compared with those from a previous study involving in-service primary mathematics teachers. Implications from the findings of this study will be drawn with respect to the focuses and challenges of teacher education in mathematical modelling.

1 Mathematical Modelling in the Singapore Mathematics Curriculum

In this chapter, mathematical modelling refers to a "process of representing real-world problems in mathematical terms in an attempt to understand and find solutions to the problems" (Ang 2010). Problem solving has been the central theme in the Singapore mathematics curriculum framework since 1990. Singapore teachers are encouraged to use a wide range of problems (non-routine, open-ended, and real world) in mathematical teaching and learning. The mathematics syllabus encourages the use of mathematical modelling where students may formulate and improve mathematical models to represent and solve real world problems. Through mathematical

K.E.D. Ng (✉)
Mathematics and Mathematics Education Academic Group, National Institute of Education, Nanyang Technological University, Singapore
e-mail: dawn.ng@nie.edu.sg

G.A. Stillman et al. (eds.), *Teaching Mathematical Modelling: Connecting to Research and Practice*, International Perspectives on the Teaching and Learning of Mathematical Modelling, DOI 10.1007/978-94-007-6540-5_28,

modelling, students can learn to "deal with ambiguity, make connections, select and apply appropriate mathematics concepts and skills, identify assumptions and reflect on the solutions to real-world problems, and make informed decisions based on given or collected data" (CPDD 2012, p. 18).

Curriculum planners (Balakrishnan et al. 2010) suggested four key elements of the mathematical modelling cycle are: *mathematisation*, *working with mathematics*, *interpretation* and *reflection*. These are activities students engage in during the modelling process contributing towards the completion of a task. Mathematisation refers to the process of representing a real-world problem mathematically using models (i.e., drawings, graphs, functions, or algebraic equations) by making appropriate assumptions about, and simplifications of, the problem. Working with mathematics involves students choosing appropriate mathematical methods and tools to solve the real world problem which has been represented mathematically. Interpretation activities encourage students to make sense of their mathematical solutions in terms of relevance and appropriateness to the real-world situation. Through reflection, students examine the assumptions made and subsequent limitations of their models. This may result in the development of more sophisticated models. Good modelling tasks could consist of four features: a real world context, open-endedness, unstructuredness, and complexity, hence providing platforms for a complete experience of the entire modelling cycle proposed by Balakrishnan et al. A good modelling task should also connect with the solvers meaningfully.

The potentials of mathematical modelling have been recognized in the Singapore mathematics curriculum for the development of processes such as mathematical reasoning, communication, and connections. Arguments for the use of modelling tasks in the teaching and learning of mathematics in Singapore echoed those of the proponents of mathematical modelling (e.g., Blum 2002; English 2007; Kaiser and Sriraman 2006; Lesh and Doerr 2003). Nonetheless, challenges exist in fostering a positive climate towards mathematical modelling in a generally traditional, prescriptive, and structured classroom culture in Singapore which has accountability towards students' performances in high-stakes examinations. Stillman (2010) called for a three-pronged approach in working towards success in modelling experiences in schools. Among these, there is a need for teachers to adjust their mind-sets so as to make appropriate expectations and provide for student-centred learning during modelling tasks.

In Singapore, one of the main causes of a lack of teacher readiness in implementing modelling tasks in classrooms is limited exposure of teachers to modelling tasks – both as modellers and as facilitators. International research on the facilitation of modelling tasks is robust. Some (e.g., Doerr 2007) called for a pedagogical shift towards more student-centred facilitation, encouraging student-initiated explorations and discussions of ideas, as opposed to prescriptive authoritarian styles of delivery. Others (e.g., Galbraith and Stillman 2006; Schaap et al. 2011) embarked on developing frameworks to help teachers identify blockages in students' experiences during modelling (e.g., making inappropriate assumptions, not recognising a relevant variable) and proposed the use of strategies to overcome blockages (e.g., searching for a model using inductive reasoning, verifying the model through estimation). In her work with in-service primary school teachers, Ng (2010) discovered that beliefs and conceptions of mathematics have implications on whether the potentials of modelling tasks would be harnessed for teaching and learning in Singapore classrooms. Teachers who become modellers for the first time had difficulties accepting the open-ended nature of the modelling task because there is no single solution and model development may not involve the direct use of algorithms or algebraic equations. Many even took time to accept that graphs and drawings are also mathematical models. Arguably, there is a need therefore for teachers to first become modellers in order to appreciate the potentials of modelling, identify student difficulties in their attempts, and most of all empathise with blockages.

Yet, would there be differences in how pre-service and in-service teachers engage with the same modelling tasks? In what ways could teacher educators work with both groups of teachers to help set the stage for fostering a positive climate towards mathematical modelling in their own classrooms? A study investigating the differences between Singapore primary in-service and pre-service teachers during their engagements with the same model-eliciting tasks was embarked upon from 2009 to 2010. Results from the participation of the in-service teachers in one of the three tasks implemented have been reported earlier (see Ng 2010). This chapter will focus on the findings related to pre-service teachers and report preliminary comparisons of the two teacher samples (i.e., in-service and pre-service) in terms of the variety of (1) mathematical approaches used, (2) task interpretations, and (3) challenges faced.

2 Method

The study was conducted first in a 3 h session involving 48 in-service primary school teachers with teaching experiences ranging from 1 to 10 years. A similar session was later conducted with 57 pre-service teachers in primary mathematics education who had only a maximum of 10 weeks of teaching practice. Both teacher samples had no experience implementing modelling tasks with their classes. Neither had they engaged in modelling tasks as modellers previously. In each session, the teachers were initially given a 20-min introductory lecture on what mathematical modelling is and the key features of modelling tasks. They were then requested to take on the role of a modeller and work in groups of three to four on an assigned model-eliciting task designed for primary school students within 70 min. There were a total of three tasks used in each session. The findings presented in this chapter focus on one of the tasks entitled *Youth Olympic Games Task* (YOG) (Fig. 28.1). Although the task was designed for implementation at primary levels, the participants were asked to use their existing mathematical knowledge and skills as adults to work on the task. Given their first experience as modellers in this study and the short interaction time with the task, it was not expected that the teachers had time to embark on in-depth discussions nor produce concrete evidence on the validity and applicability testing attempts for their models. Each group recorded their

Women's 100m freestyle Results recorded (seconds)*

Competition No.	Time in 100 m Freestyle (Minutes and Seconds)				
	Swimmer A	Swimmer B	Swimmer C	Swimmer D	Swimmer E
1	00:56	00:49	01:02	00:57	00:55
2	00:55	DNC	00:59	01:05	DNC
3	DNC	00:56	DNC	00:59	00:55
4	00:58	01:01	00:57	DNC	01:04
5	DNC	00:57	DNC	00:58	00:49
6	00:59	DNC	DNC	00:56	00:57
7	01:00	DNC	00:59	00:56	00:57
8	00:59	00:56	00:55	00:58	00:57
9	00:59	00:56	DNC	DNC	00:58
10	00:59	00:57	00:57	DNC	00:58

*Note: *Best time across heats, semi-finals and finals. DNC: Did not compete.*

(1) Decide on which two female swimmers should be selected.
(2) Write a report to the YOG organizing committee in Singapore to recommend your choices. You need to explain the method you used to select your swimmers. The selectors will then be able to use your method to select the most suitable swimmers for all other swimming events.

Fig. 28.1 The *Youth Olympic Games* task

mathematical model on a flip chart along with the assumptions, conditions, and variables considered. In the final portion of the session, each group member took turns to orally present their work to others during a gallery walk. Here, participants visited stations to view and hear about all three tasks and the solutions. They then chose two solutions from groups which attempted the same task as they did to critique the models presented in written form based on three criteria on a given template: *representation* (i.e., how well does this model solve the problem?), *validity* (i.e., can you suggest how to improve the model so as to address the problem more appropriately?), and *applicability* (i.e., can this model be used in other similar contexts?). Five groups of in-service and ten groups of pre-service teachers participated in the YOG task.

The *YOG task* (Fig. 28.1) was adapted from English (2007). The Youth Olympic Games was an appropriate context to work with given that Singapore hosted the inaugural YOG in August 2010. In this task, problem solvers were to come up with a mathematical model to help the Singapore Sports School select two female swimmers for participation in the Women's 100 m freestyle event using the data provided (i.e., previous competition records). Some background information about the purpose of YOG, age group of participants and types of events was included. Model-eliciting tasks were chosen for the study because they articulate the "contextual modelling perspective" (Kaiser and Sriraman 2006, p. 306) of mathematical modelling where models are perceived as "purposeful conceptual systems" (Lesh 2003, p. 44) which describe, explain, or predict real world phenomena. It was the intention of the researcher to investigate how the two teacher samples mathematised a realistic situation during their model development, activating the key elements in the mathematical modelling process by Singapore curriculum planners mentioned briefly above.

During the process of mathematising and subsequent discussions on the representation, validity, and applicability of the models, insights could be drawn about the chosen mathematical approaches, task interpretations, and challenges. Data collection involved the use of field notes and work done by the two teacher samples, including the verbal and written comments made during the session. Data were analysed qualitatively by doing cross-group comparisons based on the identified focuses (1)–(3) above.

3 Findings

In each sub-section here, the findings derived from pre-service teachers will be presented first followed by comparisons between the two teacher samples.

3.1 *Mathematical Approaches*

All ten groups of pre-service teachers used a variety of mathematical representations and concepts in their models. Five groups used *tables* (Fig. 28.2a) and one group used a *graph* (Fig. 28.2b) to present their interpretations of the problem and its solution. In their written reports, all the groups described briefly how they determined the two proposed swimmers using some of the data provided and their calculations. Seven out of ten groups worked with *average competition time* of each swimmer. In particular, Groups 1 and 7 showed more sophisticated mathematical arguments. Figure 28.2a shows that Group 1 had based their decision to select Swimmers A and E because of consistency in performances as supported by their smaller range of *standard deviation* scores. Interestingly, their solution was echoed by Group 5 who also argued for the selection of the same two swimmers based on consistency in results but instead presented a *timeline* marking the trend in competition times over ten competitions for the five swimmers (Fig. 28.2b) without using mean scores. On the other hand, Group 7 used average as a "sieve" to select Swimmers B and E after two rounds of comparisons. This group first eliminated competition times of more than one minute (identified as outliers). Then, they took an average of the fastest (55.57 min) and slowest (59.99 min) timings across all ten swimmers (i.e., 57.78 min rounded down to 57 min) as the guide to grade the swimmers, awarding one point every time the swimmer achieved a timing of less than 57 min. Thus, Swimmer B had the best score of four, indicating that she took less than 57 min of competition time on four occasions. Finally, to decide between Swimmers D and E, the group used the mean competition times of the two

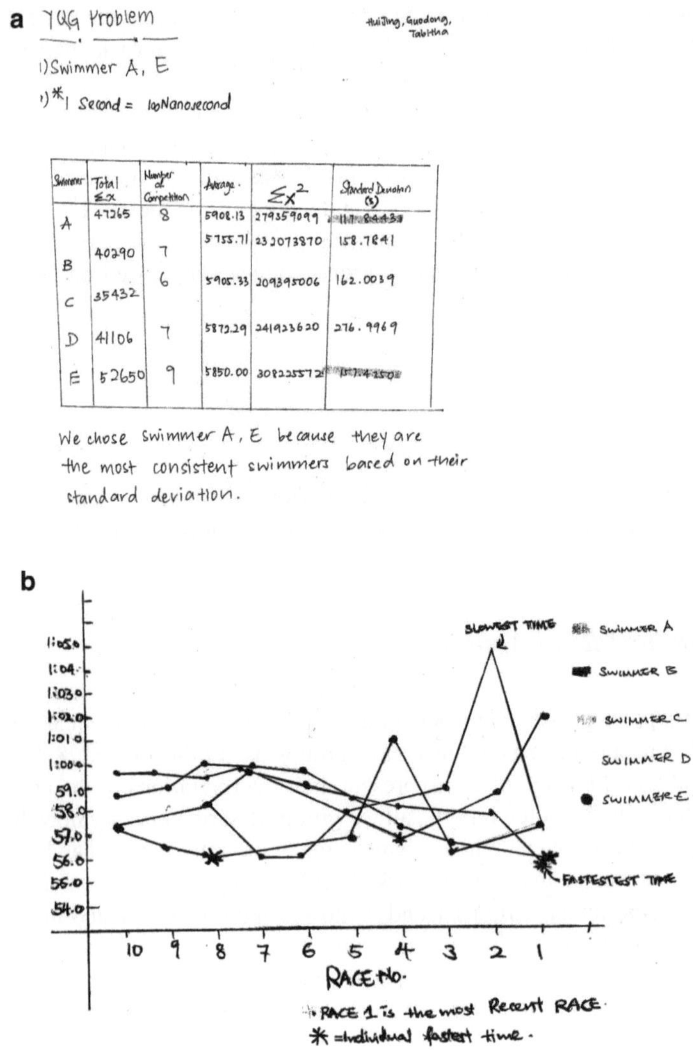

Fig. 28.2 Mathematical representations used (**a**) a table, and (**b**) a graph (colour coded in original)

swimmers. In contrast, Groups 6 and 8 worked with average or mean competition times, along with other concepts such as *probability* (i.e., the likelihood of the swimmer not competing) and *ranking* based on average timings.

3.2 *Task Interpretations*

Pre-service teachers generally interpreted the task based on the data given establishing criteria for selection using competition times and whether the swimmers had participated in key competitions. One group discussed briefly the importance of professionalism in the swimmers as one criterion but this information was not provided in the data. Although the assumptions made by the pre-service teachers had revolved around the information provided (e.g., the best timing refers to the swimmers giving their best in their performance), there were some references to *real-world constraints and expectations* placed on the swimmers. For example, Group 2 assumed that there were genuine reasons for swimmers not being able to compete in some of the competitions as opposed to non-professional tactical planning. In another example, Group 7 made assumptions about the height and weight of the swimmers as being in comparable range and conditions of each competition are kept consistent for each swimmer.

3.3 Challenges Faced

Researcher observations and verbal comments from the participating pre-service teachers seem to suggest that they faced several challenges. Firstly, some groups experienced *blockages right at the beginning* as they could not figure out how to get started because of the open-ended nature of the task. They were apprehensive as there was no apparent direct application of mathematical algorithms and straightforward solution pathways guided by formulae use. In addition, there were many variables to consider as introduced in the data (e.g., competition timings over 2 years, evidence of swimmers not competing) as well as others related to real world constraints. Secondly, there were feelings of *frustration and loss* in a few pre-service teachers who were used to more routine mathematical tasks as they were not aware that time spent on decision making to select the variables to focus upon and to discuss how to move on from the data working towards the goal was a necessary part of the modelling process. There appeared to be a need for constant reassurance by others that they were on the "right track" because the solution process was not a linear one as experienced in routine mathematical tasks. Thirdly, there were also *limited discussions on how the variables could be related to each other* and which combination of factors could be used to predict the swimmers' performances for YOG. Indeed, the written work of the sample showed that many groups had concentrated on using one variable for their decision making, the average competition time, despite the fact that this could be affected by the number of competitions the swimmers had participated in. Lastly, there were pre-service teachers who needed *prompting to use other forms of mathematical representations* other than algebraic equations or formulae to work on the problem. They had difficulty "seeing the mathematics" in the task as well as accepting that graphs, drawings, tables, and even mathematical arguments presented in prose writing was mathematical.

4 Discussion and Conclusion

This chapter is a further report of a preliminary study investigating the nature of initial engagement of pre- and in-service teachers with modelling tasks. A limitation to the research design could be the diverse backgrounds of in-service teachers in the sample. Nonetheless, similarities and contrasts could be detected in the modelling experiences of the two teacher samples.

4.1 Similarities

The nature of the task appears to support the use of average in presenting arguments for the selection of the two swimmers. This was apparent in the majority of the work done by the two teacher samples. Both in-service and pre-service teachers saw the need to relate the task to real world constraints, citing the importance of factoring in professional attitude of the swimmers as a selection criterion. Furthermore, there were incidences of the teachers experiencing blockages at the beginning of the task as well as feelings of frustration and loss during the task. This could suggest that the open-ended nature of the modelling task was a little overwhelming for the participants in both sessions. In addition, there was also evidence of a fixed mind-set of what mathematics is (i.e., using algorithms and formulae) and challenges in overcoming this. Stillman et al. (2010) reported their use of a framework developed in an earlier study (Galbraith and Stillman 2006; Galbraith et al. 2007) to identify blockages faced by Year 9 students in the various phases of the modelling cycle. This study found that teacher modellers also faced blockages during their engagement with the modelling task. Comments and questions gathered during task engagement suggest that the teachers' prior experiences in problem solving and beliefs about mathematics seem to activate blockages in their initial modelling experience. The teachers generally perceived mathematics to be formula-based involving linear track solutions. Hence, these teachers could have been taken aback with the open-ended nature of modelling tasks considering the multiple interpretations and mathematical representations where the modeller also needs to engage in a cyclical modelling process.

4.2 Contrasts

Disparities between two teacher samples came mainly from the mathematical approaches used and presentation of work during their modelling attempts. Firstly, pre-service teachers appeared to have used more sophisticated mathematical approaches in the same task (e.g., standard deviation, probability) as compared to the in-service teachers. The latter group displayed mainly the mathematics their own students could have used for the YOG task (i.e., average, ranking, addition). There could be two explanations for this. For one, the in-service teachers in this study may be used to taking on the teacher role consciously or subconsciously. Thus, despite the request by the researcher for them to engage with the tasks as adult modellers, the teachers may still have worked on the task by immersing themselves in the shoes of their students and at the same time assessed the tasks for possible implementation in their own classrooms, based on the pre-requisite knowledge of the students they were teaching at that point in time. On the other hand, the pre-service teachers in the study could have engaged with the modelling tasks as adult modellers, using the abstract mathematics they knew. A second disparity came from the level of confidence in the systematic approaches the pre-service teachers displayed when they presented their arguments in written form as compared to the in-service teachers who preferred to verbally explain their approaches as they would do in their own classrooms. This could be because it was some time since the in-service teachers had to make written mathematical arguments.

The findings from this study seem to suggest that the challenges in teacher education for fostering a positive modelling climate in Singapore schools are manyfold. Work is needed to help change the mind-set of in-service teachers as to the use of open-ended tasks situated in real world contexts. Predominantly, these teachers may have to first learn to accept and then later scaffold tasks with a non-exhaustive list of solutions which can use various mathematical representations. Although Kuntze (2011) discovered that it was the in-service teachers who were more receptive to intensive modelling activities compared to pre-service teachers, the case could not be confirmed for Singapore teachers in this study as there was qualitative evidence from both samples indicative of blockages to such tasks due to beliefs about mathematics when the teachers are modellers themselves. Lastly, teacher educators would perhaps have to work with the two teacher samples in tailoring expectations for students' work on modelling tasks, albeit on two different fronts. Pre-service teachers may need guidance in anticipating the nature of mathematics applied during modelling tasks by the students. These may be less sophisticated, less organised, and less focused than what the pre-service teachers themselves can produce. In contrast, in-service teachers may have set their expectations at a lower level during modelling tasks and be content with the students meeting those expectations in terms of the quality of mathematics applied. Teacher educators may need to work with these teachers on the range of expected mathematical outcomes for a given task, including the use of more sophisticated mathematical thinking and discuss with the teachers how to scaffold the students towards moving to using more sophisticated mathematics.

References

Ang, K. C. (2010, December 17–21). *Teaching and learning mathematical modelling with technology.* Paper presented at the Linking applications with mathematics and technology: The 15th Asian Technology Conference in Mathematics, Le Meridien, Kuala Lumpur.

Balakrishnan, G., Yen, Y. P., & Goh, L. E. E. (2010). Mathematical modelling in the Singapore secondary school mathematics curriculum. In B. Kaur & J. Dindyal (Eds.), *Mathematical applications and modelling* (Yearbook 2010 of the Association of Mathematics Educators, pp. 237–257). Singapore: World Scientific.

Blum, W. (2002). ICMI study 14: Applications and modelling in mathematics education. *Educational Studies in Mathematics, 51*(1/2), 149–171.

Curriculum Planning and Development Division [CPDD]. (2012). *Mathematics syllabus: Primary.* Singapore: Ministry of Education.

Doerr, H. M. (2007). What knowledge do teachers need for teaching mathematics through applications and modelling? In W. Blum, P. Galbraith, H. W. Henn, & M. Niss (Eds.), *Modelling and applications in mathematics education: The 14th ICMI study* (pp. 69–78). New York: Springer.

English, L. (2007, June 10). Modeling with complex data in the primary school. In *Proceedings of the thirteenth international conference on the teaching of mathematical modelling and applications*. Retrieved from http://site.educ.indiana.edu/Papers/tabid/5320/Default.aspx

Galbraith, P., & Stillman, G. A. (2006). A framework for identifying student blockages during transitions in the modelling process. *ZDM – The International Journal of Mathematics Education, 38*(2), 143–162.

Galbraith, P., Stillman, G., Brown, J., & Edwards, I. (2007). Facilitating middle secondary modeling competencies. In C. Haines, P. Galbraith, W. Blum, & S. Khan (Eds.), *Mathematical modelling: Education, engineering and economics* (pp. 130–140). Chichester: Horwood.

Kaiser, G., & Sriraman, B. (2006). A global survey of international perspectives on modelling in mathematics education. *ZDM – The International Journal of Mathematics Education, 38*(3), 302–310.

Kuntze, S. (2011). In-service and prospective teachers' views about modelling tasks in the mathematics classroom: Results of a quantitative empirical study. In G. Kaiser, W. Blum, R. Borromeo Ferri, & G. Stillman (Eds.), *Trends in the teaching and learning of mathematical modelling* (pp. 279–288). New York: Springer.

Lesh, R. A. (2003). How mathematizing reality is different from realising mathematics. In S. J. Lamon, W. A. Parker, & K. Houston (Eds.), *Mathematical modelling: A way of life – ICTMA 11* (pp. 38–52). Chichester: Horwood.

Lesh, R., & Doerr, H. M. (2003). Foundations of a models and modeling perspective on mathematics teaching, learning, and problem solving. In R. Lesh & H. M. Doerr (Eds.), *Beyond constructivism: Models and modeling perspectives on mathematics problem solving, learning, and teaching* (pp. 3–33). Mahwah: Lawrence Erlbaum.

Ng, K. E. D. (2010). Initial experiences of primary school teachers with mathematical modelling. In B. Kaur & J. Dindyal (Eds.), *Mathematical modelling and applications* (Yearbook of Association of Mathematics Educators, pp. 129–144). Singapore: World Scientific.

Schaap, S., Vos, P., & Goedhart, M. (2011). Students overcoming blockages while building a mathematical modeling: Exploring a framework. In G. Kaiser, W. Blum, R. Borromeo Ferri, & G. Stillman (Eds.), *Trends in the teaching and learning of mathematical modelling* (pp. 137–146). New York: Springer.

Stillman, G. A. (2010). Implementing applications and modelling in secondary school: Issues for teaching and learning. In B. Kaur & J. Dindyal (Eds.), *Mathematical applications and modelling* (Yearbook 2010 of the Association of Mathematics Educators, pp. 300–322). Singapore: World Scientific.

Stillman, G., Brown, J., & Galbraith, P. (2010). Identifying challenges within transition phases of mathematical modeling activities at year 9. In R. Lesh, P. Galbraith, C. Haines, & A. Hurford (Eds.), *Modeling students' mathematical modeling competencies: ICTMA 13* (pp. 385–398). New York: Springer.

Chapter 29
Exploring the Relationship Between Mathematical Modelling and Classroom Discourse

Trevor Redmond, Raymond Brown, and Joanne Sheehy

Abstract This chapter explores the notion that the discourse of the mathematics classroom impacts on the practices that students engage when modelling mathematics. Based on excerpts of a Year 12 student's report on modelling Newton's Law of Cooling, it is argued that when students engage with the discourse of their mathematics classroom in a manner that promotes the communication of ideas, they employ mathematical modelling practices that reflect the cyclical approaches to modelling employed by mathematicians.

1 Introduction

Knowing what mathematics to use in a problem situation, knowing when to use it and how to use it are powerful expressions of mathematics competence (National Council of Teachers of Mathematics 2000). However, in some classrooms, students' development of mathematics competence is limited to working out of a text book where the answers are provided and little opportunity is given to making thinking visible (Boaler 2001). As a consequence, some students are not given the opportunity to see the expression of mathematics competence as being relevant to out of classroom experiences. Using mathematical modelling to explore mathematics empowers students by allowing them to engage meaningfully with multiple contexts of mathematics use (Galbraith 1995). Through the development of

T. Redmond (✉) • J. Sheehy
Mathematics and Science Education, AB Paterson College, Gold Coast, Australia
e-mail: trd@abpat.qld.edu.au; jsy@abpat.qld.edu.au

R. Brown
School of Education and Professional Studies, Griffith University, Gold Coast, Australia
e-mail: ray.brown@griffith.edu.au

G.A. Stillman et al. (eds.), *Teaching Mathematical Modelling: Connecting to Research and Practice*, International Perspectives on the Teaching and Learning of Mathematical Modelling, DOI 10.1007/978-94-007-6540-5_29,

mathematical models students are encouraged to make their thinking visible. When thinking is made visible, ideas are able to be revisited for the purpose of making them consistent with the requirements of the context and consistent with the conventions of the mathematics being employed. For the purpose of this chapter we adopt Galbraith's (1989) definition of 'open' mathematical modelling which refers to the entire process of doing the mathematics leading from interpreting the original problem situation, formulating and designing a mathematical model, through to the validation of the model in terms of the context that it represents. In short, this approach to mathematical modelling embodies the following cyclical practices: (a) making assumptions, (b) formulating questions, (c) developing and interpreting solutions, (d) verifying models, and (e) reporting, explaining, and predicting results (Galbraith 1989).

As students engage in the process of modelling they are presented with opportunities to make choices; choices about the sophistication of the mathematics they use to model the context (for a discussion of task context as it applies here, see Stillman 2012), the assumptions they make and whether they realise they have made these assumptions, the degree of verification they implement and to what level they consider consistency between the model and the context and whether they are prepared to refine the model or accept it as it is. Decisions such as these, evident within a student's solution, provide insight for the teacher into a student's thinking at a time when he/she was engaged in the process of modelling.

2 Theoretical Framing

Learning in mathematics is a social activity (Schoenfeld 2002). When students are provided with an appropriate task, a suitable structure to interact with that task, and scaffolded in their interactions with each other, a form of classroom discourse may be brought about where students are provided with multiple opportunities to construct sophisticated understandings of a mathematics concept or procedure. Within such classroom discourse, the understanding that is demonstrated by an individual student becomes part of the collective understanding of the group (Wertsch 2002). However, when undertaking an assessment task it is generally accepted that the understanding that is demonstrated is that of the individual. While there are a variety of tools that can be used to assess understanding that range from formal examinations to group projects, the authoring of individual mathematical reports that can be worked on over a period of time and that encourage students to conference with their peers, teachers and others, is an important mechanism for assessing student understanding (Morgan 1998). In most instances the final product of such an authoring process is accepted as an individual representation by the student of the student's analysis of a task and a report of the student's synthesis of strategies for dealing with the task and the mathematical assumptions upon which conclusions are based. However, even though the end product is accredited to an individual student the process of authoring that product is inherently social as the student often has discoursed drafts of the work with the teacher and with peers.

The pedagogical context of the classroom referred to in this chapter focused on the implementation of Collective Argumentation (CA) (Brown and Renshaw 2000). Students in the Year 12 classes reported here were encouraged to engage in CA when doing mathematics. CA is an approach to teaching and learning that is based on five interactive principles. The first principle, the 'generalisability' principle, requires that students think about the problem that has been posed in terms of what mathematical concepts and procedures might be useful in building a solution. The second principle, the 'objectivity' principle, requires that ideas, relevant to the task are objectified and communicated to other members of the group. Third, the 'consistency' principle requires that ideas which are contradictory to each other or that belong to mutually exclusive points of view must be resolved through discussion. The fourth principle is 'consensus'. Consensus requires that all members of the group understand the agreed approach to solving the problem. Finally, in implementing the fifth principle, 'recontextualisation', students re-present the group response to the class for discussion and validation. The principles of CA are used by teachers and students to guide engagement in the discourse of their mathematics classrooms. The aim of this discourse is to enable students to analyse mathematical contexts, to synthesise strategies to mathematise these tasks, and to communicate solutions and conclusions to others. What is of interest to this chapter is how the discourse of a Collective Argumentation classroom may be evidenced in the authoring of an individual mathematical report.

3 Method

The school context referred to in this study is a metropolitan P-12 College that has a mathematics programme from Year 6 through to Year 12 with a major focus on mathematical modelling. Using the framework, 'Teaching for Understanding' (Perkins 1992), the mathematics department of this College has identified a sequence of generative topics that are explored to develop an understanding of mathematical concepts using mathematical modelling. The Year 12 classes referred to in this chapter comprised female and male students studying the Queensland Studies Authority (QSA) Mathematics B Syllabus (2008).

Task content focused on an individual mathematical report given to Year 12 classes of students. The task was comprised of two parts. The first part asked students to use the practices of mathematical modelling to investigate Newton's Law of Cooling. The task was an out-of-class task enacted over a 6-week period designed to provide summative feedback to students regarding their competence in developing a mathematical model.

Investigation of Newton's Law of Cooling (Year 12 Task)

Newton's Law of Cooling states: "The rate at which a body cools is proportional to the difference between the temperature of the body and the temperature of the surroundings."

Investigate a situation that directly relates to this statement by looking at an experiment that contains a 'hot mass', and using the N-Spire calculator and a temperature probe or thermometer collect data of a mass cooling.

Analyse the data collected, decide the most appropriate model for this data set and build an equation for that model. Use the model to investigate Newton's Law of Cooling; identify and document a set of attributes necessary for Newton's Law of Cooling to hold true.

This chapter explores how the discourse of Year 12 classrooms may be evidenced in the authoring of one student's report that required her to model generalisations developed from investigating Newton's Law of Cooling.

4 Analysis of Segments of Student Work

The analysis focuses on Jane's first draft of her mathematical report. As Jane had been exposed to the principles of CA to engage in the discourse of her classroom for a number of years, the analysis looked for evidence of her incorporating the principles of CA into her mathematical report. We will not be considering the report in total but, due to word constraints, consider sections which link to the principles – Generalisability, Objectivity, Consistency, Consensus, and Recontextualisation.

4.1 Generalisability

Jane begins her draft report by detailing what she understands the task to involve (see Fig. 29.1). She outlines the process she is going to employ to investigate Newtons' Law of Cooling: that is, Jane evidences the principle of Generalisability, as she clarifies her understanding of the task and details the approach she is going to adopt. She represents the task from a practical point of view. In doing so, Jane identifies the variables of room temperature, temperature of the probe, and the temperature of the cup, as being important. In the report, Jane does not disclose why she considers these variables to be the ones that need to be controlled, out of the many that are available, but she does indicate procedures that she is going to adopt in an attempt to control them.

As seen in Fig. 29.1, Jane has chosen to use graphing calculator technology and a temperature probe to collect data. The use of technology and probe were identified

Newton's Law of Cooling states: *The rate at which a body cools is proportional to the difference between the temperature of the body and the temperature of the surroundings.* The purpose of this investigation is to explore this statement from a practical point of view. This will be done through the analysis of data collected from the cooling of hot water. The data were collected through the use of a temperature probe and a Ti N-Spire calculator. This data will then be analysed and modelled so as to use this as an example to explore Newton's Law of Cooling.
The controlled variables in this experiment were the outside temperature (24ºC), the probe (which was left in the water for 20 seconds before data collection to ensure it was the same temperature as the water) and the mug (the mug was rinsed out with boiling water a few times before data collection, so as to ensure that the mug was not cooling down the water).

Fig. 29.1 Identifying variables

in the task as one method Jane could use to collect data, but, in the end, was a choice she made. Jane's choice of technology provides a clue as to the thinking that Jane is employing. In identifying the variables that need to be controlled and by indicating her choice of the technological tool that she is going to use to assist her investigation, Jane has established a way of thinking and operating about the task. This way of thinking and operating is consistent with the principle of Generalisability that she has been encouraged to use, over a number of years, to direct her initial thinking about a task so that she can engage in the discourse of her mathematics classroom.

Through identifying variables and indicating her preferred tool of representation, Jane is able to enter into discussion with others about the task, ask questions of others, share ideas with others, and to monitor her understanding. Jane has also provided a representation of her thinking that can be compared with others who may have highlighted or de-emphasised the same variables and use of technology. As such, Jane's way of initially thinking and operating is consistent with the practice of mathematical modelling that requires a mathematician to *formulate assumptions and procedures* upon which to base an investigation. This consistency is again demonstrated as Jane objectifies her thinking about the task (see Fig. 29.2).

4.2 *Objectivity*

As seen in Fig. 29.2 Jane has contextualised her thinking about the task to the context of a 'fair experiment'. That is, Jane situates her thinking and the procedures within the practices of a 'fair experiment' comparing what she does to the scientific assumptions upon which a 'fair experiment' is based so as to keep her investigation on track. In many ways Jane compares her ideas with the practices of a scientific experiment and explains those comparisons using the practices of mathematics. In this way, Jane takes her work from the individual to the social plane of reasoning, allowing herself to see what is the same and what is different about her ideas and submitting her ideas to practices that may assist her to view her thinking and procedures from an objective perspective.

Variables:
Let x be time in seconds starting where 0 seconds is x=0, increasing by 1, e.g. 1= 1 second, 30=30 seconds.
Let y be temperature in degrees Celsius, increasing by 1, e.g. 1= 1 degree Celsius, 20= 20 degrees Celsius.

It is assumed that the external temperature, although monitored, was infact kept at a constant rate. If this were incorrect it would mean that the temperature which the water was cooling to would be different at different times, creating an unfair experiment. It is assumed that the properties of the container the water was kept in (mug) did not influence the cooling of the water. If this were an incorrect assumption it would mean that the experiment was unfair, and would not be an accurate investigation into Newton's Law of Cooling. It is assumed that no outside forces, air flow, objects with other temperatures, had any influence over the temperature of the water at any point. If this were incorrect then the experiment would be unfair, and would not be an accurate investigation into Newton's Law of Cooling.

Shown below, is a subset of the collected data. For complete data set see Appendix 1. Also shown below is the scatter plot of x vs. y.

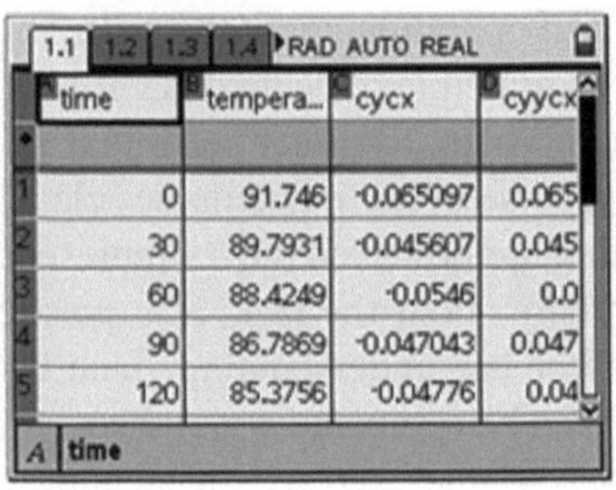

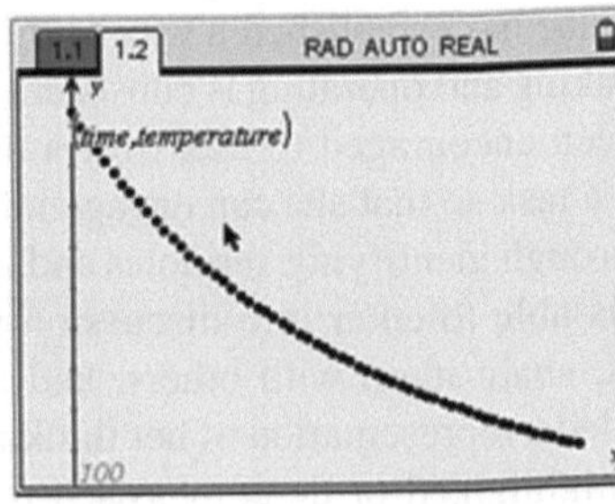

When looked at graphically, it is seen that the data would not be best modelled by a linear model as the curve of the line does not fit the shape of a linear model. This could be modelled by a quadratic equation, but this model would show that at a later time the water would increase in temperature, this does not make sense or fit the context of the data. A quadratic or higher order polynomial would not fit the context of the data as the water would not increase in temperature following the context of the situation. The data may be modelled by an exponential function; it has the same shape that an exponential function would have. The water will eventually reach the external temperature, which is similar to the asymptote of an exponential function. So contextually and mathematically, the exponential function seems to be the better choice of model.

Fig. 29.2 Contextualising thinking

This objectivity sets her up for not submitting a report that just looks authoritative, but is authoritative according to the practices of mathematics. In other words, the approach that Jane has taken speaks to the mathematical modelling practices of *formulating and solving* and requires her to engage in repeated cycles of attaining consistency within her thinking about and doing the mathematical report (see Fig. 29.3).

The first test I will do is the semi-log test. This test is done by finding the natural log (ln) of the y (temperature) data, and plotting it against the x (time) data. If this test shows a linear trend, then it is justified that the data is suitably modelled by an exponential model. Below is the graph of the semi log test, ln(temperature) vs. x.

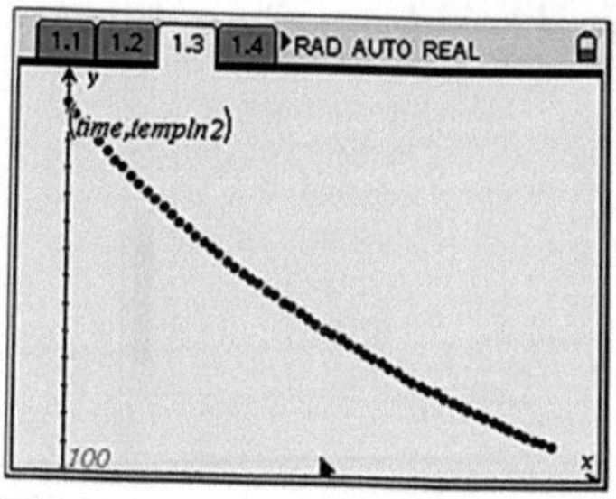

This graph shows that the semi-log test does not create a linear trend, as there is a slight curve to the data. And, therefore suggests the original data to be non-exponential.

Fig. 29.3 Testing the data

4.3 Consistency

Within Fig. 29.3 we can see the principle of Consistency operating as Jane gathers and shares evidence about the data fitting an 'exponential' model that satisfies disciplinary constraints. However, in order to satisfy the principle of consistency, Jane needs to justify her ideas and to become conscious of ways of modelling the data that may better fit task requirements. We see this happening in Fig. 29.3 as Jane becomes conscious of "a slight curve" that suggests the "original data to be non-exponential".

In the process, Jane allows her processes of thought as well as the product of her thinking to become visible and open to change. This relates to the modelling practice of *interpreting* which Jane pursues as she tries to gain consensus between her thinking and the thinking of the mathematical community as represented by her teacher (see Fig. 29.4).

4.4 Consensus

Jane sort feedback from her peers and the teacher about what this might mean. She argued that it should be exponential but the graph of the semi-log graph was inconclusive. The teacher suggested there might be an alternative test that might be performed that might lead to more information about the data and its ability to be modelled by an exponential function. At this College, students in Year 11 develop and use the test of Change in *Y* over Change in *X* vs *Y* to determine whether a data set is able to be modelled by an exponential function. So this was not new knowledge, just knowledge that was either forgotten or ignored until prompted by the teacher.

After this feedback, Jane re-represents her ideas by comparing the Change in the values of *y* divided by Change in the values of *x* versus *y* graph, noting that the data

To investigate this further I will look at the Change in Y/ Change in X vs. Y graph. This graph should model data with a linear trend. Below is the graph of Change in Y/ Change in X vs. Y, and also the linear regression on the data.

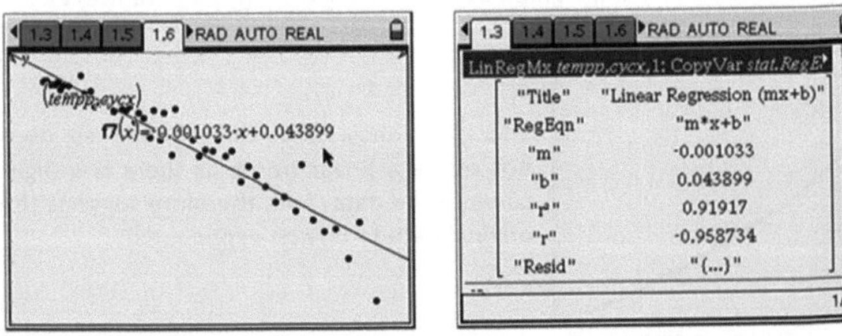

The Change in Y/ Change in X vs. Y data follows a linear trend, which verifies that the original data is exponential. This linear trend is further verified by the linear regression of the data, which has an r value of -0.95, which is extremely close to one.

Fig. 29.4 Verifying thinking

"follows a linear trend". Jane then verifies this trend by representing a "linear regression of the data". In other words, through questioning her own thinking and prompting to think about knowledge she had forgotten or ignored, Jane conducted further inquiry and attained a consensus between her thinking about the task and that based on the mathematics she was using to test the data.

In Fig. 29.2 Jane argues that given the water will cool to room temperature she expects that graph to be exponential. However, upon completing the first test, the semi-log test, she obtains a result that is inconsistent with her initial observations. Jane then looks for further evidence to support her original conjecture and finds it in the graph of Change in Y divided by Change in X vs Y graph. This graph being linear supported her original conjecture that the data and, hence the context, were able to be modelled by an exponential function. In this way, Jane takes the display of her thinking from the individual to the collective plane of functioning as she attempts to use the language of mathematics to express and to verify her thinking. This shift from the individual to the collective is supported in Jane's use of language as she uses the pronouns 'us' and 'we', attempting to regain consistency (see Fig. 29.5) in her thinking – a practice that relates to the modelling practice of *verifying*.

4.5 Consistency Revisited

In attempting to regain consistency in her thinking Jane declared that there "was some error in the semi-log test" and attributed this error to the value of the asymptote. Jane then considered the room temperature in the calculation of the semi-log test but

The Change in Y/Change in X vs. Y verifies that the original data can be accurately modelled by a linear function, and also indicates to us that there was some error in the semi-log test. This could be that the asymptote is not 0 for the data, which is not yet accommodated for, by the semi-log test. We controlled the outside temperature to be 24º C, so this should be the asymptote. However, even after manipulation to the semi-log test to accommodate for an asymptote of 24º C, the graph did not show linear trend. Therefore to find the asymptote of the data I will use the equation $d = \frac{M_1 \cdot y_2 - M_2 \cdot y_1}{M_1 - M_2}$. (For how this equation was found see appendix 2.)

When these variables are substituted for values we get the final equation to be:

$y_1 = 91.746$

$y_2 = 89.7931$

$$d = \frac{(-0.00103 \cdot 91.746 + 0.0439) \cdot 89.7931 - (-0.00103 \cdot 89.7931 + 0.0439) \cdot 91.746}{(-0.00103 \cdot 91.746 + 0.0439) - (-0.00103 \cdot 89.7931 + 0.0439)}$$

$= 42.6214$

This is done for all values of y, and all are found to have the same asymptote of 42. 6214.

This value is extremely different from the controlled value of 24º C. This however may be explained in that the water was kept in a mug, and thus only a small amount of the water was being cooled by the air. Also this air would also have been heated by the heat escaping from the water.

Fig. 29.5 Rethinking the asymptote

still was unable to find consistency between the context and the mathematics through her analysis of the semi-log graph. She then resorted to calculating a value for the asymptote and verifying its suitability by substituting values to find a value for the asymptote using the data and the graph of Change in *Y* divided by Change in *X* versus *Y*.

However, the value for this asymptote is unexpected. Jane then attempts to make this value consistent with the re-representation of her thinking about the task by explaining the value in terms of the "small amount of the water being cooled by the air" and the transfer of heat. Through re-engaging the principle of consistency to guide her investigation, Jane is regulating her attention to the requirements of the task, conceptualising the task within a different set of assumptions and integrating her ideas with scientific understandings. This integration assists Jane to re-gain consensus among her thoughts (see Fig. 29.6).

4.6 Consensus Revisited

Jane's incorporation of the new asymptote into the semi-log test allows Jane to gain consensus between her original representation and her re-representation of the task. In the process, Jane verifies her original approach to doing the task and surrenders

So, to verify that the original data is exponential, we incorporate this in the semi-log test.

Below is the scatter plot of ln (temperature-42.6214) vs. Time with the linear regression for this data, and the regression analysis.

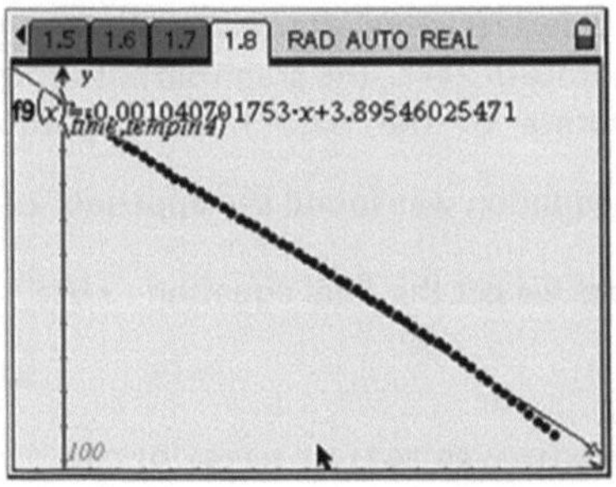

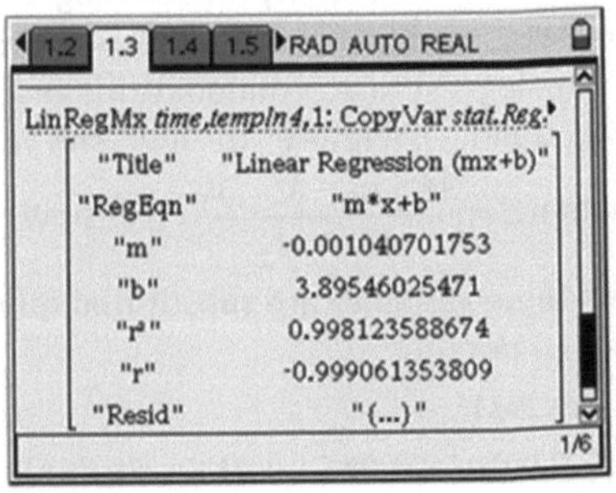

This shows that the semi log test follows a linear trend, and this is justified through the regression and that the r of the regression in -0.9991. Through this we can conclude that the original data is best modelled by an exponential function y=a*b^{t} or y=ae^{kt}.

Fig. 29.6 Validating thinking

the idea that the data are non-exponential. The re-representation of the temperature versus time plot and the construction of the regression analysis reflect her developing understanding of the task. In the process, Jane, again uses the pronoun 'we', acknowledging that her report goes beyond herself to incorporate the discourse of her Year 12 classroom.

5 Conclusion

This chapter explores the notion that the discourse of a mathematics classroom impacts on the practices that students engage with when doing mathematics. The student whose report was analysed represented her approach to doing mathematical modelling as being one that displayed practices that were congruent with those employed by mathematicians. However, in the process of informing those practices, Jane situated herself within a collective, that is, the discourse of her Year 12 classroom. Through situating herself within this discourse, Jane produced a report that not only represented her thinking about the investigation in an objective manner, but a report that evidenced engagement in repeated cycles of consistency and consensus between the mathematics and the context. As such it could be expected that through providing students with access to classroom discourses (e.g., Collective Argumentation, Brown and Renshaw 2000) that privilege the cyclical practices of mathematical modelling as articulated by Galbraith

(1989), that is the mathematisation of a context including making, reviewing and refining assumptions and verification of the mathematical model, such practices would reinforce these modelling processes. Further, these modelling processes and ways of operating should be evidenced in the reviewing and editing that students make public when authoring drafts of assessment reports. Such reviewing and editing processes may not only provide insights into student competency in mathematical modelling, but also insights into the way students represent their engagement with the discourse of their mathematics classroom. For Jane, her draft report clearly demonstrates repeated cycles of looking for consensus between the context and the mathematics. This is what applied mathematicians do on a regular basis (Clements 1989). To find consensus Jane was persistent and willing to consider alternative processes and ways of thinking. Having found consensus she then had to construct a mathematical model to represent the context taking into account the necessary changes found using the mathematical processes. Jane was thoughtful and attempted to provide a reason for the difference between room temperature and the final value developed mathematically for the asymptote. In going from the context to the mathematics in a number of iterations, Jane's draft report not only demonstrates a good understanding of the mathematical concepts and the modelling competencies but also indicates that she is an active participant in the discourse of her mathematics classroom. In other words, Jane was engaged in the process of abstracting 'within' not 'away' from the context of the task for the purpose of connecting her thinking with the discourse of mathematics of her Year 12 classroom (see Stillman 2012 for a discussion relating to abstraction and the task context). Whether this relationship between student authoring of assessment reports and their participation in discourse holds true for students other than Jane is a question for further research.

References

Boaler, J. (2001). Mathematical modelling and new theories of learning. *Teaching Mathematics and Its Applications, 20*(3), 121–127.

Brown, R. A. J., & Renshaw, P. D. (2000). Collective argumentation: A sociocultural approach to reframing classroom teaching and learning. In H. Cowie & G. van der Aalsvoort (Eds.), *Social interaction in learning and instruction: The meaning of discourse for the construction of knowledge* (pp. 52–66). Amsterdam: Pergamon Press.

Clements, R. E. (1989). *Mathematical modelling: A case study approach*. Cambridge: Cambridge University Press.

Galbraith, P. (1989). From applications to modelling. In D. Blane & M. Evans (Eds.), *Mathematical modelling for the senior years* (pp. 78–86). Melbourne: The Mathematical Association of Victoria.

Galbraith, P. (1995). Assessment in mathematics: Developments, innovations and challenges. In L. Grimison & J. Pegg (Eds.), *Teaching secondary school mathematics* (pp. 289–314). Sydney: Harcourt Brace.

Morgan, C. (1998). *Writing mathematically: The discourse of investigation*. London: Falmer Press.

National Council of Teachers of Mathematics (NCTM). (2000). *Principles and standards for school mathematics*. Reston, VA: National Council of Teachers of Mathematics (NCTM). Retrieved from http://www.standards.nctm.org

Perkins, D. N. (1992). *Smart schools: Better thinking and learning for every child.* New York: The Free Press.

Schoenfeld, A. H. (2002). A highly interactive discourse structure. In J. Brophy (Ed.), *Advances in research on teaching series: Vol. 9. Social constructivist teaching: Its affordances and constraints* (pp. 131–170). New York: Elsevier.

Stillman, G. (2012). Applications and modelling research in secondary classrooms: What have we learnt? [Regular Lecture]. In *Proceedings of the 12th international congress on mathematical education,* held 8–15 July, 2012 at Seoul. Published on-line at www.icme12.org

The State of Queensland (Queensland Studies Authority). (2008). *Mathematics B senior syllabus.* Brisbane: Queensland Studies Authority.

Wertsch, J. V. (2002). *Voices of collective remembering.* New York: Cambridge University Press.

Chapter 30
The Role of Textbooks in Developing a Socio-critical Perspective on Mathematical Modelling in Secondary Classrooms

Gloria Ann Stillman, Jill P. Brown, Rhonda Faragher, Vince Geiger, and Peter Galbraith

Abstract One approach to teaching using mathematical modelling and applications is to take a socio-critical perspective where the goal is development of a critical understanding of the world through modelling physical and social phenomena. Several curriculum documents in Australia as well as in other countries, with a modelling and applications emphasis, promote socio-critical aspects to be developed by mathematics curricula, particularly at senior secondary level. The role of modern textbook resources in promoting this development and how teachers could use these resources to mediate such a learning goal in relation to the applications of mathematics is an under researched area. This chapter interrogates curriculum documents and textbook tasks to assess the potential role of textbooks in developing this understanding in students through mathematical modelling.

G.A. Stillman (✉)
Education VIC | Faculty of Education, Australian Catholic University, Ballarat, Australia
e-mail: gloria.stillman@acu.edu.au

J.P. Brown
Education VIC | Faculty of Education, Australian Catholic University, Melbourne, VIC, Australia
e-mail: jill.brown@acu.edu.au

R. Faragher
Education ACT | Faculty of Education, Australian Catholic University, Canberra, Australia
e-mail: rhonda.faragher@acu.edu.au

V. Geiger
Education QLD | Faculty of Education, Australian Catholic University, Brisbane, QLD, Australia
e-mail: vincent.geiger@acu.edu.au

P. Galbraith
School of Education, The University of Queensland, St Lucia, QLD, Australia
e-mail: p.galbraith@uq.edu.au

G.A. Stillman et al. (eds.), *Teaching Mathematical Modelling: Connecting to Research and Practice*, International Perspectives on the Teaching and Learning of Mathematical Modelling, DOI 10.1007/978-94-007-6540-5_30,

1 Introduction

According to Burkhardt (2006) amongst others, "Society now needs *thinkers*, who can use their mathematics for their own and society's purposes. Mathematics education needs to focus on developing these capabilities" (p. 183). Niss et al. (2007) when discussing the goals of embedding applications and mathematical modelling within curricula, point out that "since mathematics accounts for a large proportion of time in school, it needs to provide experiences and abilities that contribute to education for life after school, whether in further study, work, or in enhancing the quality of life" (p. 19). Many curriculum documents (e.g., ACT BSSSS 2007; QSA 2010; VCAA 2010) that advocate, to a more or lesser extent, the use of applications and mathematical modelling give as a rationale the developing of socio-critical capabilities with respect to use of mathematics. In a socio-critical perspective the pedagogical goal is a "critical understanding of the surrounding world and of the models and the modelling process" (Kaiser et al. 2010, p. 224). How does this translate into practice? Biembengut (2013) reports that 13 % of syllabus documents from 183 courses for training Brazilian teachers, which she identified in her study as having mathematical modelling as a subject in its own right or mathematics education subjects involving modelling, used socio-critical modelling in their pre-service teacher education courses. This was to deal with social issues allowing pre-service students to develop a critical positioning regarding the context of the modelling of physical and social phenomena. Our suspicion is that social-critical aspects are even less frequently used in secondary school settings despite curriculum document statements.

2 Widening Perspectives on Mathematics Education and Modelling in School

Kaiser et al. (2010) point out that there is no "homogeneous understanding on modelling and its epistemological backgrounds within the different groups working in the area of mathematical modelling in school" (p. 221). Should we really expect that there would be? Mathematical modelling research in school must be set against the backdrop of what has happened over the years as mathematics education has evolved as a field of study in its own right. Mukhopadhyay and Greer (2001) give a simplified overview (see Fig. 30.1) from their perspective of the evolution of mathematics education and associated influences and goals that have resulted in widening perspectives on mathematics education as the notion that "mathematics is a human activity" (p. 295) gained support over time. According to these authors, expanding viewpoints resulted as mathematics education has shifted from a very strong cognitive perspective, influenced by cognitive psychology and focussing on individual technical competence as goal, to embrace firstly social and cultural contexts and then political aspects as interpretative and ideological influences have diluted the dominant empirical influence.

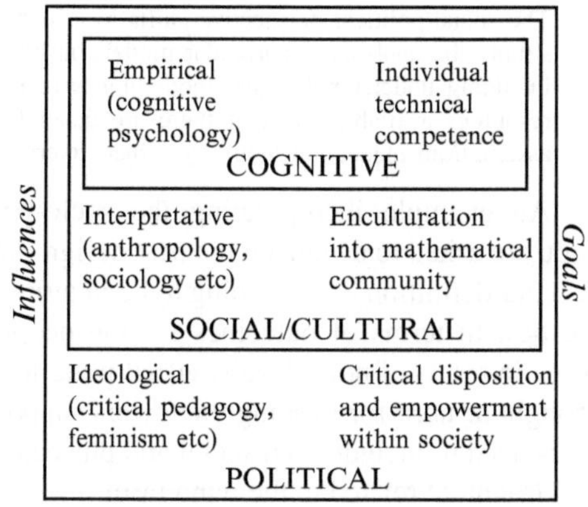

Fig. 30.1 Widening perspectives of mathematics education and associated influences and goals (Source: Mukhopadhyay and Greer 2001, p. 296)

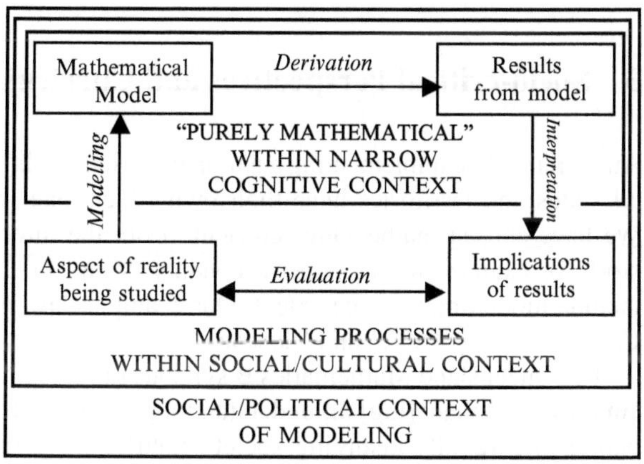

Fig. 30.2 Perspectives on what might be included in school mathematics in terms of modelling (Source: Mukhopadhyay and Greer 2001, p. 300)

Although at times educators and researchers are seen as working within particular broad perspectives, all are part of a holistic view of what currently constitutes mathematics education. It is not surprising then that elements of all three perspectives are seen in rationales or global aims of curriculum documents for schooling.

In relation to mathematical modelling in schooling, Mukhopadhyay and Greer (2001) identify three nested views of what they see as being advocated by proponents of the different perspectives on mathematics education generally as necessary to be included in school mathematics in terms of modelling (Fig. 30.2). Roughly, these correspond firstly, to development of underpinning mathematical competencies so that particular models and procedures that have been learnt can then be applied; secondly, to development of modelling competencies but within the social/cultural context so reflective knowledge is as important as technical knowledge; and thirdly, the development of a critical disposition emphasising awareness of the social/political context of modelling, that is, socio-critical competency. According to Mukhopadhyay and Greer (2001, p. 301),

> The social/political perspective shifts to a higher level, in that it is concerned with not simply the analysis of particular models but the nature of the modelling process per se, including understanding that modelling is relative to the goals of the modeler and the resources available and that there are generally alternative models, depending on the assumptions made, that should be subject to criticism.

An example incorporating the socio-political perspective is the following: You are asked to decide on the best design for a wind farm. What might best mean? Is your definition of best going to be in terms of most efficient in energy generation or best in terms of outcomes for the people who will live near this wind farm (i.e., in terms of noise levels, visual pollution) or best for those whose land will be bought or used for locating the towers supporting the turbines? The modelling will thus need to include both social and physical phenomena from the natural and built environment related to the wind farm.

3 Socio-critical Perspectives and Curricula

Australia will be implementing a national curriculum in the near future but currently all states and territories have their own educational jurisdictions and curricula. We have chosen mathematics curricula from the three jurisdictions in which we work for examination as upper secondary (Years 11–12) curricula in these states include modelling to some extent. Our examples are from Queensland, Australian Capital Territory (ACT) and Victoria.

The Queensland Mathematics B syllabus rationale, for example, presents mathematics as "enhanc[ing] an understanding of the world and the quality of participation in a rapidly changing society [and] a truly international system for the communication of ideas and concepts" (QSA 2010, p. 1). A syllabus global aim is development of "an awareness of the uncertain nature of the world and be[ing] able to use mathematics to help make informed decisions in life-related situations" (p. 3). General affective objectives include appreciation of "the contribution of mathematics to human culture and progress [and the] power and value of mathematics" (p. 5). The ACT Mathematical Methods course guide states that students' capabilities as "critical thinkers", "informed and ethical decision-makers", and "environmentally and culturally aware citizens" should be developed and students provided with "personal attributes enabling effective participation in society" (ACT BSSS 2007, p. 2). "Environmentally and socially aware citizens…consider the implications of problem solutions on the natural and constructed world and the society around them" (p. 15). Thus, both curriculum documents espouse aspects indicative of developing a critical disposition.

The Victorian Study Design for Mathematical Methods (2006–2014), on the other hand, states that "Mathematics provides a means by which people can understand and manage their environment". Modelling is listed as one of the "essential mathematical activities". "This study is designed to promote students' awareness of the importance of mathematics in everyday life in a technological society, and confidence in making effective use of mathematical ideas, techniques and processes" (VCAA 2010, p. 7). Rationale statements in this document place the least emphasis on the socio/political perspective.

Before we began delving into how these statements might be influencing what was potentially done in classrooms, we listed what we perceived as potential sites for contexts that might serve these purposes. These are not meant to be exhaustive but rather to demonstrate accessibility for curriculum writers, textbook authors and classroom teachers. For enhancing an understanding of the world, population growth, food production, spread of disease, climate change, diet, quality of life, poverty and measures of poverty and socio-economic indices are suggested. In order to develop a critical disposition of the surrounding world, measures of humankind's impact on the environment such as the IPAT equation (i.e., Impact = Population × Affluence × Technology, see Chertow 2001) are suggested as having potential. All are opportunities that could be chosen.

4 Textbook Analysis

Bearing in mind the different emphases in these curriculum documents with respect to our focus, we began an analysis of textbooks and associated resources used in the different states to see what role textbook tasks might be able to play in developing a socio-critical perspective on mathematical modelling. We chose three textbooks each for Year 11 and 12 that were being used in the three educational contexts (see Table 30.1 for details) in similar pre-tertiary mathematics courses.

Taking into account that textbooks cannot really do modelling for you but they can be a platform for broadening teaching horizons by the nature of tasks provided, the following questions were framed to guide examination of tasks in each textbook with respect to opportunities for socio/political perspectives to be pursued by the teacher or student.

(1) What are the contexts being used? (2) What are students asked to do? (3) To what extent are the selected texts able to provide a foundation for teaching modelling? (4) To what extent are the selected texts able to provide a foundation for developing a socio/critical competency/perspective in students with respect to (a) enhancing an understanding of the world and (b) developing a critical disposition towards the surrounding world?

Table 30.1 Textbook details

Author	Level	Where used	Format	Code
Dooley (2001)	11	Queensland	Student text, website	MIC
Dowsey et al. (2009)	11, 12	Victoria	Student text, CD, teacher text	MW11, MW12
Evans et al. (2009)	11, 12	Victoria, ACT	Student text, CD	EM11, EM12
Hodgson et al. (2010)	12	Victoria	Student text, ebook, website	MQ12

In reporting findings, the answers to the first two questions will be subsumed under the sections dealing with the latter two questions.

5 Findings

5.1 Texts as a Foundation for Teaching Modelling

The *Mathematics Quest* (MQ) and *Essential Mathematical Methods* (EM) texts give quite an impoverished view of modelling. For MQ, Hodgson et al. take a curve fitting view of modelling summarising their perspective as: "Modelling is the process of finding the rule (mathematical model) that fits the given data" (p. 122). They advise students "the best way to start is to plot the data, as the shape of the graph might suggest the type of relationship between the variables" (p. 111). In an extended example using the context of a cyclone (MQ12, p. 131), students are led to first translate given real data into a linear model of the relationship between wind speed and time, then two quadratic models using two data points then three, and finally a cubic model. Clearly the message is that "modelling" is a vehicle to display mathematical understanding. Evans et al. (2009) merely introduce the term "modelling" in the Year 11 version of EM without explanation in a heading "Functions and modelling exercises" (p. 197) following the introduction of the terms "Linear models" (p. 43) and "Quadratic models" (p. 135) in headings also without explanation. The subsequent "modelling" exercise (EM11, p. 199) includes a purely mathematical area maximisation task then three applications involving postal charges, phone call costs, and car rental. In a later section, the heading "Exponential models and applications" is qualified by the sub-heading "Fitting data" (EM11, p. 433). In the Year 12 text neither the word "model" nor "modelling" is mentioned although the term "applications" is used on occasions.

Both the *Mathematics in Context* (MIC) and *MathsWorld* (MW) texts provide a foundation for teaching modelling to some extent. Dooley (2001) gives a three page explanation of modelling in MIC including a five stage pentagonal model building diagram with the processes of analyse, create, solve, translate and evaluate being the steps between stages. A mathematical model is defined as "a representation of a problem situation that contains the essential characteristics of an object or event, often in the real world" (MIC, p. 1–27). In keeping with the nature of the Queensland Mathematics B course for which the text is written, Dooley claims that as "modelling plays a central part in this course", students "will be given many opportunities to develop skill in modelling" (MIC, p. 1–29). Here, modelling appears to be seen as content in its own right. Dowsey et al. (2009) are less explicit about the nature of modelling when introducing the term in the Year 11 MW text (pp. 171, 229) in the context of functions. In addition, several modelling tasks for School Assessed Coursework are introduced in the text at various points and extended versions appear on a teacher CD. The first, *Energy Balance,* dealing with the energy value of packaged and canned foods appears much earlier than any attempt at explanation (MW11, p. 128). These tasks are described as "innovative Modelling tasks" but were prepared by others, not the authors. The teacher version of the textbook does not explain modelling more explicitly for the uninitiated. There are some additional comments related to formulation but these are related to translation of worded problems (MW11, p. 229, teacher edition). This vagueness about the nature of modelling reflects mentions of "modelling" in the study design (VCAA, 2010).

5.2 *Enhancing Understanding of the World*

Several extended investigative, modelling or analysis tasks which are meant to model physical phenomena and ostensibly enhance students' understanding of the world are present in MIC and MW. Only one appears in MQ12 and there are none in EM. Contexts include age of cats, growth of weeds, crops, bacteria and crystals, biorhythms, sagging of chains, quality of frozen food, wind speed in cyclones, spread of disease, decay of drug levels in blood and length of daylight. Contexts from the constructed environment include design of a fish pond, volume of water in a dam, speed humps and ferris wheels.

In the *Length of Daylight Investigation* (MIC, p. 6–34 to 6–39), for example, some interesting activities are given to cover most of the usual trigonometric function work in a practical context. As part of an extensive investigation, data are fitted to $y = A \cos B(t + C) + D$ to find A, B, C, D. The function is then used to obtain theoretical answers to questions such as when there are 12 h of daylight. Data are given for Brisbane. Students are then asked to work in groups using data from where they live and any other location to find a general model that will determine the length of the day at any location on the Earth, given the time of the year and the latitude and longitude. Application to their own location and other world locations adds some more applied flavour. Much communicative work is required in asking for qualitative explanations of phenomena so in this sense students could be expected to be enhancing their understanding of the world.

Social phenomena are less often used as contexts for extended tasks where the one context is sustained throughout. These included community market stall profitability, currency exchange, cricket bowling averages and gambling on horse races, card games and at the Casino. For the uninitiated, these tasks could extend students' understanding particularly of Australian culture and social life.

There were many examples of standard applications in MIC and MW of all sorts of functions in simple plausible contexts (e.g., exchange rates, simple interest, internet usage, break even analysis to determine viability of a business venture, shape of a roller coaster track, skate board track profile, bridge arch, and composition of tennis club committee).

Simple Interest Investigation (MIC, pp. 4–35 to 4–36)

Jenny has earned $2000.00 cotton-chipping over her holidays. She wants to keep the money for a holiday in a year's time. Her older sister Alice is starting a new career at a solicitors firm and needs new clothes. Jenny agrees to loan the $2000.00 to Alice so long as Alice can return the money to her as soon as possible as a lump-sum, and pay her interest of 1 % per month flat (simple interest).

There were fewer in the other texts (MQ12, EM) and these were often influenced by the format of tasks in high stakes examinations held at the end of Year 12. The latter were at best "pretend" genuine contexts (*Tide Height*) or judged by the researchers as contrived (*Weight*). There were, however, examples of standard applications in extended or short written contexts that the textbook authors saw as clearly contrived situations (e.g., *The Dog ate my Homework*, MW11, p. 544) and this was conveyed to students by how the task was presented. Although some of these contexts are authentic to real life many are not and instead of enhancing students' understanding of the world are more likely to give them a rather strange view of reality or, as has been demonstrated previously, a cynical view of the applicability of school mathematics.

Tide Height (EM Yr 11, p. 487)

It is suggested that the height $h(t)$ metres of the tide above mean sea level on 1 January at Warnung is given approximately by the rule $h(t) = 4\sin\left(\frac{\pi}{6}t\right)$, where t is the number of hours after midnight.

(a) Draw the graph of $y = h(t)$ for $0 \leq t \leq 24$.
(b) When was high tide?
(c) What was the height of the high tide?
(d) What was the height of the tide at 8 am?
(e) A boat can only cross the harbor bar when the tide is at least 1 m above mean sea level. When could the boat cross the harbor bar on 1 January?

Weight (MQ12, p. 26)

The weight of a person t months after a gymnasium program is started is given by the function: $W(t) = \frac{t^2}{2} - 3t + 80$, where $t \in [0, 8]$ and W is in kilograms. Find:

(a) the minimum weight of the person
(b) the maximum weight of the person.

5.3 *Developing a Critical Disposition Towards the Surrounding World*

The modelling of social phenomena is clearly a candidate for the development of a critical disposition towards the world. The modelling task, *Supersize Me!* (MW12, p. 235) inspired by the Morgan Spurlock film (Spurlock 2004) of the same name is such an example where the habits of modern American lifestyles (i.e., poor diet, inadequate exercise) and obesity can readily be explored mathematically and from a critical perspective. The MW student text hints at this: "It is worth pointing out that the film created hostile as well as positive reaction. It was pointed out by some columnists that there was no compulsion to consume 5,000 cal per day and healthier choices are available from McDonalds and other outlets. This is relevant to discussing real-world aspects of the problem of obesity".

In an extension of an analysis task, *You win some you lose some* (MW11, p. 316) profitability of a Casino is explored in a theoretical sense at least as students are instructed: "If the probability that a player will win a game is $\frac{a}{b}$ and the player is charged $\$x$ to play the game, give a general formula in each case for what the casino would need to pay a player to achieve various profit levels." Justification for such profit levels could easily explore social/political issues such as part of the Casino's potential profit being the source of governmental funding for programs to assist problem gamblers and their families and the limitations of the proposed model in taking these considerations into account. This may not have been the intention of the authors in setting the task but scope is there for a teacher or student so inclined.

5.3.1 Missed Opportunities or Launching Pads

In the wealth of contexts that appeared in these texts, there were opportunities that seemed to have been missed. In *World Population Investigation* (MIC, p. 7–4 to 7–6), for example, the investigation begins by examining the world population from 1950 to 2000. A series of questions are posed estimating population for future years and the inverse problem – when will the population reach ten billion, when will it double its 2000 value. Assumptions are given but are contrived (e.g., "Assume population has continued to grow in same way since 2000"). In the process of viewing closeness of fit of the equation to data, however, there is a continuing discussion about relying on assumptions that might change in the future (e.g., "cultural norms such as the number of children desirable in a family"). A warning about extrapolating is discussed usefully but held at the theoretical level rather than used to review some actual outcomes (e.g., ten billion between 2025 and 2030). The contribution of the investigation to students' understanding of natural phenomena is dubious and a critical perspective is not taken. This investigation appears to be an opportunity lost.

A major social issue in Australia is gambling. In an analysis task involving casino games, *You win some, you lose a lot more* (MW Units 3&4, pp. 470–471), the clear social message in the title is not explored. The task as presented in the text stays at the mathematical level being really just a wrapper for probability exercises.

However like many other tasks on gambling at the horse races (*Weighing Up Odds*, MW11, p. 302) or the casino (*Wheel of Fortune*, MW11, p. 291) these are contexts that could be the launching pad for modelling and critique of social phenomena.

6 Possibilities

At this point it is time to pause to reflect on what could have been included in these textbooks that was readily accessible. Proposed taxes, for example, come and go over the years but political expediency means these are many and varied (e.g., currently carbon tax) and the effects of these on particular groups (e.g., self funded retirees) could be modelled. Tax cuts or reducing financial support by government raises many questions: How are families affected? Are there disadvantaged groups? Is the redistribution fair? What is the rationale for an unfair distribution? What is the political gain? Any change of law disadvantages some and not others and the political advisors have done their own modelling ahead of time to determine effect on votes. Data are readily available in newspapers so an astute teacher wishing to take a socio/critical stance could use these in preference to a textbook. Socio-critical questions about these situations could then be readily orchestrated.

7 Discussion and Conclusions

In terms of Mukhopadhyay and Greer's nested views (2001) of what to include in school mathematics, in the main all texts provided ample opportunity to develop underpinning mathematical competencies for modelling so particular learnt models and procedures could be applied. The development of modelling competencies is restricted by the authors' perspectives on the nature of modelling although one text, MW, was supplemented with modelling tasks. Where modelling is promoted by the textbook authors in its richness, technical knowledge usually holds sway over reflection on social or cultural concerns regarding the situation being modelled. Enhancing students' understanding of the world around them does seem to be a goal of many of the tasks in the texts but at times this goal is lost as didactical considerations (Hickman 1986) overwhelm authenticity. The development of a critical disposition emphasising awareness of the social/political context of modelling, (i.e., socio-critical competency) is certainly not the intention of any of the textbook authors; however, there are opportunities for teachers to take up in both MW and MIC.

In the ICMI 14 study volume, Niss et al. (2007) saw as imperative for our field "to articulate the relationship between applications and modelling, its values, methods and skills and the world we live in" (p. 18). This was in pursuit of the epistemological goal of "linking of the field of mathematics with some aspects of the world, with the purpose of enhancing knowledge, but also ensuring or advancing the sustainability of health, education and environmental well-being, and the reduction of poverty and disadvantage" (pp. 17–18). Although the analysis in this chapter has been carried out in the context of Australian state curricula, it does have implications for educators and researchers in other parts of the world where similar goals are expressed in curriculum documents. Clearly, regular textbooks will have only limited impact in broadening teaching horizons for this purpose especially if the teachers have little knowledge of socio-critical perspectives in a mathematics context and/or little inclination to take up the challenge to incorporate into modelling activities.

Acknowledgements A Faculty of Education, ACU, grant for establishment of the Australasian Mathematical Modelling and Applications Research Interest Group facilitated commencement of this research study.

References

Australian Capital Territory Board of Secondary School Studies (ACT BSSS). (2007). *Mathematical methods: Type 2 T course*. Canberra: Author.

Biembengut, M. (2013). Modelling in Brazilian mathematics teacher education courses. In G. A. Stillman, G. Kaiser, W. Blum, & J. Brown (Eds.), *Teaching mathematical modelling: Connecting to research and practice* (pp. 507–516). New York: Springer.

Burkhardt, H. (2006). Modelling in mathematics classrooms: Reflections on past developments and the future. *ZDM, 38*(2), 178–195.

Chertow, M. R. (2001). The IPAT equation and its variants: Changing views of technology and environmental impact. *Journal of Industrial Ecology, 4*(4), 13–29.

Dooley, P. (2001). *MIC mathematics in context: An investigative approach. Queensland mathematics B* (2nd ed.). Toowoomba, QLD: Mathgym.

Dowsey, J., Swedosh, P., Caruso, N. Flynn, P., & D. Tynan (2009). *MathsWorld mathematical methods CAS units 1 & 2* (2nd ed.). Melbourne: Macmillan [MW11].

Dowsey, J., Swedosh, P., Caruso, N., Flynn, P., Wayth, P., & D. Tynan (2009). *MathsWorld mathematical methods CAS units 3 & 4* (2nd ed.). Melbourne: Macmillan [MW12].

Evans, M., Lipson, K., & Wallace, D. (2009). *Essential mathematical methods 1 & 2 CAS* [TI-Nspire/Casio ClassPad version]. Melbourne: Cambridge University Press. [EM11].

Evans, M., Lipson, K., Jones, P., & Avery, S. (2009). *Essential mathematical methods 3 & 4 CAS* [TI-Nspire/Casio ClassPad version]. Melbourne: Cambridge University Press. [EM12].

Hickman, F. R. (1986). Mathematical modeling in physics. *Physics Education, 21*, 173–180.

Hodgson, B., Karanikolas, N., Langsford-Smith, B., Duncan, M., Herft, T., Kempton, L. Nolan, J., & Phillips, G. (2010). *Maths quest 12: Mathematical methods CAS* (TI-Inspire 2.0 ed.). Brisbane: Wiley. [MQ12].

Kaiser, G., Lederich, C., & Rau, V. (2010). Theoretical approaches and examples for modeling in mathematics education. In B. Kaur & J. Dindyal (Eds.), *Mathematical applications and modeling* (pp. 219–246). Singapore: World Scientific.

Mukhopadhyay, S., & Greer, B. (2001). Modeling with purpose: Mathematics as a critical tool. In B. Atweh, H. Forgasz, & B. Nebres (Eds.), *Sociocultural research on mathematics education* (pp. 295–312). Mahwah, NJ: Erlbaum.

Niss, M., Blum, W., & Galbraith, P. (2007). Introduction. In W. Blum, P. L. Galbraith, H.-W. Henn, & M. Niss (Eds.), *Modelling and applications in mathematics education: The 14th ICMI study* (pp. 3–32). New York: Springer.

Queensland Schools Authority (QSA). (2010). *Mathematics B senior syllabus 2008*. Brisbane: Author.

Spurlock, M. (Director). (2004). *Supersize me* [Film] New York: Kathbur Pictures.

Victorian Curriculum and Assessment Authority (VCAA). (2010). *Mathematics: Victorian certificate of education study design*. Melbourne: Author.

Chapter 31
Pre-service Secondary School Teachers' Knowledge in Mathematical Modelling – A Case Study

Tan Liang Soon and Ang Keng Cheng

Abstract This paper reports an exploratory study which examined how a group of pre-service secondary school mathematics teachers' knowledge in mathematical modelling was shaped by their independent modelling experiences. It also determined the pre-service teachers' views on teaching mathematical modelling. The findings show that independent modelling experiences with reflection activities had enabled the pre-service teachers to acquire knowledge on the important elements of modelling activities in the various stages of the modelling process. The pre-service teachers' views on teaching mathematical modelling were mainly limited to the scientific-humanistic perspective. Implications for these findings on mathematical modelling instructional practice were discussed.

1 Introduction

Application and modelling has been a component in the process domain of the Singapore Mathematics Curriculum since 2006. However mathematical modelling primarily has existed as a topic in the pre-university curriculum. Only recently, plans were made to integrate mathematical modelling learning experiences into the primary and secondary mathematics classrooms via the "island approach" (Blum and Niss 1991, p.60). Although Blum and Niss (1991) noted that some of the frontline developments and findings in mathematical modelling are gradually being translated into mainstream mathematics education practice, mathematics education in Singapore still lags behind in terms of the development in mathematical modelling instruction due to the recency of the explicit inclusion of mathematical modelling in the curriculum. Also, mathematics teachers' practices are guided by syllabuses that are

L.S. Tan (✉) • K.C. Ang
Mathematics and Mathematics Education, National Institute of Education, Nanyang Technological University, Singapore
e-mail: liangsoon.tan@stdmail.nie.edu.sg; kengcheng.ang@nie.edu.sg

G.A. Stillman et al. (eds.), *Teaching Mathematical Modelling: Connecting to Research and Practice*, International Perspectives on the Teaching and Learning of Mathematical Modelling, DOI 10.1007/978-94-007-6540-5_31,

topical, whereas mathematical modelling will need to connect several mathematics topics concurrently. It may therefore be expected that many secondary school teachers have limited background and confidence in mathematical modelling and would experience discomfort when faced with teaching mathematical modelling in the mathematics classroom (Ang 2010).

2 Mathematical Modelling Process

As the modelling process is central to the teaching of mathematical modelling, we seek to clarify the mathematical modelling process suggested by Ang (2009) and how its components relate with one another. The mathematical modelling process is characterised by the iterative negotiation of learning between the real and mathematical world. A typical mathematical modelling learning task traces the following trajectory: The mathematical modelling process begins with a problem motivated by practical concerns oriented in the real world. The real world problem is then formulated into a mathematical problem through which a mathematical solution is obtained. The mathematical solution is then interpreted into some plausible real world solution. The reasonableness in the real world solution will then determine if the process needs to be iterated by re-examining how the mathematical problem has been formulated.

Such a mathematical mode of thought as reflected by the mathematical modelling process can be a challenge for pre-service teachers, especially the process of formulating the mathematical model. The notion of real world model as identified by Blum and Leiss (2007) is central in helping to bridge this process, in that it highlights the importance of understanding and simplifying the real world situation. Whilst an expert modeller formulates the real world problem into a mathematical problem directly, understanding and simplifying the real world situation by stating the assumptions first can help pre-service teachers identify and relate the variables of interest in the mathematisation process.

3 Teachers' Knowledge for Teaching Mathematical Modelling

Teachers' knowledge for mathematical modelling instruction can determine, to a large extent, how pre-service teachers perceive and respond to curriculum innovation efforts to infuse mathematical modelling learning experiences into the secondary mathematics syllabus. To think about knowledge of mathematical modelling instruction requires going beyond knowing the mathematical modelling process. It may require understanding the complex interplay among aspects of other forms of teacher knowledge in the mathematical modelling teaching and learning environment.

Shulman (1986) represented pedagogical content knowledge as the blend of content and pedagogy for the understanding of how particular aspects of the subject matter are organised, adapted and represented to teach for understanding by considering students' conceptions, abilities and interests. Following the work of Shulman (1986), Ball (2000) has deconstructed mathematical content knowledge and mathematical pedagogical content knowledge into its key components. Like Shulman (1986), Ball (2000) argued that both content and pedagogical content knowledge are critical for effective teaching.

Stillman et al. (2007) developed a framework to support implementation of mathematical modelling in the secondary classroom. Essentially, this framework comprises elements of modelling activities that correspond to the respective stages of the modelling process, and serves as a guide for teachers, researchers, and curriculum designers to anticipate possible student blockages as they transit between the stages of the modelling process. Such understanding can then lead to the identification of particular prerequisite modelling skills and competencies required by students to complete a modelling task.

The framework developed by Stillman et al. (2007) is akin to what Ball et al. (2008) pointed out as the *specialized content knowledge* for the unpacked modelling process, which in turn can be useful for developing teachers' *knowledge of content and students* in identifying possible students' blockages and its corresponding prerequisite modelling skills and competencies. While the framework can be useful to experienced teachers of mathematical modelling for this purpose, novice Singapore mathematics teachers in the teaching of mathematical modelling may need to gain independent modelling experiences first. This is so that they can better make sense of the various aspects and nuances of the modelling process as outlined by the framework and to experience the possible blockages students may have during the modelling process.

4 Mathematical Modelling Perspectives

The different views of mathematical modelling in education have been discussed extensively by researchers and curriculum planners (e.g., Barbosa 2006; Kaiser and Sriraman 2006; MOE 2012). One view regards mathematical modelling as an end by itself that focuses on students developing modelling competencies needed to solve extra-mathematical problems. A second perspective sees mathematical modelling as a way to motivate and develop the learning of particular mathematics content. These views are similar to the categorisation in the pragmatic and scientific-humanistic perspectives respectively as discussed by Kaiser and Sriraman (2006). Our curriculum planners have adopted the eclectic approach of incorporating both of these perspectives in their curriculum purpose for applications and modelling (MOE 2012).

Aligned to this curriculum purpose for applications and modelling, our own orientation for teaching mathematical modelling is shaped by the perspective where the pragmatic encompass the scientific-humanistic. These perspectives can serve as a lens for us to examine the views of our pre-service teachers on teaching mathematical modelling. The pre-service teachers' decisions on future mathematical modelling teaching situations can then be explored and negotiated with regards to these views.

5 Pre-service Teachers' Knowledge and Views on Teaching Mathematical Modelling

In order to develop their knowledge and determine their views on teaching mathematical modelling, it has become very much a pedagogical imperative to explore how the pre-service secondary school mathematics teachers can be engaged in mathematical modelling learning experiences. The questions of interest are:

- How is the pre-service teachers' knowledge in mathematical modelling being shaped by the mathematical modelling learning experiences?
- What are the pre-service teachers' views on teaching mathematical modelling?

We next report an exploratory study which examines the development of a group of pre-service secondary school mathematics teachers' knowledge and their views in mathematical modelling during the course of their mathematics teacher education programme.

6 The Study

Mathematical modelling learning experiences were integrated into the mathematics teacher education programme for 6 weeks, as two course hours per week, by the researcher as the tutor. The mathematical modelling tasks presented to the pre-service teachers were selected by taking into account the mathematics content topics covered previously. The pre-service teachers completed two mathematical modelling tasks based on common modelling approaches suggested by Ang (2009) over the course of the semester. They began with empirical modelling and moved on to build deterministic models.

The empirical mathematical modelling learning task examining the relationship between the stopping distances of cars and their corresponding speeds is as follows:

Task 1: Stopping Distances. Highway Code books often recommend stopping distances for a car driven at various speeds. (The data set showing the typical stopping distances of a car driven at different speeds was presented). Construct an empirical model for the stopping distances of a car driven at different speeds to explain this. You may consider the following questions: List down the variables in the problem, write down a possible model for the stopping distances of the car, write down the solution method for the proposed model, carry out your solution method and interpret the solution, and suggest ways to refine the model or extend the problem.

The deterministic mathematical modelling learning task determining how the volume of a fuel tank is calibrated against its dipstick height is as follows:

Task 2: Fuel Tank Calibration. Petrol stations very rarely run out of fuel due to efficient stock control. Each type of fuel is stored in an underground tank and the amount in each tank is regularly monitored using a dipstick. Develop a mathematical model to explain how such efficient stock control can take place. You may consider the following questions: Write down your assumptions in simplifying the problem, list down the variables in the simplified problem, construct a possible model for the calibration of the dipstick to the volume of fuel in the tank, write down the solution method for the proposed model and carry it out, explain how efficient stock control is done, and suggest ways to refine the model or extend the problem.

The mathematical modelling process is structured slightly differently in these two mathematical modelling learning tasks. Notwithstanding this, the main stages of the modelling process are still retained and serve to scaffold the pre-service teachers in their solutions as well as reveal the ways they may think about and experience the various aspects and nuances of the modelling process.

The pre-service teachers worked in groups of four to solve the modelling problems. They alternated between periods of group work, individual work and classroom discussion. As part of their course assessment for mathematical problem solving, they were asked to make oral and written presentations of their solutions to the modelling tasks in the classroom. After some didactic instruction in the modelling process, the pre-service teachers received neither additional resources nor engaged in any discussion of their solutions with the tutor except after they had experienced solving the modelling tasks themselves.

Of the 24 pre-service secondary mathematics teachers who participated in this study, 14 were male and 10 female. Ages ranged from 23 to 39 years old. None of the pre-service teachers had a mathematics degree. The pre-service teachers expressed concerns about their lack of understanding in teaching mathematical modelling and welcomed the opportunity to explore mathematical modelling learning experiences. None of the pre-service teachers had prior experience with the type of modelling tasks implemented in the study.

The data were obtained from the pre-service teachers' performance of the mathematical modelling tasks and the questions they were asked in an open-ended questionnaire. The pre-service teachers' learning actions and thoughts about the modelling process were noted through in-class discussions and post-class reflections via the course blog, respectively. Their written reports to the modelling tasks also served as artefacts to be examined in detail for aspects of the modelling process discussed in class. After completing the modelling tasks, the pre-service teachers responded to open-ended questions that pertained to their modelling experiences and their views on teaching mathematical modelling.

The pre-service teachers' modelling learning experiences was analysed in light of Ang's (2009) modelling process as well as some elements of modelling activities in Stillman et al.'s (2007) framework for mathematical modelling instruction. Analysis of the pre-service teachers' responses to the open-ended questionnaire was based on the interpretations of teaching mathematical modelling in the pragmatic and scientific-humanistic perspectives.

7 Results

7.1 Pre-service Teachers' Knowledge in Mathematical Modelling

Table 31.1 shows the shift in development of the pre-service teachers' knowledge in the various stages of the modelling process as they progressed from the *Car Stopping*

Table 31.1 Pre-service teachers' learning and blockages in the modelling process

Mathematical modelling stages	Car stopping distance task	Fuel tank calibration task
Understand and simplify the real world situation	Assumed flat road condition, standard category of cars	Assumed fuel tank is uniformly cylindrical in shape and located on a flat surface
Formulate the real world problem into a mathematical model	VARIABLES: CAR SPEED (km/h), BRAKING DIST. (m), REACTION DIST. (m); INCORRECT DETERMINATION OF INDEPENDENT VARIABLE IN THE NUMBERING OF DATA SET	Independent variables: dipstick height of fuel h, radius of cross sectional circle r, angle of sector θ, dependent variable: cross sectional area of fuel tank; represented these variables and their relationships in a circular measure diagram
Obtain the mathematical solution	MODEL SOLUTION DID NOT SATISFY INITIAL DATA CONDITION	Cross sectional area of fuel $=\frac{1}{2}r^2(\theta-\sin\theta)$, where $\theta = 2\cos^{-1}\left(1-\frac{h}{r}\right)$
Interpret the mathematical solution in the context of real world situation	NOTED GRADIENT OF GRAPH INCREASES WITH CAR SPEED, BUT DID NOT RELATE IT TO HIGHWAY CODE RECOMMENDATIONS	Noted nature of gradient of graph at different h values and related it to efficient stock control of fuel
Validate the mathematical model	DID NOT COMPARE MODEL WITH REAL DATA	Justified difference between model generated data and real data

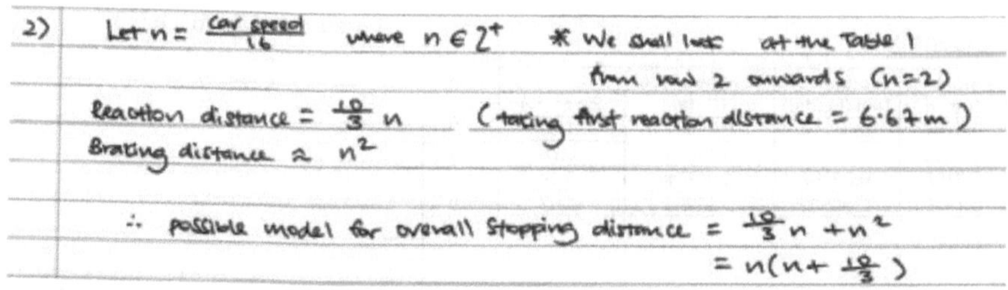

Fig. 31.1 Pre-service teacher's performance on the *Car Stopping Distances Task* (artefact 1)

Distance Task to the *Fuel Tank Calibration Task*. The pre-service teachers' blockages (Stillman et al. 2007) and learning during the modelling stages are elucidated (in capitals) and (in regular type) with reference to the modelling tasks respectively. Following this, data from selected examples of the pre-service teachers' modelling learning experiences and their reflections on the modelling process are presented in the analysis of these blockages and learning at the various stages of the modelling process.

Understand and simplify the real world situation. No blockages were noted from the pre-service teachers' written and oral presentations of the two modelling tasks for this stage of the modelling process. The pre-service teachers were able to interpret the real world situations for the modelling tasks, identify important quantities involved, and make realistic assumptions (see Table 31.1) for more tractable formulation of the mathematical model.

Formulate the real world problem into a mathematical model. Four out of the six groups of pre-service teachers made basic errors like ignoring unit conversion of speed from km/h to m/s in their formulation of the car stopping distance empirical model. One group had difficulty determining the variables of interest and were observed applying known techniques like number pattern generalisation to derive the empirical model from the numbering of the given data set n (see Fig. 31.1). The pre-service teachers were asked to consider the consequences of incorrect determination of variables of interest on the predictive and retrospective functions of empirical modelling. As they progressed to the *Fuel Tank Calibration Task*, it was noted that not only were the variables of interest determined correctly, but also these variables and their corresponding relationships were "represented mathematically so formulae can be applied" (Stillman et al. 2007) via the circular measure diagrams. In fact, from the pre-service teachers' reflective comments that mathematical modelling is about "developing a model that can relate two quantities" and the need to "know how the model behaves over time", they might have come to realize that mathematical modelling is about uncovering relationships between variables of interests that can help us better understand the real world situation.

Obtain the mathematical solution. The Least Squares Regression derivation of the quadratic function for the stopping distances of cars against their corresponding speeds was done in a black box fashion and could not plausibly satisfy the given

Appendix A

height of fuel, h	Volume of fuel (Model)	Volume of fuel (Data)	Error (%)
0.0254	0.05010	0.04921	-1.82
0.0508	0.14104	0.14006	-0.70
0.0762	0.25787	0.25741	-0.18
0.1016	0.39509	0.38990	-1.33

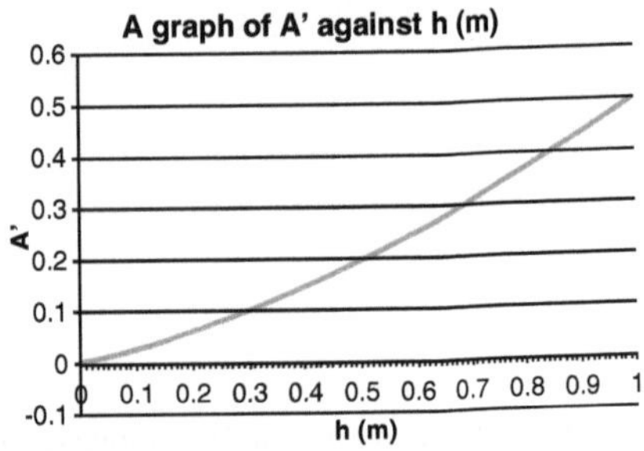

To further interpert the solution, the rate of increase of A' with respect to h is slower when h is low (less than 0.5 m). This is because the gradient is gentle for small values of h. For larger values of h (between 0.5 m to 1 m), the rate of increase of A' is faster and approximately constant as the gradient now increase to become almost constant.

Hence, as the fuel starts to be consumed, te rate of decrease of fuel volume with respect to the fuel height is almost constant and faster. When the fuel is about to be depeted, the rate of decrease of fuel volume is much slower due to the gentle gradient when the fuel height is small. This slow rate of volume consumption when fuel height is small allows the station attendant to have ample time to top it up.

Fig. 31.2 Pre-service teacher's performance on the *Fuel Tank Calibration Task* artefact

initial data condition of zero car speed and stopping distance. The pre-service teachers were asked to consider how the manner in which the mathematical model was being solved could impact its interpretation and the learning of mathematics. Using the Solver tool in Microsoft EXCEL (refer to Lawson and Tabor 2001 for details on how to use the Solver Tool), the pre-service teachers were then engaged in the process of studying the pattern of the car stopping distances data before mindfully conjecturing the general form of a quadratic function that could possibly fit the data with the given initial condition. For the *Fuel Tank Calibration Task*, it was observed that the pre-service teachers had been "applying appropriate formulae" (Stillman et al. 2007) in circular measure and trigonometry to determine the functional relationship between the cross sectional area of fuel and the dipstick height of fuel for a fixed radius length.

Interpret the mathematical solution in the context of real world situation. Although the pre-service teachers noted the gradient of the quadratic function graph for the stopping distances of cars increases with car speed, they were not able to interpret this mathematical model's result in the context of the Highway Code recommendations for increasing car stopping distances with respect to its car speeds. On the other hand, the pre-service teachers were integrating mathematical arguments to augment their explanations for effective fuel stock control in the *Fuel Tank Calibration Task*. As shown in Fig. 31.2, they argued that, since the gradient of the function graph in the fractional cross sectional area of fuel A' is more gradual for smaller dipstick height of fuel, it would mean that when the fuel supply is low, there would be a smaller decrease of fuel volume for every unit decrease in dipstick height, hence allowing more time for the station attendant to top up the fuel.

Validate the mathematical model. In contrast to the *Car Stopping Distance Task* where pre-service teachers did not evaluate their empirical model by comparing it with real data, they went about seeking online real data on the variation of fuel volume with dipstick height of fuel to be compared with the fuel tank calibration model data (see Fig. 31.2). They also attempted to explain the percentage error difference between the model generated data and the real data by suggesting that the fuel tank used in the real data set could be elliptical in shape. Such learning of the modelling process was also highlighted in the pre-service teachers' reflective comments that they would need to "check their model with any real data (if available) and justify for any difference". In addition, from the open ended questionnaire which had a response rate of 87.5 %, about 80 % of the respondents were of the view that they could understand and identify with the stages of the modelling process.

7.2 *Pre-service Teachers' Views on Teaching Mathematical Modelling*

When asked to comment on what they think the possible benefits in carrying out mathematical modelling lessons are, the pre-service teachers by and large expressed a scientific-humanistic perspective (Kaiser and Sriraman 2006) for it. Many of them opined consistently that teaching mathematical modelling can be beneficial to students' learning in terms of "helping students develop higher order thinking skills, see the involvement of real world context, allowing for highly meaningful exploration of concepts"; "bringing on the real world context to maths learning". They were also of the view that "a benefit for implementing mathematical modelling would be a deepening of students' understanding and application of concepts".

According to the pre-service teachers, learning mathematical modelling not only demands time and effort on the part of the students, but also requires them to have the basic mathematics knowledge for meaningful engagement. This was inferred from their comments that mathematical modelling learning "really induces students to think and come up with their own solution. This requires a lot of understanding on their part instead of just application of formulae and theories".

8 Discussion

It could be seen that the pre-service teachers' initial mathematical modelling learning experience was limited to applying particular mathematics topics with rather narrow and known techniques. Hence they were not used to integrating various concepts and techniques in their first mathematical modelling task on car stopping distances and their modelling process was often isolated from the real world situation. The tutor's introspections during the mathematical modelling learning task discussions provided the basis for forming case stories that could have enabled the pre-service teachers to reflect and pick up nuances about important elements of modelling activities in the various stages of the modelling process (Stillman et al. 2007). The coupling of such case stories with the learning scaffold structured in the mathematical modelling learning tasks was paramount to the development and transfer of mathematical modelling competencies to other problem contexts. This observation was supported by the shift in the development of the pre-service teachers' knowledge in the various stages of the modelling process as they progressed from the *Car Stopping Distance Task* to the *Fuel Tank Calibration Task*.

We argue for the necessity that novice teachers in teaching mathematical modelling should perform the novel modelling tasks themselves. This is so that the experience and knowledge gained in mathematical modelling can help teachers explicate aspects and nuances of the modelling process with respect to novel modelling tasks. It is noted that explicating such aspects and nuances of the modelling process is important later when we need to translate them into features of modelling tasks for teaching and learning purposes. Some features of the modelling learning tasks needed for its successful implementation may include identification of the pre-requisite modelling skills and competencies required and the employment of appropriate teaching strategies to help resolve possible student blockages in the modelling process (Stillman et al. 2007).

The pre-service teachers' views on teaching mathematical modelling were mainly limited to the scientific-humanistic perspective. This is to be expected to some extent as the pre-service teachers had not carried out modelling learning experiences in their classroom and hence may not be able to appreciate the benefits and importance of developing modelling competencies in students yet. Notwithstanding this, we note that teachers need to have wider perspectives of mathematical modelling in education as that will guide and orientate the organisation and implementation of modelling instruction in classrooms.

9 Conclusion

Through this study, we have seen that independent modelling experiences with reflection activities enabled a group of pre-service secondary school mathematics teachers to acquire knowledge of the important elements of modelling activities in the various stages of the modelling process. Future work is needed to examine the necessity and influence of such *specialized content knowledge* for the unpacked modelling process on teachers' development of *pedagogical content knowledge* in mathematical modelling instruction. The pre-service teachers' future practice in mathematical modelling instruction may also need to be shaped in part by the pragmatic perspective on teaching mathematical modelling in order for their students to have more meaningful mathematical modelling experiences.

References

Ang, K. C. (2009). *Mathematical modelling in the secondary and junior college classroom*. Upper Saddle River, NJ: Prentice Hall.

Ang, K. C. (2010). Mathematical modelling in the Singapore curriculum: Opportunities and challenges. In A. Araújo, A. Fernandes, A. Azevedo, & J. F. Rodrigues (Eds.), *Conference proceedings of EIMI 2010: Educational interfacecs between mathematics and industry* (pp. 53–61). Lisbon/Bedford: Centro Internacional de Matemática/COMAP.

Ball, D. L. (2000). Bridging practices: Intertwining content and pedagogy in teaching and learning to teach. *Journal of Teacher Education, 51*(3), 241–247.

Ball, D. L., Thames, M. H., & Phelps, G. (2008). Content knowledge for teaching: What makes it special. *Journal of Teacher Education, 59*(5), 389–407.

Barbosa, J. C. (2006). Mathematical modelling in classroom: A socio-critical and discursive perspective. *ZDM – The International Journal on Mathematics Education, 38*(3), 293–301.

Blum, W., & Leiss, D. (2007). How do students and teachers deal with modeling problems? In C. Haines, P. Galbraith, W. Blum, & S. Khan (Eds.), *Mathematical modeling: Education, engineering and economics* (pp. 222–231). Chichester: Horwood.

Blum, W., & Niss, M. (1991). Applied mathematical problem solving, modelling, applications, and links to other subjects: State, trends and issues in mathematics instruction. *Educational Studies in Mathematics, 22*(1), 37–68.

Kaiser, G., & Sriraman, B. (2006). A global survey of international perspectives on modelling in mathematics education. *ZDM – The International Journal on Mathematics Education, 38*(3), 293–301.

Lawson, D. A., & Tabor, J. H. (2001). Stopping distances: An excellent example of empirical modelling. *Teaching Mathematics and Its Applications, 20*(2), 66–74.

Ministry of Education (MOE). (2012). *Mathematical modelling resource kit*. Singapore: Ministry of Education. http://library.edumall.sg/cos/o.x?ptid=92&c=/library/linktree&sid=237. Accessed 5 Apr 2012.

Shulman, L. S. (1986). Paradigms and research programs in the study of teaching: A contemporary perspective. In M. C. Witrock (Ed.), *Third handbook of research on teaching* (pp. 3–36). New York: Macmillan.

Stillman, G., Galbraith, P., Brown, J., & Edwards, I. (2007). A framework for success in implementing mathematical modelling in the secondary classroom. In J. Watson, & K. Beswick (Eds.), *Mathematics: Essential research, essential practice*. (Proceedings of the 30th annual conference of the Mathematics Research Group of Australasia, Hobart, Vol. 2, pp. 688–707). Adelaide: MERGA.

Chapter 32
How Students Connect Descriptions of Real-World Situations to Mathematical Models in Different Representational Modes

Wim Van Dooren, Dirk De Bock, and Lieven Verschaffel

Abstract The translation of a problem situation into a mathematical model constitutes a key – but not at all obvious – step in the modelling process. We focus on two elements that can hinder that translation process by relating it to the phenomenon of students' overreliance on the linear model and their (lack of) representational fluency. We investigated: (1) How accurate are students in connecting descriptions of realistic situations to "almost" linear models, and (2) Does accuracy and model confusion depend on the representational mode in which a model is given? Results highlight that students confuse linear and non-linear models, and that the representational mode has a strong impact on this confusion: Correct reasoning about a situation with one mathematical model can be facilitated in a particular representation, while the same representation is misleading for situations with another model.

1 Theoretical and Empirical Background

The translation of one's understanding of a problem situation into a mathematical model constitutes a key step in the process of mathematical modelling (Verschaffel et al. 2000). However, as research at different levels of schooling has shown, this step is far from obvious. An element that can hinder that translation process is students' tendency to assume a linear model in non-linear situations (De Bock et al. 2007).

W. Van Dooren (✉) • L. Verschaffel
Centre for Instructional Psychology and Technology, Katholieke Universiteit Leuven, Leuven, Belgium
e-mail: wim.vandooren@ped.kuleuven.be

D. De Bock
Centre for Instructional Psychology and Technology, Katholieke Universiteit Leuven, Leuven, Belgium

Hogeschool-Universiteit Brussel, Brussels, Belgium

G.A. Stillman et al. (eds.), *Teaching Mathematical Modelling: Connecting to Research and Practice*, International Perspectives on the Teaching and Learning of Mathematical Modelling, DOI 10.1007/978-94-007-6540-5_32,

An example of this tendency was documented by Van Dooren et al. (2003): Many upper secondary students respond proportionally ($2\times 1/6=2/6$) to probabilistic problems such as "The probability of getting a six in one roll with a die is 1/6. What is the probability of getting at least one six in two rolls?" Students' overreliance on linear models has been evidenced in a variety of other mathematical domains as well (e.g., elementary arithmetic, algebra, (pre)calculus, and geometry).

Another element that can hinder the translation from a situation to a mathematical model is students' (lack of) mastery of different external representations of a function – such as graphs, tables and formulas – and their use in relation to the solution of a given problem situation (Verschaffel et al. 2010). External representations are notational systems that help to express, describe, manipulate, and communicate mathematical ideas. Vergnaud (1997) argues that external representations are inherent to the discipline of mathematics, since a characteristic of mathematical concepts is that they can *only* be accessed through their external representations (Duval 2002). For quite some time already, mathematics educators emphasize the (stimulating) role of "multiple" external representations in mathematics. External representations have also been shown to facilitate mathematical problem solving (Duval 2002; Kaput 1992). The NCTM Standards (1989) therefore hold a strong plea for establishing "mathematical connections" through the use of multiple external representations:

> Different representations of problems serve as different lenses through which students interpret the problems and the solutions. If students are to become mathematically powerful, they must be flexible enough to approach situations in a variety of ways and recognize the relationships among different points of view (p. 84).

However, research has shown that students are not always fluent in using all the external representations that are available to solve a problem, and in translating between representations (e.g., Bieda and Nathan 2009; Grüber et al. 1995; Tabachneck et al. 1994; Yerushalmy 1991). Another reason why students do not always benefit from solving problems using multiple external representations is that they are unable to make flexible representational choices, before engaging in *using* the representations as such (Acevedo Nistal et al. 2009).

In this chapter, we relate students' overreliance on the linear model with their (lack of) mastery of external representations because we think that the role of external representations so far is not sufficiently acknowledged in research on the overreliance on the linear model. More concretely, we investigate: (1) How accurate are students in connecting descriptions of realistic situations to linear and "almost" linear models,[1] and (2) Does accuracy and model confusion depend on the representational mode[2] in which a function is given?

---

[1]In this chapter we use the term linear to determine models that can be described by a formula of the form "$y=ax$". Models that can be described by a formula of the form "$y=ax+b$" with $b\neq 0$ are labelled as affine models. The term "almost" linear is used as a collective term for inverse linear and affine models, that is models that share some but not all characteristics with the linear model.

[2]In the rest of this chapter, we will use the term "representation" to denote "external representation".

Students' difficulties in selecting a model were also addressed by Frejd and Ärlebäck (2011) who investigated Swedish secondary students' mathematical modelling competencies. Selecting a model was one of the seven sub-competencies these researchers focused on. In two items of their instrument a "realistic" situation was described and students had to link this situation to a mathematical model, once to a graphical and once to a formula representation of the model. The results indicated selecting a model was one of the sub-competences students showed least proficiency with, when this model was given in either a graphical or formula representational mode.

Students' tendency to adhere to characteristics or representations of linearity has been observed in several studies. Particularly the straight line prototype proved to be very appealing for many students (De Bock et al. 2007). Leinhardt et al. (1990) show that students of different ages have a strong tendency to produce a linear pattern through the origin when asked to graph non-linear situations, such as the growth in the height of a person from birth to the age of 30. The tendency towards linearity is not limited to attempts to graph real-life situations. For example, when Markovits et al. (1986) asked 14- to 15-year-old students to draw a graph of a function that passes through two given points, they typically drew straight lines. Similarly, Karplus (1979) found that when students interpolate between two graphed data points in a science experiment, they strongly tend to connect the points using a straight line.

2 Method

Sixty-four students in the first year of Educational Sciences of the University of Leuven participated. These students have successfully finished secondary education and typically also 3 years of non-university higher education. Although they all followed the obligatory mathematics courses in secondary school, this was in most cases not the core of their curriculum. In these mathematics courses, solving realistic problems and the applicability of basic functions such as the ones central in our study receive quite some attention.

Participants were confronted with a written multiple-choice test consisting of 12 descriptions of realistic situations they had to connect with an appropriate mathematical model. For each situation the appropriate model was either linear (i.e. of the form $y=ax$) or "almost" linear: inverse linear ($y=a/x$), affine with positive slope ($y=ax+b$ with $a>0$ and $b\neq0$), or affine with negative slope ($y=ax+b$ with $a<0$ and $b\neq0$). These models were given either in graphical, tabular or formula form (each representation was provided in one third of the cases). Three examples of descriptions of situations to be connected to the appropriate model in one of the three representations as used in the study follow. Situations were chosen so that there would be a strong and clear fit with the provided models, although we are aware that models never perfectly fit to a realistic situation. In example 1, for instance, students are expected to start from the assumption that one of the four options provided is an adequate representation of the situation that is stated. Since every option has the number 150 in it, the only possible interpretation is that it points to the amount of butter.

Responses were statistically analysed by means of a repeated measures logistic regression analysis, using the generalized estimating of equations (GEE) approach within SPSS (Liang and Zeger 1986). This procedure allows analysis of repeated (and thus possibly correlated) categorical observations within series of individuals, and to appropriately correct for the inferences that can be drawn from such correlated measures. Given the dichotomous nature of the dependent variable (i.e., is a particular response alternative chosen or not), a logistic regression (modelling the probability that an observation is made in terms of some explanatory variables) was appropriate.

Example 1 (Inverse Linear Situation/Alternative Formulas)

During the war, butter was rationed. Each week, butter was delivered and fairly distributed among the people. Which formula properly represents the relation between the number of people who want butter and the amount of butter everybody receives?

- ☐ $y = 150\,x$
- ☐ $y = 150/x$
- ☐ $y = 150\,x + 30$
- ☐ $y = -150\,x + 30$

Example 2 (Linear Situation/Alternative Tables)

Jennifer buys minced meat at the butcher's shop. Which table properly represents the relation between the amount of minced meat that Jennifer buys and the price she has to pay?

☐	x	y	☐	x	y	☐	x	y	☐	x	y
	0	8		0	0		0			0	8
	1	−4		1	12		1	12		1	20
	2	−16		2	24		2	6		2	32
	3	−28		3	36		3	4		3	44
	4	−40		4	48		4	3		4	56

Example 3 (Affine Situation with Negative Slope/Alternative Graphs)

A chemical concern has a big cistern with hydrochloric. This morning they started to pump with a constant pace all hydrochloric out of this cistern. Which graph properly represents the relation between the time elapsed and the amount of hydrochloric that is still in the cistern?

(continued)

(continued)

❐
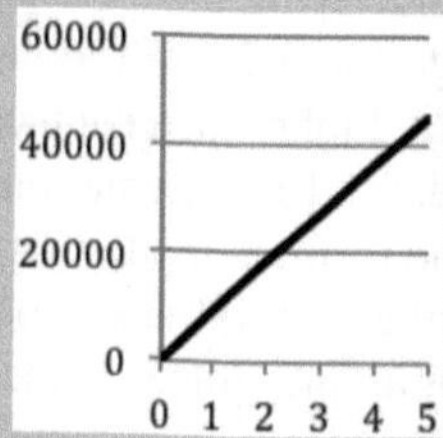

❐
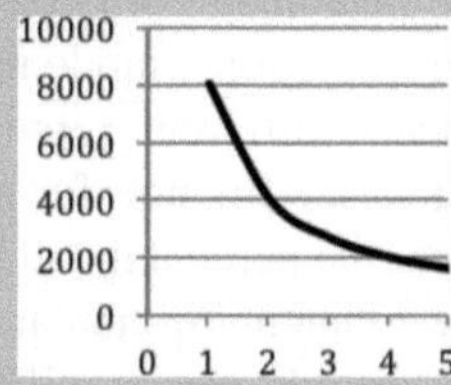

❐
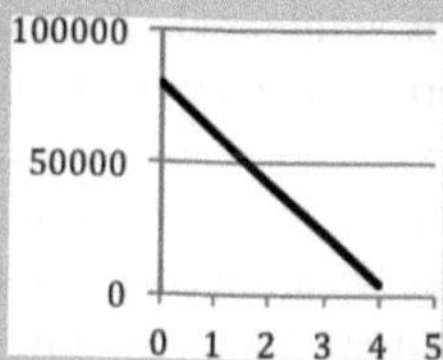

❐
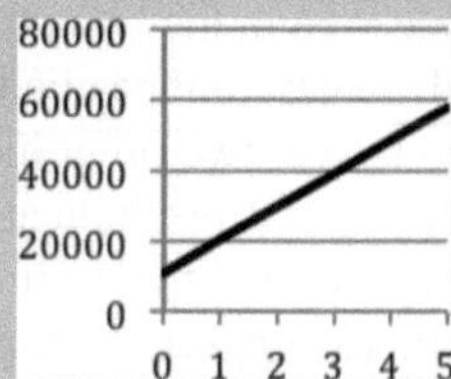

Results

2.1 Accuracy

Table 32.1 shows the percentage of correct assignments for the different models underlying the verbal descriptions and for the different representational modes. The logistic regression analysis first of all revealed a main effect of model, indicating that students' accuracy in connecting situations to the appropriate mathematical model depended on the type of model involved: The percentage of correct matches for an underlying linear model was significantly higher than for an inverse linear, affine with positive slope, and affine with negative slope model. The analysis did not reveal a main effect of representation: The percentages of correct assignments for

Table 32.1 Percentage of correct assignments for the four models underlying the verbal descriptions and for the three representational modes

	Representational modes			
Model	Graph	Table	Formula	Average
Linear	98	95	89	94
Inverse linear	70	69	92	77
Affine with positive slope	81	55	66	67
Affine with negative slope	81	81	48	70
Average	83	75	74	77

the three representational modes did not differ significantly. Finally but most importantly, an interaction effect between model and representation was found: The percentage of correct matches for a given underlying model depended on the representational mode in which that model was given, and for different underlying models, different representational modes led to higher percentages of correct matches. Although for an underlying linear model, all representations were quite good, the percentage of correct assignments in the formula mode was significantly lower than in the tabular and graphical mode. For the inverse linear model, the result for formula was significantly higher than for table and graph. Underlying affine models with positive slope elicited significantly more correct matches in the graph mode than in the two other representational modes, while underlying affine models with negative slope elicited significantly fewer correct matches in the formula mode than for the two other modes.

These results indicate that the graph is the "best" representation in all cases, except for the inverse linear relationships. For these relations the formula proved to be more supportive. A possible explanation is that students strongly tend to over-associate characteristics of linear graphs (straight lines) and linear tables (equal distances) with graphs and tables in general. Absence of these prototypical characteristics in graphs and tables of inverse linear relationships might have retained students to choose these representations, leading to a decrease of correct matches for the inverse linear model when given in graphical or tabular mode. Another more general observation is that formulas seem to be misleading for affine relations. This might have been a side effect of the fact that the situations to be modeled were given in a "$y = a \pm bx$" form while the formulas to be matched with were given in a "$y = ax + b$" form.

Based on the number of correct responses, we conclude that the students could interpret all representations (correct matches for all representations varied between 83 and 74 % and mutual differences were not significant). Students were also able to detect underlying mathematical models (more than 80 % of the students detected the underlying model in at least one of the three representations), but some representations support one underlying model better than other underlying models, while others put the students on the wrong track. To obtain a better understanding of these findings, a quantitative error analysis was conducted.

2.2 Error Analysis

Table 32.2 shows the percentage of erroneous assignments for the different models in the different representational modes. The error analysis revealed that in situations for which the underlying model was inverse linear, (direct) linear errors were most frequently made, especially in the tabular and graphical representational mode. This finding confirms results of previous studies on students' overreliance on the linear model (De Bock et al. 2007) when an "almost" linear model is appropriate. Students are most likely misled by characteristics of representations of the linear

Table 32.2 Percentage of erroneous assignments for various models in the different representational modes

	Representational modes											
Model	Graph			Table			Formula			Total		
L	IL	A+	A−	IL	A+	A−	IL	A+	A−	IL	A+	A−
	0	2	0	2	3	0	9	0	2	4	2	1
IL	L	A+	A−	L	A+	A−	L	A+	A−	L	A+	A−
	14	3	11	27	2	3	5	3	0	15	3	5
A+	L	IL	A−	L	IL	A−	L	IL	A−	L	IL	A−
	17	2	0	41	0	2	31	2	0	30	1	1
A−	L	IL	A+	L	IL	A+	L	IL	A+	L	IL	A+
	8	11	0	6	11	2	19	23	3	11	15	2

Note. Linear (L), the inverse linear (IL), the affine with positive slope (A+), and the affine with negative slope (A−) models

model (i.e., equal distances and straight lines), which are prominently present in respectively tabular and graphical representations.

Also in situations for which the underlying model was affine with positive slope, linear errors were most frequent, but now these errors mainly occurred in the formula and tabular representational mode. Apparently, for this model that comes closest to the linear model, students *see* the *Y*-intercept in the graphical representation, while this seems to be more difficult for the other representations.

Finally, in situations for which the underlying model was affine with negative slope, inverse linear errors were most frequent, especially in the formula representational mode. A straightforward explanation refers to the decreasing nature of both models. Apparently, for many students, the independent variable in the denominator is more appealing than the negative sign in the numerator. Another explanatory element refers to the attraction of the "doubling/halving" prototype in situations of decrease which is most prominent in the formula representation. Students likely experience this prototype both in daily life and from school word problems that are typically found in textbooks for teaching inverse proportionality such as "For 4 painters, it takes 12 days to paint a bridge. How many days does it take 8 painters to do that job?" (Van Dooren et al. 2005) and tend to over-generalise it to situations of decrease which are not inverse proportional.

3 Conclusions and Discussion

Our results show that students are very proficient in relating descriptions of realistic situations to models in cases where the situation described is a linear one. In cases where the situation is "almost" linear (i.e., affine with a positive or negative slope, or inverse linear), there is, however, a strong tendency to connect the situation also to linear models (and for inverse linear situations, to some extent, also to affine models with a negative slope). These results are in line with several other studies showing the "default" role of the linear model, this time in situations in which an "almost" linear model is appropriate (inverse linear, affine with positive slope, and affine with negative slope models).

Results also indicate that the representational mode has a strong impact on students' modelling accuracy and on the tendency to inappropriately connect non-linear situations to linear models. Apparently, a particular representation may highlight aspects of non-linearity that are easily noticed by students and therefore facilitate correct reasoning, but be misleading when representing a situation with another model. Further research should also draw out students' ability to link given representations to other representations and their ability to link given representations to properties of these representations.

An implication for mathematics education is the need for drawing sufficient instructional attention to representations, to match representations with each other and to link them to realistic situations, and for explicitly discussing differences between linear and different types of "almost" linear models. Modelling exercises in prototypically clean situations as used in the current study can be applied to develop students' ability to more easily recognize mathematical models and their characteristics, and to fluently apply related procedures (or strategies), but, evidently, this type of exercise should be regularly alternated with tasks relating more authentic real-world situations to mathematical models and with reflections on this relation (Verschaffel et al. 2000).

References

Acevedo Nistal, A., Van Dooren, W., Clarebout, G., Elen, J., & Verschaffel, L. (2009). Conceptualising, investigating and stimulating representational flexibility in mathematical problem solving and learning: A critical review. *ZDM – The International Journal on Mathematics Education, 41*, 627–636.

Bieda, K. N., & Nathan, M. J. (2009). Representational disfluency in algebra: Evidence from student gestures and speech. *ZDM – The International Journal on Mathematics Education, 41*, 637–650.

De Bock, D., Van Dooren, W., Janssens, D., & Verschaffel, L. (2007). *The illusion of linearity: From analysis to improvement* (Mathematics education library). New York: Springer.

Duval, R. (2002). The cognitive analysis of problems of comprehension in the learning of mathematics. *Mediterranean Journal for Research in Mathematics Education, 1*(2), 1–16.

Frejd, P., & Ärlebäck, J. B. (2011). First results from a study investigating Swedish upper secondary students' mathematical modeling competencies. In G. Kaiser, W. Blum, R. Borromeo Ferri, & G. Stillman (Eds.), *Trends in teaching and learning of mathematical modelling* (pp. 407–416). New York: Springer.

Grüber, H., Graf, M., Mandl, H., Renkl, A., & Stark, R. (1995, August). *Fostering applicable knowledge by multiple perspectives and guided problem solving.* Paper presented at the 6th conference of the European Association for Research on Learning and Instruction, Nijmegen.

Kaput, J. (1992). Technology and mathematics education. In D. A. Grouws (Ed.), *Handbook of research on mathematics teaching and learning* (pp. 515–556). New York: Macmillan.

Karplus, R. (1979). Continuous functions: Students' viewpoints. *European Journal of Science Education, 1*, 397–413.
Leinhardt, G., Zaslavsky, O., & Stein, M. K. (1990). Functions, graphs, and graphing: Tasks, learning, and teaching. *Review of Educational Research, 60*, 1–64.
Liang, K.-Y., & Zeger, S. L. (1986). Longitudinal data analysis using generalized linear models. *Biometrika, 73*, 13–22.
Markovits, Z., Eylon, B.-S., & Bruckheimer, M. (1986). Functions today and yesterday. *For the Learning of Mathematics, 6*(2), 18–24, 28.
National Council of Teachers of Mathematics. (1989). *Curriculum and evaluation standards for school mathematics*. Reston: Author.
Tabachneck, H. J. M., Leonardo, A. M., & Simon, H. A. (1994). How does an expert use a graph? A model of visual and verbal inferencing in economics. In A. Ram & K. Eiselt (Eds.), *Proceedings of the 16th annual conference of the Cognitive Science Society* (pp. 842–847). Hillsdale: Erlbaum.
Van Dooren, W., De Bock, D., Depaepe, F., Janssens, D., & Verschaffel, L. (2003). The illusion of linearity: Expanding the evidence towards probabilistic reasoning. *Educational Studies in Mathematics, 53*, 113–138.
Van Dooren, W., De Bock, D., Hessels, A., Janssens, D., & Verschaffel, L. (2005). Not everything is proportional: Effects of age and problem type on propensities for overgeneralization. *Cognition and Instruction, 23*(1), 57–86.
Vergnaud, G. (1997). The nature of mathematical concepts. In T. Nunes & P. Bryant (Eds.), *Learning and teaching mathematics: An international perspective* (pp. 1–28). Hove: Psychology Press.
Verschaffel, L., Greer, B., & De Corte, E. (2000). *Making sense of word problems*. Lisse: Swets & Zeitlinger.
Verschaffel, L., De Corte, E., de Jong, T., & Elen, J. (Eds.). (2010). *Use of representations in reasoning and problem solving: Analysis and improvement*. Oxon: Routledge.
Yerushalmy, M. (1991). Student perceptions of aspects of algebraic function using multiple representation software. *Journal of Computer Assisted Learning, 7*, 42–57.

Chapter 33
Pre-service Teacher Learning for Mathematical Modelling

Mark Winter and Hamsa Venkat

Abstract Evidence has shown that teachers in South Africa often lack the capacity to both connect their mathematics to real-life contexts and struggle to see the internal connections between mathematical concepts. Situating our argument within the 'critical competence' and 'utility' perspectives, we focus on pre-service teachers' initial mathematical modelling competencies in a professional development course. Using the notion of modelling competencies with specific reference to the didactic modelling process, we argue that the pre-service teachers' initial mathematical modelling competencies are at early stages of development.

1 Introduction

Mathematical modelling is one of the key goals of mathematics education in most countries across the globe despite differences in emphases and purposes. The relevance of pushing for a modelling agenda in schools is evident in the numeracy demands contained in a wide range of real world situations. The need for students to develop their capacity to use mathematics in modelling situations in their present and future lives has been strongly emphasised in the PISA studies (OECD 2003, 2009). The PISA intentions imply the need for teachers to have a deep and connected understanding of mathematical modelling in order to provide adequate guidance to students in modelling classrooms. Pre-service teachers in particular need to have the modelling-based understanding of mathematics developed in order to reduce the barriers related to integration of modelling examples in modelling classrooms (Kaiser and Maaß 2007).

M. Winter (✉) • H. Venkat
School of Education, University of the Witwatersrand, Private Bag 3,
WITS 2050 Johannesburg, South Africa
e-mail: Mark.Winter@wits.ac.za; Hamsa.Venkatakrishnan@wits.ac.za

G.A. Stillman et al. (eds.), *Teaching Mathematical Modelling: Connecting to Research and Practice*, International Perspectives on the Teaching and Learning of Mathematical Modelling, DOI 10.1007/978-94-007-6540-5_33,

In line with the global curricula reforms, the South African government introduced mathematical literacy as a subject in the Further Education and Training (FET) band (Years 10–12) to provide more opportunities for learners who intend to pursue tertiary degree programs which are not mathematically based. The subject focus on real-life situations coupled with the use of elementary mathematics to solve the situations within the citizenship domain contributes to this cause. The National Curriculum Statement (NCS) (Department of Education 2003) for mathematical literacy defines the subject in the following terms:

> Mathematical Literacy provides learners with an awareness and understanding of the role that mathematics plays in the modern world. Mathematical Literacy is a subject driven by life-related applications of mathematics. It enables learners to develop ability and confidence to think numerically and spatially in order to interpret and critically analyze everyday situations and to solve problems (p. 9).

South Africa is one of the countries whose educational policy explicitly emphasise development of learners' skills and abilities to model real life contexts which contain mathematical features. The notion of mathematical modelling is captured in both the mathematics and mathematical literacy curricula. Within the mathematics curriculum statement, the rhetoric suggests that mathematics modelling skills would be developed across all levels (Years R-12) (Department of Basic Education 2011). It states that "Mathematical modelling is an important focal point of the curriculum" and further stresses the need for use of realistic rather than contrived contexts (p. 8). Since mathematical literacy is introduced at FET level (Years 10–12) the modelling skills learnt in school mathematics from R to Year 9, are further developed within the context of citizenship agenda in this subject. The subject is focused largely on modelling real life situations using elementary mathematics as demonstrated in its purposes: "Mathematical literacy will develop the use of basic mathematical skills in critically analysing situations and creatively solving everyday problems" (Department of Education 2003, p. 9). The above intentions in both mathematics and Mathematical literacy curricula, point towards the need for teachers who are competent in mathematical modelling to teach either mathematics or mathematical literacy in the schools. However, evidence from research studies conducted in South Africa has revealed that teachers often struggle to successfully engage with real life contexts, an aspect which is more manifested in mathematical literacy classrooms (Brombacher 2003; Venkat 2007). The need to further develop the pre-service teachers' modelling-based knowledge provided a strong rationale for this study.

Our main agenda in this chapter is to document the mathematical literacy of pre-service teachers' learning in the context of a university mathematics course focused on modelling real world contexts. We consider initial modelling competencies of the pre-service teachers at the beginning of the 2-year course as they solve some contextual tasks which in our view present an entry point into complex modelling. We situate our discussion within the *"critical competence"* and *"utility"* arguments. The critical competence argument is related to the utility argument in that the former focuses on empowering students to "see and judge" situations in life independently in order to make informed decisions, and the latter emphasises the need for students to use mathematical ideas in making sense of extra-mathematical situations (Blum and Niss 1991, p. 43). The citizenship perspective adopted by mathematical literacy informed the choice or design of contexts which fell within the citizenship domain. The pre-service teachers' responses are analysed in terms of how they engaged with contextual problems across the didactical modelling process (Kaiser and Schwarz 2006). We therefore intend to address the following specific question: What kinds of initial modelling competences could be seen in pre-service teachers' performance in the first 3 months after participating in a university course?

2 Modelling with Elementary Mathematics

The idea of modelling with elementary mathematics has been extensively documented in school mathematics (English and Watters 2005; Verschaffel et al. 1997) and less so in mathematical literacy (e.g., Geiger et al. 2013). There is a strong push globally towards introducing modelling to young children to ensure development of modelling specific competences across all levels of schooling. Within Mathematical literacy, modelling focuses on the use of basic mathematics (Years 0–9 mathematics), implying that knowledge of elementary mathematics is sufficient for solving real life situations (Department of Education 2003). The elementary mathematics base and use of simple word problems, provide some links between young children modelling and mathematical literacy modelling. In South Africa, the advocacy to have contexts for mathematical literacy drawn from either personal, educational/occupational, or public situations, coupled with its citizenship perspective provides a good rationale for the use of elementary mathematics. Greer and colleagues (2007) suggest that elementary mathematics is useful in situations where modelling of social issues focusing on critical analysis is concerned, an aspect closely linked with the citizenship view. Such modelling, they argue, provides "a sense of agency through recognising the potential of mathematics as a critical tool for analysis of issues in their lives" (Greer et al. p. 89).

3 Learning for Mathematical Modelling

The demand for new ways of structuring classroom instruction in mathematical literacy so that real life contexts are starting points, provided a strong rationale for introducing a university mathematics content course which aimed at developing the pre-service teachers' modelling skills. Being a new course, research was needed to explore the nature of knowledge development in terms of their understandings of mathematics and how this knowledge could be applied in modelling situations. As highlighted above, we intend to explore the initial modelling competences of the pre-service teachers through a focus on solutions to some contextual test items.

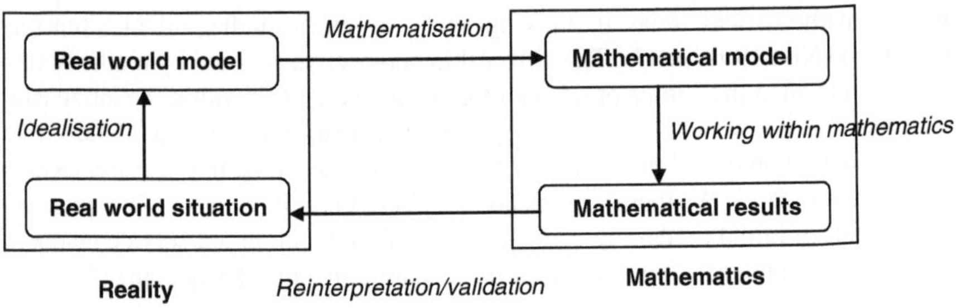

Fig. 33.1 Didactical modelling process (Kaiser and Schwarz 2006, p. 197)

3.1 Theoretical Frame

Within the mathematics didactical debate, modelling abilities and competencies have provided theoretical lenses for understanding students' modelling processes (Blomhøj and Jensen 2003; Kaiser 2007). By analysing pre-service teachers' solutions to contextual tasks, we were able to gain insight into the nature of initial modelling competencies across the modelling cycle. Kaiser and Schwarz (2006) suggest a four-stage didactical modelling process (Fig. 33.1) which is useful in terms of understanding thought processes as students engage with modelling tasks. We found these theoretical lenses to be useful in understanding the pre-service teachers' initial modelling competencies. Their notion of didactical modelling process requires students' demonstration of competencies related to four major stages:

- *Idealisation*: simplifying, structuring the real world situation to get a real world model.
- *Mathematisation*: transforming the real world model into a mathematical model which includes identifying relevant quantities and their relationships.
- *Solving mathematical model*: using symbols and mathematical language to solve a problem.
- *Reinterpretation and validation*: making sense of the mathematical solution in the context of the original situation. Here, we are answering the question; does this mathematical solution make sense in the context of the problem situation?

Although the modelling process suggests a linear relationship between stages, students' thought processes involve going back and forth during problem solving. The modelling competencies which are at play during this process have been given adequate attention in Maaß (2006). Her conception of modelling competencies in relation to the didactic modelling process was found to be useful in this study and has helped us to understand the pre-service teachers' modelling processes.

3.2 Methodological Approach

This study is qualitative in nature and adopts a case study approach (Yin 2009). We present results where data were drawn from the first 3 months of a 2-year longitudinal study (2011–2012). Eight second-year Bachelor of Education (BEd) students were sampled purposively. They were enrolled in a new 3-year (2011–2013) Concepts and Literacy in Mathematics (CLM) course. The course was aimed at developing modelling skills within a citizenship perspective among the pre-service teachers. All the participants had a pass in Mathematical literacy at the end of their high school education. We use data from the first 3 months of the CLM course. In the first year of their BEd degree programme they had enrolled for, and passed, 'mathematical routes', a university course aimed at both developing the pre-service teachers' numerical skills and consolidating their mathematical content knowledge.

The first 3 months of the course focused on further developing the pre-service teachers' knowledge related to elementary mathematics content and contextual problem solving. At the end of the first 3 months the participants were then given a test which consisted of both pure mathematics ($N=5$) and contextual ($N=5$) tasks. The main purpose of the test was to assess pre-service teachers' content knowledge and their modelling abilities. We have selected the contextual tasks for this chapter because our focus is to understand the pre-service teachers' initial competencies relating to modelling situations. The tasks were designed or selected in such a way that the teachers would draw from their elementary mathematical understandings. The course coordinator was responsible for selecting or designing of the test items and the researchers were not involved. The test scripts were then collected for analysis. Table 33.1 presents the test items. As argued above these items provide an entry point into modelling situations within a citizenship perspective.

Table 33.1 Contextual test items

Item number	Items
Q1	My daughter wants to paint her bedroom pink. I mixed three tins of red paint with five tins of white paint and she says the pink it makes is perfect. I figure we need about 12 tins of paint to paint her bedroom (a) If I add two tins of white paint and two tins of red paint to the perfect pink mix will it be too red, too white or still perfect? (b) How many tins of red paint and white paint must I add to the original perfect pink mix to make 12 tins worth of perfect pink mix?
Q2	At Pizzaz, the pizza with a 10 cm radius costs R30. The pizza with a 15 cm radius costs R45. Which is the better deal or is there no difference? Explain fully and clearly why you say so.
Q3	I have 8.2 m of material. I need 0.4 m of material to make a doll's dress (a) How many complete dresses can I make from the material? (b) How much material will I have left over?
Q4	The only sports offered at Burg High School are soccer and netball. The principal loves soccer so he allocates the sports budget so that for every R2 spent on netball, R3 will be spent on soccer (a) If R450 is allocated to soccer, how much will be allocated to netball? (b) If the school gets R8,000 to spend on sport, how much will be allocated to netball?
Q5	You buy a car for R85,000. If each year the value of the car depreciates by 10 % of its value the previous year, what will its value be at the end of 3 years?

4 The Nature of Pre-service Teachers' Mathematical Work

We focus our analysis on the nature of initial modelling competencies across the didactic modelling process. By adopting the deductive approach (Creswell 2009; Yin 2009), we use the four indicators adopted from the theoretical framework, namely: idealisation, mathematisation, use of mathematical language, and interpreting mathematical solutions in the context of the real life situation.

4.1 *Idealisation in Modelling*

Idealisation involves competencies related to transforming real world problems into a real model. Although the pre-service teachers' solutions do not show how the situations have been simplified or structured to get a real world model and then a mathematical model, there is evidence around the teachers' identification and selection of "quantities that influence the situation" (Maaß 2006, p. 116). The results show that the pre-service teachers tend to rush into constructing a mathematical model without explicitly showing how the real world situation has been simplified or structured. Furthermore, the results suggest that pre-service teachers do not seem to have problems with idealisation at this level of question, as evidenced in the way relevant quantities from the situations were identified. Evidence of pre-service teachers' work which does not explicitly show idealisation is shown in Fig. 33.2.

4.2 *Mathematisation in Modelling*

Mathematisation competencies include establishing relationships between quantities and choosing appropriate notations to be represented in a mathematical model (Maaß 2006). It involves setting up a procedure. As the teachers move from the "reality space" into the "mathematical space" there seem to be a breakdown in terms of choosing the correct operation to be used. For instance, after dividing 8.2 m by 0.4 m in item Q3a, 20.5 was obtained. However, instead of using a multiplication operation in item Q3b (i.e. 0.5×0.4), subtraction was used as shown in Fig. 33.3.

4.3 *Using Mathematical Language to Solve Modelling Tasks*

After setting up a mathematical model, mathematical content knowledge becomes fundamental in solving the problem mathematically. Those with strong mathematical content knowledge would have different ways of solving the problem, weigh up

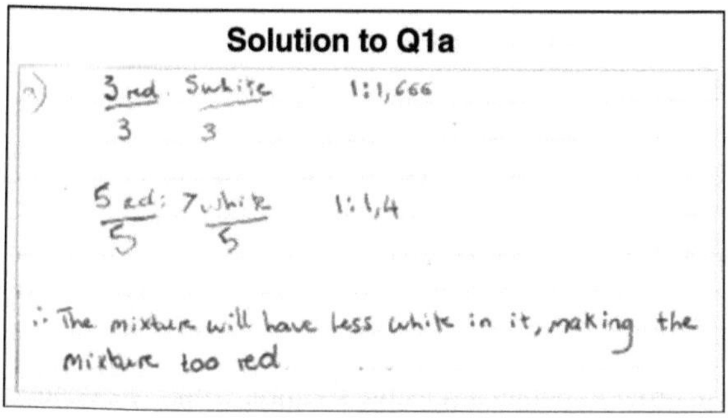

Fig. 33.2 Evidence of solution not showing idealisation

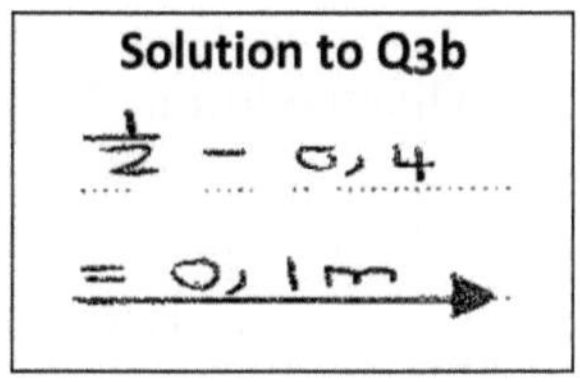

Fig. 33.3 Evidence of incorrect choice of operation

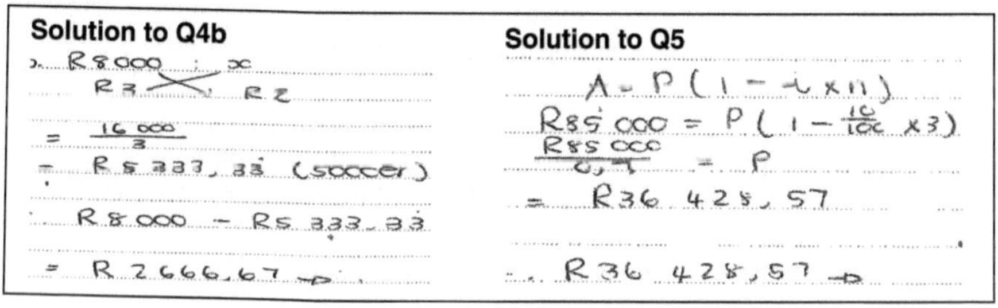

Fig. 33.4 Evidence of incorrect mathematical work

the different ways and choose the strategy that best solves the problem. The results show that some of the pre-service teachers in the study found it difficult to work within mathematics, suggesting that the mathematical content knowledge was not strong enough, as shown in Fig. 33.4.

4.4 *Interpretive Aspect of Modelling*

Interpretation can sometimes be the most challenging stage of modelling because it draws from insight in relation to the problem situation. It involves making realistic

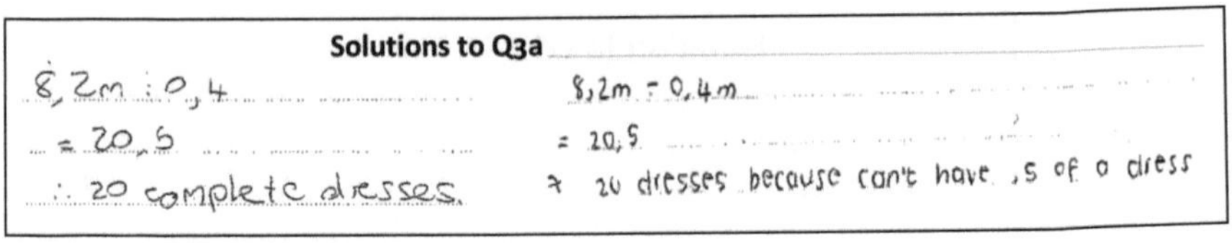
Solutions to Q3a

8,2m : 0,4
= 20,5
∴ 20 complete dresses.

8,2m ÷ 0,4m
= 20,5
→ 20 dresses because can't have ,5 of a dress

Fig. 33.5 Evidence of realistic considerations

considerations in order to obtain a sensible solution within the context of the problem situation. The evidence given in Fig. 33.5 illustrates the importance of making considerations which make sense within the context of the problem situation.

5 Discussion of Results

Analysis of the pre-service teachers' solutions to contextual problems which in our view provide an entry point into modelling complexities involved in dealing with real life problems has revealed two critical issues.

First, the pre-service teachers' abilities to reason within the realistic space was a critical issue. This includes transforming the real-world context into some statement(s) which can be subjected to mathematical treatment. It also involves insight related to interpretation of mathematical solutions in the context of the situation. Although there is no evidence pointing towards specificity of competencies related to idealisation from the data, the pre-service teachers' initial stages of their solutions suggest that they were able to identify quantities that influenced the situations. Thus the correct quantities in the solution strategies can be linked to some recognition processes which took place prior to the setting up of the procedure. Reasoning within the realistic space is also evident in the way mathematical solutions were interpreted in the context of the problem situation. Realistic considerations were employed which assisted in providing more sensible solutions. For instance, instead of rounding up the answer to test item Q3a to obtain 21 as is the case in a purely mathematical context, the results show evidence of insight related to the context.

Second, there was evidence of lack of deep and connected understanding of elementary mathematics ideas. Mathematics content knowledge is central in problem solving. This involves identifying relationships between variables and representing these relationships using some mathematical operations. As the problem is mathematised, some abstraction using mathematical language is needed so that mathematical principles can be applied. The study results have revealed some difficulties relating to enacting a procedure particularly when choosing an operation that best represents the relationship between the quantities. For example, the solution to Q3b required multiplication of 0.5 by 0.4 to obtain the left over material. Instead some teachers interpreted it as a subtraction problem (0.5–0.4) which resulted in 0.1 as an answer. Furthermore, in some cases, correct quantities were identified but incorrect procedures were used, which resulted in an incorrect answer. Difficulties in constructing mathematical models and finding solutions to such models such as these are associated with what Kaiser and Maaß call "mathematics-distant modellers" (Kaiser and Maaß 2007).

6 Conclusion and Recommendations

In this chapter, we have presented preliminary results from a larger study which broadly focuses on pre-service teachers' development in knowledge and practice related to modelling real-life situations within the citizenship domain. By focusing our analysis on solutions to realistic word problems using modelling competencies as lenses, we have argued that the reasoning within the realistic space among the sample of pre-service teachers is, in some ways, developed but comprehension of fundamental mathematics is fragmented. This means that the pre-service teachers' initial mathematical modelling competencies are at early stages of development. In view of this, we recommend further development of the pre-service teachers' modelling competencies with a specific focus on mathematics content understandings.

References

Blomhøj, M., & Jensen, T. H. (2003). Developing mathematical modelling competence: Conceptual clarification and educational planning. *Teaching Mathematics and Its Applications, 22*(3), 123–139.

Blum, W., & Niss, M. (1991). Applied mathematical problem solving, modelling, applications, and links to other subjects – State, trends and issues in mathematical instruction. *Educational Studies in Mathematics, 22*, 37–68.

Brombacher, A. (2003). *AMESA submission to the Department of Education on the National Curriculum Statement Grades 10–12 (Schools) and in particular on the mathematics and mathematical literacy subjects statements* (pp. 1–12). Johannesburg: AMESA. Retrieved 25th June, 2011, from http://amesa.org.za/Hearing.htm.

Creswell, J. W. (2009). *Research design: Qualitative, quantitative, and mixed methods approaches*. Thousand Oaks: Sage.

Department of Basic Education. (2011). *Curriculum and Assessment Policy Statement (CAPS): Mathematics*. Pretoria: Department of Basic Education.

Department of Education. (2003). *National curriculum statement grades 10–12 (general): Mathematical literacy*. Pretoria: Author.

English, L. D., & Watters, J. J. (2005). Mathematical modelling with 9-year-olds. In H. L. Chick & J. L. Vincent (Eds.), *Proceedings of the 29th conference of the International: Group for the Psychology of Mathematics Education* (Vol. 2, pp. 297–304). Melbourne: PME.

Geiger, V., Goos, M., & Dole, S. (2013). Taking advantage of incidental school events to engage with the applications of mathematics: The case of surviving the reconstruction. In G. Stillman, G. Kaiser, W. Blum, & J. Brown (Eds.), *Teaching mathematical modelling: Connecting to research and practice* (pp. 175–184). New York: Springer.

Greer, B., Verschaffel, L., & Mukhopadhyay, S. (2007). Modelling for life: Mathematics and children's experience. In W. Blum, P. Galbraith, W.-H. Henn, & M. Niss (Eds.), *Modelling and*

applications in mathematics education: The 14th ICMI study (Vol. 10, pp. 89–98). New York: Springer.
Kaiser, G. (2007). Modelling and modelling competences in school. In C. Haines, P. Galbraith, W. Blum, & S. Khan (Eds.), *Mathematical modelling (ICTMA12): Education, engineering and economics* (pp. 110–119). Chichester: Horwood.
Kaiser, G., & Maaß, K. (2007). Modelling in lower secondary mathematics classrooms – Problems and opportunities. In W. Blum, P. Galbraith, W.-H. Henn, & M. Niss (Eds.), *Modelling and applications in mathematics education* (pp. 99–108). New York: Springer.
Kaiser, G., & Schwarz, B. (2006). Mathematical modelling as bridge between school and university. *ZDM, 38*(2), 196–208.
Maaß, K. (2006). What are modelling competences. *ZDM, 38*(2), 113–142.
OECD. (2003). Mathematical literacy. In *Assessment frameworks: Mathematics, reading, science and problem solving knowledge and skills* (pp. 23–102). Paris: OECD.
OECD. (2009). *Learning mathematics for life: A perspective from PISA*. Paris: OECD.
Venkat, H. (2007). Mathematical literacy-mathematics and/or literacy: What is being sought? *Pythagoras, 66*, 76–84.
Verschaffel, L., De Corte, E., & Borghart, I. (1997). Pre-service teachers' conceptions and beliefs about the role of real-world knowledge in mathematical modelling of school word problems. *Learning and Instruction, 7*(4), 339–359.
Yin, R. K. (2009). *Case study research: Design and methods*. London: Sage.

Chapter 34
Initial Perspectives of Teacher Professional Development on Mathematical Modelling in Singapore: Conceptions of Mathematical Modelling

Chan Chun Ming Eric

Abstract In the Singapore context, mathematical modelling is often equated with the Model – Drawing Approach – a key pedagogical tool in the Singapore mathematics curriculum. Though related, the two concepts are not identical. The participants in a professional development workshop program on mathematical modelling were initially facilitated through a 1 day workshop to experience what mathematical modelling entailed. This chapter captures the participants' understanding of mathematical modelling and the value they attached to it as part of the total mathematical learning experience for their students after having undertaken this professional development. Concerns the teachers had about mathematical modelling in the Singapore context are shared.

1 Introduction

Mathematical modelling has been advocated by some researchers to be an important goal in mathematics education (Lesh and Sriraman 2005). The inclusion of mathematical modelling as part of the teaching and learning of mathematics in school curricula has become more prominent recently. This transition from an emphasis on mathematical problem solving in the 1980s towards mathematical modelling apparently stems from the need to recognise the importance of modelling in dealing with real-world situations and solving problems in an ability-driven economy. To function well in the twenty-first century, learners will need to acquire knowledge and skills to manage complex systems and solve real world problems (English and Sriraman 2010) and mathematical modelling is perceived to provide students with the opportunity to develop those competencies (Maaβ 2006).

Chan C.M.E. (✉)
Mathematics and Mathematics Education Academic Group,
National Institute of Education, Nanyang Technological University, Singapore
e-mail: eric.chan@nie.edu.sg

G.A. Stillman et al. (eds.), *Teaching Mathematical Modelling: Connecting to Research and Practice*, International Perspectives on the Teaching and Learning of Mathematical Modelling, DOI 10.1007/978-94-007-6540-5_34,

The Singapore Ministry of Education revised its curriculum in 2007 to include *applications and modelling* as part of the process component in alignment with international reform movements (Ministry of Education 2007). As mathematical modelling is in its formative stage in Singapore, the lack of detailed description about what mathematical modelling entails has not helped to actualise the intended curriculum (Chan 2008). Most primary and secondary school mathematics teachers tend to relate modelling to the model-drawing approach, a heuristic that is popularly used in Singapore mathematics classrooms to solve structured word problems. That is one form of direct modelling but it does not adequately fit the description of what mathematical modelling means in the curriculum. The curriculum document defines mathematical modelling as "the process of formulating and improving a mathematical model to represent and solve real-world problems" (Ministry of Education 2007, p. 8). It requires students to "use a variety of representations of data, and to select and apply appropriate mathematical methods and tools in solving real-world problems" (p. 8), including dealing with empirical data and using mathematical tools for data analysis. In this regard, without a good understanding of mathematical modelling, it is only reasonable that teachers do not carry out mathematical modelling activities in the mathematics classroom.

2 Selected Reviews on Teachers' Conceptions of Mathematical Modelling

According to Ärlebäck (2009), having a better understanding of teachers' beliefs about mathematical modelling is important if teachers are to be encouraged to integrate mathematical modelling in their mathematics teaching. Teachers may hold different beliefs about mathematical modelling and these beliefs come from their understanding of modelling based on their experiences and what they think modelling is. As mentioned earlier, many primary school teachers in Singapore have the notion that the new process component of mathematical modelling that has been included in the revised syllabus of 2007 is about an exemplification of the model-drawing approach (Chan 2008). On the other hand, those who teach higher level mathematics in junior colleges or tertiary courses that involve applied mathematics or mechanics see mathematical modelling as formulating and using abstract formulas. Kaiser and Maaβ (2007) found in a German study that there were teachers who viewed mathematical modelling as stressing the process of creating opportunities for developing solutions while others strictly stress the establishment of formulas. These different orientations can lead to a particular emphasis in the way modelling is taught and carried out in classrooms. As well, differences in opinions about mathematics and reality and teacher-held beliefs could build resistance to introducing mathematical modelling into teaching (Kaiser 2006).

Selected literature on teachers involved in mathematical modelling professional development sessions have shown that they generally appreciate the potential of what mathematical modelling could offer as part of students' learning of mathematics but not without challenges. Sixteen Taiwanese teachers, for example, who underwent model-eliciting activities (MEAs) workshops found that MEAs were useful to enhance students' problem-solving ability and attitudes towards modelling but the MEAs were difficult to develop and design to fit the modelling design principles (Yu and Chang 2011). According to these researchers, this can be attributed to the lack of theoretical knowledge of the teachers to make the connections between MEAs and the school curriculum.

In Chan's (2011) case study of Singaporean primary six students' mathematical modelling endeavours where two mathematics teachers were the facilitators, the teacher-facilitators' journals reflected mathematical modelling as being beneficial to students in ways they had personally observed like developing creative thinking, collaborative skills, problem-solving skills and decision making skills, and independent learning. They were concerned about the tensions that exist such as classroom management issues, completing the syllabus, the high-stake examinations, and the large class sizes that needed to be reconciled before mathematical modelling can be effectively carried out or would become a prominent feature in the classroom.

In Cheah's (2008) case studies of teachers who underwent mathematical modelling sessions, he found that the teachers appreciated the creative and collaborative experience of mathematical modelling in learning mathematics that enabled their thinking to move from the real world to the mathematical world, and that learning was acquired through mathematical discourse. A concern raised by teachers was the issue of time and space in the curriculum to do mathematical modelling.

This brief review offers some preliminary information with respect to what teachers appreciate about mathematical modelling and the tensions they face in thinking about how to effectively carry out modelling activities in classrooms that have been based on a predominantly teacher-centred pedagogy.

3 Stimulating Teacher Conceptions at the Mathematical Modelling Outreach

The Mathematical Modelling Outreach (MMO) was conceived by the Mathematics and Mathematics Education Academic Group of the National Institute of Education to promote mathematical modelling as a means to support the school mathematics curriculum in Singapore. The main objective of this outreach was to put theory into practice by providing students with the experience of engaging in modelling tasks spread over 3 days. Lee's chapter in this volume describes the methodology for data collection in this exploratory study for reporting findings through surveys (Lee, Chap. 35, this volume).

By teachers experiencing modelling workshops where they were scaffolded through a modelling task, it was believed that what the teachers gained from that experience would enable them to formulate some conceptions of what mathematical modelling is. This chapter reports on the teachers' conception of mathematical modelling based on the first day's workshop where they were taken through a modelling task to experience what mathematical modelling was like.

The task, entitled *My Hometown*, was used with 31 teachers during the immersion programme because of the then current local news items pertaining to new migrants in Singapore:

My Hometown (Years 7 and 8)

According to the Urban Redevelopment Authority, more than 80 % of the 4,000 Singaporeans and permanent residents who participated in a lifestyle survey conducted in 2010 found Singapore to be a great place to live in, work and play. When interviewed, people shared what they liked about the town they live in, however new or old. The editor of "Good Living", a lifestyle magazine, is featuring an article to showcase the most suitable town for a family of new migrants. Come up with a proposal for your recommended town.

The teachers worked in five groups. They were given internet access during the session so that the required information could be sourced. The two facilitators, who are teacher educators, implemented the task using a purpose designed facilitation framework (see Ng, Chap. 36, this volume) in three sessions over the day, allowing for whole class discussions and facilitation inquiries by participating teachers between different phases of the modelling process.

To ascertain the teachers' conceptions of what modelling was a simple survey questionnaire based on a 5 point Likert scale (i.e., Strongly Disagree (1), Disagree, Neutral, Agree, Strongly Agree (5)) was administered at the end of the day's workshop for the teachers to pen their conceptions. The two main items in the survey were:

(i) I have an understanding of what mathematical modelling is.
(ii) I see the potential of mathematical modelling tasks as part of the total mathematical learning experience for students in my school.

To obtain a better understanding of the teachers' perceptions concerning their responses to the two main items, teachers were encouraged to express as open-ended responses what they knew about the features/process and the potential of mathematical modelling respectively in blank spaces below the two main items. There was also one other item not linked to the two main items where teachers were asked to write about "My concerns about mathematical modelling".

The teachers' responses to the two main survey items were reported as percentage figures while the open-ended responses were tallied based on the comments given (or similar comments) as frequency counts. For a participant who gave a single comment with two or more distinctive descriptions, the responses were treated as separate occurrences in the tallying process.

4 Responses

4.1 *Teacher Understanding of Mathematical Modelling*

Table 34.1 shows the percentage response of the teachers to the statement "I have an understanding of what mathematical modelling is" with 83 % ("agree" and "strongly agree") of the teachers responding that they had an understanding of mathematical modelling after the first day introductory workshop.

Table 34.2 presents the frequency of the open-ended responses. Responses that did not specifically fit the notions of the features or process of mathematical modelling (e.g., "the need to design tasks that are relevant to pupils" or "they are demanding") were not considered.

After the modelling experience, the response that surfaced the most was specific mention of the modelling process including the various phases that they had encountered (17 occurrences). The other most occurring response relates to the openness of the modelling endeavour (16 occurrences) in terms of the modellers' responses, goals and interpretations of task and solutions. As well, several respondents listed a few specific modelling behaviours (16 occurrences). There were 11 responses concerning mathematical modelling as relating to the real world or involving the use of real data towards solving real world problems. Six responses described the toggling between the real world and mathematical world for transforming the problem. There was only one response indicating that group work was a feature of mathematical modelling.

Table 34.1 Teachers' response towards attaining an understanding of mathematical modeling ($N=31$)

	Strongly disagree	Disagree	Neutral	Agree	Strongly agree
I have an understanding of what mathematical modelling is	0 %	3 %	13 %	77 %	7 %

Table 34.2 Teachers' conceptions of the features and process of mathematical modelling

Conceptions of the features and process of mathematical modelling	Frequency
The process includes representation of relationships, experimentation, verification and generalisation	17
They are open-ended/elicit different interpretations/achieve different goals	16
They involve (naming of some specific modelling behaviours) problem posing, explaining, describing, simplifying, classifying, exploring, conjecturing, predicting, justifying, reasoning, assuming, reasoning, communicating, revising, metacognitive thinking	16
They involve the real world/real data/real world problems	11
They involve making links to transform a real world model into a mathematical solution	6
They involve group work	1

Table 34.3 Teachers' response concerning the potential of using mathematical modelling tasks in schools ($N=31$)

	Strongly disagree	Disagree	Neutral	Agree	Strongly agree
I see the potential of mathematical modelling tasks as part of the total mathematical learning experience for students in my school	0 %	3 %	13 %	63 %	20 %

Table 34.4 Teachers' conception of the potential of using mathematical modelling tasks in schools

The potential of using modelling tasks	Frequency
It links maths concepts to real world situations/applications	6
It is enriching/enjoyable/interesting/refreshing	5
It helps to improve students' motivational level/appreciation of maths	4
It makes students think more/apply their maths learned/be creative	3
It is suitable for higher-ability students	1

4.2 *The Potential of Using Mathematical Modelling Tasks*

The percentage figures of the teachers' response to the item "I see the potential of mathematical modelling tasks as part of the total mathematical learning experience for students in my school" are presented in Table 34.3. Most teachers (63 %) agreed that there is potential to use mathematical modelling tasks in the mathematics curriculum with 20 % of the teachers strongly agreeing. The figures reflect that teachers viewed positively mathematical modelling as playing an integral part in the mathematics learning experience of students.

Table 34.4 shows the frequency of the open-ended responses in relation to the potential of having mathematical modelling in the classroom. As before, responses that did not specifically give the potentials (e.g., "it has many potentials") were not considered further. There were 6 responses that indicated the potential of using mathematical modelling as linking students' mathematical ideas to real-world applications or situations. There were five responses that suggested mathematical modelling is enriching, enjoyable, interesting or refreshing. Four responses mentioned the affective domain, namely, modelling increases students' motivational level or appreciation of mathematics. There were three responses that suggested mathematical modelling could make students think more, apply their mathematics learned or be creative (but these responses did not indicate linking to the real world). There was one response that related its potential mainly for higher-ability students.

4.3 *Teacher Concerns About Mathematical Modelling*

Table 34.5 shows the consolidated 41 responses of the teachers with respect to their concerns about mathematical modelling. It can be inferred that most teachers (19 occurrences) found the constraints of time and the completion of syllabus

Table 34.5 Teachers' concerns about mathematical modelling

Concerns about mathematical modelling	Frequency
It takes up (substantial) curriculum time/cannot complete syllabus	19
I am unsure how to facilitate/lack competence/unsure how to scaffold	7
How to differentiate the instruction for different (ability) groups of students?	6
Lacks confidence in dealing with openness of interpretation of task/solution	3
How to design (appropriate) modelling task?	2
How to incorporate into examinations/daily maths practice?	2
What support is available for teachers?	1
Will parents buy in?	1
Lack of familiarity with the task	1
How to assess student learning in mathematical modelling?	1
May discourage students when they do not get breakthrough	1
Are schools/teachers/students ready for mathematical modelling?	1

requirements to be the inhibiting factors in carrying out mathematical modelling. The concern that teachers need to be well-equipped so as to be able to teach mathematical modelling had the next highest response, garnering seven occurrences. Six responses concerned how to differentiate the instruction in terms of managing different ability groups of students. Three responses were task related, in that they might not be able to deal with the openness of the interpretation of the task and solution. Two responses concerned task design and two responses sought answers for linking the relevance of mathematical modelling to high-stakes examinations and routine classroom practice. The other concerns each raised by only one respondent included the availability of support for teachers, parents' concerns, the lack of familiarity with the task, how to assess student learning, readiness to implement mathematical modelling and whether students would be discouraged if they were not successful in completing a modelling task.

5 Discussion and Implications

From the responses of the first survey item about teachers' understanding of mathematical modelling, most of the teachers (83 % agree and strongly agree) after their modelling experience, believed that they had an understanding of what mathematical modelling entailed. The more prominent accompanying open-ended responses entailed their stating modelling phases, the open nature of making interpretations to tasks and solutions, as well as the stating of certain specific modelling and mathematising behaviours. As this modelling endeavour was an initial, but encouraging, step in promoting mathematical modelling in accord with the goal of the MMO, it must be recognised that the teachers did not have the opportunity to design, carry out and facilitate mathematical modelling sessions with students. In this regard, what might have been acquired as their conceptions of modelling was a surface level understanding about mathematical modelling, which nonetheless was a good start to address future mathematics teachers' professional knowledge in the area of modelling competencies.

Similarly, 83 % of these local teachers saw the potential of having mathematical modelling in the classroom although not many provided open-ended responses to what the potentials were. The two highest occurrences (six and five) had to do with linking mathematics to the real world and that such tasks offered something refreshing and interesting. There were more responses highlighted as concerns though. Being new to mathematical modelling, the concerns raised by teachers were not surprising. With a tight curriculum and the need to complete the syllabus (this highest response garnered 19 occurrences), this tension would need to be addressed towards reconciling how to integrate modelling activities in the mathematics classroom. Notably too, a number of teachers were unsure of how to facilitate such activities (seven occurrences) as well as in managing the differentiation in instruction for different ability-groups of students (six occurrences).

We see that training teachers for future professional knowledge and modelling competencies as a means to address some of the concerns. As well, tackling the affective aspect should not be dismissed. Lim (2002) noted similar concerns with mathematics education in Singapore in light of implementing reform classroom practices and recommended that the issue be approached from the stance of intrinsic motivation of the teachers to carry new developments through. She also believed that it takes dedicated teachers who are confident and passionate in mathematics besides being grounded in pedagogy and new developments in mathematics education to advance more contemporary pedagogies. Thus, starting small and appealing to cultivate the teachers' interest in mathematics in the area of mathematical modelling could help infuse the gradual use of modelling into the curriculum. It is indeed demanding for a facilitator to manage a class of 40 students. To circumvent this, the school could arrange to have two teachers involved in facilitating a modelling session as Chan (2011) has done in his study related to implementing model-eliciting activities. This helped the teachers to reach out to the significantly fewer groups of students on a more timely basis.

6 Conclusion

By the end of the introductory workshop on mathematical modelling, most teachers had a basic understanding with regards to what mathematical modelling entailed. Most teachers also saw the benefits of having mathematical modelling in the school curriculum. Teachers would gain greater confidence and develop modelling competencies if they are better exposed and directly involved in modelling activities with students. While generally most of the teachers saw the potential of mathematical modelling, their concerns were mainly with respect to time constraints and completing the school syllabus. There is scope therefore in developing a teacher training programme towards helping teachers acquire future professional knowledge in mathematics competencies and thereby address some of the concerns. As well, a greater promotion of using simple modelling tasks for the purpose of helping students to communicate and reason mathematical ideas towards giving meanings to problem situations can be a good starting point to integrate modelling activities into the classroom.

References

Ärlebäck, J. B. (2009, January 28–February 1). Towards understanding teachers' beliefs and affects about mathematical modelling. In V. Durand-Guerrier, S. Soury-Lavergne, & F. Arzarello (Eds.), *Proceedings of the sixth congress of the European society for research in mathematics education,* (Working Group 11, pp. 2096–2105). Lyon: Institut National de Recherche Pédagogique. Available on line http://www.inrp.fr/editions/editions-electroniques/cerme6/

Chan, C. M. E. (2008). Using model-eliciting activities for primary mathematics classroom. *The Mathematics Educator, 11*(1/2), 47–66.

Chan, C. M. E. (2011). *Investigating primary 6 pupils mathematical modelling process in a problem-based learning setting*. Unpublished doctoral dissertation, Nanyang Technological University, Singapore.

Cheah, U. H. (2008). Introducing mathematical modelling to secondary school teachers: A case study. *The Mathematics Educator, 11*(1/2), 21–32.

English, L. D., & Sriraman, B. (2010). Problem solving for the 21st century. In L. D. English & B. Sriraman (Eds.), *Theories of mathematics education: Seeking new frontiers*. New York: Springer.

Kaiser, G. (2006). The mathematical beliefs of teachers about applications and modelling – Results of an empirical study. In J. Novotná, H. Moraová, M. Krátká, & N. Stehliková (Eds.), *Proceedings 30th conference of the international group for the psychology of mathematics education* (Vol. 3, pp. 393–400). Prague: PME.

Kaiser, G., & Maaß, K. (2007). Modelling in lower secondary mathematics classroom – Problems and opportunities. In W. Blum, P. Galbraith, H. W. Henn, & M. Niss (Eds.), *Modelling and applications in mathematics education: The 14th ICMI study* (pp. 99–108). New York: Springer.

Lee, N. H. (2013). Initial perspectives of teacher professional development on mathematical modelling in singapore: Problem posing and task design. In G. Stillman, G. Kaiser, W. Blum, & J. Brown (Eds.), *Teaching mathematical modelling: Connecting to research and practice* (pp. 415–425). New York: Springer.

Lesh, R., & Sriraman, B. (2005). John Dewey revisited—Pragmatism and the models modeling perspective on mathematical learning. In A. Beckmann, C. Michelsen, & B. Sriraman (Eds.), *Proceedings of the 1st international symposium of mathematics and its connections to the arts and sciences* (pp. 7–31). Schwöbisch Gmund: The University of Education.

Lim, S. K. (2002). Mathematics education in Singapore: Looking back and moving on. *The Mathematics Educator, 6*(2), 1–14.

Maaβ, K. (2006). What are modelling competencies? *ZDM, 38*(2), 113–142.

Ministry of Education. (2007). *Mathematics syllabus – Primary*. Singapore: Curriculum Planning and Developmental Division.

Ng, K. E. D. (2013). Initial perspectives of teacher professional development on mathematical modelling in singapore: A framework for facilitation. In G. Stillman, G. Kaiser, W. Blum, & J. Brown (Eds.), *Teaching mathematical modelling: Connecting to research and practice* (pp. 427–436). New York: Springer.

Yu, S. S., & Chang, C. K. (2011). What did Taiwan teachers think of model-eliciting activities and modeling? In G. Kaiser, W. Blum, F. Borromeo, & G. Stillman (Eds.), *Trends in teaching and learning of mathematical modelling – ICTMA 14* (pp. 147–156). New York: Springer.

Chapter 35
Initial Perspectives of Teacher Professional Development on Mathematical Modelling in Singapore: Problem Posing and Task Design

Lee Ngan Hoe

Abstract The Singapore Mathematics Curriculum is often referred to as the Problem Solving Curriculum. Despite two major reviews, it has remained largely unchanged. Consequently, teachers are familiar with problem solving but not problem posing, a key feature in mathematical modelling, an addition to the curriculum since 2003. As mathematical modelling is context based, teachers expressed concern with the availability of appropriate relevant tasks and the sustainability of mathematical modelling. This chapter is the second in a series of three on initial perspectives of teacher professional development on mathematical modelling in Singapore. It captures the readiness of participants in a relevant professional development event in designing modelling tasks and the value they attached to problem posing as part of the total student mathematical learning experience.

1 Introduction

One aspect of the 1990s Singapore mathematics syllabus document that differed significantly from past syllabus documents was the articulation of the key role played by problem solving in school mathematics. The primary aim of the mathematics curriculum was to "enable pupils to develop their ability in mathematical problem solving" (Ministry of Education 1990a, p. 3, 1990b, p. 3). The buzz internationally on mathematical problem solving in the 1980s may have been the key contributing factor for Singapore to develop its national mathematical curriculum with an emphasis on mathematical problem solving. However, as Lee (2010)

N.H. Lee (✉)
Mathematics and Mathematics Education Academic Group,
National Institute of Education, Nanyang Technological University, Singapore
e-mail: nganhoe.lee@nie.edu.sg

G.A. Stillman et al. (eds.), *Teaching Mathematical Modelling: Connecting to Research and Practice*, International Perspectives on the Teaching and Learning of Mathematical Modelling, DOI 10.1007/978-94-007-6540-5_35,

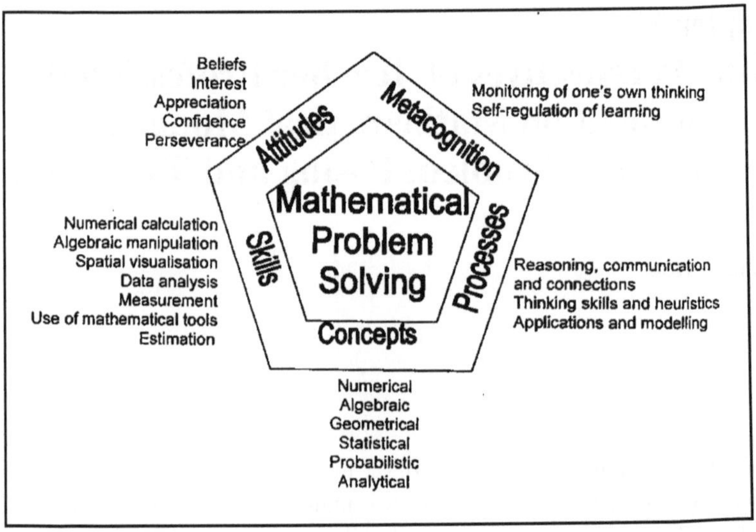

Fig. 35.1 Framework for the Singapore mathematics curriculum (Ministry of Education 2006b)

observed, the Singapore mathematics curriculum of the 1990s survived, with minor modifications, the major curriculum reviews for implementation in 2001 and 2007 (Ministry of Education 2000a, b, 2006a, b). Consequently, Singapore mathematics teachers are generally familiar and comfortable with the role mathematical problem solving plays in the mathematics classroom.

1.1 The Context

1.1.1 The Singapore Mathematics Curriculum

The Singapore Mathematics Curriculum Framework (Fig. 35.1) provides a succinct description of the philosophy of the curriculum and articulates the important aspects of learning and teaching in Singapore mathematics classrooms. The centre of the Framework, also known as the Pentagonal Framework, depicts the primary aim of the curriculum – mathematical problem solving. The five sides of the pentagon reflect the dependence of the attainment of problem solving ability on five inter-related components – concepts, skills, processes, attitudes, and metacognition. Concepts and skills refer to the relevant mathematical concepts and skills, attitudes refers to the affective aspects of mathematical learning, and metacognition refers to the awareness of, and the ability to, control one's thinking processes. The processes refer to the "knowledge skills (or process skills) involved in the process of acquiring and applying mathematical knowledge" (Ministry of Education 2006b, p. 13).

1.1.2 Singapore Mathematics Curriculum and Mathematical Modelling

As shown in Fig. 35.1, processes in the Singapore Mathematics Curriculum Framework include reasoning, communication and connections, thinking skills and heuristics, and application and modelling. As noted by Balakrishnan et al. (2010), applications and modelling were introduced in the Singapore Mathematics Curriculum Framework in 2003. This is to highlight the importance of applications and modelling in mathematics learning so as to meet the challenges of the twenty-first century (Ministry of Education 2009; Soh 2005). However, it should be pointed out that the inclusion of the applications and modelling in the Singapore Mathematics Curriculum Framework was only reflected in the Syllabus documents for implementation in 2007 (Ministry of Education 2006a, b).

1.1.3 Professional Development of Singapore Mathematics Teachers

In the Singapore Education System, all trained teachers (those deemed to have successfully completed the necessary pre-service education) are "eligible to [sic] 100 hours of professional development a year" (Lim-Teo 2009). The professional development courses range from one session workshops, whole day conference/seminars to long certification programmes. However, up to the end of 2009, there was no record of professional development courses dedicated to address mathematical modelling in the mathematics classroom.

1.2 The Issues

With a lack of professional development opportunities to avail mathematics teachers with the necessary knowledge and skills to address mathematical modelling in the mathematics classroom, this particular aspect of the Singapore Mathematics Curriculum appears to be often more neglected than others. Many issues are involved, but this chapter seeks to discuss and shed some light on two related ones: problem posing and task design.

1.2.1 Problem Posing

Although Polya (1957) considered problem posing as an inseparable part of problem solving, it is observed that only after 10 years of existence, did the Singapore Problem Solving Curriculum explicitly encourage problem posing – in the form of "extend and generate problems" (Ministry of Education 2000b). As Yeap (2002, pp. 3–4) observed, students "reason with laws that involve symbols to solve well-defined problems that have fixed conceptual meaning". On the other hand, Reed (1999) argued that one way in which real-world problem solving differs from school problem solving is that in the former the problem solver often has to formulate the problem and collect relevant data for solving it. In fact, Bonotto (2010) noted that problem posing activities lay the foundation for a mathematisation disposition – a key aspect of the modelling process (Niss 2010). However, as Bonotto (2010, p. 406) also recognised that such activities require "a radical change on the part of teachers" who are more used to using tools that are always "highly structured, rigid, not really suitable to develop alternative processes deriving from circumstantial solicitations, unforeseen interests, particular classroom situations". Such radical change could often only be effected by first impacting on the relevant teachers' belief systems. It is thus important to address and examine the value Singapore mathematics teachers attached to problem posing as part of the total student mathematical learning experience.

1.2.2 Task Design

Stillman (2010, p. 301) noted that "mathematical modelling connects from the outside world into the classroom", whilst Kaiser et al. (2010, p. 227) observed that "a real world situation is the [modelling] process' starting point". Muller and Burkhardt (2006, p. 269) also pointed out that "context-based mathematical modelling provides ideal settings to blend content and process so as to produce flexible mathematical competence". Thus, one of the key features of a mathematical modelling task is its context-based real-world nature. As mathematical modelling is new to Singapore mathematics classrooms, the availability of such context-based real-world modelling tasks that will allow Singapore students to "address problems in their world" (Galbraith et al. 2010, p. 135) has become a main concern of Singapore mathematics teachers. This in turn affects how these teachers perceive the sustainability of mathematical modelling in mathematics classrooms. However, if teachers feel that designing modelling tasks is just part of their broader repertoire of professional strategies, such concerns may no longer be an issue. It is thus useful to gather some insight into teachers' perceived level of competency in designing modelling tasks after being exposed to some relevant professional development.

2 Methodology

In June 2010, the Mathematics and Mathematics Education Academic Group of the National Institute of Education – the sole provider of pre-service teacher education and main provider of in-service teacher education in Singapore, organised two consecutive events, namely a Mathematical Modelling Outreach (MMO) and the Second Lee Peng Yee Symposium. Both events centred around the theme of mathematical modelling. In addition to invited talks on mathematical modelling for participating students and teachers, there were hands-on sessions for the students solving a mathematical modelling task. There were also hands-on workshop sessions to better prepare teachers to implement the mathematical modelling aspect of the Singapore Mathematics Curriculum. These workshops addressed issues related to conceptions of mathematical modelling, problem posing for mathematical modelling, mathematical modelling task design, and facilitation of mathematical modelling instruction. These were identified as areas of concern given the unfamiliarity of teachers with mathematical modelling, the lack of appropriate mathematical modelling tasks, and inexperience of teachers to conduct mathematical modelling lessons. This chapter, first of a series of three to provide some insights into the form and type of teacher professional development on mathematical modelling in these aspects for Singapore mathematics teachers, through an exploratory survey study, addresses the issues related to problem posing and task design for mathematical modelling. Chan (2013) deals with the issue of conceptions of mathematical modelling whilst Ng (2013) explores facilitation.

2.1 The Participating Teachers

A total of 31 teachers participated in the hands-on workshop sessions, of these, two were from an Australian high school and one from an Indonesian International primary school as the events also generated interest among educators in the Asia Pacific region. Of the 28 Singapore teachers who participated, 13 were teaching at primary levels and 15 at secondary levels. The schools that these teachers were teaching in also provided a good mix of school backgrounds in Singapore, as it was endeavoured by the organiser to ensure that schools were well-represented at both events. The teaching experience of these participating teachers was unable to be controlled as it was at the discretion of the schools which teacher to send to the events. However, from informal interaction with participating teachers, it was observed that level of teaching experience among the participants was quite varied. There were novice teachers who have just joined the schools upon completion of their pre-service education, whilst others had taught for more than 20 years. A few were also appointment holders, such as senior subject teacher or heads of department.

2.2 Workshop Sessions for Participating Teachers

The session on *Problem Posing for Mathematical Modelling* was conducted by an international expert – an invited speaker for the events. The session was of one and a half hours duration and consisted of short lectures, demonstrations, and hands-on activities. The session dealt with the role of problem posing from the perspective of its relationship to modelling and finding modelling tasks. In particular, the session exposed teachers to the use of photographs as a stimulus to task design and the kind of problems that could be posed. Examples used were contexts drawn from common daily life experiences, alerting teachers to the importance of sensitising one to the surrounding environment for suitable contexts for mathematical modelling.

The session on *Task Design* was conducted by Dr Dawn Ng and Dr Lee Ngan Hoe, the two co-Chairs for the Organising Committee. The session was of 1 h duration and consisted of a sharing of the journey taken to create eight modelling tasks that the participating students were undertaking during the MMO as well as a short hands-on session for the participating teachers to design a task. In particular, the teachers were exposed to the use of a stimulus, such as printed materials, video clips, the physical environment, and photographs, to create the tasks. The process taken to lend contexts to the designed tasks relevant to the participating students was also illustrated. Teachers were also shown how the tasks were designed to ensure that the key attributes of modelling tasks, namely openness of the tasks and the focus on mathematics, were well aligned. The participating teachers were then given an opportunity to apply these principles to design a sample task.

2.3 *Collection of Feedback from Participating Teachers*

After having attended both Workshop sessions, participating teachers were asked to respond to a survey. In relation to the session on *Problem Posing for Mathematical Modelling*, participating teachers was asked to rate on a 5 point Likert scale how they saw the importance of students posing their own problems during mathematical modelling. Qualitative comments were also sort from the participating teachers through the completion of the following statements:

(i) When encouraging problem posing by students, I would like to focus on:
(ii) When encouraging problem posing by students, I would like to be careful about:
(iii) My concerns about facilitating problem posing by students during mathematical modelling:

For the *Task Design* session, participating teachers were asked to rate on another 5 point Likert scale if they had obtained some ideas for designing modelling tasks. Qualitative comments were again sought from the teachers through the completion of the statements:

(i) When designing modelling tasks, I would like to focus on:
(ii) When designing modelling tasks, I would like to be careful about:
(iii) My concerns about task design in mathematical modelling:

At the end of the survey, the teachers were also asked to list the key learning points from the sessions.

3 Results

A total of 27 participating teachers responded to the survey. Of the 27 sets of survey forms collected, it was noted that despite the structured nature of the qualitative statements, the teachers still did not write at length – a common trait of Singapore teachers. Nonetheless, compared to the norm whereby many blanks are left by participants in qualitative surveys, numerous short sentence responses by the teachers provided some insights into their thoughts after the workshop sessions. However, teacher responses to these statements were not as focussed as it was desired. The teachers' qualitative comments with regards to problem posing and task design were often stated in general terms and not distinctly on only one of two aspects. Sometimes the qualitative comments even touched on other general aspects of mathematical modelling. This is not surprising, as the aspects though distinct, are not actually independent, but are inter-related aspects of addressing mathematical modelling in the mathematics classroom. Consequently, this section will report on two consolidated aspects of the survey data, namely quantitative versus qualitative, instead of delineating the results along the lines of problem posing and task design.

3.1 Quantitative Results

Based on a Likert scale of 1–5 (Strongly Disagree, Disagree, Neutral, Agree, Strongly Agree), the responses on the importance of students posing their own problems during mathematical modelling had a mean of 4.26 and a SD of 0.45. All 27 respondents either agreed (20) or strongly agreed (7 responses). In other words, all the responding teachers agreed that it was important for students to pose their own problems during mathematical modelling.

Similarly, using a Likert scale of 1–5 (Strongly Disagree, Disagree, Neutral, Agree, Strongly Agree), the responses of the participating teachers on their having obtained some ideas for designing modelling tasks had a mean of 4.11 and a SD of 0.42. Other than one Neutral response, the other 26 responses consisted of 22 agreements and 4 strongly agreed. So, the responding teachers generally felt they had obtained some ideas for designing modelling tasks.

3.2 Qualitative Results

Table 35.1 summarises the key qualitative comments provided by the participating teachers that are relevant to the discussion of this chapter, in order of frequency, that is, the number of participating teachers who provided the respective response.

More than half of the respondents had captured the key considerations when designing a modelling task, namely the need to use a context that is relevant to the students, and to ensure a mathematical focus of the task while retaining an openness to the problem. Eleven of 27 respondents expressed that the key learning point of the workshop sessions was to ensure that modelling tasks are real-life contexts that are of interest and relevance to students. On the other hand, teachers' competencies to provide appropriate scaffolding when encouraging problem posing without 'closing' the task remained a concern to many (more than a third of the respondents). Interestingly, the perennial concern of inadequacies of teachers in dealing with

Table 35.1 Participating teachers' qualitative comments to survey

Comments	Number of responses
Need to focus on the mathematics	16
Need to ensure relevance of task to students	14
Need to ensure openness of task	14
Concern about teachers' competencies to provide appropriate scaffolding	10
Concern with the need for teachers to trek unfamiliar grounds (lack of necessary knowledge)	6
Need to get students to examine assumptions made	6
Need for consideration of students' ability	5
Need to get students to examine the variables (reasonable number, and ensure that variables are quantifiable)	6
Time constraint when implementing modelling task	3
Need to get students to verify their models	2
Need to get students to appreciate the limitations of their models	2
Availability of resources	1

real-life tasks did not appear to concern as many of the participating teachers as was expected (6 respondents). The message that the teachers of the twenty-first century are no longer simply dispensers of knowledge but facilitators of students' learning appears to have sat well with these teachers.

It was also apparent that comments with regards to problem posing and task design from the teacher participants also centred on more detailed concerns relating to implementation of mathematical modelling in the classroom. Some respondents expressed explicitly the need for task design to include opportunities for problem posing that will help students to better identify assumptions made and variables, verify their models and appreciate limitations of their models. This reflected a greater level of immersion of the participants in a culture of mathematical modelling.

4 Discussion, Implications and Conclusion

It is evident that the short professional development, in the form of hands-on workshop sessions, on problem posing and task design in relation to mathematical modelling for the participating teachers yielded a positive, though maybe short term, outcome. More than 90 % of the participants had not implemented mathematical modelling in their classroom and had little or no experience in carrying out problem posing in their mathematics class. After the workshop sessions, not only did they generally consider it important for students to problem pose during mathematical modelling, but also they were able to identify some of the key principles for designing modelling tasks. In the latest curriculum document

(Ministry of Education 2012, p. 15), teachers were alerted to the fact that "applications and modelling allow students to connect mathematics that they have learnt to the real world, enhance understanding of key mathematical concepts and methods as well as develop mathematical competencies. Students should have opportunities to apply mathematical problem-solving and reasoning skills to tackle a variety of problems, including open-ended and real-word problems". Thus, the three most frequent comments from teachers (Table 35.1), that is, focus on mathematics, relevance of task to students, and the need for openness of the tasks, are also reflective and in line with the Singapore Mathematics Curriculum. In fact, Galbraith et al. (2010) listed six principles for identifying a "potential situation for model development":

- Principle 1: Relevance and Motivation
- Principle 2: Accessibility
- Principle 3: Feasibility of Approach
- Principle 4: Feasibility of Outcome
- Principle 5: Validity
- Principle 6: Didactical Flexibility

The participating teachers have at least been able to relate the design of a modelling task to Principles 1 and 2. In addition, the qualitative comments have touched on some other aspects of the other four Principles. Maybe a more systematic and explicit addressing of these six principles, coupled with a more generous allocation of time for professional development in these areas, could better equip teachers with the necessary skills and knowledge so as to better empower them to address the issue of mathematical modelling in the mathematics classroom.

It must also be pointed out that the receptivity of the teachers, as reflected by their responses to the survey and during casual discussions, was indeed a welcoming surprise. The lack of experience in mathematical modelling and the many initiatives that the Singapore mathematics teachers need to address (Wong and Lee 2009), often result in teachers bringing a skeptical view towards new approaches. One could, of course, attribute the positive outcome to the dynamism, charisma and effectiveness of some of the facilitators of the sessions with one presenter described as enlightening on both mathematical modelling and problem posing. Some participants also expressed their ability to link the materials presented at the sessions with real classroom situations as the tasks discussed were also being implemented simultaneously with participating students in the MMO in a classroom context. Participating teachers could, in between teacher workshop sessions, visit these classes to observe how students reacted and approached the modelling tasks. Could these demonstration classes lend more credibility to the theories behind such new pedagogical approaches to the teachers? Should Singapore teachers' professional development courses be structured around more demonstration classes? Such considerations, rather than simply making available more in-service courses on mathematical modelling might be needed to effect the "radical change on the part of teachers" that Bonotto (2010, p. 406) described.

References

Balakrishnan, G., Yen, Y. P., & Goh, E. (2010). Mathematical modelling in the Singapore secondary school mathematics curriculum. In B. Kaur & J. Dindyal (Eds.), *Mathematical applications and modelling* (pp. 247–257). Singapore: World Scientific.

Bonotto, C. (2010). Realistic mathematical modelling and problem posing. In R. Lesh, P. L. Galbraith, C. R. Haines, & A. Hurford (Eds.), *Modeling students' mathematical modeling competencies (ICTMA 13)* (pp. 399–406). New York: Springer.

Chan, C. (2013). Initial perspectives of teacher professional development on mathematical modelling in Singapore: Conceptions of mathematical modelling. In G. Stillman, G. Kaiser, W. Blum, & J. Brown (Eds.), *Teaching mathematical modelling: Connecting to research and practice* (pp. 405–413). New York: Springer.

Galbraith, P. L., Stillman, G., & Brown, J. (2010). Turning ideas into modeling problems. In R. Lesh, P. L. Galbraith, C. R. Haines, & A. Hurford (Eds.), *Modeling students' mathematical modeling competencies (ICTMA 13)* (pp. 133–144). New York: Springer.

Kaiser, G., Lederich, C., & Rau, V. (2010). Theoretical approaches and examples for modelling in mathematics education. In B. Kaur & J. Dindyal (Eds.), *Mathematical applications and modelling* (pp. 219–246). Singapore: World Scientific.

Lee, N. H. (2010). The role and nature of curriculum frameworks in mathematics curriculum development Initiatives. In Y. Shimizu, Y. Sekiguchi, & K. Hino (Eds.), *Proceedings of the 5th East Asian Regional Conference on Mathematics Education (EARCOME5): In search of excellence in mathematics education* (pp. 607–614). Tokyo: Japan Society of Mathematics Education.

Lim-Teo, S. K. (2009). Mathematics teacher education: Pre-service and in-service programmes. In K. Y. Wong, P. Y. Lee, B. Kaur, P. Y. Foong, & S. F. Ng (Eds.), *Mathematics education: The Singapore journey* (pp. 48–84). Singapore: World Scientific.

Ministry of Education. (1990a). *Mathematics syllabus (lower secondary)*. Singapore: Ministry of Education.

Ministry of Education. (1990b). *Mathematics syllabus (primary)*. Singapore: Ministry of Education.

Ministry of Education. (2000a). *Mathematics syllabus – Lower secondary*. Singapore: Ministry of Education.

Ministry of Education. (2000b). *Mathematics syllabus – Primary*. Singapore: Ministry of Education.

Ministry of Education. (2006a). *A guide to teaching and learning of O-Level mathematics*. Singapore: Ministry of Education.

Ministry of Education. (2006b). *Mathematics syllabus – Primary, 2007*. Singapore: Ministry of Education.

Ministry of Education. (2009). *The Singapore model method for learning mathematics*. Singapore: Panpac Education.

Ministry of Education. (2012). *O-Level mathematics teaching and learning syllabus*. Singapore: Ministry of Education.

Muller, E., & Burkhardt, H. (2006). Applications and modelling for mathematics – Overview. In W. Blum, P. L. Galbraith, H. W. Henn, & M. Niss (Eds.), *Modelling and applications in mathematics education: The 14th ICMI study* (pp. 267–274). New York: Springer.

Ng, K. E. D. (2013). Initial perspectives of teacher professional development on mathematical modelling in Singapore: A framework for facilitation. In G. Stillman, G. Kaiser, W. Blum, & J. Brown (Eds.), *Teaching mathematical modelling: Connecting to research and practice* (pp. 427–436). New York: Springer.

Niss, M. (2010). Modelling a crucial aspect of students' mathematical modeling. In R. Lesh, P. L. Galbraith, C. R. Haines, & A. Hurford (Eds.), *Modeling students' mathematical modeling competencies (ICTMA 13)* (pp. 43–59). New York: Springer.

Polya, G. (1957). *How to solve it?* (2nd ed.). New York: Penguin.

Reed, S. K. (1999). *Word problems: Research and curriculum reform*. Mahwah: Lawrence Erlbaum.

Soh, C. K. (2005, November). *An overview of mathematics education in Singapore*. Paper presented at the 1st International Mathematics Curriculum Conference, Chicago, USA.

Stillman, G. (2010). Implementing applications and modelling in secondary school: Issues for teaching and learning. In B. Kaur & J. Dindyal (Eds.), *Mathematical applications and modelling* (pp. 300–322). Singapore: World Scientific.

Wong, K. Y., & Lee, N. H. (2009). Singapore education and mathematics curriculum. In K. Y. Wong, P. Y. Lee, B. Kaur, P. Y. Foong, & S. F. Ng (Eds.), *Mathematics education: The Singapore journey* (pp. 13–47). Singapore: World Scientific.

Yeap, B. H. (2002). *Relationship between children's mathematical word problem posing and grade level, problem-solving ability and task type*. Unpublished PhD Thesis, Nanyang Technological University, Singapore.

Chapter 36
Initial Perspectives of Teacher Professional Development on Mathematical Modelling in Singapore: A Framework for Facilitation

Kit Ee Dawn Ng

Abstract Challenges exist for the incorporation of modelling tasks in Singapore mathematics classrooms particularly due to the open-ended, interdisciplinary nature of such tasks. This chapter presents a preliminary teacher facilitation structure developed for initial professional development in an outreach programme for mathematical modelling in Singapore primary and secondary schools. Findings on teacher readiness in facilitating students' mathematical thinking, reasoning, and communication during mathematical modelling will be discussed.

1 Introduction

Proponents of mathematical modelling have long argued for the inclusion of modelling tasks in school curricula (English 2009; Kaiser and Maaß 2007; Lesh and Doerr 2003; Stillman et al. 2008) to promote connections between school-based mathematics and real-world problem solving. Mathematical modelling refers to the process of formulating and improving a mathematical model to represent and solve real-world problems (CPDD 2012). Although mathematical modelling was introduced in the Singapore mathematics curriculum in 2003, more concerted efforts at teacher education for task design, facilitation, and evaluation of learning occurred only in recent years (CPDD 2012; Ng 2013). As Stillman (2010) observed, the teacher factor is one crucial condition for successful incorporation of modelling activities in schools. Three main challenges exist among teachers when promoting the use of modelling activities in Singapore mathematics classrooms.

K.E.D. Ng (✉)
Mathematics and Mathematics Education Academic Group,
National Institute of Education, Nanyang Technological University, Singapore
e-mail: dawn.ng@nie.edu.sg

G.A. Stillman et al. (eds.), *Teaching Mathematical Modelling: Connecting to Research and Practice*, International Perspectives on the Teaching and Learning of Mathematical Modelling, DOI 10.1007/978-94-007-6540-5_36,

Firstly, there appears to be confusion among teachers with respect to what mathematical modelling is. Mathematical modelling is often mistakenly interpreted as the Model Drawing Approach, which is often used as a problem solving heuristic and a teaching approach at primary and lower secondary levels (Chan 2013). In addition, there is also a need to clarify differences between applications and modelling tasks (Stillman et al. 2008). Applications are used when the teacher looks for possible real-world contexts where specific taught mathematical knowledge and skills can be applied. For many years, Singapore teachers have designed problem solving activities revolving around applications. A modelling task on the other hand begins with the real-world context where various mathematical knowledge and skills can be elicited to derive models representative of the context for problem solving purposes. Singapore teachers may also find it challenging to identify and subsequently weave the spectrum of possible mathematical knowledge to be drawn from the open-ended modelling task in their facilitation.

Secondly, Singapore teachers and students are bound by traditional high stakes assessment systems. These bring about constraints in time management and the nature of mathematical learning activities.

Thirdly, there is a lack of teacher readiness in implementing modelling tasks for several reasons. Singapore schools have limited, sporadic exposure to modelling activities. Modelling tasks often require problem posing, justification and articulation of decision making factors in model development and validation by students as well as critique processes to be displayed by their teachers. These challenge the mind sets and comfort levels of Singapore teachers who are generally used to prescriptive teacher-centred approaches where problems with a limited selection of approaches all converging towards an answer are discussed. Furthermore, it has been ingrained in many Singapore teachers that mathematics is about algorithms, working out, solution steps, and one final answer all occurring in a linear sequential manner. This has narrowed the views of what can comprise appropriate mathematical models (Ng 2010) with resulting implications on facilitation.

These challenges provided impetus for the development of a framework to guide teachers towards facilitation of modelling tasks hoping to bring about a pedagogical shift in approach towards a more student-centred mode but within the comfort levels of teachers who are used to prescriptive forms of instruction. This chapter is the third and last in a series of reports in this book on initial perspectives of teacher professional development in Singapore. It presents a facilitation framework for modelling in Singapore classrooms and exemplifies its use when teacher educators scaffolded teachers in their preliminary attempts as modellers in the second session of a "mathematical modelling outreach" 3-day immersion programme (see Lee 2013). Teacher readiness at facilitating modelling in their own classrooms will be examined from data collected in an exploratory study conducted during the programme, with implications drawn for future teacher education.

2 Theoretical Perspectives of Modelling for Facilitation Framework

Mathematical modelling in the context of the Singapore curriculum appears to reflect two perspectives proposed by Kaiser and Sriraman (2006): realistic modelling and contextual modelling. The former views modelling as having pragmatic-utilitarian goals to promote the solving of real-world problems for relevance of school mathematics and to nurture competencies related to the modelling process. Modelling is taken as an activity to solve authentic real-world problems from industry and science, not as the development of mathematical theories (e.g., Kaiser and Maaß 2007). The latter perspective, however, perceives models as "purposeful conceptual systems" (Lesh 2003, p. 44) which describes, explains, or predicts real-world phenomena. Here, the mathematisation of reality is placed in the foreground during modelling tasks. When mathematising a situation, students devise or select appropriate symbolic mathematical representations (i.e., model development) for the given real-world situation in meaningful ways, engaging in processes like constructing, explaining, justifying, predicting, conjecturing, and representing. This approach is espoused to contrast textbook-based mathematics where students try to make sense of symbolically described situations. Hence, an important goal of the contextual modelling perspective is to facilitate students to develop, critique, and validate their own models of mathematical representations for identified purposes based on the assumptions and conditions set in relation to specific real-world contexts. English (2003) and Lesh and Doerr (2003) posit the use of model-eliciting activities to achieve the stated goal. Model-eliciting activities focus on the usefulness of models and their development in the given situations and beyond.

The two theoretical perspectives of modelling set the foundation underlying the development of the facilitation framework. An objective of the immersion programme was for teacher participants to experience the entire journey of being modellers guided by teacher educators demonstrating the use of the facilitation framework.

3 A Framework for Facilitating the Modelling Process

The framework (Fig. 36.1) consists of six phases which are articulated based on Polya's four stages (1957) of problem solving process: (a) understand the problem, (b) plan, (c) implement plan, and (d) reflect. Singapore mathematics teachers are generally familiar with Polya's work as part of their pre-service education as well as their teaching repertoire articulated in the syllabus documents. Thus, it was a conscious decision to anchor the facilitation framework on Polya's stages.

Each phase in the facilitation framework is marked by focused questions aligned to the development of modelling competencies and crucial modelling elements highlighted by experts. Houston (2007) emphasised the importance of looking

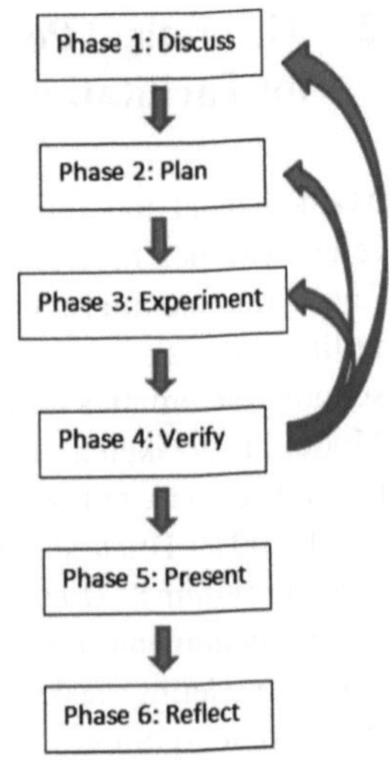

Fig. 36.1 A preliminary facilitation framework for scaffolding student learning during modelling tasks

holistically into students' overall sense making of the real-world situation presented in the problem and their efforts at generating a mathematical model. The facilitation framework in Fig. 36.1 allows for time spent discussing the problem in order to make sense of it. The framework also provides opportunities to embark on repeated attempts to derive an appropriate model for the problem situation. Responding to calls for nurturing metacognitive competencies during modelling espoused by Stillman (2009), the framework provides platforms for developing student's ability to reflect critically about their completed modelling attempts. Lesh (2003) and English (2003) brought out key elements in the modelling process echoed by Balakrishnan et al. (2010): (a) mathematisation of the problem, (b) the explicit statements of assumptions and conditions during the interpretation of the problem, (c) validity checking of the derived model, and (d) reflections about the applicability of the model in other situations. These elements are included in the framework.

Phase 1 of the framework, "discuss", is translated from Polya's first stage of "understanding the problem". The term "discuss" was specifically chosen to highlight the importance of communication and group consensus during the negotiation of meaning when developing understanding of the given real-world problem. In this phase, important modelling competencies (Niss et al. 2007) such as the following are developed: ability to perform problem posing, goal setting, defining key ideas, selecting relevant information, and identifying variables. This phase simultaneously elicits the use of processes such as explaining, describing, conjecturing, justifying, and comparing while assumptions based on interpretations are stated.

"Plan", in Phase 2 of the framework, replicates the second stage of Polya's structure because modelling encourages group collaborations. Careful planning and organisation of approaches related to set conditions of investigations during the task may well be instrumental to positive group experiences.

Phase 3 is referred to as "experiment" because it scaffolds students in devising data collection, interpretation, analysis, and representation approaches for testing of conjectures as part of preliminary model development. This phase extends Polya's "implementation" stage to encompass the fine tuning of initial macroscopic plans through sometimes repeated microscopic but focused implementation procedures guided by the goal set. Students are prompted to relook, reflect, and possibly revise their implementation attempts within this phase. An important instructional goal for teachers during this phase is to steer students towards developing meaningful symbolic mathematical representations of the given real-life problem situation (i.e., mathematisation).

Phase 4, "verify", guides students to check the accuracy of their calculations, the reasonableness of their approaches, their experimentation attempts, and the subsequent conclusions drawn based on the results obtained. It encourages students to evaluate the appropriateness of their representations (i.e., models) based on the goals set.

It is important to note the cyclical relationships between Phases 1–4 because of the need for repeated clarifications of key ideas, sharpening of focuses, and revisions of the model. By Phase 5, students finalise their models and work towards crafting a response or recommendation adhering to the requirements of the task. They should justify their responses and conclusions based on the models generated.

Lastly, in Phase 6, students reflect on the validity of their models in relation to the real-world context and comment on the appropriateness of their attempts in addressing the problem. They are also invited to discuss the possible applications of their models in other real-world scenarios by comparing the assumptions and conditions.

4 Exemplifying the Facilitation Framework

During the second session of the immersion programme, 31 teachers took on the role of modellers and worked in groups on *My Hometown Task* (for more details see Chan 2013), facilitated by two teacher educators using the framework proposed above. The goal was to recommend the most suitable town in Singapore for a family of new migrants. Although internet access was provided, the teachers were encouraged to use other sources to collect relevant information that was not given in the problem. The two facilitators implemented the framework alternating between whole class and group discussions within all phases except during Phase 5 (Present) where the groups worked on their own following brief instructions. Table 36.1 shows some of the question prompts used during each phase.

There was an initial state of excitement among the teachers. Many questions surfaced resulting from varying interpretations of the task. Each group was encouraged to find a common understanding of key terms related to the task such as "town", "migrants", and "suitability". They also had to decide what comprises "a family of migrants". Different variables affecting the "suitability" of a town and hence different sets of criteria for ranking the towns were explored. Information was collected on the needs of new migrants. The teachers applied arithmetic concepts such as percentage and fractions as well as various forms of visual data representations to

Table 36.1 Question prompts

Phase	Prompts
Discuss	Can your team state the goal clearly in a problem statement?
	What information does your team need to help determine the "suitability" of a town for a family of migrants? Were there any assumptions made?
Plan	What criteria has your team chosen to determine the "suitability" of a town?
Experiment	Using mathematics, how would your team explain the approach to determine the "suitability" of a town?
Verify	How does your team check the reasonableness of the mathematical representations?
	Does your team need to relook at the problem and form new plans of approach? Why?
Present	Which town does your team recommend to the editor of *Good Living* magazine?
	How would your team support the recommendations using mathematics?
Reflect	How successful is your team in addressing the requirements of the problem?
	Can your team use the same mathematical representations to choose the most suitable town for other residents of Singapore? Why?

support their recommendations. Some used measures of central tendencies on the identified variables (e.g., family income, expected monthly expenditure) as part of their ranking decisions. There were intense discussions on how they would determine the validity and applicability of their models under differing assumptions and conditions.

5 Teacher Readiness in Facilitating Modelling Tasks in Singapore Classrooms

Teacher readiness in facilitation was deemed a crucial aspect of investigation in this exploratory study (for details and methodology, see Lee 2013) as findings would help inform future professional development programmes towards meaningful, not just successful incorporation of modelling tasks in schools. Teacher readiness was conceptualised as having two aspects in the study in addition to teacher knowledge of facilitation phases presented in the framework: (1) *facilitation focuses* identified by the teachers themselves in view of enhancing their own repertoire and (2) teachers' awareness of *"tensions"* (i.e., concerns) they may face during facilitation.

Participating teachers were given a brief survey at the end of the session to gather feedback on their (a) facilitation focuses when incorporating modelling tasks in their own classrooms in view of their existing repertoire, (b) concerns about facilitation, and (c) key learning points from the framework. These were analysed qualitatively. Written comments were sought about the usefulness and limitations of the framework. Twenty-six of 31 participants completed the survey voluntarily. They also rated whether they had gained some good teaching points for facilitating modelling tasks on a Likert scale from 1 to 5 (representing a very positive learning experience). All except one respondent indicated productive learning experiences (i.e., scores of 4 and 5 on the Likert scale) during the session.

5.1 Facilitation Focuses

The teachers were asked to identify focus areas they would like to embark on when facilitating modelling tasks. Most listed *questioning techniques* as one of the key focuses in their facilitation. They indicated they wanted to ask the right questions at the right time so as not to be too prescriptive in their question prompts. The right questions address the entry levels of the students during the task and guide them further in their investigations without "helping too much". Another important focus was *task design*, as the teachers saw this as playing a large role in their facilitation pathways. There was a clear understanding that a modelling task should be open-ended and meaningful to the students. Some proposed using simple tasks of direct relevance to the daily lives of their students. Others wanted to help students gain a good understanding of the context presented as well as the tasks requirements. A few teachers highlighted the importance of guiding students in problem posing and goal setting. A third focus was on *mathematics*. The teachers would specifically like to help draw out the mathematics from the tasks and subsequently bridge students' approaches to provide quality mathematical representations based on their entry levels.

5.2 Facilitation Concerns

The teachers' concerns about facilitation were related to their perceived focuses in facilitation described above. Of most concern was how to achieve *a "balance" of scaffolding* during their interactions with students so as to maintain sufficient student-ownership of the task to promote enriching learning experiences. The teachers were conscious of being over-prescriptive during facilitation. At least two teachers inquired about how much guidance should be given to lower-ability students so they could still seek and expand their knowledge in an independent way. Another concern high on the priority list was the *tension* between providing just-in-time guidance through appropriate questioning techniques and presenting students with ready-made pathways of investigations when they are stuck. At least two teachers from the session indicated over-reliance of students on teacher help as an undesirable outcome of this tension. Also of concern was the *scaffolding of mathematical learning* from the task. The open-ended nature of the modelling task elicits varying mathematical concepts and skills and hence, the adequacy of students' pre-requisite knowledge for the task will be crucial. The teachers were concerned with how they could ensure that mathematical learning from the existing pre-requisites could be realised for different pockets of students during their facilitation. *Classroom management* during modelling activities was mentioned as another concern by teachers. They were worried about engaging students throughout the task and keeping discussions on task. Lastly, some teachers wondered whether mathematical modelling would form part of school *assessment* requirements. Such a decision from the Singapore Ministry of Education would be seen as having wide implications for the use of modelling tasks in mathematics classrooms.

5.3 Learning Points from Framework

The data implied that the framework had some impact on teachers' facilitation repertoires in at least three areas. Firstly, the phases of the modelling process that students could be brought through during the task as articulated in the framework provided a structure for teachers new to modelling. Secondly, the importance of questioning for thinking, mathematical reasoning and justification was emphasised when the framework was exemplified. Thirdly, the framework prompts modellers to reflect on their modelling process. Nonetheless, some teachers indicated the framework alone was insufficient to guide them through the subtle nuances of facilitating modelling, feeling they needed to be able to adapt the questions pertaining to the phases to different tasks.

6 Discussion and Implications

It appears that more can be done to ease teacher tension with respect to incorporating modelling tasks in Singapore mathematics classrooms. At this stage, whether these tasks are used and whether the attempts at these tasks are formally assessed are still decided by schools. The purpose of providing a facilitation framework for modelling tasks as presented in this chapter was a first step by teacher educators to help Singapore teachers, who are used to prescriptive instructional approaches, structure their modelling lessons. As aptly pointed out by the teachers in the study, the framework alone does not adequately inform teachers about the nature of scaffolding of mathematical learning during modelling activities. Appropriate key questions have to be asked in each phase of the framework at the right time to the right students. In addition, how this framework is used in actual classroom situations will vary depending on the teaching experiences and confidence levels of teachers. This may impact the quality of mathematics presented in the final outcomes of the modelling tasks.

Feedback from the teachers with respect to their focuses and concerns about facilitating modelling tasks echoes previous findings of Ng (2011); there needs to be a balance of scaffolding during open-ended tasks where the teacher has to be aware of when to step in and when to draw away to encourage thinking and promote competencies. Lingefjärd and Meier (2010) described the "theory of teacher as a manager" of the modelling process where the focus is on "adaptive teacher intervention" which consists of "organizational, metacognitive or content-related support" (p. 95). This means the teacher will have to know when to step in, how to respond to students' queries without being prescriptive, how to invoke students' thinking and prod them along their decision making path by making explicit justifications of the choices made. Such pedagogical repertoire would be enhanced with experience over time. In addition, the teachers' focuses and concerns during facilitation of modelling tasks suggest the development of the following two key competencies in teacher education: (a) questioning skills and (b) fostering a modelling climate.

The nature, purpose, and timing of questioning were mentioned by teachers as important considerations when scaffolding students' learning in modelling activities. Teachers have to know when and how to ask "diagnostic questions" (Lingefjärd and Meier 2010, p. 105) to assess students' progress and help them reflect on their thinking, approaches, and decisions. Such diagnostic questions include "framing" questions where teachers help students sharpen their focus and work towards an intermediate answer and "clearing frame" questions where the teacher explicitly questions towards a specific solution pathway initiated by the students.

Stillman (2010) emphasised the need to establish a modelling climate before embarking on modelling activities. This climate is co-created by teachers and students because there will be common expectations set within the classroom for modelling activities. To do this, both teachers and students have to undergo some mind-set changes towards mathematics teaching and learning, working towards a more student-centred approach. This involves time spent in class to listen to students as they interpret and explain their models (Doerr 2007).

References

Balakrishnan, G., Yen, Y. P., & Goh, E. (2010). Mathematical modelling in the Singapore secondary school mathematics curriculum. In B. Kaur & J. Dindyal (Eds.), *Mathematical applications and modelling* (pp. 247–257). Singapore: World Scientific.

Chan, C. M. E. (2013). Initial perspectives of teacher professional development on mathematical modelling in Singapore: Conceptions of mathematical modelling. In G. Stillman, G. Kaiser, W. Blum, & J. Brown (Eds.), *Teaching mathematical modelling: Connecting to research and practice* (pp. 417–425). New York: Springer.

Curriculum Planning and Development Division [CPDD]. (2012). *Mathematical modelling resource kit*. Singapore: Ministry of Education.

Doerr, H. M. (2007). What knowledge do teachers need for teaching mathematics through applications and modelling? In W. Blum, P. Galbraith, H. W. Henn, & M. Niss (Eds.), *Modelling and applications in mathematics education: The 14th ICMI study* (pp. 69–78). New York: Springer.

English, L. D. (2003). Mathematical modeling with young learners. In S. J. Lamon, W. A. Parker, & K. Houston (Eds.), *Mathematical modelling: A way of life* (pp. 3–18). Chichester, UK: Horwood.

English, L. (2009). Promoting interdisciplinarity through mathematical modelling. *ZDM – The International Journal on Mathematics Education, 41*(1–2), 161–181.

Houston, K. (2007). Assessing the "phases" of mathematical modelling. In W. Blum, P. Galbraith, H. W. Henn, & M. Niss (Eds.), *Modelling and applications in mathematics education: The 14th ICMI study* (pp. 249–256). New York: Springer.

Kaiser, G., & Maaß, K. (2007). Modelling in lower secondary mathematics classroom – Problems and opportunities. In W. Blum, P. Galbraith, H. W. Henn, & M. Niss (Eds.), *Modelling and applications in mathematics education: The 14th ICMI study* (pp. 99–108). New York: Springer.

Kaiser, G., & Sriraman, B. (2006). A global survey of international perspectives on modelling in mathematics education. *ZDM – The International Journal of Mathematics Education, 38*(3), 302–310.

Lee, N. H. (2013). Initial perspectives of teacher professional development on mathematical modelling in Singapore: Problem posing and task design. In G. Stillman, G. Kaiser, W. Blum, & J. Brown (Eds.), *Teaching mathematical modelling: Connecting to research and practice* (pp. 405–415). New York: Springer.

Lesh, R. A. (2003). How mathematizing reality is different from realising mathematics. In S. J. Lamon, W. A. Parker, & K. Houston (Eds.), *Mathematical modelling: A way of life – ICTMA 11* (pp. 38–52). Chichester: Horwood.

Lesh, R., & Doerr, H. M. (2003). Foundations of a models and modeling perspective on mathematics teaching, learning, and problem solving. In R. Lesh & H. M. Doerr (Eds.), *Beyond constructivism: Models and modeling perspectives on mathematics problem solving, learning, and teaching* (pp. 3–33). Mahwah: Lawrence Erlbaum Associates.

Lingefjärd, T., & Meier, S. (2010). Teachers are managers of the modelling process. *Mathematics Education Research Journal, 22*(2), 92–107.

Ng, K. E. D. (2010). Initial experiences of primary school teachers with mathematical modelling. In B. Kaur & J. Dindyal (Eds.), *Mathematical modelling and applications: Yearbook of Association of Mathematics Educators* (pp. 129–144). Singapore: World Scientific.

Ng, K. E. D. (2011). Mathematical knowledge application and student difficulties in a design-based interdisciplinary project. In G. Kaiser, W. Blum, R. Borromeo Ferri, & G. Stillman (Eds.), *International perspectives on the teaching and learning of mathematical modelling: Trends in the teaching and learning of mathematical modelling* (Vol. 1, pp. 107–116). New York: Springer.

Ng, K. E. D. (2013). Teacher readiness in mathematical modelling: Are there differences between pre-service and experienced teachers? In G. Stillman, G. Kaiser, W. Blum, & J. Brown (Eds.), *Teaching mathematical modelling: Connecting to research and practice* (pp. 339–348). New York: Springer.

Niss, M., Blum, W., & Galbraith, P. L. (2007). Introduction. In W. Blum, P. L. Galbraith, H.-W. Henn, & M. Niss (Eds.), *Modelling and applications in mathematics education* (pp. 3–32). New York: Springer.

Polya, G. (1957). *How to solve it.* Garden City: Doubleday.

Stillman, G. (2009, June 5). *Fostering metacognitive competencies during beginning mathematical modelling tasks.* Paper presented at the seminar for Mathematics and Mathematics Education Academic Group, National Institute of Education, Singapore.

Stillman, G. A. (2010). Implementing applications and modelling in secondary school: Issues for teaching and learning. In B. Kaur & J. Dindyal (Eds.), *Mathematical applications and modelling: Yearbook 2010 of the Association of Mathematics Educators* (pp. 300–322). Singapore: World Scientific.

Stillman, G., Brown, J., & Galbraith, P. L. (2008). Research into the teaching and learning of applications and modelling in Australasia. In H. Forgasz et al. (Eds.), *Research in mathematics education in Australasia: New directions in mathematics and science education* (pp. 141–164). Rotterdam: Sense.

Chapter 37
Teacher Professional Development on Mathematical Modelling: Initial Perspectives from Singapore

Vince Geiger

Abstract In this chapter, I will provide commentary on the symposium dedicated to teacher professional development on mathematical modelling in Singapore which was based on papers by: Chan, *Teacher professional development on mathematical modelling: Conceptions of mathematical modelling*; Lee, *Initial perspectives from Singapore: Problem posing and task design*; and Ng, *Teacher professional development on mathematical modelling: Facilitation and scaffolding*. Across these three themes emerge – the importance for teachers to understand the nature of mathematical modelling; the need to acknowledge the interconnection between teaching, learning and assessment; and the influence of teacher dispositions toward designing modelling tasks. I conclude this commentary by offering observations on the research designs employed in these studies.

1 Understand the Nature of Modelling

As the corpus of literature related to what knowledge teachers require to teach mathematics through applications and modelling indicates that subject matter knowledge alone is not enough (Doerr 2007), what teachers and students need to know and understand in addition to the relevant mathematics to solve a problem is a vital question when attempting a system wide change to teachers' instructional practices. Chan argues that one of the challenges associated with the implementation of a new mathematics curriculum that includes mathematical modelling and applications, is a lack of clarity, in the minds of teachers, about the nature of mathematical modelling. This lack of clarity appears to stem from the term used for a

V. Geiger (✉)
Education QLD | Faculty of Education,
Australian Catholic University, Brisbane, QLD, Australia
e-mail: vincent.geiger@acu.edu.au

G.A. Stillman et al. (eds.), *Teaching Mathematical Modelling: Connecting to Research and Practice*, International Perspectives on the Teaching and Learning of Mathematical Modelling, DOI 10.1007/978-94-007-6540-5_37,

heuristic employed in the well established approach to teaching problem solving in Singaporean schools – *draw a model of a problem situation*. Thus, many teachers believe they have already incorporated the modelling and applications element of the new curriculum because they were incorporating the similarly named heuristic when teaching problem solving.

In order to understand and then describe the range of teachers' beliefs about an understanding of mathematical modelling, Chan surveyed 27 teachers who attended a professional learning event – the Mathematical Modelling Outreach (MMO). From this survey, Chan concludes that teachers were able to develop an understanding of the important features of mathematical modelling after only one day's engagement at the MMO. This seems a remarkable turn around since most previous research (see, e.g., Kaiser and Maass 2007) suggests that teachers' mathematical beliefs and understanding are difficult to reconstruct.

While we must regard this finding with some caution, as I believe there is some potential for ambiguity in the survey items used to determine this finding, none-the-less, participating teachers appeared to display a positive attitude towards introducing mathematical modelling tasks with 83 % agreeing or strongly agreeing that such activity had the potential to benefit their students' mathematics education. These teachers saw the major benefits for students as the meaningfulness brought to teaching and learning through teaching in real-life contexts, and as motivation for student learning by engaging with interesting applications of mathematics. These are important conceptions for teachers to hold if they are to engage with modelling approaches to teaching mathematics (Chapman 2007). Time and the constraints of the syllabus were identified as limiting factors in any attempt to include mathematical modelling activities as part of the mainstream business of school mathematics classrooms. In conclusion, Chan suggests that teachers in the program developed a clear understanding of mathematical modelling through the MMO as they were able to identify the modelling stages of representation of relationships, experimentation, verification and generalisation. However, it is unclear if Chan established whether the teachers also understood the cyclic nature of the mathematical modelling process.

Lee also raises the issue of teachers' understanding of the modelling process. He asserts that teachers are familiar with the ideas of mathematical problem solving because of its inclusion in syllabus documents since the 1990s (problem solving is the primary aim of the mathematics curriculum) but they are less familiar with mathematical problem posing – which he argues is an essential part of the mathematical modelling process. An additional factor that may have delayed the development of teachers' understanding on mathematical modelling is that, while applications and modelling were introduced into the Singaporean Mathematics Curriculum Framework in 2003, this element of the Framework was only reflected in syllabus documents, from which teachers develop their teaching and learning plans, since 2007. Further, despite a requirement that Singaporean teachers complete 100 hours of professional development per year, Lee could find no record of professional development opportunities for teachers in mathematical modelling and applications until 2009.

Ng also reports that there appears to be confusion among teachers in regards to the nature of mathematical modelling. She poses a finer grained view of this problem by identifying the lack of understanding by teachers of the distinction between application and modelling tasks. This, in Ng's view, has impacted upon the degree of implementation of the syllabus intent in relation to mathematical modelling and applications as some teachers believe they have met requirements by engaging students in application activities alone. This means students are usually provided with a model of a real-life context and asked to make use of it to solve a problem and so are denied the opportunity to develop their own models. Ng identifies the capacity to "identify and weave the spectrum of possible mathematical knowledge and skills to be elicited from the open-ended nature of a modelling task in their teaching" (p. 428) as a significant challenge for Singaporean teachers in attempting to include effective approaches to teaching mathematical modelling in their classrooms.

In offering a possible way forward with this challenge, Ng has attempted to connect modelling with aspects of the problem solving with which teachers are already familiar, that is, Polya's four stage plan for problem solving (a) understand the problem, (b) plan, (c) implement the plan, and (d) reflect, as she sees this as a way to encourage teachers to include application tasks that have an element of open-endedness to them. This in turn, she believes, will encourage teachers towards including more modelling tasks rather than simply tasks based on applications of mathematics.

2 The Interconnection Between Teaching, Learning and Assessment

All three authors raise the issue of the interconnection of teaching, learning and assessment as a major influence in the implementation of the applications and modelling element of the Singaporean Mathematics Curriculum Framework. This is an indication of the important role teachers play in the development of students' modelling competencies (Blum 2011) and of the influence of assessment regimes on classroom teaching practices (Stillman 2010; Stillman and Galbraith 2009).

Lee comments that the nature of open-ended modelling tasks requires problem posing, justification, and articulation of decision making factors in model development. This process also challenges teachers to develop the capacity of offering relevant critique of students' work "in progress". He argues that this challenge points teachers in a very different direction from their current norms of practice which are often prescriptive, teacher-centred and textbook-based.

Ng points out that most teachers listed questioning techniques as one of the key areas they wished to develop in relation to mathematical modelling while at the MMO. The challenge, as teachers saw it, was to ask the right questions at the right time so as not to be too prescriptive in their question prompts. Teachers want to achieve a balance between scaffolding students in a way that they could begin a task while not providing a level of assistance that would trivialise a task. This was particularly pertinent when students encountered blockages to this forward progress through a task. Teachers saw this as a tension that was difficult to resolve. They also pointed out the importance of making the mathematics embedded within a modelling or application task explicit to students. Both of these issues are taken up by Blum (2011) who argues that it is important for teachers to provide sufficient freedom for students to explore their own solutions and not have them imposed by the teacher. In order to be confident in managing such an approach, it is necessary for teachers to have a knowledge of the cognitive demands of a task in order that they have the capacity to see potentially valid solutions different from their own.

Assessment is a critical issue in relation to the inclusion of modelling activities in mathematics instruction as it can both drive and constrain curriculum change (Stillman 2010; Stillman and Galbraith 2009). While it was not a focus of his study, Chan draws attention to the influence of the content and focus of high stakes examinations on the implementation of mathematical modelling. He points out that the closed-ended nature of questions on high stakes examinations sit uneasily, in the minds of some teachers, with the extended and open-ended nature of mathematical modelling tasks. Thus, they view modelling as a developmental activity, at best, that has limited value in preparing students for examinations. This is a position consistent with other curriculum contexts across the world. Antonius (2007), for example, notes that while modelling has been a key element of the Danish curricula for more than a decade, the final examination still consists of traditional written and oral components. This has resulted in teaching not yet reflecting the goals and intentions of the curriculum as teachers tend to focus on preparing students for examinations which are not suitable for assessing open-ended modelling tasks.

Ng agrees that Singapore teachers and students appear to be bounded by traditional high stakes assessment systems which bring about constraints in time management and the nature of mathematical learning activities (Ang 2010). She points out that it is difficult for teachers to harness the potentials of modelling tasks if these tasks are not integrated into the main stream curriculum foci of teaching, learning, and assessment. Whether mathematical modelling will form a part of assessment requirements in schools is still an open question, but one that has wide ranging implications for the type of activity that will receive emphasis in school mathematics classrooms.

3 Designing for Modelling – Teacher Dispositions

As pointed out by Niss (2008), one of the key ways in which mathematics is put to use in other disciplines is in order to design, create or shape objects, systems and structures. Yet one of the challenges often identified by teachers as a constraint in the implementation of application and modelling approaches to learning mathematics is the difficulty in developing accessible but challenging tasks. The availability of resources which support context based, real world tasks was identified by Lee as a serious concern for Singaporean mathematics teachers as they attempt to implement the syllabus. This reflects on how Singaporean mathematics teachers view themselves as designers of modelling tasks and has the potential to impact on the sustainability of modelling approaches as part of a mainstream mathematics education.

In an attempt to address this issue, Lee provided teacher professional learning opportunities, as part of the MMO, that focused on the development of teachers' problem posing capabilities as a means of assisting teachers to become designers of quality modelling tasks. After this program, Lee reported that more than half of the respondents captured the key considerations of designing a modelling task, namely the need to identify a context that is relevant to the students, and to ensure a mathematical focus of the task while retaining at least a degree of openness to the problem. This was despite more than 90 % of the participants self reporting as not having implemented mathematical modelling in their classrooms and having little or no experience of attempting activities that demanded the capability to problem pose. Lee believes the success of this program was, in part, due to the opportunities which were available for participating teachers to visit groups of teachers and students, in a parallel program, to see how students reacted to and carried out the modelling tasks. This observation leads Lee to speculate that teachers' professional development courses in mathematical modelling might be best structured around more demonstration classes.

Task design was a feature that Ng noted was a focus for teachers during the MMO as they saw this as playing a large role in their capability to facilitate effective mathematical modelling activity. She observes, however, that the capability to design effective tasks is strongly related to teachers' understanding of the nature and processes of mathematical modelling. Variation in this understanding was evident through the MMO as while there was a clear understanding that a modelling task should be open-ended and meaningful to students among some teachers, only a few highlighted the importance of guiding students through the processes of problem solving and goal setting.

4 Conclusion

Widespread reform in teaching and learning practice with mathematics education is always an enormous challenge. This is even more the case when the proposed reform, in this case the introduction of open-ended, context laden approaches to the teaching and learning of mathematics in the form of applications and mathematical modelling, is of a very different nature to existing traditional, teacher-centred modes of practice. The papers presented for this symposium indicate that supporting teachers to understand, and then to adopt, these new practices will be at the heart of attempts to progress this reform. Teachers will need support, not just for implementing the processes and modes of knowing and coming to know mathematics through applications and modelling approaches, but also to develop the capability to design accessible but challenging tasks. Teachers will also need to find ways to implement the intent of the syllabus within an assessment environment that appears to be at odds with the extended nature of applications and modelling. Each of these issues is an important point of departure for future research. For such research to have an impact on practice, it must be concerned with how teachers implement new knowledge and capabilities in their classrooms in addition to how they develop themselves as teachers of mathematics through programs such as the MMO.

References

Ang, K. C. (2010). Mathematical modelling in the Singapore curriculum: Opportunities and challenges. In A. Araújo, A. Fernandes, A. Azevedo, & J. F. Rodrigues (Eds.), *Educational interfaces between mathematics and industry: Proceedings of the EIMI 2010 conference* (pp. 53–62). Lisbon/Bedford: Centro Internacional de Matemática/COMAP.

Antonius. (2007). Modelling based project examinations. In W. Blum, P. Galbraith, H. Henn, & M. Niss (Eds.), *Modelling and applications in mathematics education: The 14th ICMI study* (pp. 409–416). New York: Springer.

Blum, W. (2011). Can modelling be taught and learnt? Some answers from empirical research. In G. Kaiser, W. Blum, R. Borromeo Ferri, & G. Stillman (Eds.), *Trends in teaching and learning of mathematical modelling* (pp. 15–30). New York: Springer.

Chapman, O. (2007). Mathematical modelling in high school mathematics: Teachers' thinking and practices. In W. Blum, P. Galbraith, H. Henn, & M. Niss (Eds.), *Modelling and applications in mathematics education: The 14th ICMI study* (pp. 325–332). New York: Springer.

Doerr, H. (2007). What knowledge do teachers need for teaching mathematics through applications and modelling. In W. Blum, P. Galbraith, H. Henn, & M. Niss (Eds.), *Modelling and applications in mathematics education: The 14th ICMI study* (pp. 67–78). New York: Springer.

Kaiser, G., & Maass, K. (2007). Modelling in lower secondary mathematics classroom- problems and opportunities. In W. Blum, P. Galbraith, H. Henn, & M. Niss (Eds.), *Modelling and applications in mathematics education: The 14th ICMI study* (pp. 99–109). New York: Springer.

Niss, M. (2008). Perspectives on the balance between applications and modelling and "pure" mathematics in the teaching and learning of mathematics. In M. Menghini, F. Furinghetti, L. Giacardi, & F. Azarello (Eds.), *The first century of the International Commission on Mathematical Instruction (1908–2008): Reflecting and shaping the world of mathematics* (pp. 69–84). Roma: Instituto della Enciclopedia Italiana.

Stillman, G. (2010). Implementing applications and modelling in secondary school: Issues for teaching and learning. In B. Kaur & J. Dindyal (Eds.), *Mathematical applications and modelling* (pp. 300–322). Singapore: World Scientific.

Stillman, G., & Galbraith, P. (2009). Softly, softly: Curriculum change in applications and modelling in the senior secondary curriculum in Queensland. In R. Hunter, B. Bicknell, & T. Burgess (Eds.), *Crossing divides. Proceedings of the 32nd annual conference of the Mathematics Education Group of Australasia* (Vol. 2, pp. 515–522). Adelaide: MERGA.

Part IV
Influences of Technologies

Chapter 38
Reality Based Test Tasks with Digital Tools at Lower Secondary

Gilbert Greefrath and Michael Rieß

Abstract Some results of the German CASI-Project (Computer Algebra Systems used in lower secondary schools) in terms of test tasks with real life examples are described. The CASI-Project examines the long-term use of digital tools in mathematics teaching with Years 9 and 10 secondary students. Different tests were developed for special teaching units with digital tools. This contribution brings reality based test problems into focus. The different representations of functional relationships and especially the change between the representations graph and real situation. The resulting differences between students using handheld CAS-calculators, and control group students, using simple pocket calculators, are presented with a focus on test tasks about real situations.

1 Modelling and Digital Tools

The solution of modelling problems using digital tools requires two important translation processes to take place. Firstly, the real situation of the problem has to be understood and translated into mathematical language. The digital tool (e.g., handheld CAS-calculator), though, cannot be used before mathematical expressions have been translated into the language used by the digital tool. Thus, a special technological model has to be built. The technological results then have to be translated into mathematical expressions again. Finally, the problem can be solved by relating the mathematical results to the given real situation. Using digital tools not only broadens the range of approaches that can be taken to solve certain mathematical models, but also the types of situations that

G. Greefrath (✉) • M. Rieß
Institute for Mathematics, Education and Information Technology
Education, University of Münster, Fliednerstr. 21, 48149 Münster, Germany
e-mail: greefrath@uni-muenster.de

G.A. Stillman et al. (eds.), *Teaching Mathematical Modelling: Connecting to Research and Practice*, International Perspectives on the Teaching and Learning of Mathematical Modelling, DOI 10.1007/978-94-007-6540-5_38,

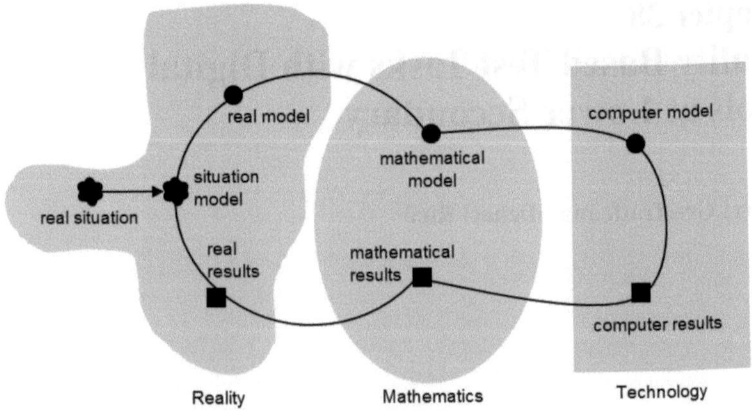

Fig. 38.1 Modelling cycle (Blum and Leiß 2007) with added technology component

can be investigated by providing the possibility of using solution strategies which would not be accessible otherwise. The use of digital tools as in Fig. 38.1 gives a restricted view of using digital tools in applications and modelling. It is also possible to use the digital tools in many phases of the modelling process (e.g., for experimenting and visualising). Therefore the use of digital tools, not only creates an important addition to the modelling cycle (see Fig. 38.1), but also influences each part of the cycle. Thus the technology is relating to the real world and mathematical world of the modelling cycle (Greefrath 2011).

2 The German School System

The findings presented are from a project conducted in German schools thus their meaning has to be evaluated considering the German school system. There are four types of secondary schools in Germany all starting at Year 5. Students showing high academic abilities during primary school attend the Gymnasium to graduate to the German Abitur, comparable to the British A-levels at Year 12. Students with moderate academic abilities attend the Realschule until Year 10 and start an apprenticeship afterwards or try to switch schools to the Gymnasium to reach Abitur later. Students showing low academic abilities in primary school attend the Hauptschule and most likely start an apprenticeship after Year 10 even though there are possibilities to switch schools and reach Abitur later. In some German states Hauptschule and Realschule are replaced by a new integrated type of school. The fourth type of school, the Gesamtschule, incorporates all three types. This is done by differentiation through special courses on various academic levels in each grade; some students learning at Gymnasium-level others at Hauptschule-level.

3 Usage of Handheld CAS-Calculators in Mathematics Education

Some empirical studies regarding the usage of handheld CAS-calculators have been undertaken in Germany, in particular the CALIMERO-Project in Lower Saxony and the M^3-Project in Bavaria. In Lower Saxony handheld CAS-calculators are employed at the Gymnasium from Years 7–10 and up to the Abitur. The Bavarian study examined students at the end of the Gymnasium (Bruder and Ingelmann 2009; Weigand and Bichler 2010). A study from Barzel (2005) showed that the results of students using handheld CAS-calculators did not fall off in quality. Ingelmann and Bruder (2007, p. 95) verified an increase in test results regarding the competence "Using symbolic, formal and technical elements of mathematics" and the big ideas *measuring* and *space and form*. Weigand and Bichler (2010) observed an improvement regarding the translation between the graph and equation of a function. Generally at the end of term, students from classes using a handheld CAS-calculator did as well in tests as the control group.

All major empirical studies regarding the use of digital tools in mathematics education in Germany took place at the German Gymnasium. An interesting aspect, which could be observed during these studies, was the particularly large increase in school achievement of the underachieving students from the Gymnasium in comparison to the examined total of students (29 classes of 6 Gymnasiums). The increase in school achievement found in this group of students was significantly greater than the average (Ingelmann and Bruder 2007). Weigand and Bichler (2010) note, in Year 10 particularly the weaker students may benefit from the use of digital tools. International meta-studies even observe substantially better test results in classes using digital tools (Ellington 2003; Hembree and Dessart 1986).

With regard to real life examples Drijvers (2005) argues, the introduction of handheld CAS-calculators facilitates handling realistic numbers and enables students to explore mathematical situations which would, due to calculation difficulties, be inaccessible without digital tools. Kendal and Stacey (2001) and Kutzler (2003) note the use of realistic data enhances the relation between mathematics and reality. In addition, handheld CAS-calculators provide new opportunities to solve real-life examples without dealing with wearisome computations, by applying graphical or numerical approaches (Lagrange 2003).

4 The CASI-Project

The Project *Computer Algebra Systems used in lower secondary schools* (CASI-Project) examines the long-term use of handheld CAS-calculators at the German Realschule and courses of the same academic level at the Gesamtschule. It aims to support, test, and examine the use of digital tools by students taught in Years 9 and 10. The focus lies on the design and evaluation of educational concepts for the use of handheld CAS-calculators by lower achieving students. At present, there are five project schools in North Rhine-Westphalia participating in the project providing a total of ten project classes, all equipped with ClassPad handheld CAS. In addition, six parallel classes from these schools serve as a control group for the competence tests. Approximately 250 project and 120 control group students participated in the project. Ten project teachers were involved not only in carrying out, but also in designing educational concepts. The project started in July 2009 and ended in June 2011 spanning two German school years.

One part of the project concept is the constant support for participating schools. During the 2 year duration of the project, five content areas (linear functions and equation systems, Pythagoras theorem, circles and the number π, quadratic functions and equations, and mathematical growth) were chosen to be of special interest and the educational concept of these was jointly planned and carried out by all project teachers. To make this possible several project meetings were held each term as well as email support at any time. The educational concept fixed for these special content areas included the definition of competencies the students are meant to achieve with digital tools, but also without any technological support at all. For this reason examination parts in which the use of any technological help is forbidden are considered obligatory whenever it makes sense. Another part of the project concept is the aim to encourage and support diverse use of digital tools. The intent is to use the handheld CAS-calculator not only to calculate, but also to experiment, visualise, create algebraic models and control (Greefrath 2010).

The test-design consists of a classical pre-test-post-test-control-group design for the competency tests with added questionnaire for the teachers in order to collect information on the prerequisites of the students. These tests take place before and after the lessons addressing the subject matter of special interest. Pre- and post-test consisted of parallel examination questions which included the newly learned knowledge in the post-test. An example for the first content area *linear functions and equation systems*, follows:

Question 2. (pre-test)	Question 2. (post-test)
Draw the graph of the function $y = 2x - 1$	Draw the graphs of the functions and determine the intersection point $y = 2x - 5$ $y = -x + 7$

Whenever reasonable the test was divided into two parts; one to be solved without technical devices and another where digital technology was allowed to be used as normal. Additionally, a third part only for project students was added to the post-tests in order to evaluate the usage of handheld CAS-calculators in a more precise way. It was possible for teachers to make the post-test a normal school examination. During the lessons on the five content areas, teachers kept a journal detailing the type and length of CAS-usage, teaching method and subject of each lesson. In addition, at three different moments (beginning of the project, after 1 year, end of the project) questionnaires dealing with attitudes towards certain aspects of mathematics, applications of mathematics, and the usage of digital tools were filled out by the students.

5 Different Types of Examination Questions Using Digital Tools

When creating examination questions many aspects should be considered. As a basic principle following Heinrich Winter, real life applications in education should only be used in sense-making real situations or if they bring advantages in understanding the problem (Henn 2007). However, due to the complexity factor examination tasks containing the whole modelling process are not possible in most cases. Consequently, we have to construct examination tasks within parts of the modelling cycle. Examination questions can be categorised into four types according to the potential use of digital tools when solving (see Table 38.1). Designing examination questions of type four proved to be rather difficult. Looking at German examinations over the last years mostly questions of types two and three are used in examinations with digital tools (Bichler 2007). In the next section we present some results from the CASI-Project regarding real-life-applications of mathematics.

Table 38.1 The four task types when using digital tools

Type	Elaboration
1	Digital tools cannot be used to solve the problem
2	Digital tools can be used, but have no potential to contribute to the solution
3	Digital tools could contribute to the solution of the examination question but their use is not required, (The *Copier Task* [post-test] can be solved by-hand, but students face difficult calculations. The calculator also provides the possibility of a graphical approach)
4	Tasks which cannot be solved without the use of digital tools (Brown 2003) (e.g., *Bridge of the Great Belt Task* (Greefrath and Mühlenfeld 2007, p. 31), part of a longer task)

Copier Task

Question 12. The old copier of the school broke down and has to be replaced. The head of school already decided to buy the model KM-C2520. Two firms offer maintenance for this model

Offer 1: 649 € per year and 0.01 € per copy

Offer 2: 749 € per year and 0.009 € per copy

Which offer is the most suitable?

Bridge Over the Great Belt Task

In mid-1998, a fixed link across the Great Belt was inaugurated. Its main part is the so-called Eastern Bridge – a road suspension bridge with a length of 6,790 m – with a free span of 1,624 m between the two pylons. The vertical clearance for ships is 65 m, at 254 m above sea level, the two pylons are the highest points on solid structures in Denmark. The lowest point of the main cable between the two pylons is at about 3 m above the road deck. (...)

The form of the cable can be approximated by the graph of the function *g* with $g(x) = a\bullet\left(e^{bx} + e^{-bx}\right)$, $a, b>0$. Determine *a* and *b*. (...)

6 Results from the CASI-Project

6.1 Translation Skills

One of the aims of the CASI-Project was to examine the influence of digital tools on the translations between described situations and graphs. This is an interesting question considering the observed improved school achievements regarding the translation between graph and algebraic expression when using digital tools (Weigand and Bichler 2010). The first observation, which has little to do with the usage of digital tools, is the importance of taking the direction of the translation into account. Figure 38.2 shows a comparison between the two directions of translations regarding the description of a situation and a graph. There is practically no

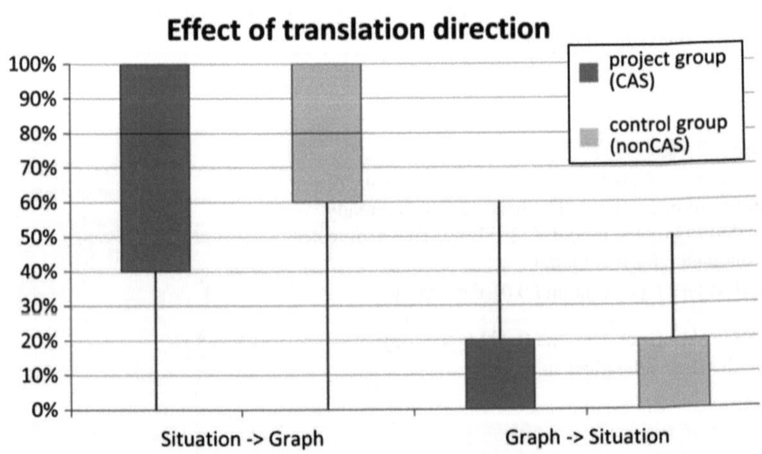

Fig. 38.2 Translations results, graph-situation description, in the context of quadratic functions

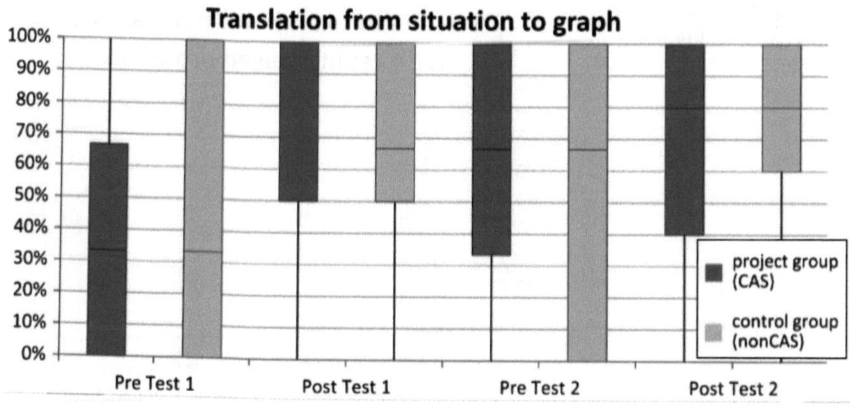

Fig. 38.3 Comparison of results on the translation from situation to graph between pre- and post-test of linear (test 1 and 2) resp. quadratic (test 3 and 4) functions and students with and without CAS

difference between the project students and the control group, but whilst one half of the students scored 80 % or more on the examination question dealing with the translation from a described situation to a graph, more than 50 % of the students did not achieve any points when translating in the other direction.

An example of an examination task on the translation from situation to graph follows:

> **Question 4.** Two water basins are connected by a pipe. The first one contains 48 l while the second one is empty. Through the pipe there is a flow of 8 l per hour from the first basin into the empty one. Draw both graphs representing the situation in the given coordinate system.

At least when the translation involves a situation, the direction has to be noted. Therefore, the translation between situations and graphs leads to two different types of tasks. Results on these show little variation regarding the difference between project students and control group, so we limit the analysis to one case. Figure 38.3 compares results of project students and control group students on four different tests during the first 18 months of the project regarding the translation from a described situation to a graph.

Although some small differences between project and control group students are noticeable if only one test is taken into account (e.g., Test 1 in Fig. 38.3), the bigger picture shows mixed results regarding the nature of these differences. The quantitative analysis is not able to show any advantage or disadvantage of the usage of digital tools for translations involving a description of a situation. The effect of the

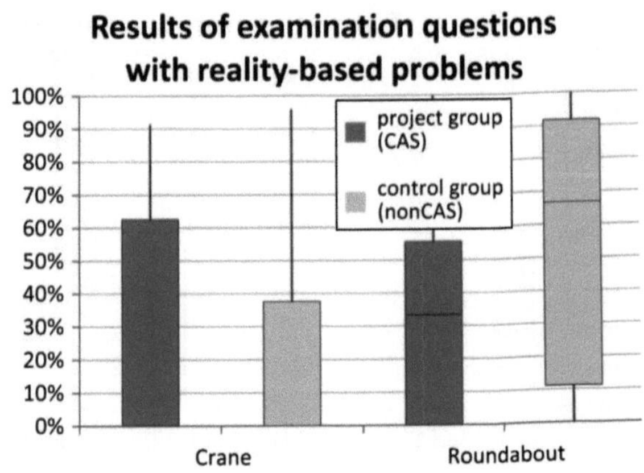

Fig. 38.4 Comparison of results on for reality-based problems

digital tools seems to be overlaid by the effect of other variables. For our project it is an important result that the use of digital tools does not lead to weaker skills of the students. Nonetheless some indicators exist that there are more subtle differences, which will be subject to qualitative studies.

6.2 *Modelling Skills in Examinations*

Questions addressing the full modelling cycle are often too complex to be treated in an examination. Nevertheless, it is possible to test certain aspects of reality-based problems. One is the *Crane Task* in which students receive a data sheet for a crane with a photograph and calculate the height the crane could reach at the maximum operating distance of 92 m above the ground. Filtering important information, considering height of the vehicle and structuring the situation are some of the modelling-related aspects students need to address. Clearly these are not sophisticated modelling problems, but involve some reality-based problem solving as well as transforming the given situation into a mathematical model. The *Roundabout Task* (Post-test on circles and π) is closer to a standard exercise whereas the *Crane Task* can be considered more challenging. Figure 38.4 compares the resulting performance of students with and without digital tools, on these two tasks.

Over half of the students were unable to score one point (out of 10) on the *Crane Task*. There was a slight advantage for project students, related to the knowledge of a similar problem involving a fire engine. Even though the project teachers shared this task with the teachers of the control group it is possible it might have been ignored. This task was very challenging for the students and solutions like the one shown in Fig. 38.5 were rare.

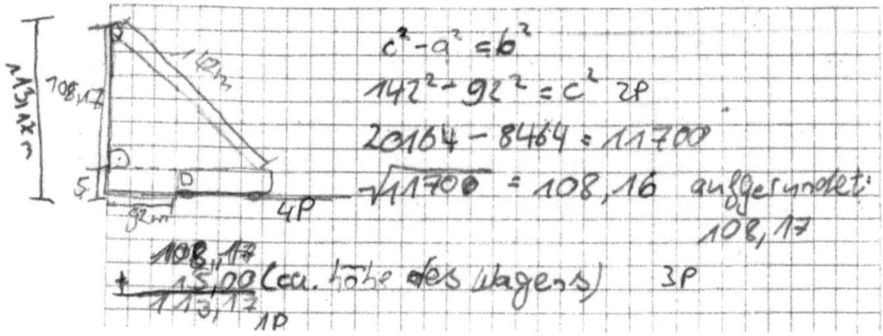

Fig. 38.5 One student's solution for the *Crane Task*

Roundabout Task

Question 4. Traffic in and around modern cities relies more and more on replacing classic junctions with traffic lights by roundabouts. This is an aerial photo of a roundabout. Naturally the asphalt of the circle is especially prone to wear and tear and has to be renewed from time to time How big is this area?

As expected, students generally did better on the *Roundabout Task* and surprisingly the control group did much better than the project students. One possible explanation might be their more intense training in recognising annuli which was agreed by project teachers to be undesirable for their students. Summing up, the qualitative results are indecisive. Note, these findings can only be applied to examination questions, whereas one of the most significant advantages in using digital tools is the possibility to deal with complex and otherwise impossible problems which clearly are not suitable for examinations.

Effects of digital tools use were visible in a more subtle way. The expectation, that students using digital tools are able to employ multiple different methods to solve a problem, proved true even in an examination environment. For example a geometrical problem which suggested the use of Pythagoras' Theorem was solved using the handheld CAS-calculator in an effective but unintended way: students drew the situation using the included dynamic geometry software and solved the problem by reading off the value in the computer-drawing. A scale drawing without digital tools would have been impossible.

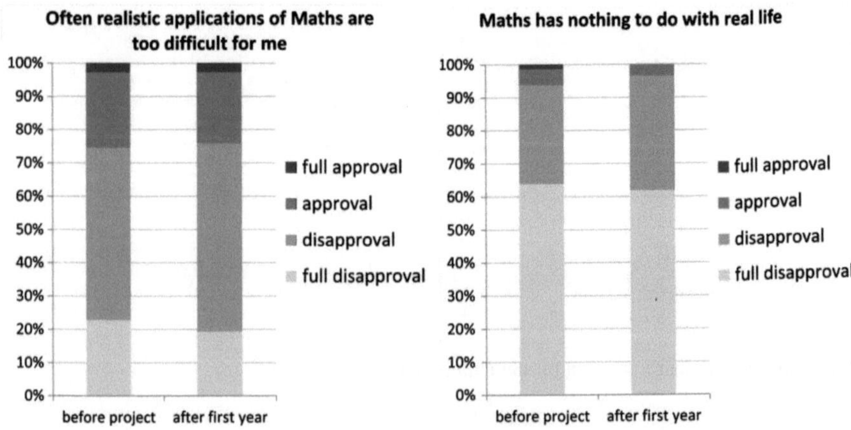

Fig. 38.6 Comparison of the students' opinions towards real applications of mathematics

6.3 *Attitude Towards Reality-Based Problems in Mathematics*

Questionnaires evaluating student attitudes towards certain aspects of mathematics, applications of mathematics and use of digital tools have been completed twice by project students: at the beginning and after the first year of the project. A four-level Likert scale was used. Figure 38.6 shows the comparison of results concerning application of mathematics in real life between these two moments. A slight trend towards a more positive attitude regarding applications of mathematics can be observed, but this is barely more than a trend to avoid giving the extreme answers. This tendency was also found for the other items.

7 Conclusion

The results produce a rather mixed picture. So far, we have been unable to duplicate the expected better test results of project students concerning translation skills involving a description of a situation. The quantitative approach seems unable to detect more subtle differences between the project students and the control group. Qualitative analysis is a promising perspective here. Reality based examination tasks showed some differences (sometimes even quite big ones) between the two groups, but these were indecisive regarding the advantages of one group over the other. The reason for these findings is not clear at this time and remains to be examined further. Methods have to take into account the different possible approaches students are now able to take when using digital tools. As mentioned, looking closely at the solutions, it was clear that students at this educational level are able to find and use these possibilities even in an examination environment.

Attitudes of the students towards reality-based problems in mathematics changed slightly, but taking into account the stronger changes in the students' attitude towards other examined aspects, this can only mean that there is a need to improve the representation of reality-based tasks when surveying the concept. However, it can be stated that the use of digital tools made more complex examination questions (e.g., *Copier Task*) available for lower secondary level students attending German Realschule or similar courses at Gesamtschule. They may still be challenging, but even diverse approaches can be verified.

References

Barzel, B. (2005). Open learning? Computeralgebra ...No time left for that.... *Zentralblatt für Didaktik der Mathematik, 37*(5), 336–342.

Bichler, E. (2007). Computer und prüfungen – geht cas auch? Erfahrungen aus dem bayrischen Modellversuch. *Beiträge zum Mathematikunterricht, 2007*, 98–101.

Blum, W., & Leiß, D. (2007). How do students and teachers deal with modelling problems? In C. Haines, P. Galbraith, W. Blum, & S. Khan (Eds.), *Mathematical modelling (ICTMA 12): Education, engineering and economics* (pp. 222–231). Chichester: Horwood Publishing.

Brown, R. (2003). Computer algebra systems and mathematics examinations: A comparative study. *The International Journal of Computer Algebra in Mathematics Education, 10*, 155–182.

Bruder, R., & Ingelmann, M. (2009). CAliMERO aus sicht der forschenden. *Mathematikunterricht (Seelze), 55*(4), 13–19.

Drijvers, P. (2005). Learning algebra in a computer algebra environment. *International Journal for Technology in Mathematics Education, 11*(3), 77–89.

Ellington, A. J. (2003). A meta-analysis of the effects of calculators on students' achievement and attitude levels in precollege mathematics classes. *Journal for Research in Mathematics Education, 34*(5), 433–463.

Greefrath, G. (2010). Mit dem computer qualitativ arbeiten? *Praxis der Mathematik i.d Schule, 52*(31), 20–24.

Greefrath, G. (2011). Using technologies: New possibilities of teaching and learning modelling – Overview. In G. Kaiser, W. Blum, R. Borromeo Ferri, & G. Stillman (Eds.), *Trends in teaching and learning of mathematical modelling: ICTMA 14* (pp. 301–304). New York: Springer.

Greefrath, G., & Mühlenfeld, U. (2007). *Realitätsbezogene aufgaben für die Sekundarstufe II*. Troisdorf: Bildungsverlag Eins.

Hembree, R., & Dessart, D. J. (1986). Effects of hand-held calculators in precollege mathematics education: A meta-analysis. *Journal for Research in Mathematics Education, 17*(2), 83–99.

Henn, H.-W. (2007). Modelling pedagogy: Overview. In W. Blum, P. Galbraith, H.-W. Henn, & M. Niss (Eds.), *Modelling and applications in mathematics education. The 14th ICMI study* (pp. 321–324). New York: Springer.

Ingelmann, M., & Bruder, R. (2007). Sinnvoller Einsatz von CAS in den Klassen 7 und 8. In Gesellschaft für Didaktik der Mathematik (Ed.), *Beiträge zum Mathematikunterricht 2007. Vorträge auf der 41. GDM Tagung für Didaktik der Mathematik* (pp. 94–97). Hildesheim: Franzbecker.

Kendal, M., & Stacey, K. (2001). The impact of teacher privileging on learning differentiation with technology. *International Journal of Computers for Mathematical Learning, 6*(2), 143–165.

Kutzler, B. (2003). CAS as pedagogical tools for teaching and learning mathematics. In J. T. Fey, A. Cuoco, C. Kieran, L. McMullin, & R. M. Zbiek (Eds.), *Computer algebra systems in secondary school mathematics education* (pp. 53–72). Reston, VA: National Council of Teachers of Mathematics.

Lagrange, J.-B. (2003). Learning techniques and concepts using CAS: A practical and theoretical reflection. In J. T. Fey, A. Cuoco, C. Kieran, L. McMullin, & R. M. Zbiek (Eds.), *Computer algebra systems in secondary school mathematics education* (pp. 269–285). Reston, VA: NCTM.

Weigand, H.-G., & Bichler, E. (2010). Symbolic calculators in mathematics lessons: The case of calculus. *International Journal for Technology in Mathematics Education, 17*(1), 3–15.

Chapter 39
On Comparing Mathematical Models and Pedagogical Learning

Janeen Lamb and Jana Visnovska

Abstract Comparing and contrasting different mathematical models of realistic situations is one way in which the relative strengths and weaknesses of these models, and the mathematics that underpins them, can become the focus of discussion in a mathematics classroom. This chapter reports on one episode from a research and development project where teachers were learning to orchestrate such classroom discussions with a view to providing opportunities for their students to apply mathematical understanding and skills in context. While the teachers did discuss a range of models they also experienced difficulty in reconciling the conflicting ideas represented in these models.

1 Introduction

Modelling and application tasks in which students are required to mathematise realistic situations and examine the usefulness of different mathematical models can be effectively used to support students' development of a variety of mathematical competencies. Niss et al. (2007) argue that "modelling problems provide contexts for coherent rather than piecemeal learning, providing a vehicle both for connecting individual pieces of mathematical knowledge and giving them purpose – a whole that is much more than the sum of the parts" (p. 21). This is evident in research where tasks based on comparisons of two data sets were successfully used to support middle years students' appreciation of different ways in which

J. Lamb (✉)
Education QLD | Faculty of Education,
Australian Catholic University, Brisbane, QLD, Australia
e-mail: Janeen.lamb@acu.edu.au

J. Visnovska
School of Education, The University of Queensland, Brisbane, QLD, Australia

G.A. Stillman et al. (eds.), *Teaching Mathematical Modelling: Connecting to Research and Practice*, International Perspectives on the Teaching and Learning of Mathematical Modelling, DOI 10.1007/978-94-007-6540-5_39,

data can be organised that ultimately led to discussions about data distributions (e.g., Cobb 1999; Cobb et al. 2003b). By making and comparing models of task situations, students developed an understanding of measures of centre and spread, and of conventional representations such as histograms, box plots, and scatter plots as tools for making sense of data distributions. This however, did not happen solely by students exploring models using a computer applet tool. The teacher's role in orchestrating classroom discussions that supported students' understandings was crucial (McClain and Cobb 2001).

Stein and colleagues (2008) outline five practices they consider crucial for facilitating productive mathematical discussions: (a) anticipating likely student solutions to cognitively demanding mathematical tasks, (b) monitoring students' responses to the tasks during the explore phase, (c) selecting particular students to present their mathematical responses during the discuss-and-summarise phases, (d) purposefully sequencing the student responses that will be displayed, and (e) helping the class make mathematical connections between different students' responses and between students' responses and key mathematical ideas. Balancing these practices to facilitate classroom discussions, which are invaluable in the context of modelling, is a complex endeavour. It demands teachers have a deep understanding of both the mathematics and the pedagogy of mathematical modelling. Since skilfully led classroom discussions are shown to benefit students' mathematical learning (Sfard et al. 1998), finding ways to support teachers' orchestration of discussions within the context of modelling is imperative.

Professional development programs often focus on helping teachers identify and interpret students' mathematical ideas in their written work as well as in their verbal contributions. The intention of such programs is to prepare teachers to better anticipate student solutions and increase their ease of monitoring these solutions. Nonetheless, it is a widely shared experience among teacher professional development researchers, that while teachers frequently become competent in interpreting and categorising students' solutions, they often continue to struggle with how these solutions can be used in classrooms to meet particular mathematical goals (Kazemi et al. 2010).

One way to support teachers to develop an appreciation of their own reasoning, and also that of their students, is through professional development that engages teachers in the kinds of learning experiences that are effective for students (Fennema et al. 1993). This helps teachers to identify and interpret one another's mathematical ideas privileging the importance of interpreting and comparing different solutions that will assist them when orchestrating discussion in their own classroom.

The questions explored in this study were: (1) What different models do the teachers use to represent the data? and (2) To what extent do the teachers compare different models?

2 Research Design

This longitudinal research and development project in a rural and remote area of Queensland, Australia, commenced in 2011. The study aims to enhance instructional practices in the teaching of statistics for mathematics teachers of 13 and 14 year old students. Participants in this project included one regional mathematics advisor and 11 teachers from six high schools, spread over an area of 500 km^2. Reported in this paper is the second full-day session that involved 8 of the 11 participating teachers.

A design research approach (Cobb et al. 2003a) was adopted to support and study teachers' learning. Prior studies that focused on students' statistics learning (Cobb 1999; Cobb et al. 2003b; McClain and Cobb 2001) and the use of technologies in mathematics classrooms (Zbiek et al. 2007) informed the design. Means of support developed in prior design studies (Dean 2004; Visnovska et al. 2012; Zhao et al. 2006) were tested and revised while working with this cohort of Australian teachers. Central to the adopted approach is the engagement of teachers as learners in comparing and contrasting different mathematical models of statistical data.

2.1 Speed Trap Task

Teachers were introduced to a road safety scenario that considered problems associated with cars regularly speeding on a local road.[1] The eight teachers collectively proposed what action could be taken to reduce the car speeds deciding that placing a visible speed camera would be the most effective deterrent. They also discussed how the effectiveness of this action could be measured deciding to record the speed at which cars travelled before and after the introduction of a speed camera. They noted that keeping the conditions the same, such as time, weather, and day in the week was important for drawing valid conclusions from data. Two data sets specifically generated to provide opportunities for maximum model development were then introduced to the teachers as data that had been generated in the manner they had proposed. The teachers were asked to use the data to support their advice to the police department about the effectiveness of the speed trap. The data were presented in the computer tool[2] pictured in Fig. 39.1, which was used by teachers for the first time.

In Fig. 39.1, each dot represents the speed of a car in kilometres per hour. The 60 dark dots represent the speeds of cars measured before and the 60 light dots represent the speeds of cars measured 3 weeks after the speed trap has been in operation. The applet permits several actions allowing users to model the speed trap data. These include structuring the data by creating one's own groups, partitioning the data into groups of a fixed size, partitioning the data into two groups with an equal number of data points in each, partitioning the data into four equal groups (precursor to the box plot), and partitioning data into groups with fixed interval width

[1] The actual speed limit was intentionally not made known to teachers. Whether or not to introduce speed limit in classrooms and how doing so may enhance or limit students' analyses later became a point of discussion (i.e., students might be tempted to introduce a cut off point at the speed limit value, thus limiting the diversity of models used).

[2] The applet tool is part of three statistical applet Minitools (© 1998 Vanderbilt University) developed in collaboration with the Freudenthal Institute in The Netherlands. Please contact the second author for more details.

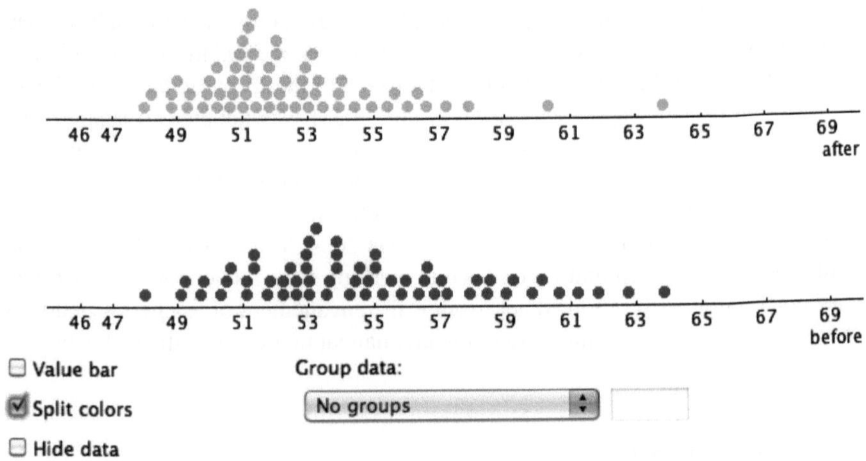

Fig. 39.1 Computer applet tool with speed trap data

(precursor to the histogram). Elaborations of the modelling options are discussed in the Results Section.

2.2 *Data Collection and Methodology*

The professional development session was video and audio recorded and subsequently transcribed. As teachers discussed various models the images they presented were saved as screen captures. The transcript and screen captures were analysed to gain insight into the research questions outlined above, and thus gauge teachers' preparedness to facilitate productive mathematical discussions that compare different models. The data were analysed using an adaptation of constant comparative method described by Cobb and Whitenack (1996) that involves testing and revising tentative conjectures while working through the data chronologically. Teacher names used in this chapter are pseudonyms.

3 Results and Discussion

Working in pairs, teachers analysed the speed trap data and reported their findings as to whether the speed trap was effective in reducing drivers' speed. During reporting, all proposed models were projected for the whole group to see. Section 3.1 documents how teachers used the tool to generate a variety of models to support their data-based arguments. Section 3.2 discusses teachers' willingness to compare these generated models.

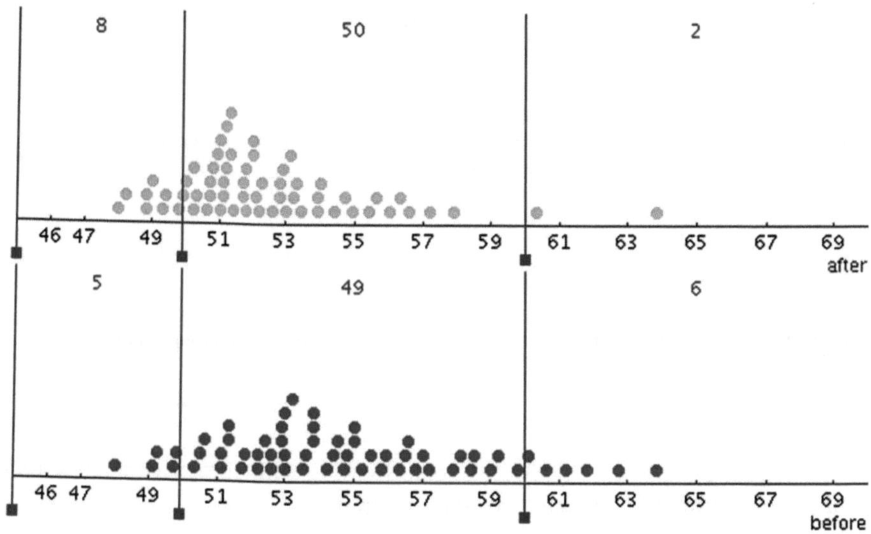

Fig. 39.2 Speed trap data with partitions at 50 and 60 km/h (create your own groups)

3.1 Diversity of Teacher-Generated Solutions

Create your own groups (see Fig. 39.2). The least sophisticated option on the computer tool involves placing one or more vertical bars at locations on the axis in order to partition the data set into groups of points. The number of points in each partition is displayed on the screen and adjusts automatically when the bars are dragged along the axis. One teacher pair, Tanya and Zoe, used this tool option when they partitioned the data, arbitrarily, at 50 and 60 km/h.

Tanya: We created our own groups and put a bar at about 50k and at about 60k… And so thinking about the speed camera and it being there we saw a general trend that the yellow [light dots] did technically move down to the left, like a lesser speed.

It transpired that Tanya and Zoe focused on position of the dots along the speed axis rather than on relative comparison of numbers of data points in different groups created by this partition. Notice that the number of cars in the middle section of their *before* and *after* partition only differed by one car. Zoe argued that *after* the speed trap was installed, most data in the middle section showed lower speeds, while the *before* group data covered the entire interval between 50 and 60 km/h. She used the tool to highlight what they saw as a shift in data towards lower speeds, manifested by no cars travelling at 58–60 km/h in the *after* group. At this point, the teachers relied on pictorial evidence and did not use the counts that the model provided to quantify and support their observations.

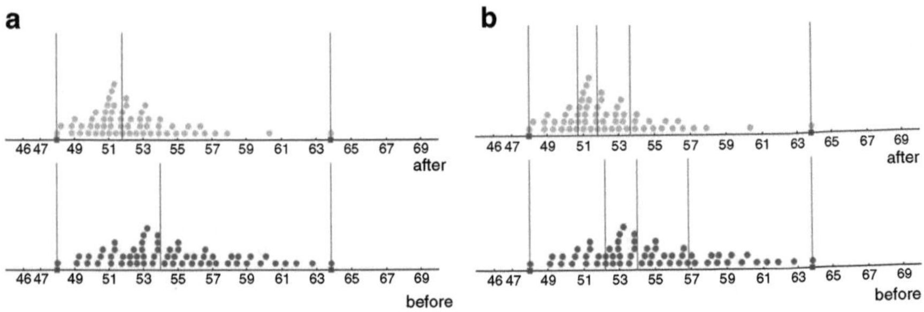

Fig. 39.3 Speed trap data with (**a**) Two equal groups and (**b**) Four equal groups option

Two equal groups (see Fig. 39.3a). John, working with Tom, used another tool where they partitioned data into two equal groups.

John: So now looking at *before*, which are blue [dark] dots, we can see that because it's two equal groups, we can say that 50 % of the cars were travelling at 54 km/h or below. So that's 50 %, whereas after the camera [was installed] we can say 50 % of the cars were travelling at 52 (km/h) or below, so I can say that, yeah, for [slower] 50 % of the cars their speed is lower [than before].

In his argument, John attempted to highlight that the shape of the distribution has changed despite the range of data being the same in both the *before* and *after* groups. He focused on speeds of the slower 50 % of the cars and used the tool to show that these cars slowed down.

Four equal groups (see Fig. 39.3b). Ben, working with Claire, shared that he would prefer his students to choose the tool that generates a model with four equal groups. He elaborated:

Ben: If you put it into the four groups, then you can look at it from a lot a higher point of view, you can look at the standard deviations away from the mean.

Tanya: Yeah. [Multiple teachers murmur in agreement, then 3 seconds pause]

Researcher: Well, you need to say more. [encouraging]

Ben: Well you can go to the mean value [sic, referring to Fig. 39.3b], which has obviously dropped like John says, but then … if you put in the next quartile, looking at coming down, obviously the quartile differences between the *after* and *before* are quite measurable and they also dropped. So this just backs up the further information based upon the mean averages.

The teachers identified the *four equal groups* option, as an equivalent of quartiles. Ben indicated that both the centre (median) and two quartiles (lines corresponding to Q_1 and Q_3 values) were lower in the *after* data, but did not explain how

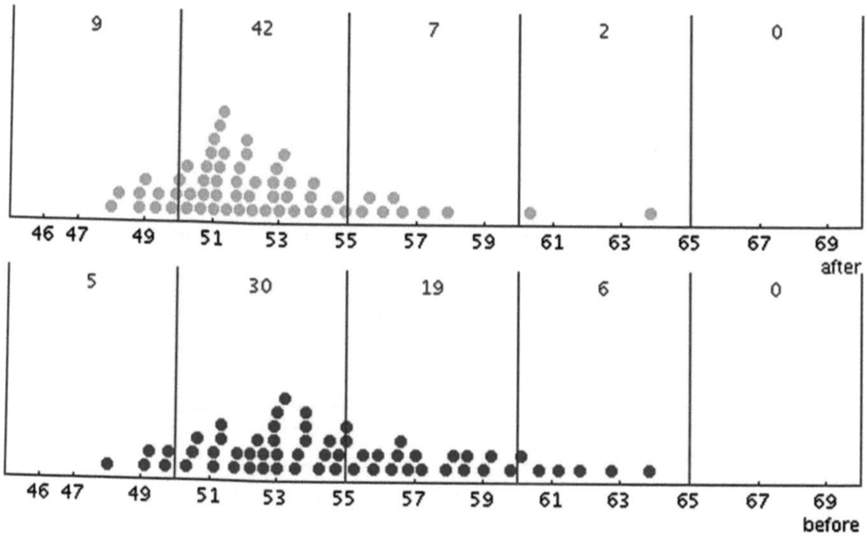

Fig. 39.4 Speed trap data grouped into 5 km/h-wide intervals

this helped him to understand what was happening with the car speeds, why these comparisons were relevant, or how they related to standard deviation. He offered that he would use the task as a follow up after the students had learned how to calculate various measures. At this stage he did not appreciate that the tool can be used to develop background understanding as to why modelling data in *four equal groups* might be useful to answering the question at hand.

Overall, despite using the computer applet tool for the first time, the teachers generated a variety of ideas, in which they made use of options afforded by the computer tool. While they did not routinely give reasons for their choices and some continued to focus on calculating statistical measures as a goal in its own right, the collection of solutions could become a basis for anticipating and recognising student approaches to the task.

3.2 Struggling to Reconcile Different Models of the Same Task Situation

Fixed interval width. Sarah and May first organised data into fixed interval width groups of 5 km/h (Fig. 39.4) and concluded that the cars slowed down after the speed camera was installed. They argued that while in the *before* data 19 cars drove at 55–60 km/h and six cars at 60–65 km/h, these numbers dropped to seven and two, respectively, after the camera was installed. Continuing to explore, they also organised data into 1 km/h intervals (Fig. 39.5). This led Sarah to question whether the speed trap had been effective.

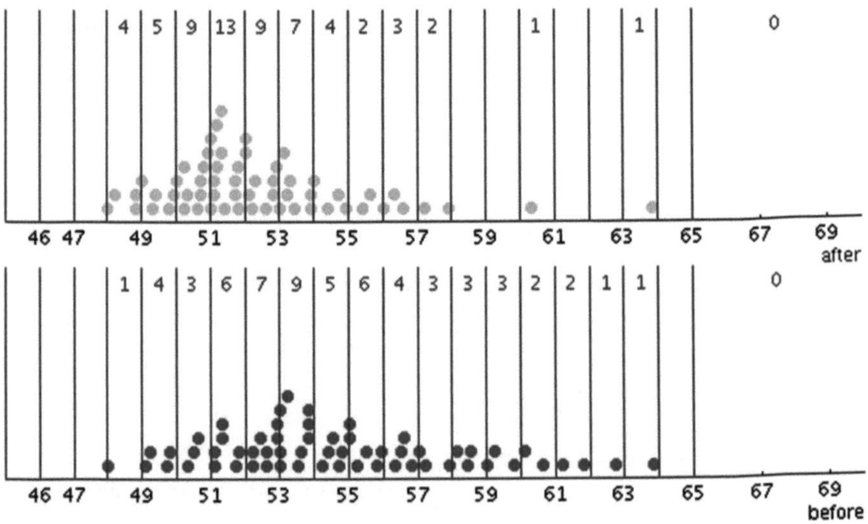

Fig. 39.5 Speed trap data grouped into 1 km/h-wide intervals

Sarah: Looking at this, if you're making comparison, is it really a great deal?

Researcher: What are you comparing?

Sarah: Well, if you're looking at each and every single kilometre, like if you look at the 57, they're all around the same sort of speed.

Researcher: So you are saying that there are 4 cars travelling at this speed *before*

Sarah: And then it's the same there [after]. Even if we look down at 53 k, there's 7 and 9, and 7 and 9 [16 cars in both the *before* and *after* groups travelled between 52 and 54 km/h]. There's not really [a difference], not until you get through 13 and 6, there was a difference [in the interval between 51 and 52 km/h]. That's where those people who were up the top in the dark [*before*] were sort of down in the bottom [in the *after* group]. So is that enough?

Researcher: So you are saying, if we look at this representation, it seems like both of them are the same. Is that what you were saying?

Sarah: Not the same but-

May: Still might be.

Sarah: Not close enough for me to have a definite yes, the camera worked, to me.

Sarah identified sections in her partitioning at 1 km/h intervals, which had the same number of cars in the *before* and *after* data. This seemed to have conflicted with her idea of a convincing trend in the data, which should be always identifiable, possibly irrespective of whether the tool used to model the situation is suitable for the purpose or not. While some teachers seemed to be convinced by the argument built with equal intervals of 5 km/h, others (e.g., May) shared Sarah's concern and struggled to reconcile what they viewed as conflicting solutions.

4 Conclusions

In this project, teachers have just commenced learning how to engage in activities where they genuinely analyse data. The ideas about purposes, goals, and acceptable ways of *doing statistics* in classrooms were not yet established in the group. Research demonstrates that as teachers initially try to make sense of professional development activities, they draw on practices established in their classrooms and attempt to incorporate new ideas into their teaching usually without profound changes to their practice (Franke et al. 2008). Evidence of this was present in Ben's suggestion to only provide modelling opportunities to his students after they had *learnt* the mathematical content. This suggests that supporting him and the entire group to experience how statistical understandings can emerge from the activity of modelling and analysing data, without being taught to students in advance (e.g., Hiebert et al. 1997), will be of key importance in future professional development sessions.

To respond to the research questions the analysis indicates that teachers did consider a range of computer tool options, and used these in different ways, as they modelled the task situation. For instance, Tanya and Zoe used the *create your own groups* option to make a gap in the data more visible. In contrast, John used both the visual impact of his partition and quantitative comparison provided by the *two equal groups* option. These generated solutions and group discussion may help these teachers to *anticipate likely student solutions* when planning to teach this task in their own classrooms. In addition, teachers' awareness of different possible approaches and how these relate to specific arrangements of the computer tool could be helpful in *monitoring students' work* during the task exploration. The PD activities thus provided the teachers with opportunities to elaborate on practices (a) and (b) identified by Stein et al. (2008).

Data also indicate that teachers, with the exception of Sarah and May, did not attempt to compare different solutions or explain how they connect with specific statistical ideas. Moreover, when Sarah pointed out what she viewed as conflicting solutions (the data split into 5 km/h and the 1 km/h intervals), the group did not come up with a way to reconcile this conflict. At least some teachers seemed to expect that if there is a *real* trend in data, it should be equally evident in all models.

Teachers' discussion of different models and their difficulty in reconciling the conflicting ideas reflected their lack of prior experience with engaging in mathematical discussions grounded in comparison of such models. While the activity supported them in making successful steps towards developing some of the skills identified by Stein et al. (2008), other skills, in particular (d) and (e), remained in the background. This episode illustrates that preparing teachers to "provide contexts for coherent rather than piecemeal [mathematical] learning … [and to create] a whole that is much more than the sum of the parts" (Niss et al. 2007, p. 21) is a nontrivial endeavour. Developing adequate skills will require supporting teachers to work flexibly with key statistical ideas as well as with the tools through which student learning can be built. Model comparisons will be further explored as a means to ground professional development discussions of teaching and learning principles (Cobb et al. 2008; Gravemeijer 2004).

Acknowledgements The research reported here was supported by Queensland Association of Mathematics Teachers and by Education Queensland, Central Queensland.

References

Cobb, P. (1999). Individual and collective mathematical development: The case of statistical data analysis. *Mathematical Thinking and Learning, 1*(1), 5–43.

Cobb, P., & Whitenack, J. (1996). A method for conducting longitudinal analyses of classroom videorecordings and transcript. *Educational Studies in Mathematics, 30*, 213–228.

Cobb, P., Confrey, J., diSessa, A. A., Lehrer, R., & Schauble, L. (2003a). Design experiments in education research. *Educational Researcher, 32*(1), 9–13.

Cobb, P., McClain, K., & Gravemeijer, K. (2003b). Learning about statistical covariation. *Cognition and Instruction, 21*(1), 1–78.

Cobb, P., Zhao, Q., & Visnovska, J. (2008). Learning from and adapting the theory of realistic mathematics education. *Éducation et Didactique, 2*(1), 105–124.

Dean, C. (2004). Investigating the development of a professional mathematics teaching community. In D. E. McDougall & J. A. Ross (Eds.), *Proceedings of the 26th annual meeting of the PME-NA* (Vol. 3, pp. 1057–1063). Toronto: PME.

Fennema, E., Carpenter, T. P., Franke, M., & Carey, D. A. (1993). Learning to use children's mathematics thinking: A case study. In R. B. Davis & C. A. Maher (Eds.), *Schools, mathematics, and the world of reality*. Boston: Allyn & Bacon.

Franke, M. L., Carpenter, T. P., & Battey, D. (2008). Content matters: Algebraic reasoning in teacher professional development. In J. J. Kaput, D. W. Carraher, & M. L. Blanton (Eds.), *Algebra in the early grades* (pp. 333–359). New York: Lawrence Erlbaum.

Gravemeijer, K. (2004). Local instruction theories as means of support for teachers in reform mathematics education. *Mathematical Thinking and Learning, 6*, 105–128.

Hiebert, J., Carpenter, T. P., Fennema, E., Fuson, K. C., Wearne, D., Murray, H., et al. (1997). *Making sense: Teaching and learning mathematics with understanding*. Portsmouth: Heinemann.

Kazemi, E., Elliott, R., Lesseig, K., Mumme, J. E., & Carroll, C. (2010, May). *Researching mathematics leader learning*. Paper presented at the annual meeting of the American Educational Research Association, Denver, CO.

McClain, K., & Cobb, P. (2001). Supporting students' ability to reason about data. *Educational Studies in Mathematics, 45*, 103–129.

Niss, M., Blum, W., & Galbraith, P. (2007). Introduction. In W. Blum, P. Galbraith, H.-W. Henn, & M. Niss (Eds.), *Modelling and applications in mathematics education: The 14th ICMI study* (pp. 3–32). New York: Springer.

Sfard, A., Nesher, P., Streefland, L., Cobb, P., & Mason, J. (1998). Learning mathematics through conversation: Is it as good as they say? *For the Learning of Mathematics, 18*(1), 41–51.

Stein, M. K., Engle, R. A., Smith, M. S., & Hughes, E. K. (2008). Orchestrating productive mathematical discussions: Five practices for helping teachers move beyond show and tell. *Mathematical Thinking and Learning, 10*, 313–340.

Visnovska, J., Cobb, P., & Dean, C. (2012). Mathematics teachers as instructional designers: What does it take? In G. Gueudet, B. Pepin, & L. Trouche (Eds.), *From text to 'lived' resources: Mathematics curriculum materials and teacher development* (pp. 323–341). Dordrecht: Springer.

Zbiek, R. M., Heid, M. K., Blume, G. W., & Dick, T. P. (2007). Research on technology in mathematics education. In F. K. Lester (Ed.), *Second handbook of research on mathematics teaching and learning* (Vol. 2, pp. 1169–1207). Greenwich: Information Age.

Zhao, Q., Visnovska, J., Cobb, P., & McClain, K. (2006, April). *Supporting the mathematics learning of a professional teaching community: Focusing on teachers' instructional reality.* Paper presented at the annual meeting of the American Educational Research Association conference, San Francisco, CA.

Part V
Assessment in Schools

Chapter 40
Formative Assessment in Everyday Teaching of Mathematical Modelling: Implementation of Written and Oral Feedback to Competency-Oriented Tasks

Michael Besser, Werner Blum, and Malte Klimczak

Abstract The interdisciplinary research project Co²CA investigates how assessing and reporting students' performances in mathematics can be arranged in every-day teaching in such a way that teachers are able to analyse students' outcomes appropriately and organise further learning as well targeted as possible. In this context, 39 classes of German middle track schools were observed for several weeks while dealing with mathematical tasks focusing on technical and modelling competencies. Based on the assumption that assessing and reporting students' outcomes regularly will foster learning processes, students from some classes were given individual, task-related feedback, in some classes several times in a written form, in some classes in addition permanently accompanying the students' solution processes. In this chapter, we describe the study and report some preliminary results.

1 Introduction

Following ideas of the Danish KOM-project (Blomhoj and Jensen 2007; Niss 2003) and of activities in the context of the development of national education standards for mathematics in several countries (Deutsche Kultusministerkonferenz 2003; National Council of Teachers of Mathematics 2000), the discussion of how to improve competency-oriented teaching and learning of mathematics is of central interest in

M. Besser (✉)
Fachbereich Mathematik, Institute of Mathematics, University of Kassel, Kassel, Germany
e-mail: mbesser@mathematik.uni-kassel.de

W. Blum
Department of Mathematics, University of Kassel, Kassel, Germany

M. Klimczak
Department of Educational Quality and Evaluation, Deutsches Institut für Internationale Pädagogische Forschung, Frankfurt am Main, Germany

G.A. Stillman et al. (eds.), *Teaching Mathematical Modelling: Connecting to Research and Practice*, International Perspectives on the Teaching and Learning of Mathematical Modelling, DOI 10.1007/978-94-007-6540-5_40,

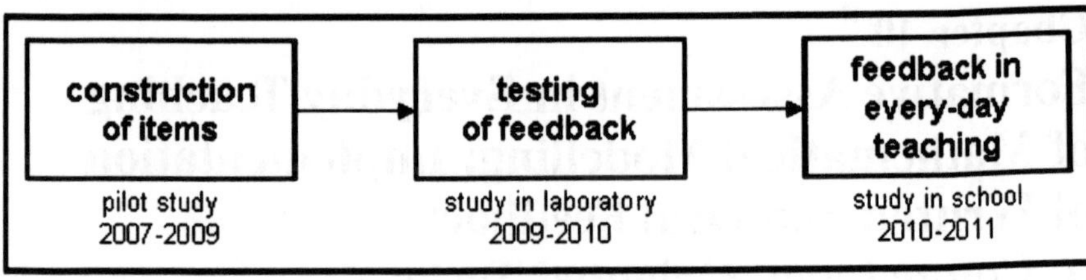

Fig. 40.1 Stages of the research project Co^{2}CA (2007–2011)

mathematics education. Considering the tension between 'unguided learning' on the one hand and 'instructional learning' on the other hand (DeCorte 2007; Hoops 1998; Kirschner et al. 2006; Mayer 2004), several studies have tried to find out how every-day teaching of mathematics could be arranged so as to foster students' learning as well targeted as possible (among many others, see e.g. Dekker and Elshout-Mohr 2004; Leiss 2010; Teong 2003).

The interdisciplinary research project Co^{2}CA (*Conditions and Consequences of Classroom Assessment*)[1] aims at investigating the impact of different kinds of feed-back in competency-oriented mathematics teaching on students' performances, emotions and attitudes. In a first step, starting in 2007, competency-oriented tasks (modelling tasks and technical tasks) that were to assess students' outcomes reliably have been constructed successfully. In a second step, special kinds of feedback to students' responses on the constructed items have been developed and tested in the laboratory (Besser et al. 2010; Bürgermeister et al. 2011; Klieme et al. 2010). Here the effect of feedback on performance tests based on marks has been compared to criteria-based feedback (students who are as good as you are generally able to deal with the following topics) and feedback directly based on students' working pro-cesses (as can be seen from your answers to the test, you are able/not able to deal with the following topics). In a third step, from October 2010 to March 2011, the items as well as the feedback that had been developed were implemented in a 13 lesson teaching unit in 39 Year 9 classes of German middle track schools (see Fig. 40.1 for a timetable of the Co^{2}CA-project; for a short overview of this special part of the study see Besser et al. 2011). In relation to this last step, one of the main research questions is: Will students in classes with an optimized kind of written and oral feedback outperform their counterparts who are not given such feedback, especially concerning their modelling competency?

In this chapter we will present the design of the Co^{2}CA study in school as well as some very first results of this study that hint at challenges we have to deal with in further steps.

[1]Supported by the German Research Society (DFG) as part of the current priority programme "Kompetenzmodelle zur Erfassung individueller Lernergebnisse und zur Bilanzierung von Bildungsprozessen" (SPP 1293); principal researchers: E. Klieme, K. Rakoczy (both Frankfurt), W. Blum (Kassel), D. Leiss (Lüneburg).

2 Implementation of Feedback in Every-Day Mathematics Teaching: Design of the Co²CA Study in School

According to results of pedagogical and psychological research (Hattie and Timperley 2007), it is reasonable to assume that assessing and reporting students' outcomes regularly in short intervals will foster students' learning. Such so called "formative assessment" (in contrast to ideas of "summing up" students' results only once at the end of a unit; for a general discussion about formative assessment see for example Black and Wiliam 2009) is said to be even more successful if the students are continuously offered feedback that is informative, individual and task-related (Deci et al. 1999; Kluger and DeNisi 1996) and if the assessment tries to answer some central questions concerning the students' learning processes: "Where am I going?", "How am I going?", and "Where to next?" (Hattie and Timperley 2007, p. 88).

The Co²CA project tries to implement written and oral feedback into teaching that sticks closely to the above-mentioned principles, that means it is given individually to students in short intervals (written feedback: three times during the 13 lessons; oral feedback: on the fly whenever possible), refers to students' solution processes, points out students' strengths as well as difficulties and offers strategies for students on how to improve themselves – especially feedback that helps students to concentrate on individual weaknesses and strengths on their own. In contrasting three different groups of students, the main question of how such feedback influences students' performances is pursued by the following research design (see Fig. 40.2).

2.1 *A Teaching Unit Dealing with Pythagoras' Theorem*

Altogether 39 Year 9 classes from 23 middle track schools (Realschule) in the state of Hessen (Germany) with 978 secondary students participated in this study. This sample can be regarded as fairly representative for this ability and age group. The classes were assigned randomly to either a control group CG: no special kind of feedback is given to the students, or one of two experimental groups, that is EG 1: students are given

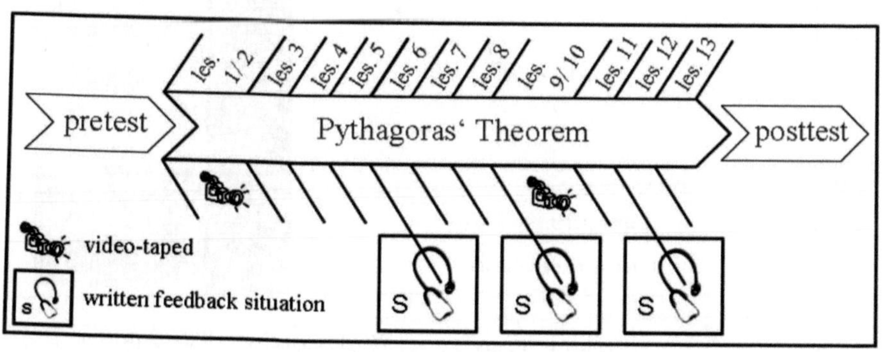

Fig. 40.2 Design of the Co²CA study in school (2010–2011)

written feedback three times within the 13 lessons and EG 2: in addition to written feedback students are supported by a special kind of oral feedback. Before starting the study, all teachers participated in half a day training to conduct a 13 lesson unit dealing with the topic area of Pythagoras' theorem. These 13 lessons comprised an introduction to Pythagoras' theorem (including a proof of the theorem), a phase with technical items, a phase with dressed up word problems and finally a phase with more demanding modelling problems. Referring to Kaiser (1995) and Maaß (2010) these modelling problems can be characterized in such a way that students' have to pass through the whole modelling cycle but that they only have to hark back to standardized, familiar ways of calculating. To control for the quality of teaching, every teacher was given a so called "logbook" with obligatory and optional tasks to use during the lessons. In addition, 4 of the 13 lessons were video-taped in all the classes.

2.2 *Written Feedback in Both Experimental Groups*

In the classes of the two experimental groups (EG 1 and EG 2) the students had to work on special short tasks on three occasions (at the end of lessons 5, 8 and 11). At the beginning of the next lesson, all students got back their solution, corrected by the teacher, together with an individual, process-oriented, written feedback and a suitable exercise to work on. The teachers were prepared to do so on a second half day of training. To ensure that all participating students worked on the aforementioned special short tasks, these were integrated into the regular lessons of the control group. An example of such a written feedback can be seen in Fig. 40.3. This example shows feedback given to the following modelling item that students were given at the end of lesson 11.

The rope of the cable car *Ristis* has to be replaced. 1 m of the rope costs 8 €. How much does a new rope cost approximately? Write down your solution process.

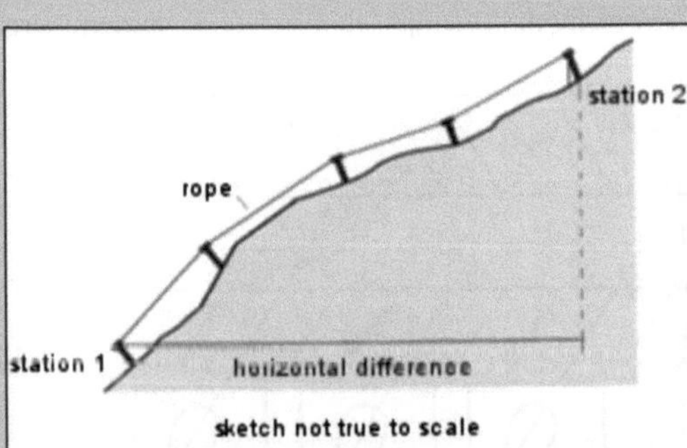

Name:	Cable car "Ristis"	Weight capacity:	132 × 3 persons
Station 1:	1,600 m above sea level	Haul capacity:	1,200 pers. per hour
Station 2:	1,897 m above sea level	Speed:	1.5 m/s
Horizontal difference:	869 m	Time of travel:	10 min

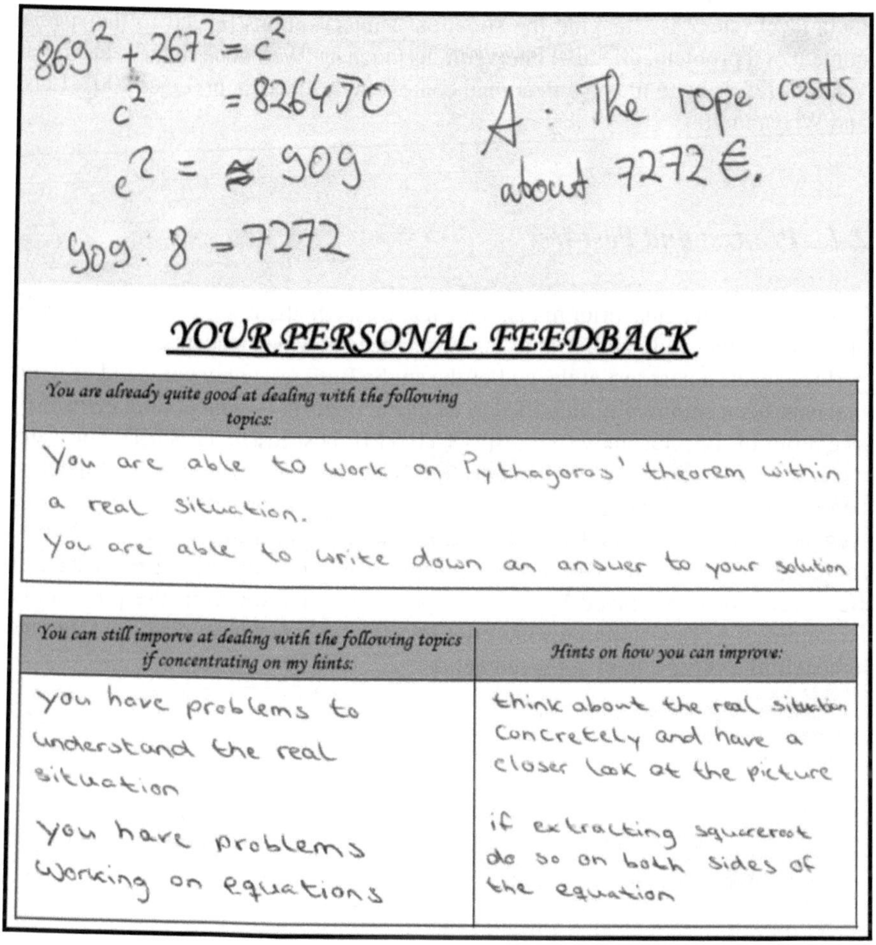

$869^2 + 267^2 = c^2$

$c^2 = 826450$

$c \approx 909$

$909 \cdot 8 = 7272$

A: The rope costs about 7272 €.

YOUR PERSONAL FEEDBACK

You are already quite good at dealing with the following topics:
You are able to work on Pythagoras' theorem within a real situation. You are able to write down an answer to your solution

You can still improve at dealing with the following topics if concentrating on my hints:	Hints on how you can improve:
you have problems to understand the real situation	think about the real situation concretely and have a closer look at the picture
you have problems working on equations	if extracting squareroot do so on both sides of the equation

Fig. 40.3 Example of written feedback

2.3 *Oral Feedback in One of the Two Experimental Groups*

In addition to the written feedback, the teachers of experimental group 2 (EG 2) were trained on a third half day to implement a special kind of oral feedback that copes with the requirements of competency-oriented tasks in every-day teaching of mathematics, similar to the so-called "operative-strategic" teaching method developed in the DISUM project (here students mainly have to deal with mathematical modelling tasks in groups and with only little support by the teacher; for details see Blum 2011 and Schukajlow et al. 2011). According to ideas of the DISUM project, the teachers were trained to orally intervene into students' working processes only by minimal-adaptive support in order to let the students work on their own as much as possible (Leiss 2005). The participating teachers were informed about different ways of intervening and supporting. Here we distinguish between four categories of teacher interventions: metacognitive interventions that give hints on a meta-level (such as 'Imagine the situation'), interventions related to the special content of a problem, affective interventions (such as 'Well done so far'), and interventions referring to the organizational context in the classroom (Leiss 2007; Leiss and Wiegand 2005).

2.4 Pre-test and Post-test

To control for students' prior mathematical knowledge there was a pre-test immediately before the study and, to find out differences between students' mathematical performances, a post-test at the end of the study. Both tests only consisted of items that have been empirically identified as technical items (TI) or modelling items (MI) as a result of the pilot study (pre-test: 13 TI, 6 MI; post-test: 9 TI, 8 MI). Since students normally cannot solve items dealing with the topic of Pythagoras' theorem before this topic is explicitly taught, here only 'prior knowledge' – elements that were necessary to work on Pythagoras' theorem in the following weeks – was asked for (e.g., finding the square root of a number or naming characteristics of a triangle). Both tests could be linked by the item-parameters known from the pilot study. Examples of a pre-test item testing prior knowledge, a technical post-test item and a modelling post-test item are given below.

Prior-knowledge pre-test item:

A broom is rested against a wall as shown below.
Broom, wall and bottom form a triangle. Mark the triangle in the picture and give names to the sides.

Technical post-test item:

Calculate the length of the side a = |BC|.

a = ____________

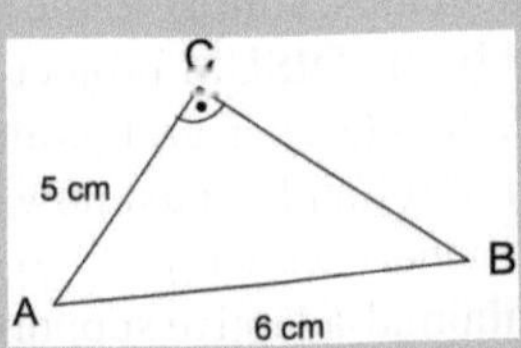

Modelling post-test item:

On May 1st people in Bad Dinkelsdorf dance around a so called "Maibaum". This is a tree which has a height of 8 m. While dancing, the people hold bands in their hands. These bands are 15 m long. How far away from the "Maibaum" are the people at the beginning of the dance?

3 Some Preliminary Results of the Field Study

Both pre-test and post-test have been rated and first analyses can be reported concerning the test results. The reported results are deduced from scores which have been given to the students' answers by trained raters, and these scores have been used for scaling the tests based on the Rasch model.

3.1 Test Results

Inter-rater reliability: The rating has been successful since the inter-rater reliability for the five trained raters can be said to be very good (pre-test: Cronbach's alphas between 0.829 and 1.000; post-test: Cronbach's alphas between 0.947 and 1.000).

Test reliability: Whether linked to the results of the pilot study or not, the *wle* (weighted likelihood estimation) and *eap* (expected a-posteriori) reliability of the tests[2] (as a one-dimensional mathematical construct) are acceptable (0.571–0.735). However, a two-dimensional scaling – separately for TI and MI – points to some problems concerning the MI dimension of the pre-test. First factor analyses hint at two out of six items not fitting sufficiently to this dimension.

Difficulties of the tests: One-dimensional as well as two-dimensional scaling illustrate bigger differences in the difficulty of the pre-test depending on whether the item-parameters of the pre-test are linked to the pilot study or not. If linked, the pre-test becomes much harder. Further analyses highlight that these differences

[2]Both *wle* reliability and *eap* reliability are computed in ConQuest as indicators for the reliability of a latent variable/construct. In general values greater than 0.6 are expected to be acceptable. For more details see also Rost (2004).

Table 40.1 Results of pre-test and post-test separated for CG, EG 1 and EG 2

	Test-results (one-dimensional-scaling)			Significance in differences	
Test	CG	EG 1	EG 2	CG vs. EG 1	CG vs. EG 2
Pre-test	−0.152 $SD=1.106$ $n=340$	0.205 $SD=0.929$ $n=303$	−0.420 $SD=0.918$ $n=206$	No	Yes ($p<0.01$)
Post-test	−0.140 $SD=1.274$ $n=334$	−0.230 $SD=1.211$ $n=302$	−0.332 $SD=1.219$ $n=220$	No	No

seem to be caused by differences in technical abilities between the populations of the pilot study and of the field study. Since there were also higher track (Gymnasium) students in the population of the pilot study, these differences are apparently caused by these higher ability students (interestingly, there are no such differences concerning the modelling dimension of the pre-test and the TI or MI dimension of the post-test).

Differences in performance: One of the main questions of the study obviously is: Are there significant differences in the post-test performance between the three groups (control group and two experimental groups)? Unfortunately, this question cannot be answered satisfactorily yet – too many variables are not yet evaluated, and too little control for appropriate treatment implementation was possible to date. Nevertheless, some very first results concerning the students' performances in the control group and the two experimental groups shall be reported here, taking into account that these results have to be dealt with very carefully (here we only refer to results of a one-dimensional scaling of the tests since the reliability of the MT-dimension of the post-test is not really acceptable) (see Table 40.1).

Table 40.1 shows that there are no significant differences between CG and EG 1 or CG and EG 2 in the post-test. The control group performed significantly better in the pre-test than experimental group 2 (−0.152 vs. −0.420) and these differences are no longer visible in the post-test. Since analyses of covariance do not show any influences of the experimental condition either, we have to know a lot more about the quality of the implementation of the treatment to explain these effects – especially we have to know in detail what really happened in the 13 lessons.

3.2 Challenges for the Future

The main research question of Co²CA is whether special kinds of formative assessment – theoretically based and optimised forms of written or oral feedback – can help teachers to improve students' learning processes when dealing with competency-oriented mathematical tasks (here: with technical and modelling tasks) and whether an implementation in everyday teaching can foster students' performances. Except for one special case (the reliability of the MI dimension of the pre-test), the performance tests have worked quite well. Within the next few months, further analyses have to be carried out in order to answer the main question stated above, that is to find out whether there are differences in students' outcomes between different groups and whether these differences are really caused by our treatments. Therefore, the big challenge is to control both for the overall quality of teaching (by analysing about 160 h of video-taped lessons; see Lipowsky et al. 2009 for some relevant variables) and for the quality of written and oral feedback given by the teachers (by developing adequate coding schemes for both written and oral feedback). We will report about these analyses in the near future in particular at the next ICTMA.

References

Besser, M., Leiss, D., Harks, B., Rakoczy, K., Klieme, E., & Blum, W. (2010). Kompetenzorientiertes Feedback im Mathematikunterricht: Entwicklung und empirische Erprobung prozessbezogener, aufgabenbasierter Rückmeldesituationen. *Empirische Pädagogik, 24*(4), 404–432.

Besser, M., Klimczak, M., Blum, W., Leiss, D., Klieme, E., & Rakoczy, K. (2011). Lernprozessbegleitendes Feedback als Diagnose- und Förderinstrument: Eine Unterrichtsstudie zur Gestaltung von Rückmeldesituationen im kompetenzorientierten Mathematikunterricht. In *Beiträge zum Mathematikunterricht* (pp. 103–106). Münster: WTM Verlag.

Black, P., & Wiliam, D. (2009). Developing the theory of formative assessment. *Educational Assessment, Evaluation and Accountability, 21*(1), 5–31.

Blomhoj, M., & Jensen, T. H. (2007). What's all the fuss about competencies? Experiences with using a competence perspective on mathematics education to develop the teaching of mathematical modelling. In W. Blum, P. L. Galbraith, H. W. Henn, & M. Niss (Eds.), *Modelling and applications in mathematics education. The 14th ICMI study* (pp. 45–56). New York: Springer.

Blum, W. (2011). Can modelling be taught and learnt? Some answers from empirical research. In G. Kaiser, W. Blum, R. Borromeo Ferii, & G. Stillman (Eds.), *Trends in teaching and learning of mathematical modelling. ICTMA 14* (pp. 15–30). New York: Springer.

Bürgermeister, A., Klimczak, M., Klieme, E., Rakoczy, K., Blum, W., Leiss, D., Harks, B., & Besser, M. (2011). Leistungsbeurteilung im Mathematikunterricht – Eine Darstellung des Projekts "Nutzung und Auswirkunen der Kompetenzmessung in mathematischen Lehr-Lern-Prozessen". http://www.schulpaedagogik-heute.de/index.php/artikel-331.html. Accessed 12 Sept 2011.

Deci, E. L., Koestner, R., & Ryan, R. M. (1999). A meta-analytic review of experiments examining the effects of extrinsic rewards on intrinsic motivation. *Psychological Bulletin, 125*(6), 627–668.

DeCorte, E. (2007). Learning from instruction: The case of mathematics. *Learning Inquiry, 1*(1), 19–30.

Dekker, R., & Elshout-Mohr, M. (2004). Teacher interventions aimed at mathematical level raising during collaborative learning. *Educational Studies in Mathematics, 56*, 39–65.

Deutsche Kultusministerkonferenz. (2003). *Bildungsstandards im Fach Mathematik für den Mittleren Schulabschluss*. München: Luchterhand.

Hattie, J., & Timperley, H. (2007). The power of feedback. *Review of Educational Research, 77*(1), 81–112.

Hoops, W. (1998). Konstruktivismus. Ein neues paradigma für didaktisches design? *Unterrichtswissenschaft, 3*, 229–253.

Kaiser, G. (1995). Realitätsbezüge im Mathematikunterricht – Ein Überblick über die aktuelle und historische Diskussion. *Materialien für einen realitätsbezogenen Mathematikunterricht, 2*, 66–84.

Kirschner, P. A., Sweller, J., & Clark, R. E. (2006). Why minimal guidance during instruction does not work: an analysis of the failure of constructivist, discovery, problem-based, experimental, and inquiry-based teaching. *Educational Psychologist, 41*(2), 75–86.

Klieme, E., Bürgermeister, A., Harks, B., Blum, W., Leiss, D., & Rakoczy, K. (2010). Projekt Co²CA. Leistungsbeurteilung und Kompetenzmodellierung im Mathematikunterricht. *Zeitschrift für Pädagogik, 56*, 64–74.

Kluger, A. N., & DeNisi, A. (1996). The effects of feedback interventions on performance: A historical review, a meta-analysis, and a preliminary feedback intervention theory. *Psychological Bulletin, 119*(2), 254–284.

Leiss, D. (2005). Teacher intervention versus self-regulated learning? *Teaching Mathematics and Its Applications, 24*(2–3), 75–89.

Leiss, D. (2007). *"Hilf mir es selbst zu tun". Lehrerinterventionen beim mathematischen Modellieren*. Hildesheim: Franzbecker.

Leiss, D. (2010). Adaptive Lehrerinterventionen beim mathematischen Modellieren – empirische Befunde einer vergleichenden Labor- und Unterrichtsstudie. *Journal für Mathematik-Didaktik, 31*(2), 197–226.

Leiss, D., & Wiegand, B. (2005). A classification of teacher interventions in mathematics teaching. *ZDM, 37*, 240–245.

Lipowsky, F., Rakoczy, K., Pauli, C., Drollinger-Vetter, B., Klieme, E., & Reusser, K. (2009). Quality of geometry instruction and its short-term impact on students' understanding of the Pythagorean theorem. *Learning and Instruction, 19*, 527–537.

Maaß, K. (2010). Classification scheme for modelling tasks. *Journal für Mathematik-Didaktik, 31*(2), 285–311.

Mayer, R. E. (2004). Should there be a three-strikes rule against pure discovery learning? The case of guided methods of instruction. *The American Psychologist, 59*(1), 14–19.

National Council of Teachers of Mathematics. (2000). *Principles and standards for school mathematics*. Reston, VA: NCTM.

Niss, M. (2003). Mathematical competencies and the learning of mathematics: The Danish KOM project. In A. Gagatsis & S. Papastavridis (Eds.), *Mediterranean conference on mathematical education* (pp. 115–124). Athens: 3rd Hellenic Mathematical Society and Cyprus Mathematical Society.

Rost, J. (2004). *Lehrbuch Testtheorie – Testkonstruktion*. Bern: Huber.

Schukajlow, S., Leiss, D., Pekrun, R., Blum, W., Müller, M., & Messner, R. (2011). Teaching methods for modelling problems and students' task-specific enjoyment, value, interest and self-efficacy expectations. *Educational Studies in Mathematics, 79*(2), 215–237.

Teong, S. K. (2003). The effect of metacognitive training on mathematical word-problem solving. *Journal of Computer Assisted Learning, 19*, 46–55.

Chapter 41
Assessment of Modelling in Mathematics Examination Papers: Ready-Made Models and Reproductive Mathematising

Pauline Vos

Abstract In the Netherlands modelling is integrated into the mathematics curriculum. This chapter describes a study of modelling characteristics in recent mathematics examination papers. Results show that the tasks convey the message that mathematics can be found in unexpected situations. In many tasks a situation and a ready-made model are given and students are asked to work with the model. For mathematising, two design principles were: to finalise the parameters of a given model, and to reconstruct a given mathematical model from verbal descriptions and diagrams of a situation. These formats were coined as *mechanistic* and *reproductive mathematising* respectively. These formats have been introduced to cater for test reliability at the expense of test validity.

1 Introduction

Standardised, large-scale assessment adheres to several criteria, such as validity and reliability. Validity means that the test really assesses the knowledge, skills and insights that are meant to be evaluated. Reliability means that a test yields results independent of factors external to the assessed candidate, such as skills and moods of the examiner. For large-scale tests that include mathematical modelling, these same criteria are at stake.

Tasks that are valid with respect to mathematical modelling simulate the work of professional modelers, or parts thereof; therefore, they require students to undertake

---

This work was completed when I was on the staff of Amsterdam University of Applied Sciences, The Netherlands

P. Vos (✉)
Faculty of Engineering and Science, University of Agder, Kristiansand, Norway
e-mail: pauline.vos@uia.no

G.A. Stillman et al. (eds.), *Teaching Mathematical Modelling: Connecting to Research and Practice*, International Perspectives on the Teaching and Learning of Mathematical Modelling, DOI 10.1007/978-94-007-6540-5_41,

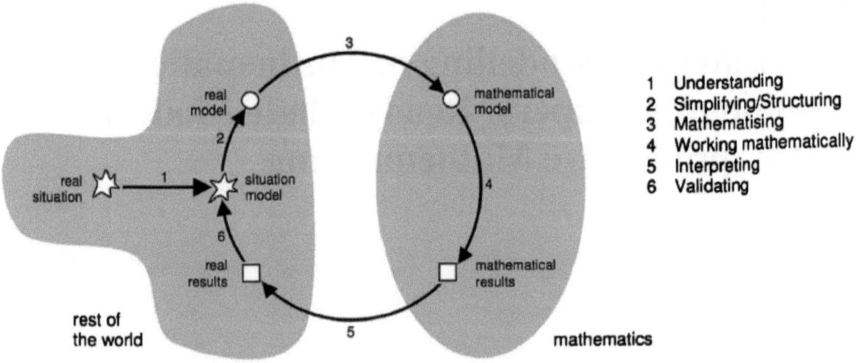

Fig. 41.1 The modelling cycle by Blum and Leiss (2005)

activities as described in the modelling cycle by Blum and Leiss (2005), see Fig. 41.1. Additionally, valid tasks with respect to modelling may require students to work for a long stretch of time, and even collaboratively. For example in the modelling competition *A-lympiad*, groups of students work for 2 days on a single task (Vos 2010), and in a modelling project at the University of Hamburg groups of students work for a week on a single problem (Kaiser et al. 2011). Tasks that are valid with respect to modelling may ask for creativity, problem solving, and collaborative, communicative and metacognitive skills. 'Open' tasks based on complex situations have a good validity with respect to mathematical modelling. However, open tasks have a lower test reliability. Reliability is a major criterion for large-scale standardised assessment, and it often leads to closed questions, to which there is only one correct answer and this answer can be reached by only a limited number of well-defined methods within a few steps. Often the mathematical skills evaluated in standardised tests are reproductive knowledge and algorithmic skills. By contrast, mathematical modelling tasks allow multiple methods and diverse answers. Modelling skills based on creativity and higher order thinking are hard to assess separately and evaluators need to be flexible in their judgement as they should be prepared for surprising answers from students.

A number of studies on the assessment of mathematical modelling have been carried out. Haines and Crouch (2001) developed a 20-min test with multiple choice questions on the mathematisation phase in modelling (e.g., on identifying variables and making assumptions). They found that students with higher modelling skills outperformed students without modelling experiences, implying that a reliable assessment format (multiple choice) can have validity to modelling, despite not covering the full range and complexity of modelling activities. Also, Houston (2007) found that separate phases in the modelling cycle can reliably be assessed. With respect to reliability, Palm (2007) and Vos (2007) found that small changes in the assessment environment of modelling tasks can significantly affect students' performance. For example, slightly altering the wording or adding a photograph can support students' mental activities or hinder these. Their research supports findings on how students construct mental models while reading the task's text and end up with individualised interpretations (Bower and Morrow 1990). Therefore, validly and reliably assessing mathematical modelling competences is still an area for further research, not only with respect to question formats, but also with regard to the descriptions of scoring guidelines to ensure that independent coders deliver equivalent evaluations.

In a number of countries modelling has been integrated into the mathematics curriculum. To sustain this integration, educational authorities have included modelling into paper-and-pencil tests, such as national examinations. Examination tasks are designed by authorities outside the schools, and they concretise the aims and objectives stated in curriculum documents. These tasks may guide teachers in lesson preparation, but also they may be used in teacher training, and they guide authors in textbook writing. As such, exam questions and standardised tests play a wider role than being merely instruments to assess students' learning. It is therefore important to study what message assessment developers put into the tasks they design and how they deal with issues such as validity and reliability. In this chapter, I will present a case study of examinations in the Netherlands. Frejd (2011) carried out a similar study for Sweden.

2 The Dutch Context

From the 1970s onwards, mathematics education in the Netherlands was shaped by Hans Freudenthal and his colleagues, developing a curriculum guided by the question: *"How can we teach mathematics so as to be useful?"* In consecutive interventions, the mathematics curriculum was reformed towards applications and modelling, based on the idea that mathematical models can be used as a vehicle to develop mathematical conceptual knowledge (see also Vos 2010). The mathematics curriculum for primary schools and vocational secondary schools (indicated as *vmbo*, see Fig. 41.2) builds on situations, emphasising *numeracy* skills needed by future citizens who should be able to gather, filter and handle information quantitatively. In the other secondary streams, the 'general' (*havo*) and the 'pre-university' (*vwo*) streams, the mathematics curricula are split into two separate subjects: *Mathematics A* for future students in the social and economical sciences and *Mathematics B* for future engineers. Mathematics A has a strong modelling component, based on Statistics, Probability, and Discrete Mathematics, whereby all models originate from *non-technical* areas. The subject of *Mathematics B* is more similar to the traditionally shaped mathematics (calculus and geometry) taught in most countries.

In the Netherlands the exit examinations of secondary schools are pivotal: a pass is the entry ticket into tertiary education. These examinations are administered nationally. At all schools extensive training is offered, often based on the papers of prior years. For the present chapter modelling characteristics of Dutch mathematics examination tasks are studied, taking these as curricular messengers.

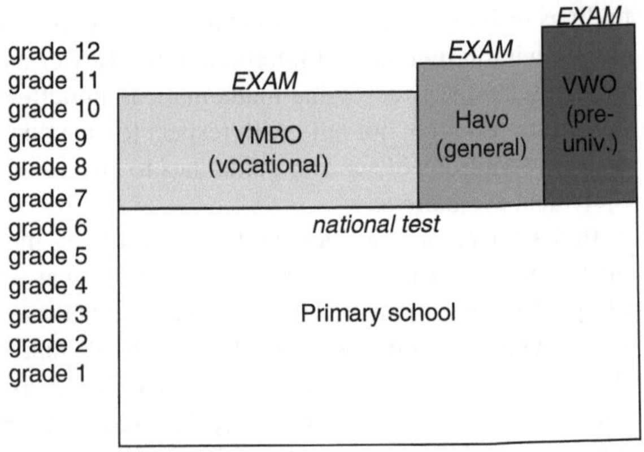

Fig. 41.2 Dutch school system for primary and secondary education

3 Method

The sample consisted of all mathematics examination papers for all levels: vocational (examination at the end of grade 10), general (examination at the end of grade 11), and pre-university (examination at the end of grade 12), see Fig. 41.2 for a scheme of the Dutch school system. Primary education starts at the age of 4 and ends at the age of 12. When students enter secondary school they still can switch between levels.

For mathematics each secondary school level (vocational, general and pre-university) has its own examination. At vocational level, there are three different papers for different streams (*KB*, *BB*, and *GLTL*). At general level, there are three different papers (*Mathematics A12*, *Mathematics B1* and *B12*). At pre-university level there are four different papers (*Mathematics A1* and *A12*, *Mathematics B1* and *B12*). Excluding the experimental computer examinations and the repeat examination for referred students, each year has a total of ten mathematics papers. To obtain a spread over recent years, I studied the papers from 2006, 2008 and 2010.

For all tasks, I coded the modelling characteristics of the given information in the tasks' text and the steps that students were asked to undertake. At first I coded in terms of the modelling cycle from Blum and Leiss (2005). For example, when a task asked for the translation from a real model to a mathematical model, I indicated this as a mathematisation task. However, after a few rounds of coding, I needed additional categories and I switched to a Grounded Theory approach (Strauss and Corbin 1997), a method recommended for exploring descriptors of observed phenomena. In this way, I could discern different kinds of tasks, that at first were all coded as mathematisation tasks. A subsample was coded by a colleague for another study (Schaap et al. 2011).

4 Results

In Table 41.1 general characteristics of the mathematics papers are presented. Generally, a paper consists of 10–12 pages. The larger number of pages at the lower level can be explained by the many space-consuming photographs and diagrams that are used to present information to the students. Also, at vocational level more often tables are used while at the higher levels more often symbolic (algebraic) models are offered.

I calculated the percentage of points that students could earn from tasks that were situated in a non-mathematical context relative to the maximum number of points in an examination. Almost all tasks at vocational level and all tasks in the papers for *Mathematics A* (for future students of social/economic disciplines) were set in real-life situations. Also in the papers for *Mathematics B* (for future engineers) many points could be earned in situated tasks: at the general level ¾ of the tasks were situated in real contexts and at pre-university level ½ of the tasks were situated.

In all papers the tasks were grouped under a theme. Each theme had a clear title. The tasks connected by a title were independent of one another (i.e., if students cannot complete one task, they still can complete the others). For example in the year 2006 the following titles were used:

- At vocational level (*Vmbo GLTL 2006*): Road signs–Tiling pattern–Food waste–Wishing well–Buying a new bicycle–Moons of Jupiter.
- At general level (*Havo Mathematics A12 2006*): Do women earn less? – Batteries – Packages – Mortgage – Daniël Education costs.
- At pre-university level (*VWO Mathematics B1 2006*): Sauna – Covering–In a square – Knock-out system–Area of trapezium.

The themes demonstrate that mathematics is everywhere, and students can expect to encounter new areas of mathematics application in the examination. This is a valid aspect with respect to mathematical modelling.

Three exemplary tasks for different levels are shown below. In these tasks students are offered a real-life situation together with a ready-made mathematical

Table 41.1 General characteristics of Dutch mathematics examination papers

School type	Average number of Pages (*SD*)	Pictures, diagrams	Tables	Graphs	Symbolic models	% score points in situated tasks
Vocational (vmbo)	12.2 (1.6)	10.6	2.8	1.9	2.0	98
General (havo)	9.9 (1.8)	3.8	1.3	4.4	5.8	100 (in Math A) 77 (in Math B)
Pre-university (vwo)	9.5 (2.1)	3.3	1.9	3.9	5.4	100 (in Math A) 47 (in Math B)

model (a diagram, a table, a graph, a formula) and then asked to work with that model. The *Food Waste Task* offers two circle diagrams that model data as if from a magazine. In the *Wind Speed Task* a model is offered: data presented in a table, as if from research measurements, and a linear equation is offered, albeit with parameters still to be calculated. In the *Trivet Task* the algebraic (trigonometric) model is offered. All mathematical models come out of the blue.

Food waste (vocational, KB, 2006)

In the Netherlands each day much of the bought food is thrown away. A large part of it disappears untouched into the waste bin. In the diagram below you can read how much food is bought by an average household in 1 year, and how much of that is thrown away.

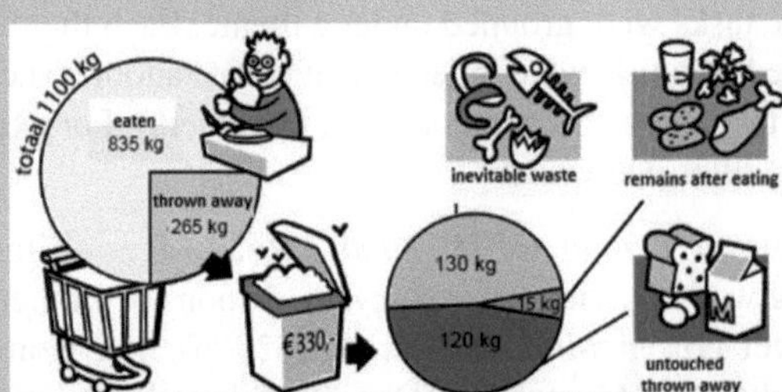

- Calculate the average price in Euro of 1 kg of thrown away food.
- In the Netherlands there are approximately 16 million inhabitants. On average a household consists of 2.4 people. Calculate the total amount of money in Euro that is thrown away as food in the Netherlands in 1 year.

Wind speed and altitude (general, Wiskunde B1, 2006)

In Vlaardingen (*a small Dutch city*) on a certain day the wind speed was measured at different altitudes. The measurements show approximately a linear relation between wind speed W in m/s and altitude h in m, when the altitude is between 10 and 80 m (see Table 41.1). The formula $W = a \cdot h + b$ gives this linear relation.

h	10	20	30	40	50	60	70	80
W	1,2	1,6	2,1	2,5	3,0	3,4	3,9	4,3

- Calculate a and b with help of the table. Round off a and b to two decimals.

Ensuing tasks use another mathematical model (logarithmic).

Trivet (pre-university, Mathematics B, 2010)

A certain trivet consists of bars that can hinge. This trivet has 19 equal rhombuses, see the photo.

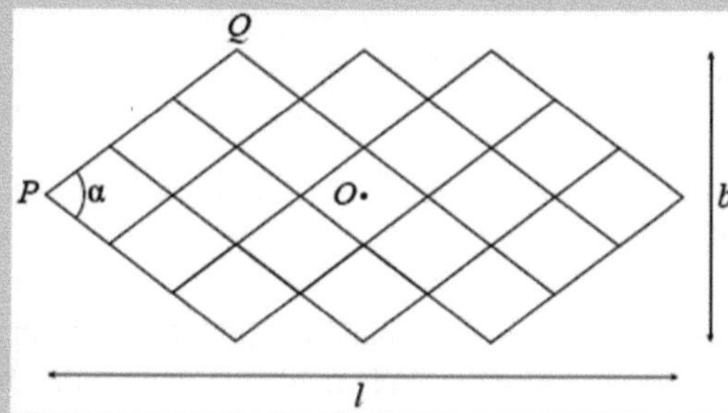

In a mathematical model for this trivet we ignore the thickness of the bars. We indicate the leftmost hinging point with P, the midpoint of the middle rhombus with O. The inner angle at P is α (in radians), see the diagram. We choose length 1 for the side of a rhombus. Length l and width w of the model are functions of α, whereby $0 \leq \alpha \leq \pi$. We have: $l = 10\cos(\frac{1}{2}\alpha)$ and $w = 6\sin(\frac{1}{2}\alpha)$.

- Show that the formula for l and w are correct.

In the ensuing questions the above mathematical model has to be used for calculating angle α at which w increases with the same rate as l decreases, for reconstructing a given formula for distance OQ, and for calculating angle α at which the trivet fits within a circle.

When a problem situation is offered together with a ready-made mathematical model, the structuring, simplifying and mathematising is skipped from the modelling cycle. The message from such tasks is that the creation of a mathematical model is not a learning aim. An advantage of this format is that all students start from the same formula; if they would create their own mathematical model and continue to work with it, the ensuing modelling activities may have a different mathematical demand between students, because of differing complexities of the mathematical models created. The presence of a ready-made model raises the reliability of the task. A disadvantage is, that students are not *owners* of the model and they cannot demonstrate competencies such as simplifying and structuring.

The ready-made mathematical model does not serve to solve a practical problem, but it is merely a starting point for 'working mathematically'. See the *Food Waste Task:* the diagrams offer information that has to be transformed into number answers. These answers do not solve the problem of food waste, although they may assist in understanding the immensity of food waste. Primarily, these tasks convey the message, that answers in mathematics examinations are numbers. More than

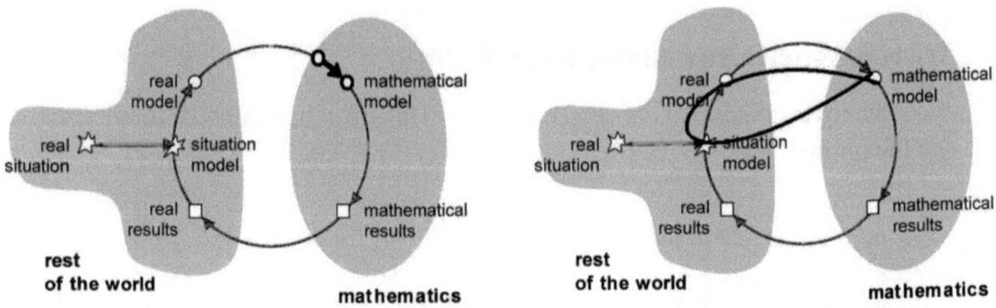

Fig. 41.3 Illustration of mechanistic mathematisation (*left*) and reproductive mathematisation (*right*)

90 % of the situated tasks had this format: ending in a number answer. In other tasks there was also a ready-made mathematical model, but then the students had to reason on model properties (e.g., whether it meets certain assumptions), compare between models, or rewrite the model.

Tasks that start from *ready-made models* are 'wrapped mathematical tasks', but this terminology covers a broad range of tasks. I therefore use additional terminology.

In the *Wind Speed Task* a ready-made table is offered to the students and they are asked to translate this table into a linear model, thus to transform the numerical model into a symbolical model. The advantage for students is that mathematising becomes easier; the advantage for examiners is that the question becomes a closed question. This raises the test's reliability. The disadvantage is that students are no longer aware that they are building a model representing phenomena: they do not need to look at the situation, but only at the table to fit points into a given parametric formula. With this question format, the transformation of numerical data into a symbolical representation becomes an algorithmic exercise that ends in two numbers (parameter values) instead of a formula. Thus, the validity of the test with respect to modelling is low. This task format in the examination papers aligns with Dutch textbooks, where the formulation *"calculate a and b"* is a reoccurring mantra to make students calculate parameters that will make a formula match a given pair of data (Vos 2011). In the textbooks the most frequently used parameters are *a* and *b*, so in the examination these letters may trigger memorisation without understanding. To categorise this task format of calculating parameters of a mathematical model prompted by the question *"calculate a and b"*, I coin it as: *mechanistic mathematising*.

Figure 41.3 illustrates how mechanistic mathematising fits the modelling cycle by Blum and Leiss (2005). In the examination papers I also observed a second format, for which the *Trivet Task* is an example. In this format a situation is extensively described by photographs and diagrams and then the ready-made model needs to be *reproduced* from the situation. In this question format, the modelling activities are asked in reversed order. I coin this as *reproductive mathematising* (see Fig. 41.3): "show that the formula fits the situation". The advantage in examinations is that a range of competencies from the beginning of the modelling cycle are covered. This raises test validity. Also, all students can continue with the correct mathematical model in ensuing tasks, even if they were not able to reproduce the model. This narrows down the variety in answers, thus raising test reliability. However, a disadvantage of this format is that students adopt *smart strategies* to reach the ready-made model, such as reasoning backwards using surface features of the model (e.g., a *cosine* indicates a horizontal projection) as found by Schaap et al. (2011).

5 Conclusions and Recommendations

The Dutch mathematics examination papers of 2006, 2008 and 2010 demonstrate that certain aspects of modelling can well be framed into assessment tasks that meet the strict criteria of reliability as needed in large-scale, standardized assessment. In this study, the tasks were scrutinized in light of validity with respect to modelling. The studied test items show a wide variety of situations, demonstrating that mathematics is everywhere; this is a valid aspect of modelling. On the other hand to increase test reliability, the complexity of tasks was reduced and students' creativity was limited. This was achieved in several ways.

First, despite the tasks being situated in real life contexts, students were not asked to solve problems related to the situation, but merely to perform mathematical activities loosely related to the situation described. Second, the examination designers broke the modelling cycle into separate steps, reducing the range of answers. Third, modelling activities such as structuring, simplifying and mathematising were avoided. These three reductions led to tasks that start from a situation together with a mathematical model (table, graph, diagram). The models were offered ready-made in the task. More than 90 % of these tasks merely asked for calculations leading to a number answer. When students are trained on such tasks, they will not have ownership over the mathematical model and the message is conveyed that, starting from a problem situation, the necessary steps to reach a mathematical model are not to be tested. This reduces the validity with respect to modelling.

The examination designers used two formats to emulate the steps of structuring, simplifying and mathematising. These formats recurred each year at general and pre-university level. The format of *mechanistic mathematising* asks students to find parameters of a ready-made model (*"calculate a and b"*); in this way students are helped into the mathematical world. Instead of a formula, they only need to produce number values. The second format, *reproductive mathematising*, asks students to deduce a ready-made model from a situation. This format asks students to give meaning to the mathematical model and to make considerations on its creation, but in not one examination paper were the students asked to introduce variables or make assumptions. Nevertheless, from research we know that the translation of a narrative description into a mental model, and thereafter into draft sketches, diagrams and all other means needed to develop a mathematical model, is not a straight-forward step (e.g., Bower and Morrow 1990). The Dutch examination papers thus may meet criteria of reliability, but with respect to modelling they are limited in validity. Therefore, it is recommended that more assessment research be carried out, in particular on the first steps in the modelling cycle. Additionally, we do not need to throw out the baby with the bathwater and abolish project-based or portfolio assessment: these formats have a high validity in relation to the creative aspects of modelling; research may assist in improving these formats with respect to test reliability.

References

Blum, W., & Leiss, D. (2005). "Filling up" – The problem of independence-preserving teacher interventions in lessons with demanding modelling tasks. In M. Bosch (Ed.), *Proceedings of 5th Congress European Research Mathematics Education.* Gerona: Universitat Ramon Llull.

Bower, G. H., & Morrow, D. G. (1990). Mental models in narrative comprehension. *Science, 247*, 44–48.

Frejd, P. (2011). *An investigation of mathematical modelling in the Swedish national course tests in mathematics*. Paper presented at 7th Congress of European Research in Mathematics Education. University of Rzeszów.

Haines, C., & Crouch, R. (2001). Recognizing constructs in mathematical modelling. *Teaching Mathematics and Its Applications, 20*(3), 129–138.

Houston, K. (2007). Assessing the "phases" of mathematical modelling. In W. Blum, P. L. Galbraith, H.-W. Henn, & M. Niss (Eds.), *Modelling and applications in mathematics education: The 14th ICMI study* (pp. 249–256). New York: Springer.

Kaiser, G., Schwarz, B., & Buchholtz, N. (2011). Authentic modelling problems in mathematics education. In G. Kaiser, W. Blum, R. Borromeo Ferri, & G. Stillman (Eds.), *Trends in teaching and learning of mathematical modelling* (pp. 591–601). New York: Springer.

Palm, T. (2007). Features and impact of the authenticity of applied mathematical school tasks. In W. Blum, P. L. Galbraith, H.-W. Henn, & M. Niss (Eds.), *Modelling and applications in mathematics education: The 14th ICMI study* (pp. 201–208). New York: Springer.

Schaap, S., Vos, P., & Goedhart, M. J. (2011). De vertaalslag van een situatie naar een wiskundige formule; vraagstellingen en leerlingprestaties op het centraal examen [The translation of a situation into a mathematical formula]. *Tijdschrift voor Didactiek der Bètawetenschappen, 28*(1/2), 21–44.

Strauss, A., & Corbin, J. (1997). *Grounded theory in practice*. Thousand Oaks, CA: Sage.

Vos, P. (2007). Assessment of applied mathematics and modelling: Using a laboratory-like environment. In W. Blum, P. L. Galbraith, H.-W. Henn, & M. Niss (Eds.), *Modelling and applications in mathematics education: The 14th ICMI study* (pp. 441–448). New York: Springer.

Vos, P. (2010). The Dutch maths curriculum: 25 years of modelling. In R. Lesh, P. L. Galbraith, C. R. Haines, & A. Hurford (Eds.), *Modelling students' mathematical modeling competencies* (pp. 611–620). New York: Springer.

Vos, P. (2011). *Modelling and task formats: Do we include mechanistic, reproductive modelling into modelling?* Paper presented at 7th Conference on European Research in Mathematics Education. University of Rzeszów.

Part VI
Applicability at Different Levels of Schooling, Vocational Education, and in Tertiary Education

Part VI
Applicability at Different Levels of Schooling, Vocational Education, and in Tertiary Education

Chapter 42
Complex Modelling in the Primary and Middle School Years: An Interdisciplinary Approach

Lyn D. English

Abstract The world's increasing complexity, competitiveness, interconnectivity, and dependence on technology generate new challenges for nations and individuals that cannot be met by continuing education as usual. With the proliferation of complex systems have come new technologies for communication, collaboration, and conceptualisation. These technologies have led to significant changes in the forms of mathematical and scientific thinking required beyond the classroom. Modelling, in its various forms, can develop and broaden students' mathematical and scientific thinking beyond the standard curriculum. This chapter first considers future competencies in the mathematical sciences within an increasingly complex world. Consideration is then given to interdisciplinary problem solving and models and modelling, as one means of addressing these competencies. Illustrative case studies involving complex, interdisciplinary modelling activities in Years 1 and 7 are presented.

1 Introduction

In recent decades our global community has rapidly become a knowledge driven society, one that is increasingly dependent on the distribution and exchange of services and commodities, and one that has become highly inventive where creativity, imagination, and innovation are key players. At the same time, the world has become governed by complex systems—financial corporations, the World Wide Web, education and health systems, traffic jams, and classrooms. These are just some of the

L.D. English (✉)
Faculty of Education, Queensland University of Technology, Brisbane, QLD, Australia
e-mail: l.english@qut.edu.au

G.A. Stillman et al. (eds.), *Teaching Mathematical Modelling: Connecting to Research and Practice*, International Perspectives on the Teaching and Learning of Mathematical Modelling, DOI 10.1007/978-94-007-6540-5_42,

complex systems we deal with on a regular basis. For all citizens, an appreciation and understanding of the world as interlocked complex systems is critical for making effective decisions about one's life as both an individual and as a community member (Bar-Yam 2004; Jacobson and Wilensky 2006; Lesh 2006).

Complexity—the study of systems of interconnected components whose behavior cannot be explained solely by the properties of their parts but from the behavior that arises from their interconnectedness—is a field that has led to significant scientific methodological advances. With the proliferation of complex systems has come new technologies for communication, collaboration, and conceptualisation. These technologies have led to significant changes in the forms of mathematical and scientific thinking that are needed beyond the classroom, such as the need to generate, analyse, operate on, and transform complex data sets (English and Sriraman 2010).

Educational leaders from different walks of life are emphasising the importance of developing students' abilities to deal with complex systems for success beyond school. Such abilities include: constructing, describing, explaining, manipulating, and predicting complex systems; working on multi-phase and multi-component projects in which planning, monitoring, and communicating are critical for success; and adapting rapidly to ever-evolving conceptual tools (or complex artifacts) and resources (Gainsburg 2006; Lesh and Doerr 2003; Lesh and Zawojewski 2007; Pellegrino and Hilton 2012).

The purpose of this chapter is threefold: First, it argues for the need to build a stronger foundation in the mathematical sciences, one that will equip students for the challenges of the twenty-first century. Second, as one means of achieving this aim, the chapter recommends an increased focus on interdisciplinary problem solving that engages students in complex modelling with challenging, life-based scenarios. Third, as an illustration of such modelling, consideration is given to selected cases from two design-based studies in Years 1 and 7.

2 Future Competencies in the Mathematical Sciences

The Australian *Mathematics, Engineering & Science in the National Interest* report (Office of the Chief Scientist 2012) highlighted the universal perspective that an education in these disciplines is essential to a nation's future prosperity. Numerous concerns have been expressed over students' achievements in the sciences, both in Australia and internationally (e.g., Australian Academy of Science 2006, 2010; Business Council of Australia 2007; The National Academies, USA 2009). Indeed, the Business Council of Australia stressed that "too many young Australians are being left behind by our school education system," beginning with early learning in mathematics and science, and that many aspects of our school system "have not changed since the 1960s." Likewise, other nations are highlighting the need for a renaissance in the science, technology, engineering, and mathematics (STEM) fields. For example, the first recommendation of The National Academies' *Rising above the Gathering Storm* (2007, 2009) was to vastly improve K-12 science and mathematics education.

With the advent of digital technologies have come changes in the future world of work for our students. As Clayton (1999) and others (e.g., Hoyles et al. 2010; Jenkins et al. 2006; Lombardi and Lombardi 2007; Pellegrino and Hilton 2012) have stressed, the availability of increasingly sophisticated technology has led to changes in the way mathematics and science are being applied in workplace settings. These technological changes have led to both the addition of new competencies and the elimination of existing skills, together with increased application of interdisciplinary knowledge in solving problems and communicating results. Although we cannot simply list a number of competencies and assume these can be automatically applied to the workplace setting, there are several that employers generally consider essential to productive outcomes (e.g., Doerr and English 2003; English 2008; English et al. 2008; Gainsburg 2006; Hoyles et al. 2002; Lesh and Zawojewski 2007; Pellegrino and Hilton 2012). In particular, the following are some of the core competencies that have been identified as key elements of productive and innovative workplace practices.

- Problem solving, including working collaboratively on complex problems where planning, overseeing, moderating, and communicating are essential components for success;
- Techno-mathematical literacy, "where the mathematics is expressed through technological artefacts" (Hoyles et al. 2002, p. 14);
- Applying numerical and algebraic reasoning in an efficient, flexible, and creative manner;
- Generating, analysing, operating on, and transforming complex data sets;
- Applying an understanding of core ideas from ratio and proportion, probability, rate, change, accumulation, continuity, and limit;
- Constructing, describing, explaining, manipulating, and predicting complex systems;
- Thinking critically and being able to make sound judgments, including being able to distinguish reliable from unreliable information sources;
- Synthesising, where an extended argument is followed across multiple modalities;
- Engaging in research activity involving the investigation, discovery, and dissemination of pertinent information in a credible manner; and
- Flexibility in working across disciplines to generate innovative and effective solutions.

3 Interdisciplinary Problem Solving

These future competencies alert us to rethink the nature of the learning tasks we implement in our classrooms. I argue that we need a greater focus on future-oriented, interdisciplinary problem-solving experiences, ones that mirror problem solving beyond the classroom. Such a focus is especially needed, given that "problems themselves change as rapidly as the professions and social structures in which they are embedded change" (Hamilton 2007, p. 2). For example, experiences that draw upon the broad field of engineering provide powerful links between the classroom and the real world, enabling students to apply their mathematics and science learning to the solution of authentic problems (English and Mousoulides 2011; Kuehner and Mauch 2006).

Our challenge then, is how to promote creative and flexible use of mathematical and scientific ideas within an interdisciplinary context where students solve substantive, authentic problems that address multiple core learnings. One approach is through modelling involving cycles of model construction, evaluation, and revision, which is fundamental to mathematical and scientific understanding and to the professional practice of mathematicians and scientists (English 2010b; Lesh and Zawojewski 2007; Romberg et al. 2005). Modelling is not just confined to mathematics and science, however. Other disciplines including engineering, economics, information systems, social and environmental science, and the arts have also contributed in large part to the powerful mathematical models we have in place for dealing with a range of complex problems (Beckmann et al. 2005). Unfortunately, many mathematics and science curricula do not capitalise on the contributions of other disciplines. A more interdisciplinary and unifying model-based approach to students' mathematics learning could go some way towards alleviating the well-known "one inch deep and one mile wide" problem in many of our curricula (Sabelli 2006, p. 7; Sriraman and Steinthorsdottir 2007). There is limited research, however, on ways in which we might incorporate other disciplines within the mathematics curriculum.

4 Models and Modelling

Modelling is increasingly recognised as a powerful vehicle for not only promoting students' understanding of a wide range of key mathematical and scientific concepts, but also for helping them appreciate the potential of the mathematical sciences as a critical tool for analysing important issues in their lives, communities, and society in general (Greer and Mukhopadhyay 2012; Greer et al. 2007; Romberg et al. 2005). In acknowledging the efforts made in many curricular documents to place a greater focus on mathematical applications and modelling, Greer and Mukhopadhyay (2012) expressed concerns that insufficient attention is being given to critical, political, and ethical considerations. These considerations warrant significant attention in the design and implementation of modelling experiences, beginning with the primary school years.

Perspectives on models and modelling are broad and have been addressed extensively in the literature. These have included solving word problems, conducting mathematical simulations, creating representations of problem situations (including constructing explanations of natural phenomena), and creating internal, psychological representations while solving a particular problem (e.g., English and Halford 1995; Gravemeijer 1999; Greer and Verschaffel 2007; Lesh and Doerr 2003;

Romberg et al. 2005). Furthermore, these perspectives appear to be ever-evolving as research in the field continues to expand. Lesh and Fennewald (2010) provide a useful "first-iteration definition of a model," one that is apt for the present case studies, namely, "A model is a system for describing (or explaining, or designing) another system(s) for some clearly specified purpose" (p. 7).

A significant component of modelling is the crossing of disciplinary boundaries, where there is an emphasis on the structure of ideas, connected forms of knowledge, and the adaptation of complex ideas to new contexts (Hamilton et al. 2008). The modelling problems I have implemented in classrooms are realistically complex situations where the problem solver engages in mathematical and scientific thinking beyond the usual school experience and where the products to be generated often include complex artifacts or conceptual tools that are needed for some purpose, or to accomplish some goal (Lesh and Zawojewski 2007). I refer to these problems as *complex modelling* because students are faced with multidimensional situations where they need to make reasoned and sophisticated choices in their application of mathematical and scientific ideas and processes. Furthermore, these problems require students to present their models to their peers for questioning and constructive feedback, which, in itself presents a complex situation requiring critical and empathetical thinking (cf. Greer and Verschaffel's third level of "critical modelling" 2007).

To illustrate some of the ideas I have advanced to this point, I give consideration to two case studies of interdisciplinary modelling, one that I implemented in Year 1, and the other, in Year 7. The former involved data modelling experiences that drew upon the children's curriculum studies of caring for the environment. The latter introduced students to the fields of engineering, specifically civil engineering, where the extensive civil constructions taking place in their city formed the problem context.

5 Case Studies of Models and Modelling

In both the Year 1 and Year 7 projects, design-based research specifically a design experiment, involving the learning of students, teachers, and researchers (Kelly et al. 2008) was adopted. Comparative case studies (focus groups) were also included (Yin 2003). The activities were designed in collaboration with the teachers, who implemented them in their classrooms across the duration of each of the 3-year, longitudinal studies.

Data collection included copies of all student artefacts and the videotaping and audiotaping of the work of the focus groups and whole-class discussions, with all taping transcribed. Data analyses involved ethnomethodological interpretative practices (Erickson 1998), with data progressively reviewed, transcribed, coded, and examined for patterns and trends in the students' developments using constant comparative strategies (Creswell 2012).

5.1 Data Modelling in Year 1: Background

From 2009 to 2011, I conducted a longitudinal study of data modelling in three classrooms in a state school situated within a middle socio-economic area of Brisbane, an Australian capital city. The study followed the children from Year 1 to Year 3 (6–8 years). The children and their teachers engaged in multiple, life-based experiences that incorporated other disciplines, such as health and nutrition, and environmental studies. The children investigated meaningful phenomena, decided what was worthy of attention (identifying complex attributes), and then progressed to organising, structuring, visualising, and representing data (English 2010a; Lehrer and Schauble 2006). Given that identifying variation, drawing inferences, and making predictions are also important components of data modelling in the early years (Watson 2006), these components were incorporated within the learning experiences.

In designing the classroom activities, literature was used as a basis for the problem context. It is well documented that storytelling provides an effective context for mathematical learning, with children being more motivated to engage in mathematical activities and displaying gains in achievement (van den Heuvel-Panhuizen and Van den Boogaard 2008). Picture story books that addressed the overall theme of *Looking after our Environment*, a key theme in the teachers' curriculum at the time, were selected.

To illustrate the data modelling activities that were implemented, the second and third of four activities of the first year of the study are addressed, namely, *Fun with Michael Recycle* and *Litterbug Doug*. The Australian picture story books that served as the basis for these activities were *Michael Recycle* (Bethel 2008) and *Litterbug Doug* (Bethel 2009). The former tells the story of Michael Recycle who came from the sky to clean up a very dirty town, with his motto, "I'm green and I'm keen to save the planet." Litterbug Doug was originally a dirty creature who lived in a pile of rubbish in a very clean town. A "green-caped crusader" then swooped to the Earth to reform Litterbug Doug. As a consequence, Litterbug Doug became the Litter Police for the town and enthusiastically monitored the town's environment. Children worked in small groups for each activity.

5.2 Data Modelling in Year 1: Fun with Michael Recycle

Fun with Michael Recycle involved two lessons (lesson one, average duration of 30 min and lesson two, 60 min). The activity involved posing questions, identifying and generating attributes, organising and analysing data, and displaying and representing data in different ways.

Prior to the lessons, the storybook, *Michael Recycle*, was read and discussed, and the classroom was set up with collections of reusable/recyclable and waste items. Next, each child in each group was given two Post-It notes and the group was directed to explore the classroom for these various items. Each group member was to draw and name an item on each Post-It note. The groups subsequently returned to their group desk and proceeded to discuss the attributes of their items, then organise, analyse, and represent their data however they chose (on a large sheet of paper provided). On completion, the groups reported back to the class on how they represented their data. A brief whole class discussion followed on the nature of the attributes the children had identified and how they had organised and represented their data.

Following this, the children were advised that Michael Recycle "really likes the different ways you have represented your recyclable/reusable and waste items but would like you to represent them in a different way on your chart paper." The children were given a second sheet of paper to do so and were to leave their initial representation sheet intact. On completion, the groups reported back to the class, during which they were encouraged to explain their new representation and indicate how it differed from their first.

5.2.1 Selected Findings

Most groups (of 15 in total) focused on the attributes of recycle/junk/waste/reuse in structuring and representing their data on their first attempt, which is not surprising given the task context. On their second attempt, however, over half of the groups changed their attributes along with their representations. Although not directed to do so, children's generation of new attributes demonstrated their ability to switch their attention from one item feature to another. That is, they needed to consider what was worthy of attention and what needed to be placed in the background, reflecting Lehrer and Lesh's (2003) notion of "lifting away from the plane of activity" (p. 377).

Children's representations for *Fun with Michael Recycle* were predominantly pictographs, which was likely influenced by the task design. Task presentation and context can create both obstacles and supports in developing children's statistical reasoning (Cooper and Dunne 2000; Pfannkuch 2011). Here, the initial use of Post-It notes likely limited the forms of representation the children created. Nevertheless, the children did display an awareness of the structure of their pictographs, making effective use of rows and columns, and appropriate inscriptions. These early representations are important in assisting young children in abstracting or simplifying information they have gathered from their data collection (Konold and Higgins 2003).

Divergent ways of creating re-representations were observed, with children again displaying a repertoire of inscriptions including drawings, written text, numerical symbols, and other referents (ticks and crosses). The children also displayed meta-representational competence (di Sessa 2004; Lehrer and Lesh 2003), an important factor in young children's development of data modelling. Such competence, albeit emerging, was evident in the children's use of inscriptions, their structuring and displaying of data, their detection of redundant information, their awareness of the need to eliminate unnecessary features, and their conservation of ideas and quantities of items. The children had not received direct instruction on these components; their seemingly naturally developing metarepresentational competence appeared to play a substantial role in shaping their learning and reasoning in working the activity (Lehrer and Lesh 2003).

5.3 Data Modelling in Year 1: Litterbug Doug

The second activity, *Litterbug Doug*, was designed to engage the children in interpreting tables of data, identifying variations in the data, posing questions, and making predictions. The activity was implemented in one lesson, average duration of 75 min. Prior to the lesson, the children read and discussed the storybook, *Litterbug Doug*. The lesson began with the teacher explaining that, "Now that Litterbug Doug has become the Litter Police, the townsfolk are interested to see what he collects in Central Park during his first three days. They also want to know if Litterbug Doug is doing a good job of collecting litter in Central Park." The children were then shown a table displaying how many of each of five items Litterbug Doug collected on day 1, with the explanation that, "As a start, the town's mayor asked Litterbug Doug to show him what he collected on his first day, Monday. Litterbug Doug showed the mayor what he saw and what he collected in the park." Next, the children were posed questions to explore their interpretation of the table, given that they had had almost no exposure to such a table. It was then explained to the children that, "Litterbug Doug has now collected litter in Central Park for three days and the townsfolk are keen to see how much he has collected." The children were next presented with a second table showing how many items had been collected from Monday through to Wednesday, with the Thursday column left blank. In their groups, children were to explore the second table, first noting the numbers of items collected on the second and third days, then how the data varied across the first 3 days and why this might be the case. Their final task was to consider the blank Thursday column. The children were to predict how many different items Litterbug Doug might have collected on Thursday. On completion, the groups reported back to the class on the variation they noticed in the data and on their predictions for Thursday.

5.3.1 Selected Findings

Children's predictions for the numbers of items Litterbug Doug might have collected on the Thursday suggested that they had an informal awareness of the range and variation in the existing data. Twelve groups (out of 13 [student absenteeism reduced the number of groups for this activity]) recorded predictions of values ranging from 0 to 10. All but one of these 12 groups explicitly recognised that wild outliers (e.g., 56, 45) would be unlikely.

In their class presentations, five groups indicated that they considered the frequencies of the values across the rows of the table. They avoided repeating a quantity, or repeated a quantity, or gave a quantity that was not in the existing row. For example, when asked why a group recorded 7 cans for Thursday, they explained, "Because there was no seven in that one. We did a number that wasn't in that line." One group justified their recording of four newspapers as, "'Cause he found six but then he didn't find that much on the other two days so I thought to do four 'cause he didn't find that much on the other two days."

Two other groups displayed a more sophisticated awareness of trends in the data, for example, "They went up and down (indicating Monday to Tuesday to Wednesday for the apple cores), then it kept counting down" (referring to the 3 they added to the Thursday column). One child in this group actually did a corresponding hand motion to illustrate the trend. Other approaches (3 groups) to predicting the values for the Thursday column included the use of patterns, numerical sequences (e.g., 4, 3, 2, 1), and odd and even numbers.

5.4 *Engineering-Based Modelling Experiences: Background*

This 3-year longitudinal study (2009–2011) across Years 7–9 was conducted in five classrooms across three Brisbane private schools (one co-educational, and two single-sex schools). The study was undertaken in collaboration with an engineering educator and a science educator (e.g., English et al. 2010). The project team worked with the classroom teachers in developing a range of engineering-based modelling problems, which engaged student groups in devising, testing, and evaluating life-based engineering problems that demonstrated the link to mathematical principles and science inquiry. We also developed comprehensive teaching notes, which supplemented the teachers' existing knowledge of engineering ideas and principles.

5.4.1 Bridge Design and Construction

In the first year of the study, we introduced the students to the varied world of engineering, including the different roles of engineers (two lessons of approximately 45 min duration). This was followed by 5–7 lessons that explored bridge designs and their construction. These lessons entailed: learning about the work of civil engineers; exploring bridge structure with a focus on the main types of bridges in Brisbane; recognising features/constraints of the main bridge types; and investigating the concepts of tension, compression, load distribution, reinforcement, and strength, and their importance in bridge designs. Next, the students were to plan, design, model, and construct a truss bridge with given constraints and materials. The students documented and reflected on the engineering design processes they used in constructing their bridge.

The activity was set within the context of the students assisting two engineer graduates, as follows:

Truss Bridge Project

Remember our engineer graduates, Ben and Jane, and our need for a bus bridge across the Brisbane River from Adelaide St to Southbank? Well you are now going to help them design and build a model bridge that will solve the transport problem near Victoria Bridge. Engineers always consider their design objective when creating their models. Also, they often have many design constraints or limitations they have to take into consideration. Ben and Jane's design objective is to make a truss bridge that:

1. Can span a distance of 150 m.
2. Must support the most weight for the vehicles that will pass over it.
3. Must not disturb the river's fish.
4. Must not obstruct normal watercraft, such as the City Cat.
5. Must be at least 12 m above water level.

A limited number of resources to build their bridge (drinking straws, sticky tape, scissors, rulers, small containers, metal washers) was supplied. The students were advised to spend time generating ideas, planning and designing their bridges, and taking measurements before the actual construction. On completion of their bridge, the students reported back to the class explaining their steps to designing, modelling, and building their bridge. They were to explain how they used engineering design processes, such as, define the problem, brainstorm, select the most promising design, communicate the design, create and test a model, and then evaluate and revise their design. Finally, the students were to indicate how they might have improved their design to strengthen their bridge.

5.4.2 Selected Findings

Students' responses revealed a number of iterative design processes, where the problem goals, including the constraints placed on bridge construction, played a major role in students' planning of their model and its construction. For example, one focus group from the all-female school reported on how they decided between two possible designs, on the importance of initially documenting their chosen design, and on how they worked at improving their evolving bridge model:

Jenny: Well, okay, everyone knows we had to design a bridge that was 150 m long and environmentally friendly and 12 m above the water, and so first of all we like brainstormed and figured out all the details and um we shared all our ideas, like we each sort of brainstormed individually and then we put them together.

Shelley: We came up with two designs and we weren't sure which one to use so we had input from every single member and from there we worked out which one would sustain the most weight.

Jilly: Um, then we drew a diagram to make sure everyone understood what we were doing and then we started to make the thing by first making the criss-crosses in our design and then putting on the layers kind of and then adding the sticks (straws) in between.

Shelley: Initially, we had, it was kind of different; we altered it a bit though, when we found some last minute materials so...

Jenny: And then when we saw it like collapsing, we realised if we had more straws we could have like put more members in but we probably could have used them more wisely maybe, in some places...

Lainie: We also had this theory, if all the straws were 1.5 cm longer our bridge would have been perfect.

It is also interesting to note the students' awareness of the environmental concerns (as Jenny mentioned above) and a consideration of minimising the costs involved, as illustrated in the report from a member of one of the co-educational school's focus groups:

The bridge with no piers and trusses supports was chosen because nothing was allowed to disturb the aquatic life in the water, meaning there were no piers allowed in the design. The trusses were placed in the bridges [sic] design to support the beam bridge because trusses can make the bridge span from a possible 80 m to a possible 180 m. Using the trusses also makes the bridge thicker meaning more strength without having so much wind resistance and without using tonnes of material.

The student's recommendations for improving his model included "extending the trusses on the bridge right to the end of the bridge, making the design more symmetrical on either side, [and] having better joints where the trusses were connected..." With his concern for the environment, the student recommended another type of bridge:

If another type of bridge was allowed, the chosen bridge would have been a suspension bridge because it would not have needed piers or anything else touching the water or interfering with the aquatic life. It would have easily spanned the 150 m distance the bridge needed to be because suspension bridges can span up to 200 m and the suspension bridge would have lasted longer and would be a lot stronger.

6 Discussion and Concluding Points

This chapter has argued for an increased focus on future-oriented learning experiences, ones that engage students in the kinds of mathematical and scientific thinking needed for challenges beyond the classroom. It is important that we design problems that cut across disciplines and encourage students to apply their learning from multiple sources because such challenges do not come neatly packaged in one-dimensional domains. As noted, students face a working future where they will need competencies that are not given adequate attention in existing curricula, and indeed, competencies that we have yet to determine.

One means of preparing students for existing and future challenges is the inclusion of complex modelling problems within the curriculum. Such problems place students in multidimensional situations that require them to make reasoned and sophisticated choices about the knowledge they will apply and how, and ways in which they might communicate and share their products. The importance of diverse forms of thinking and working cannot be underestimated here. The ability to deal with complex, life-based situations requires flexibility, creativity and innovation, critical analysis, empathetical considerations, political awareness, and a commitment to knowledge generation and refinement.

In the first case study, young children displayed abilities to generate, analyse, operate on, represent, and transform data, as well as to draw inferences from given data—competencies that provide foundations for navigating effectively through the complex data they will increasingly face from multiple directions. In completing the activities, the children were also dealing with important environmental issues in their considerations of how they might better protect their surroundings. Basing the data modelling experiences on the children's existing studies of society and the environment enhanced their engagement with the problem scenarios. Opportunities for the diverse forms of thinking identified above were ever-present.

The engineering-based modelling experiences likewise targeted future competencies in the mathematical sciences, connecting students' learning across disciplines and involving them in planning, designing, constructing, testing, and refining a life-based bridge model to solve one of their city's transport problems. In applying their learning from mathematics and science, the students also needed to deal with the problem constraints including environmental concerns and cost factors (limited materials); safety issues also emerged in their group discussions.

In both case studies, students' sharing of ideas with others, both in a group situation and in peer reporting, encouraged awareness, respect, and tolerance of the thinking and feelings of others. However, a more overt focus on these underrepresented aspects of modelling would likely have broadened and enriched the students' learning. As previously argued, modelling activities need to include these critical elements, commencing in the early primary school years, if we are to fully develop the future competencies highlighted in this chapter. Certainly, a good deal more research is warranted here, as noted by Greer and Mukhopadhyay (2012), Jablonka and Gellert (2012), and many others.

In concluding my argument for the inclusion of complex modelling in the primary and middle school years, I reiterate the words of Greer and Mukhopadhyay (2003): "The most salient features of most documents that lay out a K-12 program for mathematics education is that they make an intellectually exciting program boring," a feature they refer to as "intellectual child abuse" (p. 4). Clearly, we need to rise above this situation and look towards nurturing students' abilities to work creatively and innovatively in generating mathematical and scientific knowledge, as distinct from working creatively on tasks that provide the required knowledge (Bereiter and Scardamalia 2006).

Acknowledgements The research projects addressed in this paper have been supported by Discovery and Linkage grants from the Australian Research Council (ARC) [DP0984178 and LP0989152, respectively]. Any opinions, findings, and conclusions or recommendations expressed in this paper are those of the author and do not necessarily reflect the views of the ARC. I wish to acknowledge the enthusiastic participation of all the classroom teachers and their students, as well as the excellent support provided by my senior research assistant, Jo Macri.

References

Australian Academy of Science. (2006). *Critical skills for Australia's future*. www.review.ms.unimelb.edu.au. Accessed June 2011.

Australian Academy of Science. (2010). Academy 2010 election statement: *Empower science, power the future*. http://www.science.org.au/reports/election-statement.html. Accessed June 2011.

Bar-Yam, Y. (2004). *Making things work: Solving complex problems in a complex world*. Cambridge, MA: NECSJ, Knowledge Press.

Beckmann, A., Michelsen, C., & Sriraman, B. (2005). *Proceedings of the 1st international symposium of mathematics and its connections to the arts and sciences*. Schwöbisch Gmund: The University of Education.

Bereiter, C., & Scardamalia, M. (2006). Education for the knowledge age: Design-centered models of teaching and instruction. In P. A. Alexander & P. H. Winne (Eds.), *Handbook of educational psychology* (pp. 695–714). Mahwah, NJ: Lawrence Erlbaum.

Bethel, E. (2008). *Michael recycle*. Sydney: Koala Books.

Bethel, E. (2009). *Litterbug Doug*. Sydney: Koala Books.

Business Council of Australia (2007, April 26). *Action required to make education Australia's advantage*. http://www.bca.com.au/Content.aspx?ContentID=101154. Accessed 10 July 2010.

Clayton, M. (1999). What skills does mathematics education need to provide? In C. Hoyles, C. Morgan, & G. Woodhouse (Eds.), *Rethinking the mathematics curriculum* (pp. 22–28). London: Falmer Press.

Cooper, B., & Dunne, M. (2000). *Assessing children's mathematical knowledge: Social class, sex and problem solving*. Buckingham: Open University Press.

Creswell, J. W. (2012). *Educational research: Planning, conducting, and evaluating quantitative and qualitative research* (4th ed.). Upper Saddle River: Merrill Prentice Hall.

di Sessa, A. A. (2004). Metarepresentation: Native competence and targets for instruction. *Cognition and Instruction, 22*(3), 291–292.

Doerr, H. M., & English, L. D. (2003). A modeling perspective on students' mathematical reasoning about data. *Journal for Research in Mathematics Education, 34*(2), 110–136.

English, L. D. (2008). Setting an agenda for international research in mathematics education. In L. D. English (Ed.), *Handbook of international research in mathematics education: Directions for the 21st century* (Vol. 2, pp. 3–19). New York: Routledge.

English, L. D. (2010a). Young children's early modelling with data. *Mathematics Education Research Journal, 22*(2), 24–47.

English, L. D. (2010b). Interdisciplinary modelling. *Mediterranean Journal for Research in Mathematics Education, 9*(1), 73–90 (Special Issue).

English, L. D., & Halford, G. S. (1995). *Mathematics education: Models and processes*. Mahwah: Lawrence Erlbaum.

English, L. D., & Mousoulides, N. (2011). Engineering-based modelling experiences in the elementary classroom. In M. S. Khine & I. M. Saleh (Eds.), *Dynamic modeling: Cognitive tool for scientific enquiry*. New York: Springer.

English, L. D., & Sriraman, B. (2010). Problem solving for the 21st century. In B. Sriraman & L. D. English (Eds.), *Theories of mathematics education: Seeking new frontiers* (Advances in mathematics education series, pp. 263–285). New York: Springer.

English, L. D., Jones, G. A., Bartolini Bussi, M. G., Lesh, R., Sriraman, B., & Tirosh, D. (2008). Moving forward in international mathematics education research. In L. D. English (Ed.), *Handbook of international research in mathematics education: Directions for the 21st century* (2nd ed., pp. 872–905). New York: Routledge.

English, L. D., Hudson, P. B., & Dawes, L. (2010, November 26–27). *Middle school students' perception of engineering*. Paper presented at the STEM in education conference, Brisbane. http://stem.ed.qut.edu.au/index.php/conference-proceedings.html#m. Accessed 2 Aug 2012.

Erickson, F. (1998). Qualitative research methods for science education. In B. J. Fraser & K. G. Tobin (Eds.), *International handbook of science education Part 2* (pp. 1155–1173). Dordrecht: Kluwer.

Gainsburg, J. (2006). The mathematical modeling of structural engineers. *Mathematical Thinking and Learning, 8*(1), 3–36.

Gravemeijer, K. (1999). How emergent models may foster the construction of formal mathematics. *Mathematical Thinking and Learning, 1*, 155–177.

Greer, B., & Mukhopadhyay, S. (2003). What is mathematics education for? *The Mathematics Educator, 13*(2), 2–6.

Greer, B., & Mukhopadhyay, S. (2012). The hegemony of mathematics. In O. Skovsmose & B. Greer (Eds.), *Opening the cage: Critique and politics of mathematics education* (pp. 229–248). Rotterdam: Sense Publishers.

Greer, B., & Verschaffel, L. (2007). Modelling competencies—Overview. In W. Blum, P. L. Galbraith, H.-W. Henn, & M. Niss (Eds.), *Modelling and applications in mathematics education: ICMI study 14* (pp. 219–224). New York: Springer.

Greer, B., Verschaffel, L., & Mukhopadhyay, S. (2007). Modelling for life: Mathematics and children's experience. In W. Blum, P. L. Galbraith, H.-W. Henn, & M. Niss (Eds.), *Modelling and applications in mathematics education: ICMI study 14* (pp. 89–98). New York: Springer.

Hamilton, E. (2007). What changes are needed in the kind of problem solving situations where mathematical thinking is needed beyond school? In R. Lesh, E. Hamilton, & J. Kaput (Eds.), *Foundations for the future in mathematics education* (pp. 1–6). Mahwah: Lawrence Erlbaum.

Hamilton, E., Lesh, R., Lester, F., & Brilleslyper, M. (2008). Model-eliciting activities (MEAs) as a bridge between engineering education research and mathematics education research. *Advances in Engineering Education, 1*(2), 1–25.

Hoyles, C., Wolf, A., Molyneux-Hodgson, S., & Kent, P. (2002). *Mathematical skills in the workplace*. London: Science, Technology and Mathematics Council. Retrieved May 24, 2006, from http://www.ioe.ac.uk/tlrp/technomaths/skills2002/

Hoyles, C., Noss, R., Kent, P., & Bakker, A. (2010). *Improving mathematics at work: The need for techno-mathematical literacies*. London: Routledge.

Jablonka, E., & Gellert, U. (2012). Potential, pitfalls, and discriminations: Curriculum conceptions revisited. In O. Skovsmose & B. Greer (Eds.), *Opening the cage: Critique and politics of mathematics education* (pp. 229–248). Rotterdam: Sense Publishers.

Jacobson, M., & Wilensky, U. (2006). Complex systems in education: Scientific and educational importance and implications for the learning sciences. *The Journal of the Learning Sciences, 15*(1), 11–34.

Jenkins, H., Clinton, K., Purushotma, R., Robinson, A. J., & Weigel, M. (2006). *Confronting the challenges of participatory culture: Media education for the 21st century*. Chicago: MacArthur Foundation.

Katehi, L., Pearson, G., & Feder, M. (2009). *Engineering in K-12 education: Understanding the status and improving the prospects*. Washington, DC: The National Academies Press.

Kelly, A. E., Lesh, R. A., & Baek, J. Y. (Eds.). (2008). *Handbook of design research methods in education*. New York: Routledge.

Konold, C., & Higgins, T. L. (2003). Reasoning about data. In J. Kilpatrick, W. G. Martin, & D. Schifter (Eds.), *A research companion to principles and standards for school mathematics*. Reston, VA: National Council of Teachers of Mathematics.

Kuehner, J. P., & Mauch, E. K. (2006). Engineering applications for demonstrating mathematical problem-solving methods at the secondary education level. *Teaching Mathematics and Its Applications, 25*(4), 189–195.

Lehrer, R., & Lesh, R. (2003). Mathematical learning. In W. Reynolds & G. Miller (Eds.), *Comprehensive handbook of psychology* (Vol. 7, pp. 357–390). New York: Wiley.

Lehrer, R., & Schauble, L. (2006). Cultivating model-based reasoning in science education. In R. K. Sawyer (Ed.), *The Cambridge handbook of the learning sciences* (pp. 371–386). New York: Cambridge University Press.

Lesh, R. (2006). Modeling students modeling abilities: The teaching and learning of complex systems in education. *The Journal of the Learning Sciences, 15*(1), 45–52.

Lesh, R., & Doerr, H. (2003). Foundation of a models and modeling perspective on mathematics teaching and learning. In R. A. Lesh & H. Doerr (Eds.), *Beyond constructivism: A models and modeling perspective on mathematics teaching, learning, and problem solving* (pp. 9–34). Mahwah: Erlbaum.

Lesh, R., & Fennewald, T. (2010). Introduction to Part I modeling: What is it? Why do it? In R. Lesh, P. L. Galbraith, C. R. Haines, & A. Hurford (Eds.), *Modeling students' mathematical modeling competencies (ICTMA 13)* (pp. 5–10). New York: Springer.

Lesh, R., & Zawojewski, J. S. (2007). Problem solving and modeling. In F. Lester (Ed.), *The second handbook of research on mathematics teaching and learning* (pp. 763–804). Charlotte, NC: Information Age Publishing.

Lombardi, J., & Lombardi, M. M. (2007, June 11–14). *Croquet learning space and collaborative scalability*. Paper presented at DLAC-II symposium, Singapore.

Office of the Chief Scientist. (2012). *Mathematics, engineering & science in the national interest*. Canberra: Author.

Pellegrino, J. W., & Hilton, M. L. (2012). *Education for life and work: Developing transferable knowledge and skills in the 21st century*. Washington, DC: The National Academies Press.

Pfannkuch, K. (2011). The role of context in developing informal statistical inferential reasoning: A classroom study. *Mathematical Thinking and Learning, 13*, 1–2.

Romberg, T. A., Carpenter, T. P., & Kwako, J. (2005). Standards-based reform and teaching for understanding. In T. A. Romberg, T. P. Carpenter, & F. Dremock (Eds.), *Understanding mathematics and science matters*. Mahwah: Lawrence Erlbaum.

Sabelli, N. H. (2006). Complexity, technology, science, and education. *The Journal of the Learning Sciences, 15*(1), 5–9.

Sriraman, B., & Steinthorsdottir, O. (2007). Research into practice: Implications of research on mathematics gifted education for the secondary curriculum. In C. Callahan & J. Plucker (Eds.), *Critical issues and practices in gifted education: What the research says* (pp. 395–408). Austin: Prufrock Press.

The National Academies. (2007). *Rising above the storm: Energizing and employing America for a brighter economic future*. www.national-academies.org. Accessed 24 Feb 2010.

The National Academies. (2009). *Rising above the gathering storm: Two years later*. http://www.nap.edu/catalog/12537.html. Accessed 24 Feb 2010.

Van den Heuvel-Panhuizen, M., & Van den Boogaard, S. (2008). Picture books as an impetus for kindergartners' mathematical thinking. *Mathematical Thinking and Learning: An International Journal, 10*, 341–373.

Watson, J. M. (2006). *Statistical literacy at school: Growth and goals*. Mahwah: Lawrence Erlbaum.

Yin, R. K. (2003). *Applications of case study research*. Thousand Oaks, CA: Sage.

Konold, C., & Higgins, T. L. (2003). Reasoning about data. In J. Kilpatrick, W. G. Martin, & D. Schifter (Eds.), *A research companion to principles and standards for school mathematics*. Reston, VA: National Council of Teachers of Mathematics.
Kuehner, J. P., & Mauch, E. K. (2006). Engineering applications for demonstrating mathematical problem-solving methods at the secondary education level. *Teaching Mathematics and Its Applications, 25*(4), 189–195.
Lehrer, R., & Lesh, R. (2003). Mathematical learning. In W. Reynolds & G. Miller (Eds.), *Comprehensive handbook of psychology* (Vol. 7, pp. 357–390). New York: Wiley.
Lehrer, R., & Schauble, L. (2006). Cultivating model-based reasoning in science education. In R. K. Sawyer (Ed.), *The Cambridge handbook of the learning sciences* (pp. 371–386). New York: Cambridge University Press.
Lesh, R. (2006). Modeling students modeling abilities: The teaching and learning of complex systems in education. *The Journal of the Learning Sciences, 15*(1), 45–52.
Lesh, R., & Doerr, H. (2003). Foundations of a models and modeling perspective on mathematics teaching and learning. In R. A. Lesh & H. Doerr (Eds.), *Beyond constructivism: A models and modeling perspective on mathematics teaching, learning, and problem solving* (pp. 9–34). Mahwah: Erlbaum.
Lesh, R., & Fennewald, T. (2010). Introduction to Part I modeling: What is it? Why do it? In R. Lesh, P. L. Galbraith, C. R. Haines, & A. Hurford (Eds.), *Modeling students' mathematical modeling competencies (ICTMA 13)* (pp. 5–10). New York: Springer.
Lesh, R., & Zawojewski, J. S. (2007). Problem solving and modeling. In F. Lester (Ed.), *The second handbook of research on mathematics teaching and learning* (pp. 763–804). Charlotte, NC: Information Age Publishing.
Lombardi, J., & Lombardi, M. M. (2007, Jun. 11–14). *Croquet learning space and collaborative technology*. Paper presented at [illegible] symposium, Singapore.
Office of the Chief Scientist. (2012). *Mathematics, engineering & science in the national interest*. Canberra: Author.
Pellegrino, J. W., & Hilton, M. L. (2012). *Education for life and work: Developing transferable knowledge and skills in the 21st century*. Washington, DC: The National Academies Press.
Pfannkuch, M. (2011). The role of context in developing informal statistical inferential reasoning: A classroom study. *Mathematical Thinking and Learning, 13*, 1–2.
Romberg, T. A., Carpenter, T. P., & Kwako, J. (2005). Standards-based reform and teaching for understanding. In T. A. Romberg, T. P. Carpenter, & F. Dremock (Eds.), *Understanding mathematics and science matters*. Mahwah: Lawrence Erlbaum.
Sabelli, N. H. (2006). Complexity, technology, science, and education. *The Journal of the Learning Sciences, 15*(1), 5–9.
S[illegible] (2011). [illegible] mathematics [illegible] for the secondary [illegible]. In [illegible] (Eds.), [illegible].
The National Academies. (2007). *Rising above the gathering storm: Energizing and employing America for a brighter economic future*. www.nationalacademies.org. Accessed 24 Feb 2010.
The National Academies. (2009). *Rising above the gathering storm, two years later*. http://www.nap.edu/catalog/12537.html. Accessed 24 Feb 2010.
Van den Heuvel-Panhuizen, M., & van den Boogaard, S. (2008). Picture books as an impetus for kindergartners' mathematical thinking. *Mathematical Thinking and Learning: An International Journal, 10*, 341–373.
W[illegible], J. A. (2006). *Scientific inquiry in schools: [illegible]*. [illegible]: [illegible] Publishing.
Yin, R. K. (2003). *Applications of case study research*. Thousand Oaks, CA: Sage.

Chapter 43
Modelling in Brazilian Mathematics Teacher Education Courses

Maria Salett Biembengut

Abstract This chapter presents the mapping of the Brazilian mathematics teacher education courses that have the subject of mathematical modelling (MM) or mathematics education dealing with MM. The goal is to identify the conceptions and tendencies of modelling in these subjects. This study is a document analysis since the data come from the course syllabuses and a questionnaire directed at professors. This questionnaire sought to provide information about the development of MM teaching. In Brazil, out of the 413 courses, 183 have the subject of mathematical modelling or mathematics education dealing with MM. The analysis indicated three conceptions of MM and as a consequence, three tendencies, which represent the contributions of professors that aim at enhancing knowledge.

1 Mathematical Modelling in Brazilian Education

The international movements concerning mathematical modelling (MM) in education influenced Brazil almost simultaneously due to the collaboration of Brazilian educators representing the country in the international academic community involved in mathematics education in the 1970s. The first proposals came from mathematics professors of higher education courses, mainly from engineering courses, who tried to meet the constant questions from students about 'what is mathematics for' and also, business's criticism of mathematical training for students. Since many of those professors were aware of Applied Mathematics, in which MM is part of the syllabus, they began to propose activities for students with the objective of enabling them to identify the application of mathematical concepts

M.S. Biembengut (✉)
Faculdade de Matemática – FAMAT, Pontifícia Universidade Católica do Rio Grande do Sul – PUCRS, Porto Alegre, Brazil
e-mail: maria.salett@pucrs.br

G.A. Stillman et al. (eds.), *Teaching Mathematical Modelling: Connecting to Research and Practice*, International Perspectives on the Teaching and Learning of Mathematical Modelling, DOI 10.1007/978-94-007-6540-5_43,

and definitions, so that these students will be able to solve problems in the area they will be acting in their future careers and lives (Biembengut 2003).

MM in education sprang from lectures or presentations given in courses and conferences by those professors and their respective enthusiasts or supervised students. The lectures in education conferences aimed at improving education, since many professors became interested in implementing the proposals in their classes because they had already had contact with work that encouraged them to use the ideas in the classroom. Indeed, many classroom activities or research projects have begun with an interest provoked by a talk, research or activities that at another time, in a cyclical process, stimulated new interests.

This educational effort that aimed at providing better mathematical teaching and learning has culminated with the development of research and has gained significant space in the discussions and in the official educational documents in Brazil, especially in the last two decades. MM has become part of the methodological processes in official educational documents in Brazil, generating curricular reformulations and pedagogical proposals. Among them, the National Curriculum Guidelines for mathematics teacher education courses present orientations related to the inclusion in the curriculum, of subjects such as MM. Thus, various mathematics teacher education courses have sought to include modelling in their course syllabuses as a subject or as part of the subject of mathematics education. Along with the number of professors who have become interested in modelling through extension courses or publication, the amount of research and reports of classroom experience presented at mathematics education conferences has increased significantly.

These lectures in courses and academic events, in particular, are resources for development and improvement in education. Many professors became interested in making use of MM in their classes due to contact with work encouraging the usage of MM in education. In a feedback cycle those who are motivated and practice in the classroom or research, are willing to share, in order to advance to a better understanding of processes and results. This comprehension encompasses the conception of the person or of the group who conducted the study or the activity within the classroom with students.

According to Thompson (1992), *conceptions* about several entities come from beliefs, from knowledge acquired through experience and interactions with the surrounding environment. This set of entities and individual and social relations, closely associated, form a functional unit more or less interrelated and interdependent and it usually tends to spread. In various social activities, from the conceptions of several groups, there are tendencies that are manifested in different ways, which are renewed by the cohesion of its elements, by the education and (re)education of the integrated people (Linton 1971).

As MM in recent decades has reached a significant place in discussions of education and official documents in several countries, and considering the different approaches and consequently, different understandings, at the present time, different conceptions and tendencies can be identified. *A MM tendency* in education is understood as all action and practice in MM performed by professors/teachers, based on knowledge and the understanding they demonstrate in their work departing from the conception they have. *A MM conception* is understood as the knowledge or the understanding the professor/teacher has about MM that, in turn, comes from experience.

Kaiser and Sriraman (2006, p. 304) developed a system to classify the approaches to modelling and applications in mathematical education based on an analysis of a sample of international literature. In this system, the authors indicate five *perspectives*: (1) Realistic or applied modelling; (2) Contextual modelling; (3) Educational modelling; (4) Socio-critical; and (5) Epistemological or theoretical modelling. For the first, *Realistic or applied modelling*, pragmatic-utilitarian goals may be found, such as solving real world problems, understanding the real world, and promoting modelling competencies. For the second, *Contextual*, subject-related and psychological goals may be encountered, such as solving word problems. For the third perspective, *Educational,* the goal is pedagogical, aiming at structuring learning processes from an introduction and developing of mathematical concepts and motivating mathematical learning. The fourth perspective, the *Socio-critical*, involves pedagogical goals such as promoting critical understanding of the surrounding world; being the problem-situation departing points to analyse the nature and relationship of the mathematical model within society. The final perspective, *Epistemological or theoretical modelling* involves promoting connections among modelling and mathematics; in this way, problem-situations aim at leading the student to understand mathematical theory. The authors also add a meta-perspective, *Cognitive*, that is restricted to research and aims at analysing and understanding cognitive processes that take place during modelling, as well as promoting mathematical thinking processes by using models as mental images, pictures and emphasising modelling as a mental process of abstraction or generalisation.

With the purpose of explaining how MM has been introduced and developed in Brazilian mathematics education, I have been mapping, since 2003, the pedagogical actions with MM. I have also built a reference map of both experience and research conducted up to now. The mapping consists of raising data, organising it according to some criteria, describing and analysing the experience reports, the pieces of research, the publications, the courses, in short, the MM facts as they stand in Brazilian education. For example, in 2003, I analysed 40 pieces of research, published in the period between 1976 and 2002. Through a careful analysis, it was possible to find different MM conceptions, when compared with the 1970s professors' proposals (Biembengut 2003). According to Kaiser and Sriraman (2006) there seems not to be a homogeneous understanding of modelling and its epistemological backgrounds within the international discussion on modelling.

In this chapter, I present the mapping of Brazilian mathematics teacher education courses which have as part of their syllabus a specific modelling subject or any subject that deals with modelling. This mapping is in constant update and is also part of a greater project. For the sake of this work, the guiding research question is: *How has MM been conceived in mathematics teacher education courses in Brazil*? This research had as its main purpose the identification of conceptions and tendencies in mathematical modelling adopted in the MM subject or in the one dealing with MM in Brazilian mathematics teacher education courses.

2 Material and Methods

The aspects related to the surveyed issues were organised according to a theoretical study about conception and tendency, as well as in relation to the identification, description and understanding of the mathematical modelling subject syllabuses or the mathematics education subject that deals with modelling. It is research based on document analysis, since the data come from the syllabuses and the information about the development of these syllabuses presented by the related professors of the discipline, by means of a questionnaire. It is important to highlight that the syllabuses and the data collected in the questionnaires are considered documents in this research. The research had two parallel stages: *implementation* and *analysis of the results*.

2.1 *Implementation of Document Analysis and Questionnaire*

The first stage was divided into three sub-stages. In the first sub-stage, by using data collected from the Government Ministry of Education websites and university websites, 413 courses were identified that aimed at training mathematics teachers. Sequentially, by means of e-mail contact with the course coordinators or even information about the programs available on the Internet page, it was possible to identify 183 courses, from which 122 have the subject MM and 61 have mathematics education subjects that deal with MM. At this stage, I had access to the subjects' syllabus as in Brazil, it is guaranteed by law, that all university courses must present a syllabus, in which it must have the program of the subject, the content, methodology and bibliography used. In the second sub-stage, this early data enabled me to identify and describe the MM subject syllabus or the mathematics education syllabus that addresses MM during the teaching period. Having these data, I could classify the 183 courses by region and by the topics presented in the syllabuses: *content*, *methodological procedures* and *bibliography*. In the third sub-stage, in order to better clarify the requirements in the programs of the MM subject or the one that deals with MM, a questionnaire (Fig. 43.1) was sent by e-mail to the professors of those subjects (183 courses). In response, 78 out of 183 professors answered the questionnaire.

a) How and where did you get to know about Mathematical Modelling?
b) What types of activity have you developed with students in the Mathematical Modelling subject or the one dealing with Mathematical Modelling?
c) What are the main difficulties in using/teaching Mathematical Modelling to students (future professors of mathematics)?

Fig. 43.1 Email questionnaire questions

2.2 *Analysis of Results*

In the second stage, I identified the MM conceptions and tendencies in these subjects and attempted to explain the meanings of the data based on the syllabus and on the answers given by the 78 professors. To reach the proposed goal, I employed the five perspectives presented by Kaiser and Sriraman (2006), namely, (1) Realistic or applied modelling; (2) Contextual modelling; (3) Educational modelling; (4) Socio-critical; and (5) Epistemological or theoretical modelling, with the objective of analysing tendencies in the course programs.

3 Results and Discussion

In the first analysis of those 183/413 courses that have the MM subject or another subject that deals with MM, I could verify that the majority of mathematics teacher education courses remain divided into subjects composed of rigid plans, without any connection between each subject. In Brazil, the courses follow a national syllabus guidance document provided by the Ministry of Education. It states that the subjects must have at least 2,800 h of classes in a course. In general, the classes do not go beyond the mere transposition of content, exercise handouts, teaching of techniques or even exposing students to theorems and statements devoid of any meaning, any practical result, any link with previous subjects (Biembengut 2009).

Professors of subjects, such as Teaching Methodology, Teaching Practice, Tendencies in Mathematics Education and Mathematical Modelling, hold the responsibility for the 'preparation' of future professors. These subjects are generally part of the final stage of the course, and have about 400 h of classes, which represent less than 15 % of the total hours of the course. These subjects are commonly found at the end of the courses and the number of class hours is not always enough to adequately prepare the future teachers. For example, the MM subject of the majority of courses takes 60 h of classes (less than 3 %). This perception was confirmed by 78 professors, since they answered that unless individual experiments are undertaken, the specific subjects are treated without any link. However, these interdisciplinary links must be provided in Basic Education when these students become teachers.

According to these 78 professors, the time available, for the MM subject or the one that deals with MM, is not sufficient to cover all possible loopholes left by training: for them, there is not enough time for students to devote themselves to a work in MM, nor for the professors to guide them. For the student to understand the meaning of a fact or phenomenon studied s/he needs to have knowledge learned from experience and understanding generated by mathematical proofs. It takes time, not only for the students but also for the teacher, who should guide the time available during class hours (Osawa 2007).

In general, the 183 course syllabuses analysed present three main parts: *content*, *methodological procedures* and *bibliography*. In the *content* section, concepts and definitions of models and modelling are presented, as well as a review of literature on mathematics education, mathematical applications, problem solving situations and MM in education. The *methodological procedures* section presents the reading of texts on MM and/or related topics for reflection and discussion in class; exercises on mathematical models; a MM work based on the students' topics of interest; and especially, how the work on MM in education will be done. In the *bibliography* section, several articles and books about mathematics education are presented, in addition to books with mathematical applications and three books written by Brazilian authors on MM education.

Analysing the *content*, it can be seen that mathematical applications are prioritised. Historically, mathematics has developed from applications not previously studied: some of these situations were natural practices while others were abstract. It is known for certain that these situations encouraged those who were involved to conduct a kind of investigation to solve the situational problem. Possibly, many of the findings that emerged from heuristic endeavour turned out to be the basis for theorems (Wheal 2007). By the still existing educational structure and with regards to the answers of the 78 professors, I may say that most professors responsible for the MM subject have mathematical training focused on this tendency: from theory to techniques and from techniques to application. It does not restrict a change in tendency departing from the MM experiences in pedagogical practices, as can be seen in the bibliography indicated in most syllabuses. The conception of MM the professor has, depends on what s/he knows about MM: from whom and from which circumstances, studies, experiments and experience s/he has gone through.

In the *methodological procedures*, the modelling process is advocated, although the 183 syllabuses analysed do not make clear whether these refer to MM per se or MM for education. The purpose of the MM per se is primarily related to teaching students how to establish a mathematical model of a problem-situation, solve it, understand it or even modify it when necessary. MM for education, in turn, aims at providing students, at any stage of schooling, with mathematical knowledge and critical awareness in relation to the reality that surrounds them; besides teaching them how to formulate and solve problems, and also think critically to understand the results.

According to the 78 professors who answered the questionnaire, the time available for the subject is not enough to teach students (future teachers) neither to do MM nor to use MM in education. As most of these students did not have any experience with, or knowledge of, MM during school, it requires more time from the professor to guide them. Despite the fact that at the beginning of the subject, these students show an interest in learning about MM, the lack of experience in searching for data, in raising and formulating questions, in solving (modelling process), and/or even in adapting to MM in education, makes most of them lose their interest. In addition, the majority of students do not know how to use the 'mathematical knowledge' acquired during their school life and further, the MM subject is usually presented at the end of the course. Thus, according to the majority of the 78 professors, what they can do in this period is to show students the concepts and definitions, making students reflect upon the teaching of mathematics, and preparing the ground for at least one work about MM in education, departing from their topics of interest.

In the *bibliography* sections of the documents, the texts on mathematics education are diverse in relation to the topics covered and the authors. The adoption of one or another subject/author depends on the conception of mathematics education and/or of MM that each responsible professor or teaching group has. The articles about modelling deal with reports of classroom teaching practices and reflections upon these practices. From a certain standpoint, this is legitimate, if one considers that MM emerged as a strategy for motivating students of all grade levels and became a consolidated method that, besides motivating the learning of mathematics, enables students to learn how to solve problems, make decisions, use critical thinking and be creative.

As regards the books with mathematical applications, 87 out of 183 courses are about Integral Differential Calculus, Differential Equations, and Operational Research. Many of these books have been in the bibliography of several mathematics subjects in courses of higher education for decades, and have contributed to the mathematics formation of several generations of mathematics professors. Three Brazilian books on modelling for teaching are indicated by 139 out of 183 courses. These books come from the results of research conducted by the books' authors, during the decades of 1980s and 1990s. These books aim at clarifying the concept of modelling in teaching and the methodological bases of classroom practice for Basic Education and Higher Education professors, presenting some modelling examples as proposals or suggestions for teaching.

3.1 MM Conceptions and Tendencies

When comparing the statements in the topics *content*, *methodological procedures* and *bibliography* from the 183 syllabuses and in the questionnaire answers of those 78 professors with the five perspectives proposed by Kaiser and Sriraman (2006), I verified that the five perspectives intersperse among the statements with greater or lesser emphasis. Based on these emphases, I consider three conceptions of MM in education and three tendencies on modelling adopted by the professors responsible for MM in teaching. I rearranged the perspectives, as follows: (a) method of teaching and research (*Realistic* and *Epistemological*), (b) alternative teaching of mathematics (*Contextual* and *Educational Modelling*) and (c) the learning environment (*Socio-critical*).

In conception (a), MM as a *method of teaching and research*, defended by the precursors of MM in education, the goal is making the student learn mathematics content from the subject and at the same time, learn to do research. That is, the teacher proposes modelling examples to teach or indicate the mathematical content found in these examples (*realistic* or *applied modelling*), and subsequently, teaches the modelling process by inducing students to raise questions and data about the subject or topic, to formulate a hypothesis and then, formulate a mathematical model, and in the final step, to solve the issues raised departing from the model and finally, to make an evaluation of the model (*epistemological*). For example, each group of students chooses a topic of interest and attempts to develop the problem situations in genuine environments, enabling students to develop skills and expertise to solve them. This conception is present in about 20 % of the documents. According to this conception, "MM is essential for the professional formation, as it uses scientific knowledge as a tool to identify and understand a problem, propose models, and discuss the results and validation of the mathematical methods. MM provides the student with all these procedures". (Professor A)[1]

In conception (b), MM is an alternative teaching of mathematics, the objective is to make the student learn mathematics. MM is proposed, since it motivates the student to learn mathematics from subjects or topics in their context. Departing from the topics chosen by the students, professors raise issues (*educational modelling*) and solve them while indicating the presence of mathematical content (*contextual*). These topics are authentic and integrated with the development of mathematical theories. This conception is present in 67 % of the documents analysed. Typically this conception sees "MM in the classroom is a strategy that can and should be used in certain moments in the classroom and in different levels of teaching". (Professor B)[1]

In conception (c), MM is the learning environment, MM focuses on the social issue. It means that the MM procedures in education are able to show mathematics as a tool for judging environmental issues. This conception (*socio-critical modelling*) is complementary in nature to ethno-mathematics (see Rosa and Orey 2013). The students seek to deal with issues involving situations related to society, making them have a critical positioning regarding the context. It is present in 13 % of the documents. According to professor C,[1] "MM supports critical mathematics in education; moreover, it provides rich moments for discussion on the role of mathematics in society".

These three conceptions, and as a consequence the three MM tendencies, indicated by the document examination, represent the sum of contributions from many professors and students interested in improving school learning, enhancing knowledge and living standards in society. The actions that each one has in the environment, the activities combined and the constant training of each person in the community, professional or productive sector of society contribute to a mutual sharing of experience. Although there are different conceptions and tendencies shown by the professors responsible for the subjects, they converge on the understanding that MM can contribute not only to the improvement of mathematics teaching and learning, but also to provoke a reaction and interaction between the faculty members and the students involved in the ongoing knowledge production.

[1]Professors A, B and C are among the 78 who answered the questionnaire.

4 Concluding Remarks

The Brazilian movement for MM in education started four decades ago by a small group of professors as a proposal to invigorate students' interest in mathematics. It has been increasing significantly, leading to the formation of research groups and development of many studies. As a result, the studies and research have contributed to MM in education so that it becomes part of the syllabuses of mathematics teacher education courses. This has led to different understandings and new conceptions influenced by the experience recognised by the dominant educational communities, thus, generating new tendencies.

The integration of MM in the syllabuses of mathematics teacher education courses shows how much this area has attracted support at official levels of education, in almost all Brazilian states because of the possibility of providing youth from this millennium, with better knowledge and skills. Despite the insufficient time to develop MM and its usual placement at the end of teacher education courses, the merit and importance of MM teaching in these courses cannot be underestimated. It is vital to acknowledge the contributions made by the professors responsible for these subjects, whatever theoretical framework they follow, improving mathematics education in Brazil.

MM in teaching aims at engaging students in combining of the existing elements regarding the actual topic. It is expected that if students learn to translate real issues, or those they imagine, into mathematical language and are interested in solutions as a means of production in understandable terms, we can expect a better training that will be reflected when they act professionally. Therefore, if the MM conception in education is a method, or an alternative environment for learning, the value of their accomplishment with students will tend to grow. The objective of MM is guiding students to a proper understanding of the environment they live in, and with MM knowledge, to be able to practice in order to bring change. According to Linton (1971), the individual learns his/her environment and acts to adapt to it. While living in a community, the individual faces nature and the nature of the other, not in isolation but as a member of a cooperative and organised group.

From analysing the syllabuses and the answers from the questionnaire, it is evident that it has been possible to disseminate modelling not just as a way of applying mathematical knowledge, but especially as a way of developing, in the students/future professors: (1) the research idea, (2) a sense of critical and creative thinking, (3) a concept of culture by studying the lifestyles of the community departing from a topic of interest, (4) the ability to distinguish conceptions and purposes of the mathematics teaching, (5) the potential usefulness of modelling, especially in dealing with problems that affect society and (6) the establishment of the modelling principles contingent with the aim of improving Brazilian education.

There is still little data available that allows for understanding the extent to which pedagogical actions, experience, values, beliefs, goals and ideals guide or form the teacher for basic education, or mathematics teacher education courses, or even how these prospective professors will match educational objectives with those of the society in which they live. Nevertheless, the conceptions and tendencies of MM identified in the 183 syllabuses suggest that the professors' classroom practice has sought to encourage students (i.e., future teachers) to engage actively in their learning, so as to produce work departing from needs, interests and personal goals in a challenging talented way. In addition, professors have sought to motivate students to commit themselves to social and environmental issues.

The possibility that MM becomes regular classroom practice lies in the interest these future teachers have in contributing to a greater development in Basic Education. The difficulties reside, mainly, in the continental dimensions of Brazil, making it difficult to provide students with enough activities, courses and events to continually serve all mathematics teachers. Furthermore, it is not easy in an educational structure that offers a syllabus with so many subjects. As already mentioned, issues of time and lack of linking between subjects is a problem faced by professors and teachers (Biembengut 2009).

The mapping of modelling in Brazilian education will continue to be the objective in the future. There is much to know and many facts to be gathered. It is worth reasserting that each event is made up of the message and each person is the messenger. The slightest perception becomes an indicator or a symbol. In other words, knowing how to generate knowledge about educational questions, knowing how to develop context-maps allows us to see the new realities that are present, but perhaps, incapable of gaining significant visibility for improving education.

References

Biembengut, M. S. (2003). *Modelagem Matemática: Mapeamento das Ações Pedagógicas dos Educadores de Matemática*. Post Doctoral Thesis, Universidade de São Paulo.

Biembengut, M. S. (2009). *Processos e Métodos de Ensino e Aprendizagem Matemática na Formação Continuada dos Professores* (Research Report). Conselho Nacional de Desenvolvimento Científico e Tecnológico – CNPq.

Kaiser, G., & Sriraman, B. A. (2006). Global survey of international perspectives on modelling in mathematics education. *Zentralblatt für Didaktik der Mathematik (ZDM), 38*(3), 302–310.

Linton, R. (1971). *O Homem: Uma Introdução à Antrologia*. São Paulo: Livraria Martins Editora.

Osawa, H. (2007). Development of applications and modelling by action research in Japanese secondary school. In H. W. Henn & W. Blum (Eds.), *ICMI study 14: Modelling and applications in mathematics education* (Pre-conference Vol., pp. 357–364). Dortmund: University of Dortmund.

Rosa, M., & Orey, D. (2013). Ethnomodelling as a methodology for ethnomathematics. In G. Stillman, G. Kaiser, W. Blum, & J. Brown (Eds.), *Teaching mathematical modelling: Connecting to research and practise* (pp. 77–88). New York: Springer.

Thompson, A. (1992). Teachers' beliefs and conceptions: A synthesis of the research. In D. A. Grouws (Ed.), *Handbook of research on mathematics teaching and learning* (pp. 127–146). New York: Macmillan.

Wheal, M. (2007). Issues in implement and sustaining a mathematics curriculum based on applications and modelling – The South Australian experience. In H.W. Henn & W. Blum (Eds.), *ICMI study 14: Modelling and applications in mathematics education* (Pre-conference Vol., pp. 357–364). Dortmund: University of Dortmund.

Chapter 44
The Development of Mathematical Concept Knowledge and of the Ability to Use This Concept to Create a Model

César Cristóbal-Escalante and Verónica Vargas-Alejo

Abstract This chapter reports progress of a study to gain insight into the development of the ability of undergraduate students to work with modelling situations. The questions that guide the research are: How to relate the development of concepts and the development of modelling abilities? What does it mean that a person has knowledge of a concept? What does it mean that a person has ability in modelling? Is the development of modelling ability independent of the development of knowledge about concepts? We discuss results from a research study with mathematics students at the university level.

1 Introduction

We understand by transfer of learning the use that a person makes of knowledge learnt in one context to analyse different situations. It also refers to the individual's ability to apply that knowledge in other situations (Mestre 2002). One of the fundamental assumptions of education is that students will use knowledge learned in the classroom to solve problems associated with situations they face outside of school. It is assumed that mathematics instruction should lead students to develop their abilities to use mathematics, that is, to develop skills to model, to make abstractions, to generalise, to make inferences, to think, and to use representations (Schoenfeld 1992).

The usual way of teaching mathematics involves the sequence: reading (and memorising), defining, demonstrating and stating of properties of concepts (theorems), exercises and applications. The first activities are embedded within mathematical knowledge. The applications are related to the use of learned concepts and properties to solve problems that arise in mathematical situations. Usually

C. Cristóbal-Escalante (✉) • V. Vargas-Alejo
Department of Science, Universidad de Quintana Roo, Chetumal, México
e-mail: cescrist@uqroo.mx; vargasalejo@uqroo.mx

G.A. Stillman et al. (eds.), *Teaching Mathematical Modelling: Connecting to Research and Practice*, International Perspectives on the Teaching and Learning of Mathematical Modelling, DOI 10.1007/978-94-007-6540-5_44,

applications are at the end of the topic or course. It is thus assumed that mathematical knowledge is required to develop students' ability to apply mathematics to situations.

Usually we identify two types of mathematics courses: 'theoretical' courses and courses on applications of mathematics. This classification assumes that the development of knowledge about concepts and the development of skills to analyse situations are independent. What does it mean that a person has knowledge of a concept? What does it mean that a person is competent to apply mathematics? Is the development of the ability to model independent of the development of knowledge about concepts? What type of activities encourages the students' ability to analyse and solve "real life" situations? We present and discuss results from a research study with university students where the purpose was to seek answers to these questions.

2 Literature Review

The use of instructional activities based on problem solving has been recommended since some decades ago (Freudenthal 1991; Mathematical Sciences Education Board, National Research Council 1990; Polya 1954). Some studies show that if the students do not analyse non-mathematical situations during the teaching process, which demand the use of mathematical knowledge, then they do not think about their knowledge and they do not perform any adaptation that allows them to use the knowledge in other situations (Greeno and MMAP 1998; Sierpinska 2000). To successfully solve the problems, a person must develop mathematical knowledge and skills, and have heuristic strategies and metacognitive strategies (Schoenfeld 1992).

Perrenet and Adan (2011) describe the experience of mathematical modelling courses, where students have to analyse situations that are different to the problems of textbooks. In these situations the students need, in addition to mathematical knowledge and problem solving skills, to be able to translate the situation to mathematically tractable problems, to have a broad view of mathematics, to have communication skills, and to use their common sense and intuition. In these situations the students have to use their skills and knowledge, and they have to learn or discover new techniques or concepts.

The Models and Modelling perspective (MMP) on problem solving, learning and teaching of mathematics (Lesh and Doerr 2003) is a framework that permits integration and proposes responses to the questions posed in this research. The MMP conceives the learning process as the development of conceptual systems. These conceptual systems are continually changing by the individual's interactions with his/her environment. The conceptual systems are used to describe, to explain, to communicate, and to predict the behaviour of other systems. The learning process involves a series of cycles of understanding. In these cycles, the conceptual systems or models are changing to more refined models. Conceptual development is a gradual process and contextualised. The early development of the knowledge of students tends to be organised around experiences; that is, two ideas are considered related, not because they are logically connected, but because they were used together in a problem-solving experience. When there are a lot of experiences, the knowledge tends to be organised around abstractions.

The students' knowledge develops along a variety of dimensions: concrete – abstract, simple – complex, situated – decontextualized, specific – general, internal – external, intuitive – formal, stable – unstable. The most useful models are not always the most abstract, complex, general or formal. During the early stages of development, the conceptual systems are characterized as being diffuse, poorly differentiated, and poorly coordinated (Lesh 2010). In assessing a person's knowledge it is not useful to ask about the concepts that he/she knows. It is more productive to ask about what he/she can do with these concepts, what problems he/she can solve, the explanations he/she can make, or what calculations he/she can perform (Lesh and Doerr 2003).

The MMP uses model-eliciting activities (MEAs). They are problem situations for which there is no single solution, they have a lot of answers. These activities, together with the individual work, in teams, and in groups, allow the students to perform the actions of representation, interpretation, speculation, evaluation, communication, and explanation of situations to other people. During the process of solving problem situations, the students pose questions and seek strategies to find one answer. They must share with other students their ideas about the implemented strategies and the outcomes. They need to make explicit the criteria used to evaluate and compare these strategies. The students go through multiple stages of understanding about the problem situation and its solution, each stage is a refinement of the previous one. In this process everyone encourages others to communicate their ideas and conceptualizations of the situation and actions for the solution. The products of this activity allows the teacher to have important elements to describe characteristics of the student's thinking about these situations and about the concepts used.

3 Methodology

We used a qualitative observation of cases during the teaching process of linear algebra courses. A workshop, designed to solve problems that involved linear algebra knowledge, was implemented in these courses. The observer was the instructor of the courses and the workshop.

The first data were collected from a group of eight students, who were in the second year of the undergraduate program in physics and mathematics, during the workshop on Problem Solving which occurred in 2007. These students had taken basic courses in mathematics, including linear algebra, calculus of one and several variables, and differential equations.

The second group of data was taken from 11 students, who were in the first year of an engineering undergraduate program. They were attending a Linear Algebra course in the autumn of 2009. They had just studied Systems of Linear Equations (SLEs). After this, they undertook the workshop to solve application problems.

The third group of data was collected from 17 students, who were in the first year of an engineering undergraduate program. They were attending a Linear Algebra course in the spring of 2010. They were just starting this course. The course began with a workshop, where the students were solving problems that could be represented by a system of linear equations. They did not study SLEs before the workshop. Mexican high school students only learn ways to solve linear equation systems of 2×2 and 3×3 dimensions, and also they solve problems that can be represented by these systems. The data were taken from one activity with the characteristics of MEAs: the *Making Granola Task*.

This activity was posed in the same way in each of the three groups. The activity was developed by undertaking one phase of individual work, a second phase of working in teams, and a third phase where each team presented their progress with the other teams. The students could use any resource, including computers and mathematical software. In the first and in the third phase, each student prepared a report, including his/her response. These reports revealed the evolution of understanding of the problem.

We analyzed the approaches developed by the students. We wanted to see if they came to write a functional expression to characterise the properties of the mixture obtained; that is, if they could write something like this:

$$a_{11}x_1 + a_{12}x_2 + \ldots + a_{in}x_n = c_1$$

where x_i represents the amount of ingredient *i* used, a_{ij} represents the unit value of and c_i represents the amount of property *i* in the mixture.

Making Granola Task

The granola is a food made of seeds, dry fruit and honey. It can be very nutritious. It is high-use among people with a vegetarian diet, or trying to eat a balanced diet. A person wants to produce and sell granola. To expand the market, the person wants to know the needs of people who should control the nutrients ingested.

What is the procedure to develop and sell the product?

The cost per kg of ingredients that could be used and the nutritional values are listed in the table below.

Ingredient	Cost pesos /kg	kcal /kg	Grams protein /kg	Grams Fat /kg	Grams HC /kg	mgs Ca /kg	mgs Fe /kg	mgs Zn /kg	UI vit A /kg	mgs vit B /kg
Oats	60	3,890	168.9	6.9	662.7	540	47.2	39.7	0	01.19
Sesame	100	5,650	169.6	48	257.4	9,890	147.6	71.6	90	8
Raisin	120	2,960	25.2	0.54	780	280	26	2	0	2
Almond	85	5,780	212.6	50.64	197.4	2,480	43	33.6	50	1,310
Peanut	35	5,940	173	51.45	253.5	700	37	38	50	2.96
Prune	76	1,130	12.3	0.24	297	240	117	2.5	5,230	2
Coconut	32	3,540	33.3	33.49	152.3	140	24.3	11	0	5.4
Strawberry	135	690	05.8	0.6	173.6	210	02.2	0	900	0
Apple	86	670	01.8	0.43	168.4	40	02.4	0.5	560	0.44

In this way we wanted to know if they came to write the equation or a similar expression:

$$x_1 + x_2 + \ldots + x_n = k$$

This equation denotes the total quantity of the mixture. We wanted to know if the students used the quantity c_1/k, that is, the characteristic a_1 contained in one unit of the mixture obtained.

4 Results and Discussion

4.1 First Group (n = 8 Students)

The first group approached the problem by following two routes. The students did not use the simplified form of representation using SLEs. Four students of this group (first team) used Excel to develop a procedure that would allow determining the characteristic values of the mixture, when they knew the quantity of each ingredient used. They calculated the characteristic of a unitary quantity dividing the result by the mixture's total quantity. They asked about the characteristics of the mixture and they used trial and error to answer the questions. They used the procedure made in Excel to make new calculations. This procedure changed continually to a more effective one. In teamwork, this team failed to construct a SLE, but the procedure in Excel was very useful for solving the problem (Figs. 44.1 and 44.2). Student A2 used the same amount of each ingredient. He observed that the mixture had the same ratio of kcal/kg. He changed the quantity and explored this ratio (Fig. 44.3).

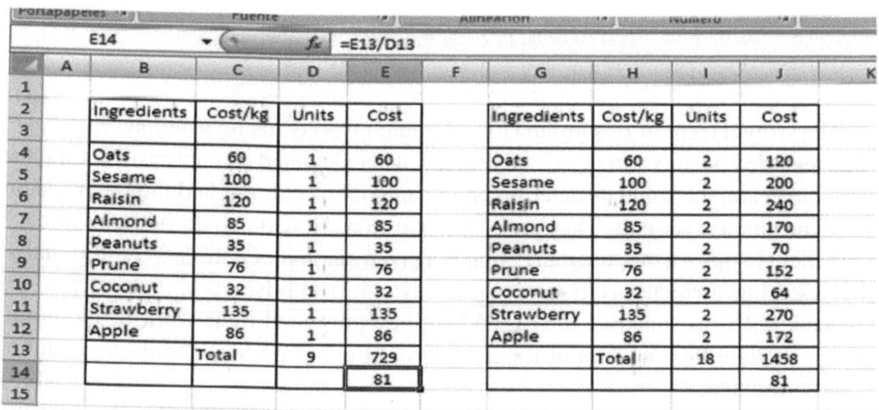
E14 =E13/D13

Ingredients	Cost/kg	Units	Cost	Ingredients	Cost/kg	Units	Cost
Oats	60	1	60	Oats	60	2	120
Sesame	100	1	100	Sesame	100	2	200
Raisin	120	1	120	Raisin	120	2	240
Almond	85	1	85	Almond	85	2	170
Peanuts	35	1	35	Peanuts	35	2	70
Prune	76	1	76	Prune	76	2	152
Coconut	32	1	32	Coconut	32	2	64
Strawberry	135	1	135	Strawberry	135	2	270
Apple	86	1	86	Apple	86	2	172
	Total	9	729		Total	18	1458
			81				81

Fig. 44.1 Excerpt from report showing Excel procedure of student A1 (first team)

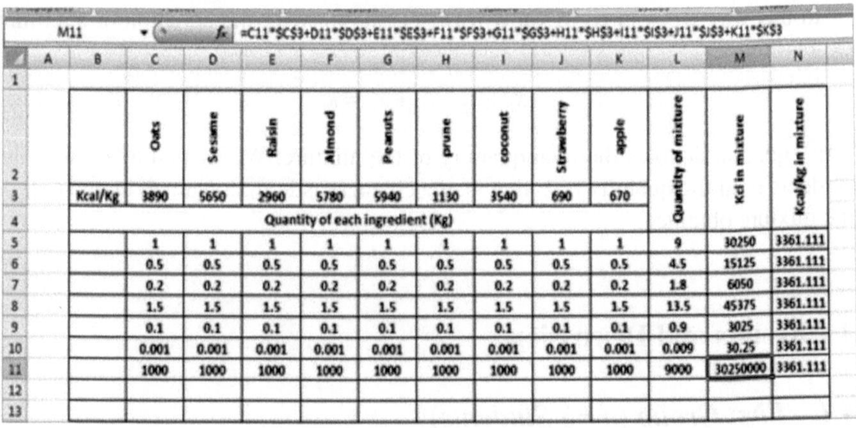

M11 =C11*C3+D11*D3+E11*E3+F11*F3+G11*G3+H11*H3+I11*I3+J11*J3+K11*K3

	Oats	Sesame	Raisin	Almond	Peanuts	prune	coconut	Strawberry	apple	Quantity of mixture	Kcl in mixture	Kcal/Kg in mixture
Kcal/Kg	3890	5650	2960	5780	5940	1130	3540	690	670			
	Quantity of each ingredient (Kg)											
	1	1	1	1	1	1	1	1	1	9	30250	3361.111
	0.5	0.5	0.5	0.5	0.5	0.5	0.5	0.5	0.5	4.5	15125	3361.111
	0.2	0.2	0.2	0.2	0.2	0.2	0.2	0.2	0.2	1.8	6050	3361.111
	1.5	1.5	1.5	1.5	1.5	1.5	1.5	1.5	1.5	13.5	45375	3361.111
	0.1	0.1	0.1	0.1	0.1	0.1	0.1	0.1	0.1	0.9	3025	3361.111
	0.001	0.001	0.001	0.001	0.001	0.001	0.001	0.001	0.001	0.009	30.25	3361.111
	1000	1000	1000	1000	1000	1000	1000	1000	1000	9000	30250000	3361.111

Fig. 44.2 Excel procedure of student A2 (first team)

FOR A SUITABLE METHOD FOR CARRYING OUT THE PROBLEM IS THE FOLLOWING:

• RAISE THE PROBLEM

We want to build a granola containing two ingredients and it also cost 50 pesos per kilo.

• TRANSFORM THE PROBLEM IN UNKNOWNS

Let x

x = Amount of oats cost of a kilogram of oats = 50 pesos

y = amount of sesame cost of a pound of sesame = 90 pesos

• ESTABLISH A SYSTEM OF EQUATIONS

$60x$ is the cost of x kilograms of oat. If $x = 1$ then $60x = 60$

$90y$ is the cost of y kilograms of sesame. If $y = 1$ then $90y = 90$

$60x + 90y = 150$

And $x + y = 2$

It is the system whose solution x=y=1 kg1 kg

Determining the number of kilograms in the mixture is obtained and the cost of the mixture divided by the total mixture is obtained and the cost of a kilogram of the mixture.

Do the same if there are three ingredients.

Fig. 44.3 Excerpt from report showing procedure of student A3 (second team)

The second team of this group (four students) worked using a simpler approach: they regarded the mixture of two ingredients and calculated characteristics. Like the previous team, the students determined the amount of nutrient and then they calculated the amount per unit. They posed the question: How much do I require of each ingredient to make a mixture that contains a given amount of nutrients? Additionally, they used linear equations to represent it (Fig. 44.3). Whilst working as a team, the students failed to construct a SLE. When they presented their work to the entire group, the fact that the sum of the amounts of the ingredients should be equal to 1 was presented.

During the problem solving process, some questions arose related to the mixture, for example, How much do I require of each ingredient to make a mixture that

Equation of the cost of the mixture
60 O at + 100Ses + 120Rai + 85Alm + 35 Pea + 76Pru + 32 Coc +135Str + 86 App = 250

Equation of the amount of calories in the resulting mixture
389 O at + 565 Ses + 296 Rai + 578 Alm + 594 Pea + 113 Pru + 354 Coc + 69 Str + 67App =1200

Fig. 44.4 Excerpt from report report showing procedure of student B2 (second group)

contains a given amount of nutrients? How much do I require of each ingredient to make a mixture that contains a given quantity of different nutrients? These questions were approached by SLEs. These two teams systematised their exploration using Excel procedures, which allowed them to characterise the mixture of ingredients roughly. They allocated quantities of ingredients, under the procedure to approach a cost, the amount of calories, and the amount of protein in the mixture. They were able to identify some characteristics of functional expressions, for example, the range of values that could be taken when assigning the amounts of ingredients.

4.2 *The Second Group (n = 11 Students)*

The procedures of the teams of the second group show that these students approached the problem in a similar way to the students of the first group: all of them sought to represent the problem using SLEs. They used two and three ingredients and determining the cost, the amount of calories and the amount of protein in the mixture. They worked the situation in a similar way to that used by team 2 of the first group (Fig. 44.3). They were able to provide a procedure for calculating the characteristics of the mixture, and determining separately the unitary characteristics of the mixture.

In this group a student (B2) proceeded to use more ingredients (Fig. 44.4). This student used nine characteristics of the ingredients and of the mixture; but he could not use a complete SLE. He could not use the total quantity of the mixture. However, he did not work with other students and he did not share the procedure. For this reason, the teams did not know of this approach.

4.3 *The Third Group (n = 17 Students)*

The third group raised equations using two ingredients (Fig. 44.5). Two students used a proportional relationship in the problem solving process; that is, they considered the fact that the sum of the ingredients was equal to 1 kg. They detected the

If you mix1 kg of oats and 2 kg of sesame we have that the mixtureis of 3 kg .
This mixture has:
60(1) + 100(2)= 260 cost
3900(1) + 5650(2) = 15200 calories
170(1) +170(2)= 510 protein
70(1) + 280(2)=630 fat

One kilogram costs 260/3 = 86.67 , 5066.67 calories, 170 protein, y 630 fat.

If we mix 2 kg of oats and 3 kg of sesame we have that the mixture is of 5 kilograms.
This mixture contains:
60(2) + 100(3)= 420 cost
3900(2) + 5650(3) = 24750 calories
170(2) +170(3)= 850 protein
70(2) + 280(3)=980 fat

One kilogram costs 420/3= 84, 4950 calories, 170 protein, and 196 fat.
If we know the amount of each ingredient we look for numbers that approximate the amount you want.

We must satisfy the sum is one kilogram, so we can get everyone.

Fig. 44.5 Excerpt of report showing procedure of team C2 (third group)

need to write additional equations associated with the characteristics of the resulting mixture. They observed that if the amounts were established in advance then they should write a number of relationships (Fig. 44.5).

During the presentation to the group, the need to use the equation that gave the total quantity of the mixture emerged (Fig. 44.5). The students of team C2 noted that they could do just 1 kg of the mixture, and then the sum of the amounts of the ingredients should be equal to one, therefore they said, "Now we have the formula to get whatever quantity of mixture".

None of the first, second or third groups wrote the SLE in a correct way when they worked individually and in teams. The SLE was written during the group discussion. The instructor used questions to highlight some of the approaches achieved by the students, and to permit the students to identify the relations among the quantities (Fig. 44.6).

5 Conclusions

The above results show that in the three groups observed, the way to approach the situation was very similar. The exploration of particular cases of the situation led to expressing the relationships among the quantities involved. The students needed to explore the particular cases of the situation to identify the functional relationship among the quantities. The students posed a question about the mixture: What characteristics would the mixture have if I take x quantity of each ingredient? Team C3

We assign unknowns to the quantities of each ingredient that we need

X1 = Required quantity of oatsin one kilogram of granola
X2= Required quantity of sesame seeds in a kilogram of granola
X3 = Required quantity of almond in a kilogram of granola
X4 = Required quantity of peanuts in a kilogram of granola
X5 = Required quantity of coconuts in one kilogram of granola
X6= Required quantity of strawberry in a kilogram of granola

We propose and equate equations for each condition

$60 X1 + 110 X2 + 85 X3 + 35 X4 + 32 X5 + 135 X6 = 76$ ---- **Eq. Cost**

$3900 X1 + 5650 X2 + 5780 X3 + 5949 X4 + 3540 X5 + 690 X6 = 5500$ ---- **Eq. Kcal**

$70 X1 + 280 X2 + 420 X3 + 454 X4 + 335 X5 + 6 X6 = 400$ ----**Eq.grams of fat.**

$560 X1 + 250 X2 + 200 X3 + 253 X4 + 150 X5 + 180 X6 = 210$ ----**Eq.grams of cbh.**

$10 X1 + 9 X2 + 50 X3 + 50 X4 + 8 X5 + 900 X6 = 80$ ----**Eq.Vit A**

$X1 + X2 + X3 + X4 + X5 + X6 = 1$ ---------------- **Eq.quantity of granola**

Fig. 44.6 Excerpt from report showing procedure of team C3 (third group)

wrote the equations in a correct way, but they could not understand them as a SLE. They did not know how to obtain the solution because they did not know methods to solve SLEs of large dimensions, and when the number of equations is greater than the number of unknowns, or when the number of unknowns is greater than the number of equations. Another question that emerged was: How much should I use of each ingredient if I want that mixture to have a characteristic value? This question led students to write the equation and ensured they identified the relationship between the concepts of function and equation.

The relationship between the development of knowledge about concepts and the ability to use them in solving problems is strong. This is what is meant when it is said that learning is a contextualized process (Lesh and Doerr 2003). The students developed different cycles to understand the situation; each one provided answers to specific questions. For some students the representations in Excel were useful to answer some questions, but later the same students recognised that the algebraic representations allowed more accurate responses.

The student work, particularly from the first group, allowed insight into what they could do with the knowledge developed about SLE. They could see some formal aspects of the SLE, for example, the SLE's conditions when it did or did not have solutions, or the range of the values of linear functions of several variables, but the students had difficulties in writing the SLE.

References

Freudenthal, H. (1991). *Revisiting mathematical education*. Dordrecht: Kluwer.
Greeno, J., & Middle School Mathematics through Applications Group. (1998). The situativity of knowing, learning and research. *The American Psychologist, 53*(1), 5–26.
Lesh, R. (2010). Tools, researchable issues and conjectures for investigating what it means to understand statistics (or other topics) meaningfully. *Journal of Mathematical Modeling and Applications, 1*(2), 16–48.
Lesh, R., & Doerr, H. M. (Eds.). (2003). *Beyond constructivism. Models and modeling perspectives on mathematics problem solving, learning, and teaching*. Mahwah, NJ: Erlbaum.
Mathematical Sciences Education Board, National Research Council. (1990). *Reshaping school mathematics: A philosophy and framework for curriculum*. Washington, DC: National Academy Press.
Mestre, J. (2002). *Transfer of learning: Issues and research agenda* (National Science Foundation Report No. NSF03-212). Washington, DC: National Academy Press.
Perrenet, J., & Adan, I. (2011). Fifteen years of experience with modelling courses in the Eindhoven program of applied mathematics. *Journal of Mathematical Modelling and Application, 1*(4), 3–19.
Polya, G. (1954). *Mathematics and plausible reasoning*. Princeton, NJ: Princeton University Press.
Schoenfeld, A. (1992). Learning to think mathematically: Problem solving, metacogniotion, and sense making in mathematics. In D. A. Grows (Ed.), *Handbook of research on mathematics teaching and learning* (pp. 334–370). New York: Macmillan.
Sierpinska, A. (2000). On some aspects of students' thinking in linear algebra. In J. L. Dorier (Ed.), *On the teaching of linear algebra* (pp. 209–246). Dordrecht: Kluwer.

Chapter 45
Problem Posing: A Possible Pathway to Mathematical Modelling

Ann Downton

Abstract Problem posing is an important component of learning mathematics as is problem solving, and it is an essential part of mathematical modelling. This chapter reports two small studies conducted in a Year 1–2 class and a Year 3–4 class, respectively. The purpose of each study was to examine the extent to which the use of real world artefacts provide a stimulus for young students to pose problems that can be investigated using mathematical modelling. The students in both studies had no prior experience in problem posing. However, they generated a range of problems linked to the real world albeit some in a superficial sense, whereas others gave genuine rise to mathematical modelling.

1 Problem Posing: What Is It? Why Use It?

Problem posing involves both the generation of new problems and reformulation of given problems (English 2003; Silver 1994; Whitin 2006). Bonotto (2010b) suggests that mathematical problem posing is the process by which "students construct personal interpretations of concrete situations and formulate them as meaningful mathematical problems" (p. 109). English (2003) goes further in suggesting that problem posing "is a fundamental part of learning and doing mathematics" (p. 187) as is problem solving to mathematical modelling. Others concur with English and argue that problem posing is a key component of mathematical exploration (Bonotto 2010b; Cai and Hwag 2002). Christou et al. (2005) argue that problem posing is "an integral part of modelling cycles, which require the mathematical idealization of real world phenomenon" p. 149. For this reason, Bonotto (2010a) considers "problem

A. Downton (✉)
Education VIC | Faculty of Education,
Australian Catholic University, Melbourne, Australia
e-mail: ann.downton@acu.edu.au

G.A. Stillman et al. (eds.), *Teaching Mathematical Modelling: Connecting to Research and Practice*, International Perspectives on the Teaching and Learning of Mathematical Modelling, DOI 10.1007/978-94-007-6540-5_45,

posing is of central importance in the discipline of mathematics and in the nature of mathematical thinking and it is an important companion to problem solving" (p. 404).

Constructing problems is only one facet of problem posing. When students are applying problem posing processes they are "actively engaged in challenging situations that involve them in exploring, questioning, constructing, and refining mathematical ideas and relationships" (English 2003, p. 197). Bonotto (2008) argues that immersing students in situations related to their real-life experience and meaningful sense–making is a way to "deepen and broaden their understanding of the scope and usefulness of mathematics as well as learning ways of thinking mathematically that are supported by mathematising situations" (p. 14).

Problem posing can be initiated using diagrams, definitions, questions, statements, equations, concrete materials, real world situations, children's literature, or objects (Brown and Walter 2005; English 1998; Whitin 2004). Bonotto (2008) suggests the use of real world artefacts that are relevant and meaningful to students, such as weekly TV guides and supermarket dockets, bridges the gap between formal school mathematics and informal out-of-school mathematics.

Although problem posing is a natural occurrence in real life it does not receive sufficient attention in the mathematics classroom. According to English (1998) educators need to

> broaden the types of problem experiences we present to children … and, in so doing, help children "connect" with school mathematics by encouraging everyday problem posing (Resnick et al. 1991). We can capitalize on the informal activities situated in children's daily lives and get children in the habit of recognizing mathematical situations wherever they might be (p. 100).

Bonotto (2008) concurs with English (1998) and argues that school mathematics should more closely relate to the experiences students have in their everyday life, to make it relevant to students. By doing so, students are enabled to view mathematics as a "means of interpreting and understanding reality" (Bonotto 2010b, p. 110), rather than purely as a set of abstract, formal structures.

As stated earlier, English (2003) indicates that problem posing underpins students' mathematical learning and involves skills such as questioning, describing, constructing, justifying, and explaining, all of which are evident in mathematical modelling. It could be argued that problem posing provides the foundation for not only students' development of problem solving skills, but also for mathematical modelling. Stillman (in press) argues, "finding and posing problems are essential ingredients in any education program in schools promoting mathematical modelling". One of the features of mathematical modelling is that "modelers find and pose their own problems to solve" (Stillman in press).

Mathematical modelling refers to a process in which a real world situation has to be "problematised and understood, translated into mathematics, worked out mathematically, translated back into the original situation, evaluated and communicated" (Bonotto 2010b, p. 108). It "involves using mathematical concepts, structures and relationships to describe and characterise, or model, a real world situation in a way that it captures its essential features" (Stillman in press). As is the case in real-life situations, "modelling activities often comprise information that might be incomplete, ambiguous, or undefined, with too much or too little data" (English et al. 2005, p. 156).

Mathematical modelling provides rich learning experiences that extend students problem solving and problem-posing abilities (English et al. 2005). Making conjectures and predictions, justifying and refuting arguments and resolving conflicts are components of mathematical modelling situations (English et al. 2005). Posing problems and questions occurs naturally throughout mathematical modelling situations because they "evoke repeated asking of questions and posing of conjectures" (p. 156).

Traditionally, mathematical modelling has been a topic for secondary school students. However, English et al. (2005) argue that younger students would benefit from rich learning experiences that could promote their problem-solving and problem-posing abilities. These abilities are necessary "to function effectively in a world that is demanding more flexible, creative, and future-orientated mathematical thinkers and problem solvers" (p. 156).

From a research perspective, little is known about students' ability to create their own problems in numerical and non-numerical contexts, or the extent to which these abilities are linked to their problem-solving competency (English 1998). English (1997) found Year 5 students who participated in specific problem posing programs exhibited greater facility in creating solvable problems than those who did not. Silver and Cai (1996) found that there was a high correlation between students' problem posing performance and their problem solving performance.

The studies reported in this chapter will attempt to demonstrate that young students (ages 6–9) can pose problems using real world contexts, such as buttons, children's literature and photographs that can be investigated using mathematics. Furthermore, these real-life situations are more conducive to young students' development of mathematical understanding and problem solving skills than formal contexts, which are quite removed from their natural real world experiences (Bonotto 2008).

2 Methods

This chapter reports on two studies in the form of teaching experiments, one in a Year 1–2 (6–7 year olds) class in their second and third year of schooling and the other in a Year 3–4 (8–9 year olds) class. Within some schools the classes are multi-age in structure, hence the notion of Year 1–2. Neither class had any formal experience with posing problems during mathematics lessons. The Year 3–4 students' problem solving experiences were limited to word problems for which they could apply a known procedure or follow a clearly defined pathway (Brown and Walters 2005). The purpose of the studies was to explore the extent to which young students, with no prior experience, could pose problems that could be investigated using mathematics, in particular mathematical modelling.

Fig. 45.1 Examples of real world photographs used for problem posing stimuli

These teaching experiments are situated within a larger project, Contemporary Teaching and Learning of Mathematics Project (CTLM), the aim of which is to improve the teaching and learning of mathematics by providing teacher professional development and in-school classroom support. These two studies are part of the in-school classroom demonstration lesson component of the CTLM project. They were conducted in two different schools, both in their second year of participation in the project.

Each teaching experiment consisted of a single lesson (approximately 90 min). The lessons were conducted by the researcher and observed by several classroom teachers. Data collection occurred in February (start of the school year) in the Year 1–2 class (12 boys, 13 girls) and in June (5 months into the school year) in the Year 3–4 class (19 boys, 7 girls). Data collected included student work samples, photographs of student engagement at various stages during the lesson, audio recording of student conversations during the task, field notes, and post lesson discussion with observing teachers.

2.1 The Tasks

In both classes real world artefacts (drawing on the work of Bonotto 2008, 2010b) were used as the main stimulus for students' problem posing activities. One task was implemented with the Year 1–2 class and two tasks with the Year 3–4 class. In the Year 1–2 lesson, the real world artefact was a very old tin containing a collection of recycled buttons of various shapes, sizes, textures, and hole configurations. In the Year 3–4 lesson the students first explored a numerical situation that drew on the work of English (1998): "What problems could you pose about the number 20?" Following English (1998) and Whitin (2006), students then selected one of their questions to compose a real-life word problem for it. Real world photographs were the stimulus for the second problem posing activity. Real world photographs used included a wood stack and a tennis stadium (see Fig. 45.1).

2.2 Implementation

As the tin of buttons was introduced to the Year 1–2 students, they were immediately engaged in problem posing and reasoning about the contents of this tin, an object that they could associate with the real world (Bonotto 2008). Fluency, flexibility, and originality (English 1998; Silver 1994) were evident in the responses. As the tin was shaken the students began to eliminate some of the suggestions with justification such as, 'If there were chocolates inside, it would not make this much noise.' Novel responses included socks, pasta, toy cars, and money. The categories included food, toys, natural objects (e.g., rocks, sand, pebbles), jewellery, money, and classroom equipment (e.g., scissors, pencils, paper clips). It took some time for buttons to be considered.

The buttons were then distributed to the tables where students worked in pairs to explore the buttons for 5 min. Students were then asked to think of questions they could pose that we could investigate as a class. During this time, the observing teachers were encouraged to observe, listen and assist with recording of questions (only if necessary) and to refrain from prompting.

The lesson in the Year 3–4 class began by indicating to the students my interest in their mathematical thinking and the type of problems they might pose for the class to investigate. Students were asked to think about some problems they could pose about the number 20. They worked individually then some of the problems posed were shared with the class. All the problems posed were in numerical form. Students were then asked if the questions could be made more challenging which led to open-ended problems such as $\_\times\_=20$ and $\_$ (+, −, ×, or ÷) $\_=20$. A brief discussion ensured about the kind of thinking they needed to do when problem posing compared to finding a solution. Some students indicated that it was more difficult to pose a problem because "you had to work backwards from the answer." Students then had to make up some word problems for $4\times5=20$. These were discussed briefly to highlight the need for the problems posed to be clear, make sense and include a question.

This activity provided the groundwork for the main part of the lesson, the use of real world photographs to pose a problem that could be investigated using mathematics. The students worked in self-selected pairs, all single gender except for one pair. Each pair had 20 min to generate a range of problems, test them out and refine them to ensure they met the following criteria: (1) it is a genuine problem to investigate using mathematics; (2) the problem is clear; (3) the problem requires more than a calculation of the given information; and (4) solving the problem requires the extra-mathematical world to be considered. The students also needed to think about how they might solve the problem and the mathematics they might use to solve it.

In sharing their problems the students were required to explain and justify how the problem posed addressed the criteria. During this time, other students were encouraged to contribute their thoughts. For example, some students did not consider *The Stamps Problem* (see Fig. 45.3) to be a real world problem because one-cent coins are no longer in circulation. Others suggested that the cost would be rounded up to the next multiple of ten like they do in the supermarket, while others said you could use a credit card!

At the conclusion of the lesson challenges faced in having to pose problems rather than solve them were discussed. Common responses included: "Making sure the problems make sense", "Trying to translate the problem that someone else has made", "Making sure the problem is clear for you and those who have to solve it", and "Coming up with a problem that is challenging".

3 Findings

3.1 Year 1–2 Problem Posing

The questions posed for the *Buttons Task* by the Year 1–2 class ranged from "How many buttons in the collection?" to "What are buttons made of?" The list of questions generated from the lesson is included in Fig. 45.2. The majority of problems posed related to quantity and exploration of number and some data representation. However, other problems led to more in-depth investigations relating to pattern, shape and size. The 'what' and 'why' problems the students posed link to the real world and provide opportunities for mathematical modelling. In contrast, the 'how many' problems posed may appear trivial from a modelling and applications point of view but are important mathematically in terms of quantifying, sorting and classifying. The 'what' and 'why' problems posed are messy and complex and require more thought to begin to solve them than the 'how' problems posed. Brown and Walter (2005) suggest that it is important in the initial stages to accept all questions even when some appear irrelevant or nonsensical.

How many holes in each button?
How many buttons have just two holes?
How many buttons have just one hole at the back?
How many holes on all the buttons?
How many buttons of each colour?
How many buttons altogether?
How many buttons fit in the tin?
How many buttons cover my page?
How far will the buttons go, if we made a line of them?
What size are the buttons?
What kinds of patterns are on the buttons?
What kinds of shapes are the buttons?
Why are most buttons round?
How do you know if something is a button?
What are buttons made of?

Fig. 45.2 Problems posed from the button exploration

Table 45.1 Examples of the word problems posed for 4×5=20

Word problem	Connection between reality and mathematical world
Patrick and Ryan go to the shops and spent $20. They bought 4 things. How much did each thing cost?	Yes, using a routine problem with result unknown
There are 20 cupcakes and five girls at a party. How many cupcakes did each girl get?	
Declan, Aidan, Luke and Dan each gave Liam 5 lollys [sic]	Yes, but assumes each gets the same amount
How many did Liam have altogether?	
Dad put 20 apples in boxes. Each box had the same amount	Yes, partially used the context to match the situation but more information is required
How many boxes did he get and how many apples in each?	
Patrick had 40 marballs [sic]. He shared 10 to Ryan and Daniel	None
How many did Patrick have left? (4×5)	
I had a box of 20 cookies and dropped it. Nine were broken and I ate 3. How many are left?	None

The observing teachers were surprised by the amount of mathematics the students used in the lesson, their engagement and the range of problems they posed particularly from some of the students who normally struggle in mathematics and are reluctant to contribute. In the follow-up lesson, which the class teacher conducted, the students chose one of the problems to work on in pairs.

3.2 *Year 3–4 Problem Posing – Story Problems for 20*

There were a range of problems generated for the equation 4×5=20. Most of the problems posed indicate students' use of the real world (contexts such as shopping, games, lollies) to demonstrate their understanding of the mathematical world. Some of the problems reflected insufficient information while others indicated little consideration of the equation. Examples of the different word problems posed are included in Table 45.1.

The students' problems posed for 20 were classified according to the connection between reality and the mathematical world and whether they reflected the original equation. There were 26 student responses and the results of the analysis of these are presented in Table 45.2.

From this analysis it is reasonable to infer that students were familiar with solving routine problems and so were able to draw on their experience and formulation of problems of this nature. It was not surprising, based on research by others (e.g., English 1998) that some students had difficulty in posing problems that related to the equation and to the real world.

Table 45.2 Frequency of word problems types posed for 20 ($N=26$)

Problem type	Frequency
Connection between reality and mathematical world using routine problem	9
Connection between reality and mathematical world with missing element	1
Partial connection between reality and mathematical world with additional information required	4
No connection between reality and mathematical world, nor does the problem match equation	8
Not formulated as a word problem	4

3.3 *Year 3–4 Problem Posing from Real World Photographs*

From an analysis of all 13 problems posed, most were based on the literal interpretation of the picture initially then as the lesson developed some students explored the possibilities of more challenging problems. This was facilitated by the problems posed being presented to the class and discussed in relation to the criteria in Sect. 2.2. Some students generated more than one problem relating to their picture while others had difficulty in moving beyond a pure computation type problem. Each pair of students was able to produce a problem that could be investigated using mathematics, regardless of the level of complexity involved. Examples of problems posed and real world photographs are included in Fig. 45.3.

Some of the problems generated could be considered too challenging for 8 and 9 year old students to solve, however they do give genuine rise to problem solving, making some assumptions and lead to mathematical modelling, such as the *Windows Problem* and the *Grains of Rice Problem*. The other problems (*Strawberries*, *Jellybeans*, and *Stamps*) are routine type word problems, which suggests the students possibly posed the problems based on a literal interpretation of the photographs. These problems do not reflect engagement with the real world. Such problems can be solved using prior knowledge and skills and do not require any problem solving. However, for some students the *Strawberries Problem* is a genuine problem, as it requires them to make some assumptions.

Two observations made by the classroom teachers watching the lesson give some insight into possible reasons for the nature of the problem posing. First, some of the boys were very competitive when the problems were shared and wanted to work them out rather than offer suggestions. This could suggest that they were focused on the solution rather than the problem posing. Second, some of the more mathematically capable students struggled initially with generating a problem beyond just the literal interpretation of the picture, possibly because of the need to think differently. It appears that the students' view of mathematics or their experiences of mathematics is as a purely abstract subject. It could be argued that from a young age students need to develop views of mathematics as being both abstract and connected to the real world.

Windows problem
How many windows in the building?
How long would it take someone to clean the windows?
How many rooms might be in the building?

Grains of rice problem
If my dad grew fifteen rice plants, how much rice would he have in a year?
If me and Flynn eat 50 grams of rice each day, how many grams of rice would we eat in a year? How many grains of rice would that be?

Stamps problems
There are 50 stamps. Tom buys 10 on Monday, 3 on Tuesday.
How many stamps does he have left? Each stamp costs 3 cents.
He has $3.00. He has until Friday to buy all 50 how many will he buy on the other three days? How much money will he have left when he buys all 50 stamps?

Jellybean candies problem
A man in a lolly shop put jellybeans in a container. There were 12 rows and 18 in each row. How many jellybeans are there?

Strawberries problem
Me and Declan picked 60 strawberries. I ate ¼ of them. How many are left?
We had to pack 60 strawberries into containers.
How many containers did we need and how many did we pack in each?

Fig. 45.3 Problems posed from photographs of real world situations

4 Discussion and Conclusions

Silver and Cai (1996) have reported that student engagement in problem posing experiences has a positive influence on their problem solving achievement or their disposition towards mathematics. The students in both teaching experiments were engaged and the teachers gained insights into their learning that they were not necessarily aware of prior to the lesson. In the Year 3–4 class students of various ability levels persevered with the challenge of formulating questions and wanted to produce something for their efforts. The observing teachers found the use of real world artefacts beneficial and some were surprised at the level of engagement of the Year 1–2 students and their understanding of number.

The findings suggest that students as young as 6 years of age, with no prior experience with problem posing, are capable of generating questions that can be investigated using mathematical modelling. As stated earlier, mathematical modelling involves skills in problem posing and problem solving so it is reasonable to argue that providing students with opportunities to develop expertise in problem posing may enhance their capabilities as mathematical modelers (English 2003; English et al. 2005; Stillman in press). Furthermore, the use of real world artefacts that are meaningful to students enabled them to construct meaningful mathematical problems as suggested by Bonotto (2008).

While acknowledging the studies reported in this chapter are small, they provide evidence to suggest that the use of problem posing and real world artefacts in the early years of primary school could provide the foundation for use of real life modelling and applications in the classroom. Further research into the extent to which the use of real world artefacts support young students' ability to pose rich mathematical problems that can be investigated using mathematical modelling, and develop their problem solving and reasoning skills needs to be conducted particularly with respect to the influence of the views of mathematics that students appear to be developing at such a young age.

References

Bonotto, C. (2008). On the relationships between informal out-of-school mathematics and formal in-school mathematics in the development of abstract mathematical knowledge [regular lecture]. In M. Niss (Ed.) *ICME10 2004, Proceedings of tenth international congress on Mathematical Education* [CD]. Roskilde: IMFUFA.

Bonotto, C. (2010a). Realistic mathematical modeling and problem posing. In R. Lesh, P. Galbraith, C. R. Haines, & A. Hurford (Eds.), *Modeling students' mathematical competencies* (pp. 399–408). New York: Springer.

Bonotto, C. (2010b). How it is possible to make real-world mathematics more visible: Some results from two Italian projects. In A. Araújo, A. Ferandes, A. Azevedo, & J. F. Rodrigues (Eds.), *Conference proceedings of EIMI 2010; Educational interfaces between mathematics an industry* (pp. 107–116). Lisbon/Bedford: Centro Internacional de Matemática/COMAP.

Brown, S., & Walter, M. (2005). *The art of problem posing* (3rd ed.). Mahwah, NJ: Lawrence Erlbaum.

Cai, J., & Hwag, S. (2002). Generalized and generative thinking in US and Chinese students' mathematical problem solving and problem posing. *Journal of Mathematical Behaviour, 21*, 401–421.

Christou, C., Mousoulides, N., Pittalis, M., Pitta-Pantazi, D., & Sriraman, B. (2005). An empirical taxonomy of problem posing processes. *Zentralblatt für Didaktik der Mathematik, 37*(3), 149–158.

English, L. D. (1997). The development of fifth-grade children's problem-posing abilities. *Educational Studies in Mathematics, 34*(3), 183–217.

English, L. (1998). Children's problem posing within formal and informal contexts. *Journal for Research in Mathematics Education, 29*(1), 83–106.

English, L. D. (2003). Engaging students in problem posing in an inquiry-oriented mathematics classroom. In F. K. Lester Jr. & R. I. Charles (Eds.), *Teaching mathematics through problem solving, Prekindergarten-Grade 6* (pp. 187–196). Reston, VA: National Council of Teachers of Mathematics.

English, L. D., Fox, J. L., & Watters, J. J. (2005). Problem posing and solving with mathematical modelling. *Teaching Children Mathematics, 12*(3), 156–163.

Silver, E. A. (1994). On mathematical problem posing. *For the Learning of Mathematics, 14*(1), 19–28.

Silver, E. A., & Cai, J. (1996). An analysis of arithmetic problem posing by middle school students. *Journal for Research in Mathematics Education, 27*(5), 521–539.

Stillman, G. (in press). Problem finding and problem posing for mathematical modelling. In N. H. Lee & K. E. D. Ng (Eds.), *Mathematical modelling—From theory to practice*. Singapore: World Scientific.

Whitin, D. (2004). Building a mathematical community through problem posing. In R. Rubinstein & G. W. Bright (Eds.), *Perspectives on the teaching of mathematics* (pp. 129–140). Reston, VA: NCTM.

Whitin, D. J. (2006). Problem posing in the elementary classroom. *Teaching Students Mathematics, 13*(1), 14–18.

Chapter 46
A Study of the Effectiveness of Mathematical Modelling of Home Delivery Packaging on Year 12 Students' Function Education

Tetsushi Kawasaki and Yoshiki Nisawa

Abstract Teaching of functions in secondary education in Japan, as in many other countries, treats only single variable functions, but in our daily life and in nature a lot of phenomena can be expressed by functions of several variables. A home delivery packaging model was provided as a typical teaching material for learning mathematical development models. Fifteen Year 12 students participated in this educational experiment. The theme given to the students was about maximising the volume of a box with a fixed outer length in order to pack the box with as large an amount as possible. This modelling will effectively help students to imagine the maximum value for three dimensional models with two variable integer functions, and it will also help them solve partial differentials. It is hoped that this study will provide additional support for Japanese mathematical activities and improve the teaching of functions in Japan.

1 Introduction

The Ministry of Education, Culture, Sports, Science and Technology (2009) mandated a new course of study for lessons in mathematics in Japanese junior high schools which began in April, 2009. In the case of high school mathematics, the new course started in April, 2012. "Learning Mathematics through Problems" and "Mathematical Activity" were described as teaching content for the first time. These new ideas entail expressing

T. Kawasaki (✉)
Gifu University, Gifu-Shi Yanagido 1-1, Japan
e-mail: tetsushi@gifu-u.ac.jp

Y. Nisawa
Kyoto Prefectural Rakuhoku Senior High School, Kyoto, Japan
e-mail: y-nisawa@kyoto-be.ne.jp

G.A. Stillman et al. (eds.), *Teaching Mathematical Modelling: Connecting to Research and Practice*, International Perspectives on the Teaching and Learning of Mathematical Modelling, DOI 10.1007/978-94-007-6540-5_46,

real case scenarios through mathematical models with the use of technological instruments such as a personal computer or a calculator. In the past, there have been few cases where mathematical modelling was practised in mathematics education in Japan. Thus, using it now may cause puzzled reactions from senior high school teachers. This revision of the course of study gives a chance to treat mathematical modelling in classes, but it is not certain whether the practice of mathematical modelling will yield positive results even if teachers understand its value. Solving many problems as fast as possible in a given time limit and repetition drills take a lot of time, so mathematical modelling seems unsuitable in the traditional Japanese university entrance examination system. It is necessary to practice mathematical modelling effectively while considering the traditional educational custom. The content of functions as a feature in high school mathematics education is an ideal starting point.

In secondary education in Japan, mathematical education treats functions of one variable. However, if natural science and social phenomena are expressed by mathematical models, expressions using functions of one variable have limitations. Some problems that can be solved by using functions of two variables are hidden in present textbooks. The importance and meaning of using these in classes will rise if such problems and mathematical modelling are effectively integrated, and it will be understood that functions of two variables are necessary to express real case scenarios through mathematics. To attempt to address this situation the following educational example was introduced to fifteen Year 12 students who are aiming at taking courses in science and engineering departments after high school.

2 Classes Based on Integral Functions of Two Variables

According to the course of study, it is thorough to plot a graph, to remember the shape of each function's graph, and to determine the intersection with the coordinate axis (the solution to an equation). We injected the essence of two variable functions into the traditional Japanese educational curriculum because one of the difficulties that should be tackled when students engage in mathematical modelling according to the modelling cycle process (Kaiser 1995, cited in Borromeo Ferri 2006) is that if new mathematics content (in this instance functions of two variables) is not given to students, they cannot form a mathematical development model (Kawasaki and Moriya 2011). The following teaching material of integral functions is a way to more easily introduce mathematical modelling to high school students.

2.1 *Maxima and Minima of Quadratic Functions*

The problem shown in Fig. 46.1 is one where Year 10 students encounter a function of two variables for the first time. The solution to this problem is called "Pre-election

(Problem) Solve for the maximum value and the minimum value.

$$y = 3x^2 - 6ax + 2 \quad (0 \le a \le 2) \text{ (quadratic function of } x\text{)}$$

(Answer) (1) The minimum and the maximum values,"m" and "M".

Change into $y = 3(x - a)^2 - 3a^2 + 2$

(2)

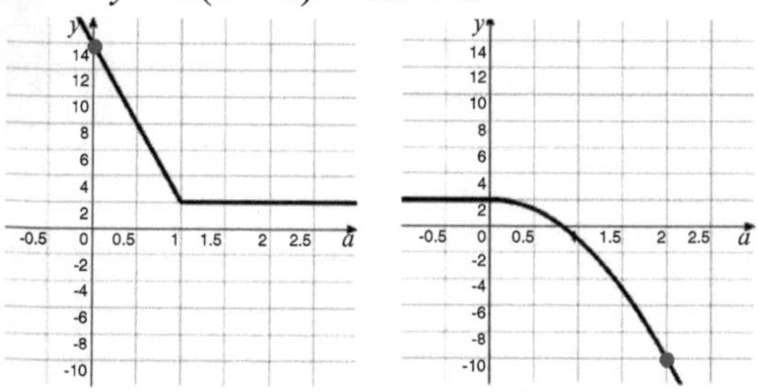

$a = 0 \Rightarrow M = 14$ $\quad$ $a = 2 \Rightarrow m = -10$

Fig. 46.1 Pre-election and final method

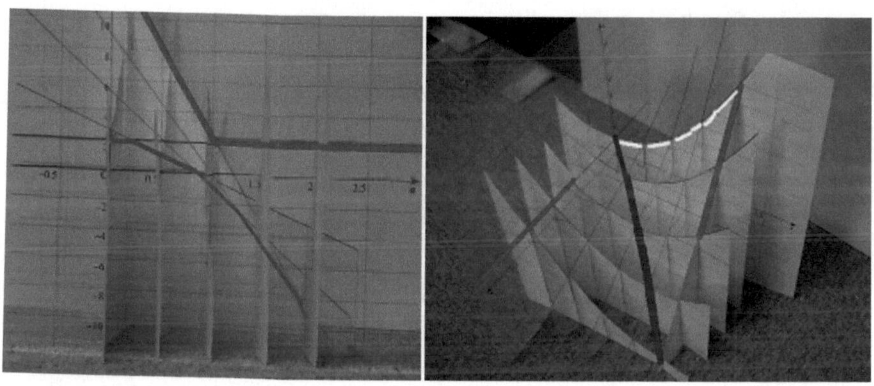

Fig. 46.2 Three dimensional model

and final method" in Japan (Fig. 46.1). In obtaining the function of x, various probable maximum and minimum values are gathered (Pre-election). The maximum and minimum values are narrowed down from these (Final). However, the meaning of a function of two variables is not described in Year 10 textbooks. The impression of using this technique or method is strong. Even if students remember the technique but do not know the meaning, they will forget it over time. In fact, 6 of the 15 students could not solve this problem. Then, it was explained to them how to make a three dimensional model as in Fig. 46.2, and they were shown the real thing. If transparent sheets are used, the two graphs can be seen at the same time. By using this model, students can observe the appearance of the change and can understand the relationship between a and x.

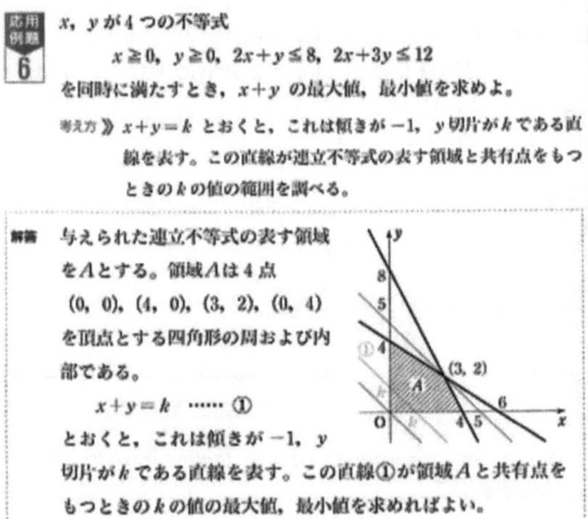

応用例題 6

x，y が 4 つの不等式

$$x \geqq 0,\ y \geqq 0,\ 2x+y \leqq 8,\ 2x+3y \leqq 12$$

を同時に満たすとき，$x+y$ の最大値，最小値を求めよ。

考え方》 $x+y=k$ とおくと，これは傾きが -1，y 切片が k である直線を表す。この直線が連立不等式の表す領域と共有点をもつときの k の値の範囲を調べる。

解答 与えられた連立不等式の表す領域を A とする。領域 A は 4 点 (0, 0)，(4, 0)，(3, 2)，(0, 4) を頂点とする四角形の周および内部である。

$x+y=k$ …… ①

とおくと，これは傾きが -1，y 切片が k である直線を表す。この直線①が領域 A と共有点をもつときの k の値の最大値，最小値を求めればよい。

Brief English translation of text.

(Problem) [Boundary condition]

Simultaneous inequality $x \geq 0$, $y \geq 0$, $2x + y \leq 8$, $2x + 3y \leq 12$

Solve for the maximum value and the minimum value of $x + y$.

(Method of thinking)

(1) $x + y = k$: This expresses a straight line that has a slope of -1 and y intercept k.

(2) Find the maximum value and minimum value of k, by plotting many straight lines in the area framed by $(0,0)$, $(4,0)$, $(3,2)$, $(0,4)$.

Fig. 46.3 Linear programming problem in textbook (*Sugaku II*, Suken Syuppan 2007, p. 91) and a contour line

2.2 Area Shown by Inequality (Maximum and Minimum)

Year 11 students are supposed to learn linear programming (Fig. 46.3), where they will find the maximum value and minimum value by plotting many straight lines in an area. The contour of the three dimensional model is represented by a group of straight lines. However, the meaning of a function of two variables is not described in the textbooks, so students must also consider this as one technique. When thinking about the maximum and the minimum of an expression like $x^2 + y$, students might similarly plot some straight lines. By making a three dimensional model students can better understand which of x or y gives major impact to the value of the expression.

3 Example of Mathematical Modelling of Home Delivery Packaging: Mathematical Development Model with Partial Differentiation

As a case of solving a problem as a real scenario, mathematical modelling of a home delivery package in Japan was done. Though this problem is a function of three variables, by using some constraints, it becomes a function of two variables. As this model can be made from an integral function, it is comparatively easy to treat. Here, the purpose is to introduce mathematical modelling that is authentic to the function for the first time in secondary education so it is important that it becomes easy for teachers and students to treat by preparing a simple and friendly scene. In addition, if teaching material that students have learnt up until this lesson can be revised from a different angle, this modelling will succeed.

According to the rules for home delivery packaging in Japan, the shipping charge is dependent on the total length, breadth and height of a box in which goods are to be packed (Table 46.1, Fig. 46.4). Note, however, that neither the weight of the goods in the box nor the delivery distances are considered. For any hexahedral delivery box, the charge will be the same within a given range. The students were given the following conditions: *"You select a size 60 box. To send as many goods as possible, what box shape must be prepared?"* Many students will decide that the answer is to find the maximum volume of the box. They will easily make the mathematical equations ($V = xyz, x + y + z = 60, 0 < x, y, z < 60$), and they will notice that V is a function of two variables ($V = xy(60 - x - y)$).

Table 46.1 List of charges (Japan Post 2011)

Size	Total (length + breadth + height)	Charge
60	Under 60 cm	¥600
80	60–80 cm	¥800
100	80–100 cm	¥1,000
120	100–120 cm	¥1,200
140	120–140 cm	¥1,400
160	140–160 cm	¥1,600
170	160–170 cm	¥1,700

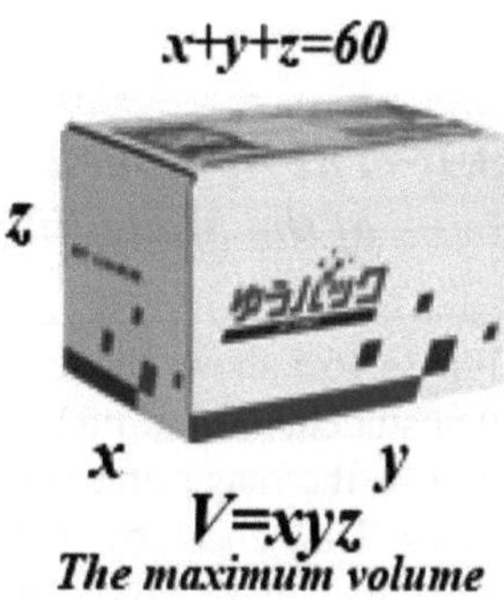

Fig. 46.4 The size 60 box

Fig. 46.5 Students creating models during a class

V=xy(60-x-y) (0<x<60, 0<y<60, 0<x+y<60)

Find the maximum

Stage 1st

Focus

y \ x	10	20	30	40	50
10	4000	6000	6000	4000	0
20	6000	8000	6000	0	-10000
30	6000	6000	0	-12000	-30000
40	4000	0	-12000	-32000	-60000
50	0	-10000	-30000	-60000	-1E+05

Zoom

Identify

y \ x	18	19	20	21	22
18	7776	7866	7920	7938	7920
19	7866	7942	7980	7980	7942
20	7920	7980	8000	7980	7920
21	7938	7980	7980	7938	7854
22	7920	7942	7920	7854	7744

By 3D model image

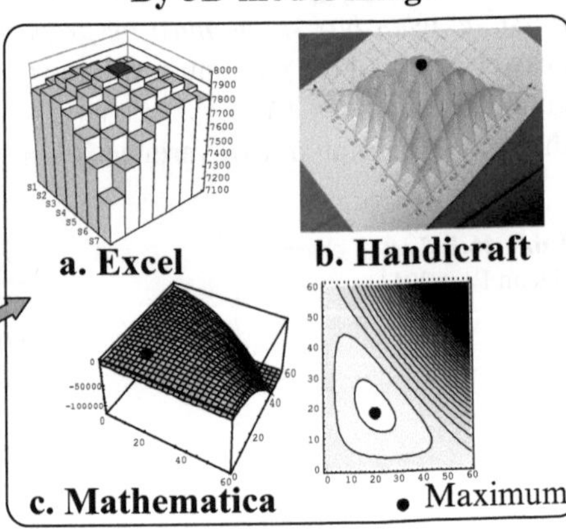

Fig. 46.6 The genealogical tree to an elementary model

3.1 *The Approach to an Elementary Model (First Stage of the Modelling Cycle Process)*

Figure 46.5 shows the lesson being facilitated. Figure 46.6 is a systematic approach that students use to make a model. First of all, a spreadsheet is prepared. Students narrow the range of *x* and *y* so that the maximum value of volume can come into focus. They check by improving the accuracy of the value for volume even a little. Next, they make a three dimensional model to observe the image of changes around the maximum value. They also use mathematical software. It can be expected that

Stage 2nd **The approximate, more exactly**

The quadratic function about x

$$V(x,y)=xy(60-x-y)=-y\left(x-\frac{60-y}{2}\right)^2+y\left(\frac{60-y}{2}\right)^2$$

$$x=\frac{60-y}{2}\Rightarrow V(y)_{\max}=y\left(\frac{60-y}{2}\right)^2$$

The cubic function about y $x=y=z=20,$

$$V_{\max}=8000$$

Fig. 46.7 The solution using the mathematics that students have studied

computer-assisted learning improves the student's performance. Students will find the maximum value of the volume of the box becoming 8,000 cm^3 at the cube ($x=y=z=20$). However, some students do not agree with the result. This is because this value is an approximate one calculated by the computer. It might have to be discussed whether this maximum value is a proper numerical value. In fact, because of the style of traditional mathematics education in Japan, this mathematical model will be dissatisfying. So, the next stage will likely be to aim to form a development model by using more advanced mathematics.

3.2 Approach to a New Model (Second Stage of the Modelling Cycle Process)

The students, who were satisfied by the previous elementary model, and even those who did not construct it must begin work on a new model using the mathematics that they have studied. Students solved for the maximum value with a "Pre-election and final method" thinking of a quadratic function concerning x as shown in Fig. 46.7.

However, this mathematical expression takes time to transform and a cubic function of y appears in the end. If they do not use differentiation, they will not be able to solve for the maximum value. There are a lot of repeated practices using differentiation of functions of one variable in the students' textbook. Students were accustomed to differentiation, and therefore were able to derive satisfaction from this maximum value using this mathematical modelling.

3.3 Approach to a Mathematical Development Model (Third Stage of Modelling Cycle Process)

Here, even if the model is the same as the elementary model, it is considered as a mathematical development model and is the reason why this modelling has improved rigour by taking on new mathematical content (i.e., partial differentials).

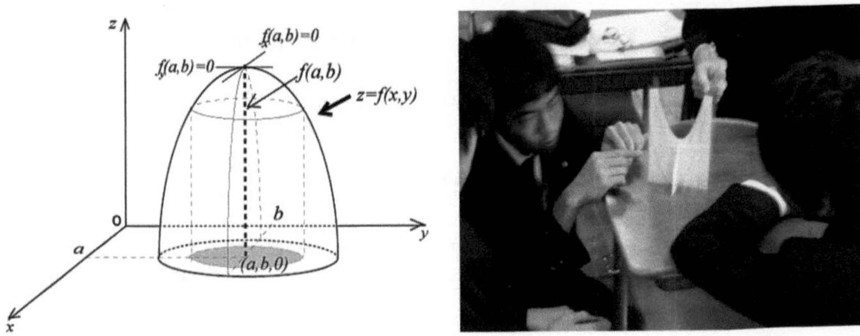

Fig. 46.8 Group discussion of a three dimensional model concerning partial differentials

Stage 3rd **More simply, evolutionally**

Partial differential

$$V_x = y(60-x-y)-xy = y(60-2x-y) = 0$$
$$V_y = x(60-x-y)-xy = x(60-x-2y) = 0$$

Only one stationary point **(Initial condition: 0<x<60, 0<y<60)**

$$\begin{cases} 60-2x-y=0 \\ 60-y-2x=0 \end{cases} \Longrightarrow x=y=z=20, \boxed{V_{max}=8000}$$

Fig. 46.9 Mathematical development model with a partial differential

This modelling does not use a delta (Δ) as the judgment condition concerning the extreme points. Moreover, a setting equal to the maximum value yields a stationary point (Fig. 46.8).

The stationary point is represented by a three dimensional model. It might be easy for students to solve simultaneous equations shown by integral expressions (Fig. 46.9). So the teacher will need to prepare other real case scenarios for students interested in partial differentials.

4 Students' Evaluation and Impressions

The fifteen Year 12 students who participated in the modelling activity were given a questionnaire (see Appendix) where they had to rate items from 1 (lowest) to 4 (highest) evaluating their understanding of the modelling lessons. Table 46.2 is the result of the students' nationwide achievement test (Sugaku Z Mondai 2009) and preliminary poll before completing this modelling practice on two questions concerning (1) preference for hand calculation over using technological instruments

Table 46.2 Year 12 students ($N=15$) academic ability (Sugaku Z Mondai 2009) and preliminary questionnaire results

Student	Mathematics T-score	(1) Preference for hand calculation	(2) Experience using spreadsheets
I	51.2	4	3
N	55.5	3	3
J	55.8	2	3
F	58.2	2	3
E	58.7	2	3
D	64.9	3	3
K	67.1	2	3
O	59.9	2	1
M	63.2	3	1
C	64.2	4	1
G	66.8	4	1
H	69.3	3	1
L	71.4	3	1
B	71.4	4	1
A	77.2	4	2

Note. Items on poll were ranked *1* (low), *2* (little), *3* (high) and *4* (much)

and (2) level of experience of analysing with spreadsheets. By the traditional educational custom of Japan, students are accustomed to studying calculations by hand, without a technological instrument. Seven students with high mathematics T-scores (see Table 46.2) were not accustomed to the use of spreadsheets or mathematical software. Rather than solving for approximate values by the personal computer, the academically-gifted students preferred hand calculation and performed remarkably. As shown in Fig. 46.10, they appeared to be satisfied with solving for an exact solution without using a computer (8); therefore, only partial differentials might have had to be taught directly to these students. The mathematical development model became more effective through each stage of the modelling cycle process. Particularly, the understanding of students improved (see Fig. 46.10, Understanding). Furthermore the students were satisfied with the effectiveness of the information technology (6). However, the first stage modelling will have to be of more interest to students (7). One student's impression was: "Until now, if I could not solve the value of a mathematical expression by hand calculation, I might have given up. I was surprised because it was easily treatable, and I want to challenge myself with other more complex equations".

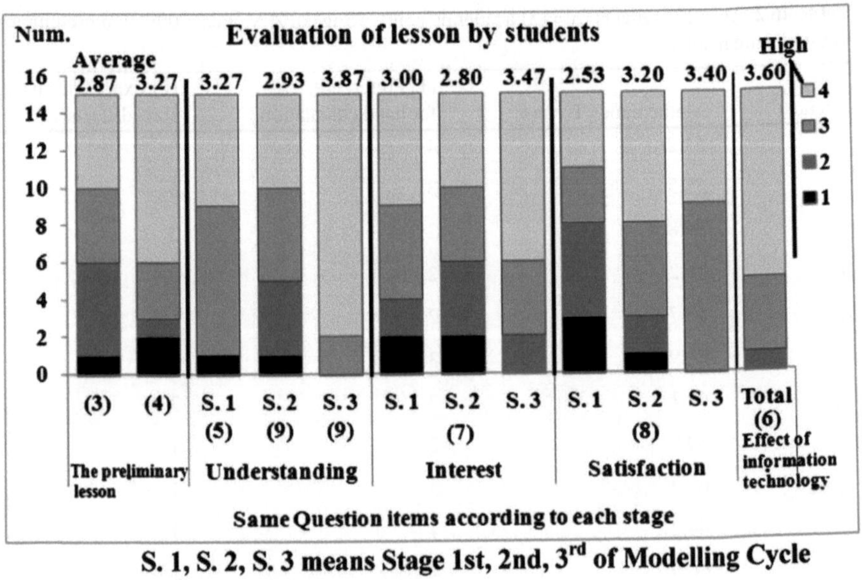

Fig. 46.10 Evaluation of each lesson by students

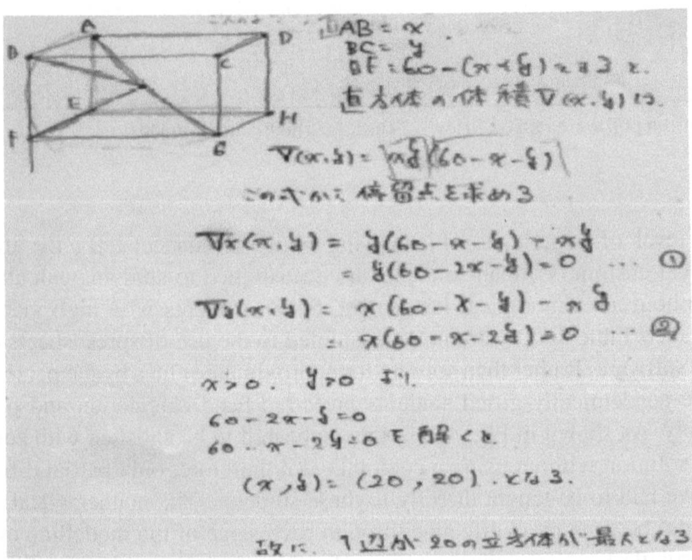

Fig. 46.11 The student's answer sheet

Some students could not use the "Pre-election and final method" in the Second Stage, but most of them made a mathematical development model with a partial differential by themselves. Figure 46.11 shows part of a student's written solution to the problem. One of the 15 students commented, "This modelling was a little easy though it was suitable

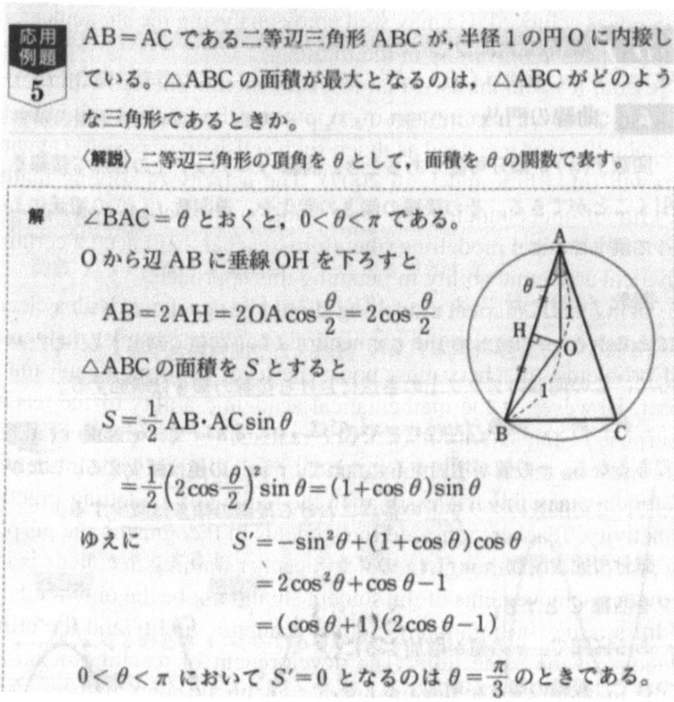
応用例題 5

AB＝AC である二等辺三角形 ABC が，半径 1 の円 O に内接している。△ABC の面積が最大となるのは，△ABC がどのような三角形であるときか。

〈解説〉二等辺三角形の頂角を θ として，面積を θ の関数で表す。

解　$\angle BAC=\theta$ とおくと，$0<\theta<\pi$ である。

O から辺 AB に垂線 OH を下ろすと

$$AB=2AH=2OA\cos\frac{\theta}{2}=2\cos\frac{\theta}{2}$$

△ABC の面積を S とすると

$$S=\frac{1}{2}AB\cdot AC\sin\theta$$

$$=\frac{1}{2}\left(2\cos\frac{\theta}{2}\right)^2\sin\theta=(1+\cos\theta)\sin\theta$$

ゆえに

$$S'=-\sin^2\theta+(1+\cos\theta)\cos\theta$$

$$=2\cos^2\theta+\cos\theta-1$$

$$=(\cos\theta+1)(2\cos\theta-1)$$

$0<\theta<\pi$ において $S'=0$ となるのは $\theta=\frac{\pi}{3}$ のときである。

Fig. 46.12 An exercise problem from a textbook (*Sugaku III*, Suken Syuppan 2006, p. 115)

for the student's level". Another student's opinion was, "With original, new and different methods, my motivation to practice mathematical modelling progressed". The same student also found a problem in a textbook (Fig. 46.12) and said, "If the following condition (AB=AC) is lost, I am sure this is a function of two variables".

5 Conclusion and Implications for Future

When mathematical modelling is introduced for the first time, it is necessary to avoid treating a sudden, difficult problem. Employing integral functions is sure to soften students' resistance. In relation to this, the researcher succeeded in developing teaching material related to the connection between high schools and universities in Japan. Students broke loose from the world of functions of one variable, and they have worked through the problems of functions of two variables easily. By producing a three dimensional model, students developed new ideas, and their motivation for learning increased. In addition, they succeeded in analysing a phenomenon in their daily life with partial differentiation. One student had the insight to recognise that if a problem sentence in the textbook was deleted or changed slightly it changed the problem into a partial differential problem. This means a significant transfer from practice was achieved. He may well enjoy analysing the phenomena in his life and natural sciences at university in the future.

In this lesson, it is emphasised that teachers should interact with their students and provide support, but a minimum of support by the teachers might be all that is needed. Especially, students need as much such self-help as possible at early stages of their growth (cf. Blum and Leiß 2007). The activity of solving mathematical problems without understanding the underlying mathematical concepts cannot be considered mathematical modelling; therefore, teachers will need a certain amount of mathematical academic ability in pursuing this approach.

Lessons using mathematical modelling should be developed with a clear purpose so that students can recognise the connections between abstract mathematics and real world problems. Teachers must boost the level of the students' mathematics achievement. However, if the mathematical academic ability of the teacher is too low, the purpose of the activity may not be realised (cf. Kawasaki et al. 2012). Of course care must be taken with mathematical modelling as well. In the worst-case scenario, students may finish a lesson with the feeling of modelling-practice as just a practice activity. Teachers must aid the students in recognising the purpose of the modelling practice. So, although the role of teacher training in college is important, the mathematics achievements of the student should not be the only focus. Teachers should address issues of both mathematical academic ability and the utilisation of that knowledge at the same time. The development of teaching-materials which increases the mathematics achievements of students through mathematical modelling lessons should be treated in teacher training colleges.

Appendix – Questionnaire

Preliminary Poll Items

(1) Do you like to solve by hand calculation better than to solve using informational instruments?
(2) Do you have experience of analyzing by use of spreadsheets or mathematics software?

Post-lesson Items

(3) Were you interested in the use of functions of two variables from what you have learned?
(4) Could you understand the use of functions of two variables from what you have learned?
(5) Did you understand the structure of the maximum value as discovered by the spreadsheet?
(6) Were the use of the spreadsheet and the production of a 3D model effective?
(7) Were you interested in this solution?
(8) Are you satisfied with this solution?
(9) Could you solve this problem using the mathematics that you have learned?

References

Blum, W., & Leiß, D. (2007). How do students and teachers deal with modelling problems? In C. Haines, P. Galbraith, W. Blum, & S. Khan (Eds.), *Mathematical modelling (ICTMA12)* (pp. 222–231). Chichester: Horwood.

Borromeo Ferri, R. (2006). Theoretical and empirical differentiations of phases in the modelling process. *ZDM, 38*(2), 86–95.

Japan Post. (2011). *Kihon Unchin*. http://www.post.japanpost.jp/fee/simulator/kokunai/parcel.html

Kawasaki, T., & Moriya, S. (2011). Using modelling experiences to develop Japanese senior high school students' awareness of the interrelations between mathematics and science. In G. Kaiser, W. Blum, R. Borromeo Ferri, & G. Stillman (Eds.), *Trends in the teaching and learning of mathematical modelling (ICTMA14)* (pp. 603–615). New York: Springer.

Kawasaki, T., Moriya, S., Okabe, Y., & Maesako, T. (2012). The problems of mathematical modelling: Introduction on mathematics education in Japanese school. *Journal of Mathematical Modelling and Application, 1*(5), 50–58.

Ministry of Education, Culture, Sports, Science and Technology. (2009). *Koutougakko gakusyuu shidou yoryou* (pp. 53–63). Tokyo: Author.

Sugaku Z Mondai. (2009). *Sougou Gakuryoku Kijyutsu Moshi 7 Gatsu. Shinken Moshi Kamoku Betsu Sousyu Hen Sugaku* (pp. 69–70). Tokyo: Benesse Corporation.

Suken Syuppan. (2006). Kansu No Saidai To Saisyo. *Kaiteiban Sugaku III* (p. 115). Toyko: Suken Syuppan.

Suken Syuppan. (2007). Futoshiki no ryoiki. *Kaiteiban Shinpen Sugaku II* (p. 91). Tokyo: Suken Syuppan.

Chapter 47
How to Introduce Mathematical Modelling in Industrial Design Education?

Geert Langereis, Jun Hu, and Loe Feijs

Abstract With competency based learning in a project driven environment, we are facing a different perspective of how students perceive mathematical modelling. In this chapter, a model is proposed where conventional education is seen as a process from mathematics to design, while competency driven approaches tend to have an inverted sequence. We assume there is a virtual barrier for on-demand learning with regards to the mathematical modelling layer under the layer of technical skills. Several successful attempts were done in the past to remove the technology skill from the chain in order to make the opportunities of modelling visible. After experiencing the modelling competency in such a setting, students can beneficially deploy it for technology. We evaluated this model based on a learning activity which was changed from traditional education into competency centred learning.

1 Introduction

The Department of Industrial Design at the Eindhoven University of Technology distinguishes itself by a focus on the design of intelligent systems, products, and related services. The user-focused application area is continued in the research and educational system by means of a competency centred learning approach. The associated reflective transformative design process (Hummels and Frens 2009) is based on learning and developing "from doing" and "by doing" and is, as such, highly dependent on action research. This is where it distinguishes fundamentally from classical design engineering approaches. In classical engineering education, a foundation of mathematical tools and thinking is taught first. Secondly, a scientific

G. Langereis (✉) • J. Hu • L. Feijs
Department of Industrial Design, Eindhoven University of Technology,
Eindhoven, The Netherlands
e-mail: geert.langereis@gmail.com

G.A. Stillman et al. (eds.), *Teaching Mathematical Modelling: Connecting to Research and Practice*, International Perspectives on the Teaching and Learning of Mathematical Modelling, DOI 10.1007/978-94-007-6540-5_47,

notion is taught, depending on the department being physics, chemistry, electronics, computer science, or any other area. After having acquired these two layers of background, the engineer can tackle practical problems in a design engineering approach. With our competency centred educational system, we start with engineering, to discover the science behind it by playing and exploring. The consequence is that the mathematical background is experienced as a deep third layer, which is not evidently a skill noted by our students as a next investment in their self directed competency development.

This chapter will explain the competency based learning system first. Next the aspects of modelling required for academic design students are discussed. In a subsequent section, some examples of teaching modelling skills are explained. In a section about balancing design engineering and modelling, a vision of the consequence of the educational system on the learning behaviour is given. With this model we can propose how teaching activities can be improved to optimise the learning outcome. The theory and proposal is evaluated based on our experience with a specific assignment on microcontrollers.

2 Competency Based Learning as Optimised for Design Education

Industrial Design (ID) is one of the nine departments of the Eindhoven University of Technology (TU/e), and was established in 2001. In consultation with industry and government, ID focuses on the development and design of user-friendly interfaces for intelligent systems, products and related services in multimedia environments. Current research topics include Human-Computer-Interaction, multi-modal interaction, perceptive user interfaces and aware environments (also known as "ambient intelligence"), as well entertainment computing research. The research aims to provide generic models and frameworks in the domains of perception, cognition, interaction and communication to the extent that these fields are relevant to the design of technical products and services.

The focal area of the Department of Industrial Design, being the creation of intelligent systems, products and related services, has resulted in an educational system that differs significantly from the other departments. For designing intelligent products at the Industrial Design Department, we are facing the problem that theory for interaction with humans is unpredictable, or at least complex. Therefore, in contrast to the knowledge driven approach of the other departments, Industrial Design has implemented an educational system that is based on action research. Education starts with practical work before students discover which scientific foundation is involved. This learning model applies particularly well for industrial designers because the industrial designer has to develop contexts of use, actively explore concepts, evaluate alternative solutions, and bring new products to the world.

The educational model is in agreement with social constructivism (Vygotsky 1978), the sociological theory of knowledge. In education, social constructivism stresses interaction over observation. As a consequence, students in our competency based system act and interact with objects, teachers and other students to learn. They need to become experts who create, apply and disseminate knowledge, and continuously construct and reconstruct their expertise in a process of life-long learning. Meeting the goals of education requires a high consistency between instruction, learning and assessment. With the new social constructivism insights in instruction and learning, the paradigm of assessment had to be changed as well (Birenbaum 2003; Segers et al. 2003). Although similar systems of Competency-Centered Learning programs are seen (e.g., Pereira et al. 2009), at the Industrial Design Department of Eindhoven University of Technology, it is implemented in every layer of the organisation, and already optimised over the past 10 years.

To assess the quality of the students, and to monitor their progress, ten competency areas are defined. The tenth competency area is *Descriptive and mathematical modelling*. This makes clear that the ability to describe models and to find the mathematical models behind concepts is seen as one of the skills of an industrial designer. The other competency areas are *Self-directed and continuous learning*, *Integrating technology*, *Ideas and concepts*, *Form and senses*, *User focus and perspective*, *Social cultural awareness*, *Designing business process*, *Design and research processes* and *Teamwork and communication*. In practice, it is expected from the students that they can integrate all ten competences into a single design process. The ten competencies are primarily learned from projects (3 days per week), and taught on-demand in classes (2 days a week). It is essential to understand that none of the classes are mandatory. Until the academic year 2012/2013, there was no class available for basic mathematics. Students who had to apply mathematics in their project, had to learn from an expert or external resources. Nevertheless, awareness of the potentials of mathematics is an intended learning outcome.

3 Interpretation of the Modelling Competency for Design

The definition of the competence "Descriptive and Mathematical Modelling" is

> Being able to create and apply descriptive and mathematical models by using formal and mathematical tools, in order to justify design decisions and support the design of complex, highly dynamic and intelligent systems.

Understanding and mastering methods and tools for descriptive modelling enables students to describe relationships between parameters resulting in system behaviour. It is the foundation of simulation and optimisation. There is a strong link to the ability to analyse complex problems: to identify structures before tackling partial problems and to work towards a solution structurally.

When mathematical modelling is put into practice, it can be subdivided into four skills for gaining system insights. The first is where models are used for *analysing a complex problem* by breaking it into pieces. In this case, the model can be a state diagram or a flow chart, and does not necessarily have to be finalised into a numerical model. It is a method of communication about the problem with others or with oneself lowering the cognitive load by drawing systems on paper. The model can be the first step to translate the problem into solutions. Based on the state diagram, a designer can evaluate options and process flow.

A second modelling deployment for design is to *identify behaviour and dynamics* of systems. The notion of feedback systems, second order dynamic systems and, for example, phenomena like friction, are typical engineering skills, almost a craftsmanship, resulting in a predictive design process. A designer can prevent oscillations in a system by identifying mass-spring systems, for example.

The *predictive power* of models becomes strongest when a numerical mathematical model is implemented. The model can consist of closed form equations, or of a finite element simulation. Closed form equations normally can be solved to find design criteria or to exclude options. However, this type of modelling, which is highly dependent on calculus and numerical mathematics, is normally not seen from our students because of the absence of basic mathematical education in our offered classes.

The fourth skill is to eliminate options before building them. This is part of *evidence based design* where calculations are used to underpin design choices. To do this, estimations or calculations are made of an envisioned implementation in order to prove that the chosen solution is correct or that the design choice is the optimum solution.

In the next section, these four skills of modelling are identified with some examples of educational innovations at our Industrial Design Department. It is worth noting, that in practice the competency "descriptive and mathematical modelling" is closely related to the competency "integrating technology". Although developing technology is not the core business of the department, hardware is needed as a substrate to explore intangible concepts.

4 Some Implementations of the Modelling Competency

Previously, a teaching method (Hu et al. 2007) was presented to teach students to understand object-oriented design principles and formal software specification methods up to a level suitable for communication with software experts. The method was based on exploring a set of simple interaction rules by means of acting. Students became software objects (or "classes" in software terminology) themselves and could so transform acted behaviour into state diagrams. Such a practical realisation of a complex concept as object-oriented programming helped students to understand contexts, evaluate design ideas, explore new ideas and communicate designs to an audience. This learning activity is an example of the first modelling skill *analysing a complex problem* and appeared to be an easy way to make state diagrams explicit. The chosen "acting-out" methods can be seen as a strategy to educate modelling without having a technological frame of reference.

In van der Vlist et al. (2008) a method to teach the abstract concept of "machine learning" to students was explained. In this case, we did not remove the technological substrate completely, but we replaced it by a platform with which most students are comfortable: Lego Mindstorms NXT. The related modelling skill is *identify behaviour and dynamics*. With the ambition to teach the students to see patterns, the mathematical background was not omitted. For both reinforcement learning using the complex Q-learning method, and voice command learning using neural networks, the underlying principles were explained to the students using equations.

In the first example, the technology was removed. In the second example, technology was replaced by a simplified vehicle as most students are confident with Lego. This was done to bring the model and the real world as close as possible. In other words, the mathematical modelling is decoupled from the hardware/software substrate. However, when the teaching activity is about hardware or software, this is not always an effective option. When teaching programming, students must write code in a commonly accepted language like C or Java. This is done by focusing on the creative part of programming (Alers and Hu 2009) and using a robotic platform for a practical approach. Although this is not a modelling nor mathematics assignment, it proves that there is room to bring students into a state where they may discover that modelling skills, *analysing a complex problem* by means of state diagrams and *identify behaviour and dynamics*, have become within reach.

The *predictive power* skill of mathematical modelling is implemented amongst others in a course about geometrical principles (Feijs and Bartneck 2009). In that course tesselations are used to create plexiglass forms. A tesselation is a collection of plane geometries with no overlaps and no gaps. Industrial Design students were asked to create tesselations by using mathematical software like Mathematica instead of the usual visual drawing tools. The didactic of this approach is that students are empowered in their success of creating when they start to express patterns in equations. Again, a setting is created where technology is not the limiting factor when students explore their thoughts.

Finally, the modelling skill of *evidence based design* is implemented in an assignment on the basics of electronics. Electronics, especially analogue electronics, is seen as the toolset to give concepts "eyes and ears" by means of sensors and actuators. The learning goals are mainly limited to (1) switching actuators with transistors and (2) placing resistors to limit currents, and (3) low-pass and high pass filtering of sensor signals. For all three learning goals, one has to calculate currents and voltages to pick the right electronic components immediately: there is no efficiency in electronic design by iterative trial and error. We empower the students by three approaches. First, we have created a low-threshold electronics studio. The electronics assignment includes a guided workshop in the studio after which students are found there on a regular basis. Assistants are always available in the atelier for answering questions. Secondly, the mathematical skills are reduced to specifically solve the three learning goals as mentioned above. Finally, specific building blocks are identified based on components available in the electronics studio (i.e., shift registers, a limited set of sensors, etc.) and are well documented on the intranet. This approach appeared to be successful.

5 Balance Between Design Engineering and Modelling

Traditional engineering education and strategies work from theory towards practice. This means that, roughly speaking, students are learning mathematics in the first year, scientific theory and practice in the second and third, before they can do the practical applied engineering work in the end or during the masters phase. In Fig. 47.1 this is represented as a three-stage approach from left to right. The mathematical foundation is seen as the base for understanding science, which in its turn is needed to create new things. In the representation, the block "science" is used in a broad sense: it includes computer code, fabrication techniques, drawing skills and electronics. The right block represents all practical work to integrate scientific knowledge into a prototype or product and to explore the impact in our society.

As already explained in the introduction, the vision of the Industrial Design Department in Eindhoven to work on intelligent systems, products and related services, resulted in the implementation of a competency centred education model. This is done by starting with practical realisations (on the right of Fig. 47.2);

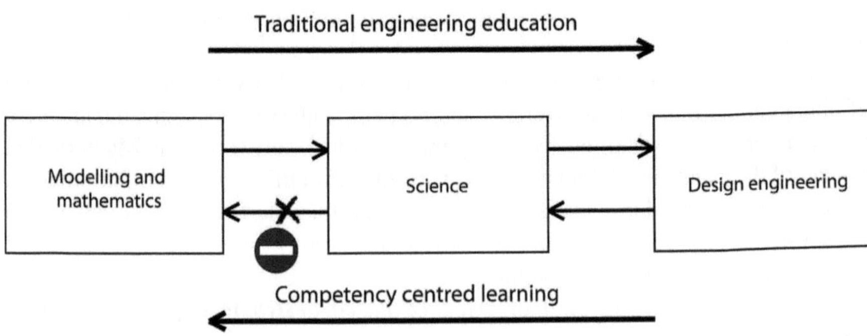

Fig. 47.1 One-way learning direction in traditional education and the opposite direction in competency centred learning

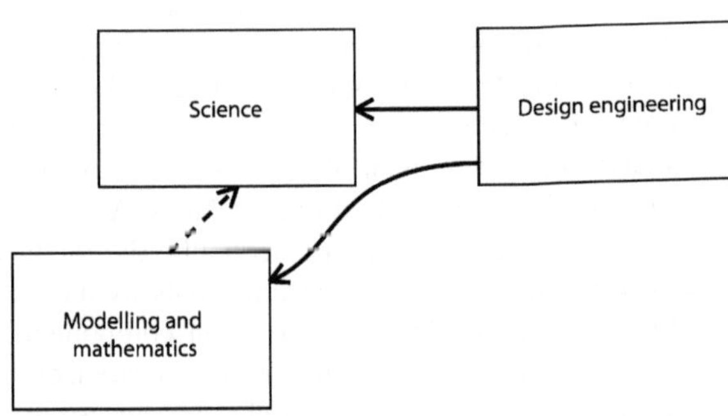

Fig. 47.2 Learning strategy for stimulating the mathematical modelling skill optimised for competency-based learning

scientific and engineering backgrounds offered on-demand. From that perspective, the mathematics and modelling question comes third, instead of being the foundation of our thinking. The inverted execution method is placing challenges on how to educate and how to do research. Design students prefer to explore using tangible artefacts, not with mathematical formulas.

For Industrial Design students in the flow of their work there is a virtual barrier when going from science to the natural need for mathematics. We experience this in their way of working and we can only guess the reason. It appears that investing in a deeper layer of abstraction is not seen as worth the effort. In the approaches discussed in the previous section, we assumed the barrier of scientific knowledge is too high to see the value in the mathematics and modelling skills; therefore, these approaches were based on lowering or removing the scientific substrate. This means in terms of Fig. 47.1 that the modelling and mathematics competency has been placed next to science as shown in Fig. 47.2.

What we are aiming at with this approach is that the opportunities of the competency *descriptive and mathematical modelling* can be experienced without being limited by a lack of technological skills. Once this is experienced, the dotted line may be put in practice in future projects, which is now in the natural direction.

6 Education Example: *Introducing Microcontrollers Assignment*

An assignment in the core of designing intelligent systems is about deploying microcontrollers for realising prototypes. At the Department for Industrial Design, assignments are learning activities offered to Bachelor students. Assignments consist of 6 weekly lessons of two contact hours, plus 36 h of self-study and/or practical work. In the *Introducing Microcontrollers Assignment* the self-study mainly consists of a design case. This assignment was originally organised by the Department of Electrical Engineering of our university. As of the academic year 2010–2011, the assignment was transferred to the responsibility of Industrial Design because "the message of the assignment did not reach the students". In the first term (Q1) of the academic year 2010–2011 the assignment was given in the old style, to discover how it could be improved. Afterwards, some adjustments were made which changed it from "traditional engineering education" (technology push) to "competency based learning" (technology pull from design perspective). Next, it was repeated in the fourth term (Q4). This is an excellent opportunity to verify the difference between the original model of Fig. 47.1 and the alternative learning strategy model of Fig. 47.2. The execution in Q1 and Q4 was the same in the sense that the offered theory was similar (microcontroller architecture, C-programming, on-chip hardware, hardware interfaces, a system design), and the design cases were the same. The cases consisted of designing an electronic game. The exact descriptions of the case for Q1 and Q4 are:

Introducing Microcontrollers Assignment (for Q1)

Measure the time between pressing button A and button B and communicate time in milliseconds over the serial bus.

- Hint 1: Use timer-interrupts to define timing
- Hint 2: A third button may be needed to reset the game
- Challenge 1: Implement as a small game. Player 1 can press a button any time, player 2 has to respond within 500 ms

Introducing Microcontrollers Assignment (for Q4)

Make a game. One of these:

- Make a "high striker": a game where you have to hit a pressure sensor. Implement such that people want to play more
- Reaction game: when person 1 presses a button, person 2 has to react within a reasonable time by pressing a second button. Test to find what a reasonable time is. Find a feedback method: a buzzer or LED for success or fail. Make it such that either player 1 or 2 can do the first press
- Colour memory game: Player 1 mixes two or three LED colours (with pot meters) into one RGB colour, presses a button and puts the pot meters in a random position. Player 2 has to memorise the colour and has to reproduce it. The microcontroller determines whether you are close enough.
  - Explain design choices, if possible by calculations
  - Show state diagram (or other used abstraction of system)
  - Evaluate functionality (user test?)

So, in Q1 the technical perspective was chosen by starting with a technical problem of measuring an interval. In Q4, the question started with describing the application, and asked for technical insights later. To substantiate the change from technology push to technology pull was further substantiated by three other modifications. First, we asked for an appropriate housing for the game to be made to focus on the user experience, rather than seeing the code plus circuit as the end result. This was accompanied by a lecture with examples of attractive functional casings. Next, one lecture was added on how to communicate about a microcontroller system. This was needed for debugging, to find effective details from experts, and to structure the problem before solving it. In that explanation there was an introduction to state diagrams to transfer concepts into programmable solutions. This is a method to create the direct link from design needs towards modelling, while bypassing the hardware state (curved arrow in Fig. 47.2). Finally, in Q1 the

Table 47.1 Scoring criteria for microcontroller assignment

Criterion	When scored?
Analysing a complex problem	State diagram, flow chart, notion of design choices
Identify behaviour and dynamics	Insight, explore, characterize a sensor, A/D input window consideration, sample rate consideration
Predictive power	Equation, FEM
Evidence based design	Eliminate options by calculations, calculate transistor operational point, power consumption calculation
Box	Electronics is packaged, integration, form and senses

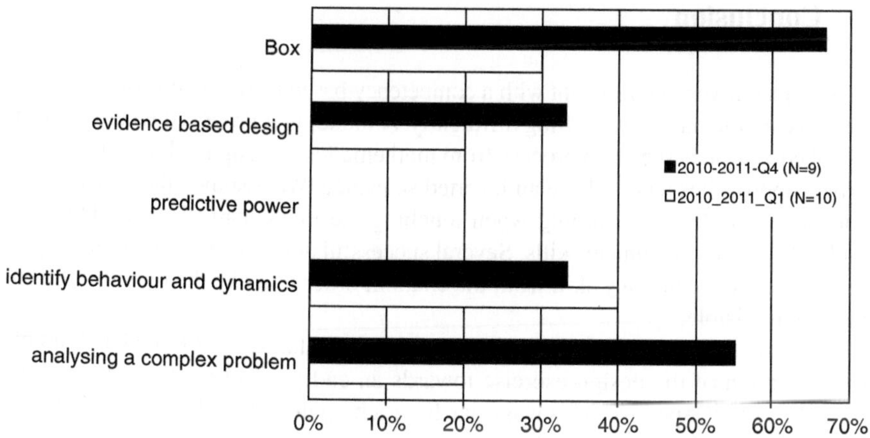

Fig. 47.3 Scores on four mathematical modelling skills before and after changing the content of *Introducing Microcontrollers Assignment*

introduction questions, before the design cases, were about specific technical functions, like "timers" and "sampling for A/D conversion". In Q4 the introduction questions were more about becoming familiar with the microcontroller: connecting it, writing subroutines, playing with communication between computer and the microcontroller board.

To compare Q1 to Q4, we assessed the end reports for the four skills of modelling for design. In addition, we evaluated whether the students worked from the perspective of the end-product; so, whether the end result has a functional shape or packaging. There was however one risk in the evaluation: the lecturer, being the first author of this chapter, did the evaluation of the outcome himself. To avoid a biasing effect, a clear scoring list was made first in order to have an objective set of criteria (Table 47.1).

In Q1 there were ten groups, in Q4 nine. The scores are shown in Fig. 47.3. It can be seen that the number of groups creating a packaged functional game increased from 30 to 66 %. This is interpreted as a perspective change from pure technology to the user or end result. In the new setting the skill of analysing a problem has been stimulated much better. This was mainly seen in the communication of students in terms of state diagrams, which immediately resulted in more structured code.

More students gave explanations of their design considerations in numbers (*evidence based design*). This was mainly done for selecting the right electronic components. The skill of *identifying behaviour and dynamics* scored less. This can be attributed to a question about timers which gave a very favourable outcome in Q1, but which was removed in Q4. Note that it is not the only assignment contributing to the competence *descriptive and mathematical modelling*; so there is no problem that not all students score on all skills. The skill of *using a predictive model* is not scored at all, because it falls outside the scope of this assignment.

7 Conclusion

In a project driven environment with a competency based learning approach, students perceive mathematical modelling differently. A model was proposed where conventional education is seen as a process from mathematics to design, while competency driven approaches tend to have an inverted sequence. We assumed there is a virtual barrier for on-demand learning when touching the mathematical modelling layer under the layer of technical skills. Several successful attempts were done in the past to remove the technology skill from the chain in order to make the opportunities of modelling visible.

We evaluated a learning activity that was changed in favour of this model. A simple reformulation of the design exercise towards an end product made students think from the user perspective. This was done by teaching them to talk about their design (e.g., with state diagrams) in order not to be confined by programming or electronics skills. Overall, there was an improvement in the ability to analyse the problem.

References

Alers, S. H. M., & Hu, J. (2009). AdMoVeo: A robotic platform for teaching creative programming to designers: Learning by playing. In M. Chang, R. Kuo, Kinshuk, et al. (Eds.), *Game-based education system design and development* (Lecture Notes in Computer Science Series, 5670/2009, pp. 410–421). Heidelberg: Springer.

Birenbaum, M. (2003). New insights into learning and teaching and their implications for assessment. In M. Segers, F. Dochy, & E. Cascallar (Eds.), *Optimising new modes of assessment: In search of qualities and standards*. Dordrecht: Kluwer.

Feijs, L. M. G., & Bartneck, C. (2009). Teaching geometrical principles to design students. *Digital Culture & Education, 1*(2), 104–115.

Hu, J., Ross, P. R., Feijs, L. M. G., & Qian, Y. (2007, June 11–13). UML in action: Integrating formal methods in industrial design education. In *Proceedings of the second international conference on technologies for E-learning and digital entertainment (Edutainment 2007)*, Hong Kong, China (Lecture Notes in Computer Science, 4469, pp. 489–498). Berlin: Springer.

Hummels, C. C. M., & Frens, J. W. (2009, April 4–9). The reflective transformative design process. In *CHI 2009 – Digital life, new world: Conference proceedings and extended abstracts; the 27th annual CHI conference on human factors in computing systems*, Boston (pp. 2655–2658). New York: ACM Press.

Pereira, A., Oliveira, I., Tinoca, L., Amante, L., Relvas, M. J., Pinto, M. C., & Moreira, D. (2009). Evaluating continuous assessment quality in competence-based education online: The case of the e-folio. *European Journal of Open, Distance and E-learning. EDEN*. http://www.eurodl.org/index.php?article=373. Accessed 25 July 2012.

Segers, M., Dochy, F., & Cascallar, E. (2003). The era of assessment engineering: Changing perspectives on teaching and learning and the role of new modes of assessment. In M. Segers, F. Dochy, & E. Cascallar (Eds.), *Optimising new modes of assessment: In search of qualities and standards*. Dordrecht: Kluwer.

van der Vlist, B. J. J., van de Westelaken, R., Bartneck, C., Hu, J., Ahn, R. M. C., Barakova, E. I., Delbressine, F. L. M., & Feijs, L. M. G. (2008, June 25–27). Teaching machine learning to design students. In Z. Pan, X. Zhang, & A. El Rhalibi (Eds.), *Technologies for E-learning and digital entertainment: Proceedings of third international conference, edutainment 2008*, Nanjing (pp. 206–217). Berlin: Springer.

Vygotsky, L. (1978). *Mind in society*. London: Harvard University Press.

Chapter 48
Rationality of Practice and Mathematical Modelling – On Connections, Conflicts, and Codifications

Lars Mouwitz

Abstract There is a tacit rationality, with a broader more qualitative mathematical essence, that has another origin and another character and function than traditional classroom mathematics. This kind of rationality is bound to personal acting in complex settings, and to our bodily interactions with the outside world. It is easily identified as a specific quality in the work of experienced craftsmen, surgeons, engineers, musicians, and sportsmen, but we will find it everywhere in ordinary working life. It is also a necessary interpretation tool when you shall bring to life and concretion abstract and general scientific models, a bridging process that needs dialogue and mutual respect.

1 Background

The role of mathematics in the workplace and in vocational proficiency has been a subject for research during the last 25 years (Noss and Hoyles 2010; Noss et al. 2000; Pozzi et al. 1998). This research is still only scratching the surface of the intricate relationships between mathematics (in a broad sense) and its use at the workplace, and has only a minor place in the mathematics education research community so far. Yet what emerges from these studies is that the relation between school mathematics and its use at the workplace is much more complicated and problematic than normally assumed (Hoyles et al. 2001; Pozzi et al. 1998; Wake 2007; Wake and Williams 2001).

In school settings there is a strong belief in the value of formal mathematics and its transferability to diverse contexts outside the school context. At the workplace however, it is not unusual, even though mathematical knowledge is highly regarded

L. Mouwitz (✉)
National Centre of Mathematics Education,
Gothenburg University, Gothenburg, Sweden
e-mail: lars.mouwitz@ncm.gu.se

G.A. Stillman et al. (eds.), *Teaching Mathematical Modelling: Connecting to Research and Practice*, International Perspectives on the Teaching and Learning of Mathematical Modelling, DOI 10.1007/978-94-007-6540-5_48,

per se, to hear that school mathematics plays only a minor role in practising diverse work tasks. A rather extreme and fascinating example of the impact from more informal kinds of knowledge was presented by Cooley (1985):

> At one aircraft company they engaged a team of four mathematicians, all of PhD level, to attempt to define in a programme a method of drawing the afterburner of a large jet engine. This was an extremely complex shape, which they attempted to define by using Coons' Patch Surface Definitions. They spent some two years dealing with this problem and could not find a satisfactory solution. When, however, they went to the experimental workshop of the aircraft factory, they found that a skilled metal worker, together with a draughtsman had actually succeeded in drawing and making one of these. One of the mathematicians observed: 'They may have succeeded in making it but they didn't understand how they did it.' ... All their knowledge of the physical world about them acquired through years of making things and seeing them break and rupture, is regarded as insignificant, irrelevant ... (p. 171)

How, indeed, was it possible for the metal worker and the draughtsman to solve the problem? What kind of knowledge did they use? What lessons should be learned about how to validate the professional knowledge of a skilled and experienced worker? Who should have the "definition power" to decide what "understanding" is, and the workers used "mathematics" without knowing it? One lesson we can learn is that there is a lot of knowledge transfer outside the school system, and also a lot of knowledge production outside academic institutions. This disturbing fact about the importance of informal learning outside the classroom has been highlighted by some researchers during the last decade, for instance Feutrie (2007):

> The main problem is that this [informal] learning is:
>
> - not formalised, less codified than traditional knowledge
> - not organised as traditional knowledge in disciplines, domains, ...
> - rather unconscious
> - hidden in action
> - contextualised, attached to a specific environment
> - built of elements more or less coherent, specific to an individual. (p. 17)

Indeed this kind of knowledge is contextualised and specific to an individual. But what about school mathematics, is it free from these kinds of "burdens"?

2 Models and Practice

A common view is that practice is, or always should be, applied theory. This, however, is only realised for one special type of practice, typical for science and technical applications. At least equally common is that practice lives its own life, without the support of scientific models and formalised language. Scientific application is, in addition, dependent on the adaptation and implementation of "tacit" practice which is not formulated in the model itself. In many cases we also try to model "regularities" (Popper 2002) on a lower level than the "laws of nature" level, for instance in social science, economy and meteorology. Therefore we have to handle in practice also that the very regularities are changing over time, and the most unpredictable effects will suddenly occur (Taleb 2007).

The knowledge ideal of our era is theoretical and intimately related to the concept of a model. A model is an abstraction formulated with the support of one or more examples. The modelling phase involves removing the concrete, which is considered to be unnecessarily complex, and also the distracting deviations. A model thus always represents a *loss of reality* and implicitly a *choice of view* since it is based on a view of what exists, its structures and connections, how these interact, and what is considered to be essential. For instance, the views of time and space are very different in Newton's and Einstein's physics. The views are both expressed in accordance with mathematical theory and therefore "mathematical models", but they are very different in structure and consequences. It is quite obvious that it is not reality itself that is quantified by mathematics, it is our *conceptions* about that reality that are quantified. A mathematical model therefore has the same weakness as the underpinning conceptions behind it. If a model is perceived as general; that is, it claims applicability to a variety of new situations, it should be able to approximate reality, and be applicable to the complexity of specific cases in the future, as well as able to explain or forecast concrete future events and the success of constructions. A model is explicitly formulated, for example mathematically formalised, to enable its meaning to be communicated through education.

Since the model should be able to explain the complexity it has been extracted from, some practical complications must arise. In many cases, successive adaptation between the model and the individual concrete case is required for its application to be possible. The strength of a general model is therefore at the same time its weakness. Sometimes a real situation creates such intensive "resistance" that the model must be revised or rejected. If the problem has to be resolved quickly, the model must be replaced by the hands-on knowledge and skills of the labour force, as when an unanticipated error suddenly occurs, for example, with a nuclear power facility. One example is a rather serious "incident" in a Swedish nuclear power setting some years ago that had not been foreseen neither by the formal operator educators nor by the simulator device, therefore nothing in the instructions explained how to handle it. A naïve belief of this kind, that nothing will happen "outside" the model, can be disastrous.

Especially when you try to connect model applications to a messy surrounding reality, there is a risk of total collapse, for instance the oil leak accident in the Mexican Gulf, and the horrible Fukushima nuclear catastrophe. A way to avoid this risk is to try to *rebuild* reality surroundings, both material and social contexts, so that the crucial part of reality will fit the model, instead of vice versa (Gellert and Jablonka 2007; Skovsmose 2005). The correctness of the model will therefore be seemingly perfect, but sometimes to tremendous economical, ecological, or human "external" costs because of the model's inherent nature of self- confirmation.

3 Analogical Thinking

Practical proficiency developed through more or less practical actions does not fit comfortably into the model domain. The boat builder, the carpenter and the sheet metal worker do not develop theoretical models in the same way as a physicist, chemist or economist does. The proficiency that a tradesman possesses is of a more analogical type. The analogies are not abstract, but rather consist of sets of concrete memorised examples, analogically connected with each other from the practitioner's repertoire of past experience. In this way knowledge becomes highly personal and often unformulated. The examples may retain their complexity. Each new concrete situation is compared and related to earlier concrete examples, in a pattern matching process, and these examples have more or less general applicability without claiming the generality of a model. Analogical proficiency is related to the person, situation and complexity, and is not easily transferred through formal education. Instead it must be *demonstrated* rather than formulated, and the classical method of conveying such proficiency is through a master-apprentice system (Dreufus 2004; Polanyi 1998; Sennet 2008). Practice will remain more complex than the representing model, otherwise the model has to be a new complete universe, in every aspect identical with the present. Practice must also be formed in other categories, presupposing a human agent continuously acting in complex and sometimes unpredictable material settings.

4 School Practice

In school situations, attempts are often made to solve mathematical problems of a purely theoretical nature, that is they deal with a world of mathematical concepts. Sometimes attempts are made to link these to reality, but these become "cosmetic" since not only the starting point, but also the aim, is to solve what is essentially a mathematical problem. Practical reality is thus used to illustrate a *theoretical* problem, as opposed to formulating a practical one. It is also common that focus is put on students demonstrating a particular theoretical method of solving a problem, a method which, in the problem context, is perhaps unnecessarily advanced and cumbersome. It is also worth considering that this practical reality does not exist as such, but only as an artificial interpretation, a virtual and theoretical "school reality". The pupils' *real* practice is to sit at their desks, and satisfy the specific requirements on how theoretical knowledge should be presented only *orally* or in *writing*. Sometimes concrete aids such as plastic or wooden cubes are used. The aim of this, however, is also abstract, and you only "touch down" in the concrete world for a brief landing. Concrete "tools" are not used to process a concrete reality, but instead to represent a theoretical activity. The actual arrangement of the cubes has no practical relevance, and they are discarded as soon as the theoretical problem is solved.

Other types of aids such as calculators and computers are different in kind, as they are not used to illustrate theoretical reasoning, but as tools to replace the *person* making the calculations. Computers can, with greater speed and precision in their calculations, carry out theoretical work that earlier required significant brainpower. In such cases, however, the theoretical school domain also determines the nature and purpose of the activity.

Despite claims for generality, it can be argued that theoretical mathematical education is just as context dependent as other mathematical activities in vocational life. A strong indication that this is the case is the relative helplessness that people, for instance new engineers, with purely theoretical backgrounds initially demonstrate at a workplace (Göranzon et al. 2006). New aims, methods, strategies and evaluations of results must be identified and internalised, much that was valued and encouraged in the school environment lacks, to varying degrees, relevance at the workplace. As a conclusion we can say that applying school mathematics is in fact about the difficult task of transferring knowledge from one practical context to another (Evans 2008; Lieberman 2000).

5 Usability of Mathematics in Practice Contexts

The aim of mathematics in practical proficiency is rather instrumental. Now mathematics is a *tool* among other tools: the problem in focus is of a practical nature, and the criteria for evaluation of the solutions are only based on practical considerations: *Is the practical problem solved?* From this, it follows that the mathematical reasoning most often is carried out in the form of simple rules and approximations. The heights of advanced mathematical theories, logical deductions, scientifically accepted methods, rigorously defined concepts, and extreme precision of results, are of little relevance or value per se in solving practical problems. The *rationality of practice* is here at work, just as "effective" as the theoretical rationality, but for another type of purpose. In many cases, problems are solved on the spot immediately, in physical interaction with the surroundings, because of your responsibility to *do* something as fast as possible. Withdrawing to a different setting, for example a computation room, to carry out calculations becomes cumbersome, time-consuming, costly and irrelevant. In the practical application of mathematics, there is no need for mathematical proof or internal theoretical consistency, practical usability is the criterion for "truth" and for the relevance of the methods used.

In addition to mathematics in the form of rules and methods in practical proficiency, there are also mathematical models incorporated in, for example, computer programs. One example is how a successful sheet metal worker today must both master a long established trade tradition, and at the same time understand how to handle a computer and various drawing and spreadsheet programs. But usability in this context is also the very point, not the underlying mathematical theory in the software. In many cases instruction can be linked to the proficiency the adult already possesses; for instance working as a carpenter with the number 1.414 "opens the window" to Pythagoras' theorem, linear models and irrational numbers in a possible theoretical education process (Gustafsson and Mouwitz 2008). Another form of "window opening" is when a *dilemma* occurs, which ordinary rules of thumb cannot resolve, but where a more general method could provide a solution (Noss 2002).

You can also find *qualitative mathematics* in the form of a familiarity with the "big ideas of mathematics" embedded in practical knowledge, for instance: counting, communicating, change, shape, symmetry, regularity, and position (Devlin 1997; Lakoff and Núnez 2000). These kinds of invisible connecting threads between practice and theory seem to be crucial for understanding both intuition and creativity in designing and the interpretation processes of symbols in science and its applications.

6 Ethics, Aesthetics, Dialogue and Cooperation

As mentioned, practical proficiency is person-related and often forms a part of the adult's identity. Knowing one's job is a source of self-esteem and vocational pride, and leads to the desire to do a "good job" which has both an aesthetic and ethical dimension: taking responsibility for ensuring that the result is good, and corresponds to the customer's or employer's quality expectations. Losing one's job can lead to an identity crisis, which is further aggravated if the adult's vocational proficiency is not identified, validated and taken advantage of in future educational or vocational situations.

Much practical work is carried out in teams where communication and the ability to co-operate is an important competence. Sometimes joint initiatives are taken putting high demands on discipline, planning and coordination. It is also possible to see that the tools used "speak to" the user and vice versa. The tool becomes an extension of the body in a continuous interplay with the work situation. In some industries, there is also a master-apprentice trainee period, or where a new employee merely functions as an observer, and the person with experience demonstrates and talks whilst the trainee imitates and puts questions.

The above are important aspects of practical proficiency, often involving some mathematical content, aspects which have very low priority in "school mathematics". In practical application in the real world, thought and action, tools and materials, quality and responsibility, identity and co-ordination together form an integral whole.

7 Skill and Technology at KTH (The Royal Institute of Technology), Stockholm

In Arbetslivscentrum (The Swedish Centre for Working Life) during the 1970s, there was an intensive discussion on the meaning of vocational proficiency in relation to contemporary research into working life at that time. The latter basically focused on research into qualifications, that is research into the qualifications an individual needed to be able to carry out a specific work task. In the first instance, the findings showed that vocational proficiency appeared to be an application of specific advanced theoretical knowledge. Attempts to theoretically describe different work tasks produced, however, only marginal success, as they were often misleading or counterproductive. The skill and familiarity typical of well-established vocational proficiency appeared to be quite different from what could be "caught" in theoretically formulated models and rule systems. This insight gradually led to the development of a completely new research area, Yrkeskunnande och Teknologi (Skill and Technology) under the supervision of Professor Bo Göranzon at KTH (The Royal Institute of Technology), Stockholm. A number of philosophers, amongst others, Bengt Molander, Tore Nordenstam and Kjell S. Johannessen, at the same time worked on trying to analyse the underlying theoretical knowledge complex. The last, in particular, has had a major impact on this research (Johannessen 1988, 1999). Typical for his work is that Wittgenstein's philosophy of language is of major interest, both the idea of a perfect logical language and the idea of language games: a craving for a perfect language will create a certain kind of silence, or "tacitness", and this sometimes suppressed knowledge can only be expressed in other forms as professional *action* and in art. A good overview of this research is Göranzon et al. (2006). Other research environments, which from completely different theoretical and methodological starting points, focusing on the relationship between theoretical and practical proficiency, have come to conclusions which essentially coincide with the analysis presented here. Examples of this are findings of narrative research, activity theory, situated learning and socio-cultural theories. There are also findings of cognitive science that point out that the human brain has two kinds of memory systems: one for declarative knowledge, and one for tacit knowledge, the latter not reachable by explicit thinking or declaration (Evans 2008; Lieberman 2000).

8 Tacit Rationality

Practical knowledge has indeed many rational elements. It relies, for instance, on doing a set of adjustments in the right order, to put together a complicated machine with all parts in the right place, or connecting an electrical network correctly. In new situations this cannot be a routinised behaviour. Instead it is a kind of rational not formalised, intentional, and often unconscious, acting: a *tacit rationality*. If we focus on a test situation in school contexts, the approved mode to articulate an underlying rationality is traditionally restricted to the "tools" pencil and paper (and sometimes orally), but in working life settings rational acting is expressed through a broad repertoire of tools and materials. If the "answer" is wrong the costs will often be extensive in many aspects. Indeed the context poor school settings are possible as an ideal because the educational system has by tradition the authority to define and decide what counts as knowledge (FitzSimons 2002; Zevenbergen and

Zevenbergen 2009). Behind this idea of the almost empty classroom is often a non-formulated idea that a poor context is good for abstraction.

Due to its character as not being clearly worded, tacit rationality becomes invisible and therefore neglected in school contexts. The ordinary way to organise knowledge tests in school settings reveals indirectly a hidden paradigm about what knowledge is, how it should be represented, how it should be tested, and how it should be learned.

Tacit rationality includes, of course, some general cognitive abilities. Partly as a consequence of its embeddedness in traditions, culture, artefacts and action it is constituted by our experiences and manifests itself as *knowledge by familiarity* and as *practical knowledge*. Some of these manifestations have resemblance to mathematics, but not as mathematical knowledge in a traditional sense. The craftsmen that tessellated the walls and ceiling of Alhambra were not mathematicians, and did not use mathematics, but their craftsmanship can be thoroughly analysed from a purely mathematical perspective. In *Finding moonshine: a mathematician's journey through symmetry,* Marcus Du Sautoy (2008) gives a striking example of the power of praxis knowledge. In Alhambra there are 17 different kinds of symmetries in the tilings – no less and no more. It took the science of mathematics 800 years to develop methods by which it was possible to strictly prove that there theoretically could not be more than these 17 different underlying symmetries. Another, perhaps even more striking, example is the discovery of quasi-crystalline Penrose patterns in Arabic tilings 500 years before they were discovered and described by mathematicians in the West (Lu and Steinhardt 2007). Finally, it is important to emphasise the fact that tacit rationality is not restricted only to traditional trades. It is, as mentioned earlier, equally significant, in highly specialised and academic occupations as well (Roth 2003; Vergnaud 2000).

Model thinking and analogical thinking represent two thinking styles which can very well come into conflict with each other. In many cases model thinking is the winner in such conflicts, and this can lead to a loss of praxis knowledge in, for example, a company. This applies particularly where there is a generational change and the importance of hidden praxis knowledge becomes evident. A new group of practitioners, even though highly educated, may not be able to replace the many years of praxis knowledge accumulated over time, currently a very problematic issue for the Swedish nuclear industry.

An interesting question is the extent to which mathematical proficiency is also embedded in the analogical thinking that typifies the more trade-like aspects of vocational proficiency. The ability of a sheet metal worker to recognise that a desired construction is similar to something he has done before requires some form of ability to understand similarities and differences between two geometrical structures in three dimensions. Such recognition gained through experience should be of great relevance, not just in the sheet metal trade, but also in many other occupations with similar demands.

If conceptual tools do not exist for making a deeper analysis of knowledge, then practical knowledge will also remain invisible and unknown. Practical knowledge is also, in contrast to propositional knowledge, personal; whilst instruments for validation and assessing qualifications are generally of a more abstract and impersonal nature. Vocational proficiency is closely connected with questions about identity and self-esteem, and an impersonal test that only recognises school knowledge may have an overwhelmingly negative impact on a person's desire and ability to develop and advance.

9 Concluding Remark

The importance of mathematical modelling and the effectiveness of its use is overwhelming; but just for that reason it is important to point out its borders and its connections to practice, both *before* and *after* the model design. If we do not recognise and highlight this, the "knowledge society" will instead rather initiate a massive societal de-professionalisation. Several young research environments are just now focusing on these crucial issues, and there is a great need for cooperation both on a national and international level. At best this chapter will be a small contribution to such future cooperative endeavours.

References

Cooley, M. (1985). Drawing up the corporate plan at Lucas Aerospace. In D. MacKenzie & J. Wajcman (Eds.), *The social shaping of technology: How the refrigerator got its hum*. Milton Keynes: Open University Press.

Devlin, K. (1997). *Mathematics: The science of patterns*. New York: Scientific American Library.

Dreufus, S. E. (2004). The five-stage model of adult skill acquisition. *Bulletin of Science Technology Society, 24*(3), 177–181.

Du Sautoy, M. (2008). *Finding moonshine: A mathematician's journey through symmetry*. London: Harper Collins.

Evans, J. S. B. T. (2008). Dual processing accounts of reasoning, judgement, and social cognition. *Annual Review of Psychology, 59*, 255–278.

Feutrie, M. (2007, March). *Validation of non formal and informal learning in Europe: Comparative approaches, challenges and possibilities*. Paper presented at Recognition of prior learning: Nordic-Baltic experiences and European perspectives congress, Copenhagen.

FitzSimons, G. (2002). *What counts as mathematics? Technologies of power in adult and vocational education*. Boston: Kluwer.

Gellert, U., & Jablonka, E. (Eds.). (2007). *Mathematisation and demathematisation: Social, philosophical and educational ramifications*. Rotterdam: Sense Publishers.

Göranzon, B., Hammarén, M., & Ennals, R. (Eds.). (2006). *Dialogue, skill & tacit knowledge*. Chichester: Wiley.

Gustafsson, L., & Mouwitz, L. (2008). *Validation of adults' proficiency – Fairness in focus*. Göteborg: National Center for Mathematics Education, University of Gothenburg.

Hoyles, C., Noss, R., & Pozzi, S. (2001). Proportional reasoning in nursing practice. *Journal for Research in Mathematics Education, 32*(1), 4–27.

Johannessen, K. S. (1988). The concept of practice in Wittgenstein's later philosophy. *Inquiry, 31*(3), 357–369.

Johannessen, K. S. (1999). *Praxis och tyst kunnande*. Stockholm: Dialoger.

Lakoff, G., & Núnez, R. E. (2000). *Where mathematics comes from: How the embodied mind brings mathematics into being*. New York: Basic Books.
Lieberman, M. D. (2000). Intuition; A social cognitive neuroscience approach. *Psychological Bulletin, 126*(1), 109–137.
Lu, P. J., & Steinhardt, P. J. (2007). Decagonal and quasi-crystalline tilings in medieval Islamic architecture. *Science, 315*, 1106–1110.
Noss, R. (2002). Mathematical epistemologies at work. *For the Learning of Mathematics, 22*(2), 2–13.
Noss, R., & Hoyles, C. (2010). Modeling to address techno-mathematical literacies at work. In R. Lesh, C. Haines, P. Galbraith, & A. Hurford (Eds.), *Modeling students' mathematical modeling competencies (ICTMA 13)* (pp. 75–86). New York: Springer.
Noss, R., Hoyles, C., & Pozzi, S. (2000). Working knowledge: Mathematics in use. In A. Bessot & J. Ridgway (Eds.), *Education for mathematics in the workplace* (pp. 17–35). Dordrecht: Kluwer.
Polanyi, M. (1998). *Personal knowledge: Towards a post-critical philosophy* (First published 1958. Corrected 1962. Reprinted 1998 ed.). London: Routledge.
Popper, K. (2002). *The logic of scientific discovery*. London: Routledge.
Pozzi, S., Noss, R., & Hoyles, C. (1998). Tools in practice, mathematics in use. *Educational Studies in Mathematics, 36*(2), 105–122.
Roth, W.-M. (2003). Competent workplace mathematics: How signs become transparent in use. *International Journal of Computers for Mathematical Learning, 8*(2), 161–189.
Sennet, R. (2008). *The craftsman*. New Haven/London: Yale University Press.
Skovsmose, O. (2005). *Travelling through education: Uncertainty, mathematics, responsibility*. Rotterdam: Sense Publishers.
Taleb, N. N. (2007). *The black swan. The impact of the highly improbable*. New York: Random House.
Vergnaud, G. (2000). Introduction. In A. Bessot & J. Ridgway (Eds.), *Education for the workplace* (pp. xvii–xxiv). Dordrecht: Kluwer.
Wake, G. (2007). Considering workplace activity from a mathematical modelling perspective. In W. Blum, P. L. Galbraith, H-W. Henn, & M. Niss (Eds.), *Modelling and applications in mathematics education: The 14th ICMI study* (pp. 395–402). New York: Springer.
Wake, G., & Williams, J. (2001). *Using college mathematics in understanding workplace practices. Summative report of research project funded by the Leverhulme trust*. Manchester: University of Manchester.
Zevenbergen, R., & Zevenbergen, K. (2009). The numeracies of boatbuilding: New numeracies shaped by workplace technologies. *International Journal of Science and Mathematics Education, 7*(1), 183–206.

Chapter 49
Extending Model Eliciting Activities (MEAs) Beyond Mathematics Curricula in Universities

Mark Schofield

Abstract This chapter recognizes the potential efficacy of MEA-related approaches in university classrooms. It explores principles associated with MEA design and raises challenges to their transfer beyond mathematical settings into natural and environmental sciences, engineering and beyond. It attempts to address the state of readiness of individuals to engage with MEA-related processes. A first stage model of progression of learners and teachers into MEA-related activities is proposed alongside consideration of the phenomenon of *flow* and the beginnings of a set of questions intended to set the scene for emerging development and research agendas. A new seventh 'Entry' principle for MEA design is proposed.

1 Introduction

The chapter originated as a vehicle to provoke discussion and debate about Modelling Eliciting Activities (MEAs) in contexts beyond the mathematics curricula in schools, which have been the prime test-bed. It is derived from consideration of the Modelling Eliciting Activity (MEA) concept in contexts outside of mathematics curricula in universities. In engaging with MEAs

> the products that students produce go beyond short answers to narrowly specified questions – which involve sharable, manipulatable and reusable conceptual tools (eg models) for constructing, describing, explaining, manipulating, predicting or controlling mathematically significant systems. (Lesh and Doerr 2003, p. 3)

A team of UK and US researchers has been in receipt of National Science Foundation (NFS) and Economic Social Research Council (ESRC) funding to

M. Schofield (✉)
SOLSTICE, Centre for Excellence in Teaching and Learning,
Edge Hill University, Lancashire, L39 4QP, UK
e-mail: Mark.Schofield@edgehill.ac.uk

G.A. Stillman et al. (eds.), *Teaching Mathematical Modelling: Connecting to Research and Practice*, International Perspectives on the Teaching and Learning of Mathematical Modelling, DOI 10.1007/978-94-007-6540-5_49,

explore extrapolation of the MEA concept and design principles to university curricula areas such as engineering, natural and environmental sciences and education, in pursuit of richer understandings of the potential pedagogic efficacy of MEA-related approaches. (For details of the proposal and work streams work see Hamilton and Hoyles 2006.) There are some fundamental questions emerging which represent a commitment to understanding the nature, design and conditions associated with implementation and researching learner and teacher engagement in MEAs.

The NSF/ESRC proposal (Hamilton and Hoyles 2006, p. 3) expressed a shared aspiration

> to enhance ways in which learners acquire capacity to adopt and synthesise multiple frameworks in solving complex problems, to anticipate new situations, to develop the ability to see both problems and the underlying systems in which they are embedded differently.

A main task was to research and theorize the design of complex problems, reflecting upon six principles for the design of MEAs (Diefes-Dux et al. 2004; Lesh et al. 2000) in an attempt to trial emerging prototype problems in a range of disciplinary contexts.

> The NSF/ESRC proposal recognises extended potential for MEAs as there is a stark gap between (mathematical) problem solving of the classroom and that encountered by those who use mathematics 'in the wild' … responders to natural disasters or unfolding terrorist scenarios. A parallel exists in activities involving management of critical incidents in medical and education settings. (p. 4)

Investigations focused on the potential efficacy of immersion of learners in professional, transdisciplinary settings in MEA-related tasks. We hypothesized, in the spirit of the MEA movement, enhancement of their ability to visualise the nature of problems and to apply metacognitive schema (models) to similar problems and to transfer elements of their models to novel situations. The focus in this chapter is based upon an assumption that knowledge and rich thinking can be elicited by 'ill-formed' problems predicated on the need for a solution in a more meaningful, contextualised and vital way than commonly presented in traditional 'transmissionist' modes in many school and university classrooms.

MEAs have been traditionally structured to enable teachers, researchers, and students to track the evolution of conceptual schema as learners produce, trial, and reconfigure solutions or models in problems. Lesh et al. (2000) designed and tested numerous iterations of MEAs in their attempts to elicit learners' rich mathematics concepts. This approach has been historically used with school students and more recently in selected university undergraduate classrooms with mathematics foci. One simplistic interpretation of the original MEA approach is that it was based upon selection of a single or close group of 'big ideas', such as need for analysis of variance, or other specific mathematical concepts which may be used to design problems to elicit rich, systematic, thinking. This, for convenience, I will codify as a design down approach. Analyses and evidence from the application of Reflection Tools (Hamilton et al. 2007) suggest that MEAs may indeed evoke concepts of a deep nature. Such tools are paper-based questions and prompts (scaffolds) to capture evidence of thinking and reflection in action. Similar outcomes can be achieved through use of digital recorders by participants during MEAs and both can give insights into learners' (and teachers) metacognitive events and developments before, during and after a MEA. These mathematical concepts, it is argued, are better understood (or can become better understood) in the context of MEAs than if they are taught as rules, algorithms and heuristics applied to de-contextualised tasks. In these latter instances there is limited motivational need in many learners to engage and persist towards a solution. There are parallels here to the distinction drawn by Biggs (2004) between deeper and surface approaches to learning. Surface approaches may be categorised as rote or report, or administration of formulae to instrumentally survive an examination or test. The former are represented by critical, higher order, thinking and enhanced potential for transferability to other contexts and present challenges to more longitudinal assessment of effectiveness beyond testing for grading.

The 'Leshian' position of construction of a MEA from a big mathematical idea, involves the nature of the MEA being controlled and organised (to a greater or lesser degree) by the chosen high order concept/s. A group of engineers and environmental scientists in the NSF/ESRC group, reflecting on their attempts to design a MEA related to a natural flooding disaster, alluded to reductionist characteristics of this type of design approach. I suggest that such natural phenomena may indeed be hyper-complex, containing numerous big ideas and that they constitute a source of potential for eliciting numerous models. In short, the design down notion is argued here as being more easily situated in the exploration of MEAs in mathematics contexts. I suggest that complex (MEA-related) problems in areas such as medical emergencies, military incidents and those related to natural disasters may be rich sources of elicitation of deep understandings through integration of existing models in addition to potentially generating new ones, by acts of creativity or coalescence of existing, incipient models in the problem solving scenario space. The use of nature's challenges and problems in scenarios provides rich test-beds for research into such theoretical positions. The complexity of such systems, when referenced to the developing status of our current command of models that may be applied in any design down approach outside of mathematics, raises key questions for research: What models could be anticipated as useful in designing or adopting our problem solving scenarios? How may we identify appropriate models? To what extent may natural scenarios align with the design legacy of MEAs in mathematics education? And, is this the right track of enquiry?

2 Reflections on the 'Six Principles of MEA Design'

Lesh et al. (2000) and Diefes-Dux et al. (2004) elaborated six principles for design of MEAs, which have influenced the emerging position and research agenda in trans-disciplinary groups of engineers, environmental scientists and educationalists. The principles are summarized below:

1. The personal meaningfulness principle – could it happen in real life?
2. The model construction principle – construction, modification, extension or refining of a model

3. The self-evaluation principle – judging whether responses are good enough
4. The model-documentation principle – revealing thinking
5. The simple prototype principle – emerging prototype or metaphor for interpreting other 'structurally similar' situations
6. The model generalisation principle – reusable, sharable, transferable.

The first principle (*personal meaningfulness*) may be considered axiomatic in that the problem foci in our interest groups were based on real case scenarios, such as Hurricane Katrina, floods and other natural occurrences. Principles 2, 3 and 4 (*construction*, *self-evaluation* and *documentation*) are currently being interpreted in such 'MEA-related' scenarios as opportunities for reflection in action and after action by individuals and groups through the use of simple Reflection Tools (Hamilton et al. 2007). These can be deployed as cognitive scaffolds to stimulate the capture of feelings, thought processes and models and as research tools to capture evidence of individual and group modelling in action. Research should necessarily focus on analyses of reflection and similar observational and ethnographic data from modelling activities in non-mathematical contexts to develop better sightlines into relationships between design, content and organization of problem scenarios. Attention should be given to understanding key features of problems (simulated or real world) that may impact upon elicitation of models, systematic thinking and transferable learning.

My personal articulation of the Reflection Tool concept is one of a 'capture device' or record of metacognitive events as consciously accessible and re-accessible language. This may be in the form of text or audio recording of reflections during and after MEA activity. These artefacts may be used to influence development of the repertoire of learners and teachers. As such they can be learning, evaluation, research and development tools. The act of articulation and 'going public' as a function of using Reflection Tools, I propose, cements and embeds existing and new learning (i.e., it enhances models). As reflection tools promote and scaffold cognitive processing they cement schema, impact upon memory for model integration, and may thus support prediction in new problems where learners are able to identify familiar, transferable, features. They have the potential to act as diagnostic assessment tools for teachers and learners as well as in research data collection. An under-investigated potential of reflection tools is their potential use in tracking affective activity before, during and after problem solving and thus the potential they may have to enhance our understanding of design for experiences to ensure the 'entry' of learners and teachers into the MEA space is appropriately prepared for. The location of reflection tools in MEA-related instructional modules is suggested here as essential for deepening learning in and from complex problem solving scenarios.

The potential of Reflection Tools for assisting in elaborating principles for good problems or scenarios; for drilling down into characteristics of models per se, and for investigation of characteristics of problem solving and metacognitive apparata is documented with clear learning and research foci by Hamilton et al. (2007). Serial immersion in rich MEA-related experiences could prove key to enhancing and deepening our current understanding through research.

> Our underlying conjecture is that as modelers become more sophisticated in the way they think about themselves as problem solvers, they become more sophisticated in thinking about problems. (Hamilton et al. 2007, p. 1)

In considering principles 5 and 6 (*simple prototype* and *model generalisation*), a challenge emerges for the use of Reflection Tools and other research enquiry methods. Such tools aim to gain insights into learners' abilities to interpret structurally similar problems and transfer of elements of models, groups of models and integration strategies to novel situations. I suggest pedagogic opportunities arise in using Reflection Tools as scaffolds to enhance articulation and consciousness of models by learners and that assisted self-articulation of application, integration and elicitation of new models adds a layer of cognitive processing that may make models more accessible and transferable for learners in addition to assisting educators' research analyses.

Notions of *internal* psychological conditions of motivation and identity as a problem solver run alongside *external* conditions and climate (in cultures, classrooms, and groups) which may affect MEAs. The former includes affective factors linked to my proposal that experts not only see problems differently from novices, but also *feel* them differently and hence handle and moderate feelings differently/effectively during problem events. Both internal and external aspects expand to culture per se; permission to engage in classroom experiences, related to group dynamics and self-permission (of teachers and learners) to operate in iterative, often messy, open-ended MEA-related scenario cycles. These cultural foci are sources of challenge and anxiety for learners and teachers in classrooms in current education systems globally, where content, curriculum-coverage, testing, public accountability and performance-related instrumentalism are arguably prevalent factors. The state of readiness of participants, learners and teachers, for MEA-related activity is thus dealt with below in the proposal of a seventh 'Entry Principle' in MEA design.

3 State of Readiness for Participation in MEA-Related Activities and the Case for a Seventh Principle

Expansion of MEA-related curriculum experiences in higher education are perhaps helpfully considered against anecdotal concerns about the discomfort experienced by some students and teachers whose previous school and cultural experiences make the human dynamics of MEA-related approaches challenging. This may preclude or limit engagement as MEA-related approaches may conflict with previous experiences and personal epistemologies. If learners' school histories or early experiences in university are focused on values of individual reward and less on collaboration and group success, then a well conceived MEA-related activity may fail in its implementation.

I have raised the debate as to a need for development of an additional 'Entry Principle' in addition to the existing six. This would focus on, and address, cultural readiness to work with MEA-related activities, on groups and role expectations, and on consideration of diversity in teachers' and learners' point of arrival at MEA related activities. In Ausubellian terms, it would focus attention on starting closer to where learners are 'at'. Not all will be in an immediate state of readiness or susceptibility to a desirable depth of engagement characteristic of highly engaged individuals on a trajectory of deeper learning through MEAs or related complex problem solving activities.

This plays forward into relatively unchartered territory in our knowledge map of MEA-related processes in university classrooms and into research opportunities. What is needed is further investigation as to the extent to which serial experiences of MEAs may enhance confidence, learner's access to and transfer of their models and potential for self-management of affective experiences? Will measurable effects related to 'practice' and 'rehearsal phenomena' emerge? Will enquiry lead to design for MEA-related activities organized on familiar principles of progression as in traditional curricula, (with induction and a sequenced approach) or will a different mindset emerge? We are yet to see, but the importance of researching such activity is pressing and aligns with this proposal for theorizing and elaborating such an 'Entry Principle' and the nature of a positive climate and conditions for learner engagement.

4 Progression into MEA-Related Activities

As a way forward towards refining an 'Entry Principle' I have reflected on pedagogic approaches associated with development of writing in different genres in school classrooms and its extrapolation to University classroom settings for academic writing (Schofield 2005). The journey towards *Independence* as a 'maximized problem solver', (and as a writer in different genres) should focus on progression. The notion of progression in designing series of MEA-related events for learners and teachers from a carefully considered point of 'Entry' which recognizes learner history and the need for sequenced orientation and preparation is emphasised here. As a starting point for debate, Table 49.1 is offered to illustrate a proposed developmental sequence. This sequence is adapted from Schofield (2005).

5 Engagement in Complex Problem Solving, the Affective Domain and the Elusive Concept of 'Flow'

The elusive and desirable concept of deep engagement, may be (in MEA-related activities) considered as a function of the interaction of a range of components such as confidence, identity as a problem solver, arousal state, motivation, situated 'comfort' in problem spaces and history of exposure to problem solving in education and life settings. All contribute to the identity of the problem solver, their management of their affective domains, and to their personal rules of engagement with problems

Table 49.1 A proposed developmental sequence for introduction of MEA-related approaches in university curricula

Progression	Detail
Immersion and modelling (Enculturation)	Introduction and orientation to MEA-related activities, acclimatisation and enculturation (inherent in the 'Entry Principle'): Teacher explains pedagogy and concepts, roles, expectations Emphasis placed on reflection, positive climate and building trust and giving permission not to have to produce the definitive answer; support; observing problem solving in action by experts and within peer groups; using Reflection Tools; building understanding of processes and appropriate confidence levels; rehearsal and seeing and feeling the art of the possible in MEA-related contexts Learners given opportunities to practice with in lower risk, 'less complex' problem solving activities
Joint construction	Engagement in group activities and articulation of learning using Reflection Tools (as scaffolds to encourage and capture reflection and learning) individually and in de-briefing activities and use of these as diagnostic tools by the teacher Emphasis on metacognition and learners' emerging cognitive 'repertoire' as problem solvers (as confidence and competence develops)
Independence	Enhanced repertoire of problem solving capacities as a Maximized Problem Solver alone and in groups, with deeper engagement

which are inextricably linked with previous schooling and cultural experiences and hence the emphasis placed on preparation and 'Entry' above.

Csikszentmihaly (2000, p. 4) posited nine characteristics of 'play and work flow situations'. Hamilton et al. (2007, p. 4) suggest these are useful in considering design of MEAs, for formation and testing, and for analyses of their affect on participants. They summarise the nine characteristics as follows:

1. There are clear goals every step of the way
2. There is immediate feedback on one's action
3. There is balance between challenges and skills
4. Distractions are excluded from consciousness
5. There is no worry of failure
6. Action and awareness are merged
7. Self-consciousness disappears
8. The sense of time becomes distorted
9. The activity becomes autotelic (having itself as its only purpose)

Assuming flow is a desirable state, some questions follow in relation to MEA and MEA-related like approaches:

- is it unique to individuals and variably experienced and triggered?
- is it ascribable to groups? What might trigger group flow?
- is it potentially allied to a 'type' description and analysis in participants like a 'learning style' 'preference' or 'attribute'?

- could there be a qualitatively describable developmental continuum of flow on which individuals/groups may be located, for example No flow ↔ maximum flow?
- if so, what predictors of movement on such a continuum do we know about currently? Are there additional principles to those of Csikszentmihaly that we may need to deploy to precipitate flow or flow-like phenomena in learning environments where MEA-related activities are situated? Also, to what extent may some form of an Entry Principle be worthy of consideration?

What is lacking is clarity in the nature of factors that may precipitate flow which may be related to individual conceptions of self and classroom cultures. For illustration, consider teachers' permission, and learners' self-permission to persist with or even enjoy dissonance in a complex problem solving environment. Is this socialised out of learners by teachers suffering the imperious constraints of public accountability in contemporary classrooms? Previous experiences must be considered. Has, in some instances, teaching for success on tests reduced the climate of risk taking and hypothesising and limited learners' and teachers' comfort with uncertain or provisional knowledge in school and university classrooms? To what extent may this be acting as an impediment to readiness to engage with MEA-related activities and a key factor in flow's elusiveness? Will flow continue as an elusive, aspirational concept without radical, research-informed re-conceptualisation of the climate and conditions prevailing within learning environments for the future? I suggest that this is worthy of further consideration in terms of teaching for deeper learning approaches if these are to be achieved and due regard given to the diversity of learners and teachers, learner histories and teaching cultures that exist within countries and around the world who are implementing MEAs and related approaches.

6 Conclusion

This paper attempts to raise questions and issues for research and theoretical pursuit as MEAs and MEA-related approaches are gaining momentum in contexts outside of mathematics education. Whilst much has been learned from the elaboration of MEA design principles in schools, many questions are arising as the ideas and approaches are shifted towards university classrooms. Challenges of individual and organizational cultures, customs and practices in curriculum design in test-oriented societies, and currently less well-formed concepts of MEAs outside of mathematics are a rich substrate for further research and enquiry.

References

Biggs, J. (2004). *Teaching for quality learning at university* (2nd ed.). Milton Keynes: SRHE, OU Press.

Csikszentmihalyi, M. (2000). *Finding flow: The psychology of engagement with everyday life*. New York: Basic Books.

Diefes-Dux, H., Moore, T., Zawojewski, J., Imbrie, P. K., & Follman, D. (2004). A framework for posing open-ended engineering problems: Model-eliciting activities. In *Proceedings of the 34th ASEE/IEEE Frontiers in Education conference*, Savannah, GA.

Hamilton, E., & Hoyles, C. (2006). *Enhancing and assessing complex reasoning through models and modeling: UK-US collaborative research.* To NSF and to the Economics and Social Research Council of England, in collaboration with the University of London. Funded Supplement to 0433373, September 2006. http://tinyurl.com/HamiltonHSDSupplement-pdf. Accessed 24 Dec 2012.

Hamilton, E., Lesh, R., Lester, F., & Yoon, C. (2007). The use of reflection tools to build personal models of problem-solving. In R. Lesh, E. Hamilton, & J. Kaput (Eds.), *Foundations for the future in mathematics education*. Mahwah: Lawrence Erlbaum.

Lesh, R., & Doerr, H. (2003). *Beyond constructivism: A models & modelling perspective on mathematics teaching, learning, and problems solving*. Mahwah: Lawrence Erlbaum.

Lesh, R., Hoover, M., Hole, B., Kelly, A. E., & Post, T. (2000). Principles for developing thought revealing activities for students and teachers. In A. Kelly & R. Lesh (Eds.), *The handbook of research design in mathematics and science education* (pp. 591–646). Mahwah: Erlbaum.

Schofield, M. (2005). Constructivist principles and conditions for learning. *Educational Developments, 4*(4), 9–16.

Chapter 50
Building Awareness of Mathematical Modelling in Teacher Education: A Case Study in Indonesia

Wanty Widjaja

Abstract Interest to teach mathematics closely connected to its use in daily life has grown in Indonesia for over the last decade (Sembiring RK, Hadi S, Dolk M, ZDM – Int J Math Educ 40(6):927–939, 2008). This chapter reports an exploratory case study of the building of an awareness of mathematical modelling in teacher education in Indonesia. A modelling task, *Re-designing a Parking Lot* (Ang KC, Mathematical modelling in the secondary and junior college classroom. Prentice Hall, Singapore, 2009), was assigned to groups of pre-service secondary mathematics teachers. All groups collected data on a parking lot, identified limitations in the current design, and proposed a new design based on observations and analyses. The nature of the mathematical models elicited during the modelling task were examined. Implications of this study suggest a need to encourage pre-service teachers to state assumptions and real-world considerations and link them to the mathematical model in order to validate if the model is appropriate and useful.

1 Background

Mathematical modelling is not formally introduced at any school level in Indonesia. At the tertiary education level, mathematical modelling is offered as either a compulsory or an elective subject at the university but it is not formally part of teacher education training programmes in mathematics education. Nonetheless, courses in

The study reported in this chapter was conducted in Universitas Sanata Dharma Indonesia, the author's former affiliated institution.

W. Widjaja (✉)
Department of Mathematics Education, School of Education,
Deakin University Faculty of Arts and Education,
221 Burwood Highway, Melbourne VIC 3125, Australia
e-mail: w.widjaja@deakin.edu.au

G.A. Stillman et al. (eds.), *Teaching Mathematical Modelling: Connecting to Research and Practice*, International Perspectives on the Teaching and Learning of Mathematical Modelling, DOI 10.1007/978-94-007-6540-5_50,

teacher education programmes encourage pre-service teachers to work on projects which open up possibilities for mathematical modelling ideas to be embedded in the course. There has been an increased awareness in Indonesia of the need to create an engaging and meaningful learning process in learning mathematics. One of the reform initiatives in Indonesia, known as PMRI (*Pendidikan Matematika Realistik Indonesia*), has adapted Realistic Mathematics Education (Freudenthal 1983) and put into classroom practice the use of real-world contexts for mathematics learning (Sembiring et al. 2010). PMRI advocates pedagogical practices that cultivate reasoning, sense making and communication skills using real-world contexts. Students are encouraged to explore multiple pathways in solving real-world tasks, and to communicate and challenge each other's ideas during the process of learning mathematics.

Potential of real-world tasks to engage students in meaningful and interactive instructions have been well documented in recent studies in Indonesia (Sembiring et al. 2008; Widjaja et al. 2010). However, teachers face challenges in facilitating students to move beyond the real-world bound setting and in establishing a different set of classroom norms to support productive discussions (Dolk et al. 2010; Sembiring et al. 2008; Widjaja 2010). Establishing a classroom practice and norms that support productive use of real-world tasks is quite demanding for both teachers and students. There is a strong impetus to better equip teachers with a broader scope of knowledge and skills to facilitate student-centred lessons which encourage multiple interpretations and diverse solution pathways. Provision of a wide range of exposure to mathematical modelling tasks during their training in teacher education will lay a foundation for effective facilitation of mathematical modelling in classrooms (Burkhardt 2006; Doerr 2007).

This chapter reports preliminary findings from a case study of an attempt to build an awareness of mathematical modelling in teacher education in Indonesia. A mathematical modelling project to re-design a parking lot (Ang 2009) was assigned to groups of pre-service secondary teachers as a mathematics education project to be completed in 3 weeks. The nature of mathematical models elicited by pre-service teachers while completing this project will be examined and discussed in order to identify key aspects that need to be emphasised in the facilitation of such projects. Hence this chapter will examine: What is the nature of the knowledge being activated by pre-service teachers while engaged in the *Re-designing a Parking Space* modelling project?

2 Mathematical Modelling at Tertiary Education

Facilitating mathematical modelling entails different beliefs from those traditionally held by mathematics teachers in Indonesia of what constitutes mathematics and mathematical learning processes as well as a major reversal in the usual role of teachers and students (Doerr 2007; Niss et al. 2007). Niss et al. (2007) underscored the significant role of teachers in ensuring the successful development of applications and modelling competencies. They argued for effective promotion of inclusion of modelling in teacher pre-service education to develop pre-service teachers' capacities in providing rich learning environments for modelling activities and scaffolding strategies that support mathematisation. It is imperative to engage pre-service teachers in first-hand experience as mathematical modellers before expecting them to facilitate mathematical modelling in their own teaching.

Teachers' own exposure to, and experience of, mathematical modelling play critical roles in developing facilitation skills in mathematical inquiry involving real world situations (see e.g., Blum 1993; Doerr 2007; Stillman 2010). Stillman (2010) emphasised the importance of teachers' first-hand experience as mathematical modelers over an extended period of time to equip them with the skills needed in mathematical inquiry and modelling. Ideally, early exposure to mathematical modelling from the schooling years, even as early as primary school, will pave a strong path for mathematical modelling skills to develop (English 2006, 2010). Lamon et al. (2003, p. ix) claim that late adoption of mathematical modelling at the tertiary level (e.g., in teacher education) is challenging due to different beliefs and ways of looking at mathematics and its role in understanding real world phenomena.

The open-ended nature of mathematical modelling tasks also present significant challenges as substantial diversity and unexpected responses are likely to occur. Doerr (2007) pointed out the importance of listening and moving away from prescriptive modes of instruction. She called for teachers to provide ample room for students to exercise their reasoning and decision making skills and to support students in making progression in their model development. According to Doerr (2007), the authority of evaluating students' work is no longer placed solely in the hands of the teacher but rather shared with the students.

3 Method

3.1 Setting and Participants

The data reported in this chapter came from 14 pre-service secondary teachers who completed a modelling task, *Re-designing a parking space* (Ang 2009, pp. 52–53), during a 3-week session in April 2010. These pre-service teachers were a subset of the second year cohort undertaking a 4-year bachelor degree program to become qualified secondary mathematics teachers in Indonesia. They did not have any prior experience working with mathematical modelling tasks because mathematical modelling was not listed as part of any mathematics education course. The inclusion of mathematical modelling tasks was the author's own initiative to build pre-service teachers' awareness of mathematical modelling and to engage them in a productive classroom discourse to justify the mathematical model.

Pre-service teachers formed their own groups consisting of 2–4 pre-service teachers to facilitate a productive working relationship. Gender was not the basis in forming groups; in fact they were encouraged to form mixed-gender groups. In the first week, the task was introduced with the guidelines that highlighted various stages of the modelling process to help groups in planning their investigations. No readings on particular modelling literature were assigned but they were required to conduct real-world investigations and online research in their own time. In the second week, each group was allocated a 10-min presentation to share the progress of their ongoing investigation and to report any challenges they faced during the process of completing the task. At the end of the third week, a gallery walk session, consisting of poster presentations of written work, was conducted followed by a whole class discussion. Written reports of the group findings, and posters were collected. During the gallery walk, one member of each group was assigned to stand and explain the poster to the visitors whilst other members observed and asked questions while visiting posters from other groups. It was expected that each group would arrange their members to take turns during the gallery walk sessions. This arrangement allowed pre-service teachers to engage in meaningful discussion of different approaches and solutions to the problem and to exercise their evaluative communication skills. Throughout the session, the author observed posters from group to group, took notes and asked questions. At the end of the gallery walk, two groups were invited to summarise and present their posters to the whole class. The analysis reported in this chapter will be based mainly on groups' written reports to examine mathematical knowledge and real-world considerations that were activated by pre-service teachers. Pre-service teachers' knowledge and difficulties will be discussed in respect to the two main didactical modelling processes, that is *mathematisation* and *re-interpretation of mathematical solutions* to the real-life situations.

3.2 The Re-designing of Parking Space Task

The original task was situated in the context of re-designing a car parking space to increase the number of lots available (Ang 2009). Taking into account that the majority of pre-service teachers involved in this study rode motorbikes as their mode of transport; a decision to change the setting to the motorbike parking space was made. Unlike the parking lots for cars, the parking lots for motorbikes are not clearly marked with lines. On the mathematical front, the task carried rich potential to bring forth relationships among pertinent variables such as the notion of angle, the relationship between the curb space and the sine of the angle, and maximum capacity interpreted within the real-world settings. The task was open-ended in nature and allowed for multiple pathways to derive a mathematical solution depending on interpretations, assumptions and choice of variables. Guidelines for the *Re-designing Parking Space Project*, (Ang 2009, pp. 52–53) follow.

Re-designing Parking Space Project

Choose a parking space of your own choice. Are the parking lots arranged or designed in a way that makes the best use of the parking space? Develop a mathematical model that optimises the use of the space for parking. Use the following steps to guide your completion of the project:

1. Sketch a rough plan of the parking space of your own choice.
2. List factors or variables in the problem.
3. Discuss possible designs for the parking lot.
4. Design a new parking lot that caters for an optimum parking space.
5. Derive a mathematical model from the initial design to the proposed design.
6. Identify the limitations of the proposed model for the parking space.

4 Findings

4.1 Variables Identified and Mathematical Approaches Used

Four groups opted for different locations, namely campus, hospital, supermarket, and traditional market parking space to complete the project. Pertinent variables such as location, a variety of brands and dimensions of motorbikes, parking positions, distance between two motorbikes, space between lots, and the area of the parking space were identified. Other factors such as time of the day which was likely to affect parking arrangements and positions (i.e., 1 stand or 2 stand) in some places was noted in the report (Fig. 50.1). These were valid variables because during the busy time of the day such as visiting hours in the hospital, the arrangement of motorbikes and their positions turned out to be incredibly messy and complex. However, an assumption to hold time as a non-affecting variable was made to simplify the task. As evident in their observations of various parking lots, there were great variations in the parking positions and dimensions of the motorbikes in one parking lot. The messiness of real-world conditions related to re-designing motorbikes parking lots was clearly recorded in the report.

All groups took into consideration a wide range of dimensions of motorbikes and parking positions. Examination of data from the parking space and dimensions of motorbikes were evident in the written reports. In dealing with a variety of brands and dimensions of motorbikes, motorbikes were classified into scooters and sport motorbikes. However, the levels of mathematics being activated in analysing the capacity of the parking space and in re-designing a new parking space showed great variation.

Fig. 50.1 Parking arrangement and position as identified variables

Group 1 conducted an investigation on the campus parking space where the lots were clearly partitioned. A methodical approach in measuring the dimension of each parking lot, the width of motorbikes, and the space between two motorbikes was evident in the report (Fig. 50.1). Different parking positions were documented and reported during their poster display but they were not taken into account in the analysis. Group 1 suggested narrowing a gap between two parallel lots to increase the capacity of the parking space in the proposed design. Whilst the calculation suggested an increased parking space capacity, this approach was impractical because in reality enough space should be reserved for motorbikes to move in and out of the parking area safely. Clearly this group did not validate their result but focused on mathematical calculations that reflected a higher parking capacity.

Deriving a new design for the parking lot proved to be not simple. Group 2 opted to re-design a parking space in one of the private hospitals and gathered detailed measurements of different sections to produce the initial sketch of the parking space (Fig. 50.2). They measured the crowdedness in the three busiest sections (Section A, Section B, and Section C) to determine the area taken for each motorbike in the three sections by dividing the area of each section by the number of motorbikes in the corresponding section. An average of the three results of 1.2438 m^2 was noted as an area for parking one motorbike. This approach was mathematically sound but not entirely appropriate in the real world setting. Group 2 noted that the parking space in the three sections was overcrowded and non-ideal. This was a direct result of their choice to focus on the most crowded sections. However, this was not articulated which raised a concern if they realised the connection between their decision and the mathematical result. Taking an average area of 1.2438 m^2 resulted in an overestimate of the parking space capacity. Group 2 decided to follow the recommended area for one motorbike by the transport authority (1.3125 m^2) and offered to eliminate the space in between two sections of the initial design of the parking lot to increase the capacity of the parking area. However, the general structure of the initial design was retained (Fig. 50.2). The proposed design was also represented in a scaled-down model during the gallery walk presentation. During the gallery display session, one member of the group explained the group's consideration:

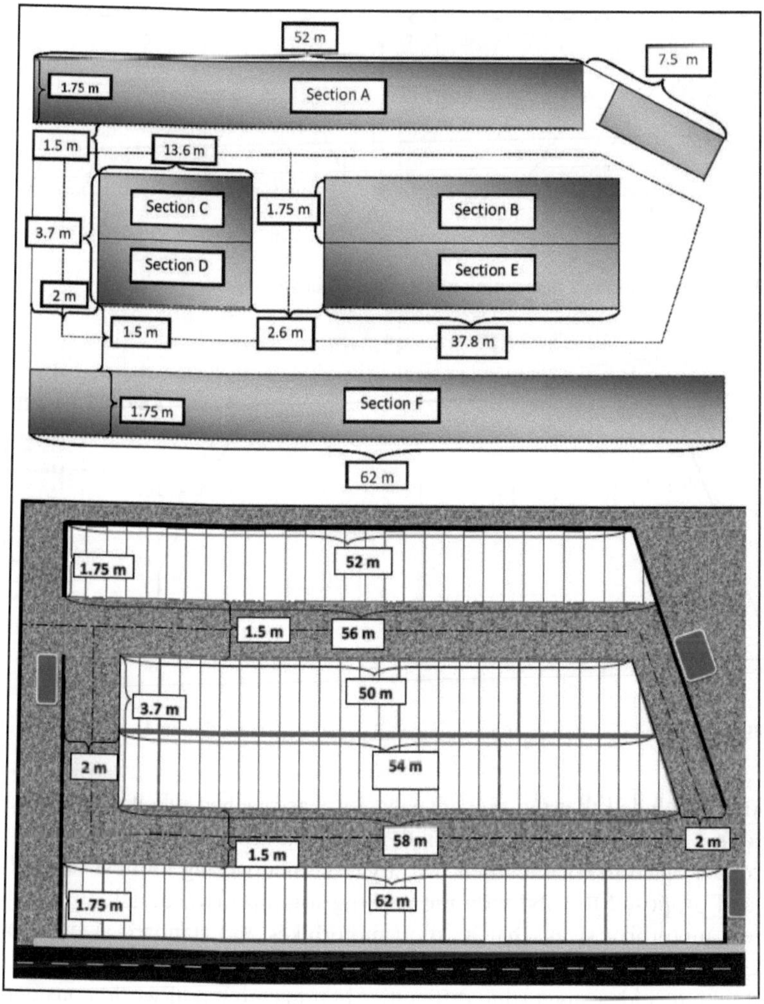

Fig. 50.2 Group 2 initial sketch of the hospital parking (*top*) and the proposed re-design (*bottom*)

"We are not only concerned about the design that offers optimum capacity but also the comfort and easy access for parking the motorbikes".

Group 3 was the only group who examined the parking positions (i.e., 1 stand or 2 stand) in the analysis even though all groups acknowledged that the parking positions affected the parking space capacity. Several possibilities by varying parking arrangements (curb angle between two motorbikes) of motorbikes and comparing the number of motorbikes that could occupy the same area were examined. However, the relationship between the curb space, and the angle the lines make with the curb space using the sine of the angle was not activated in the strategy. Instead, Pythagoras' theorem was activated when calculating the area of the parking space.

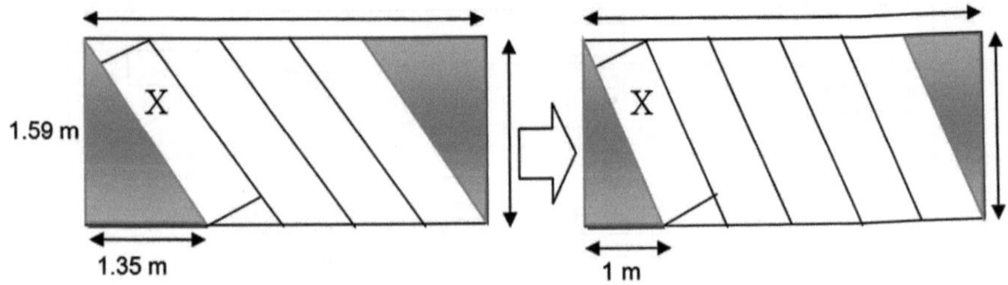

Fig. 50.3 Group 3 strategy in examining different parking arrangements

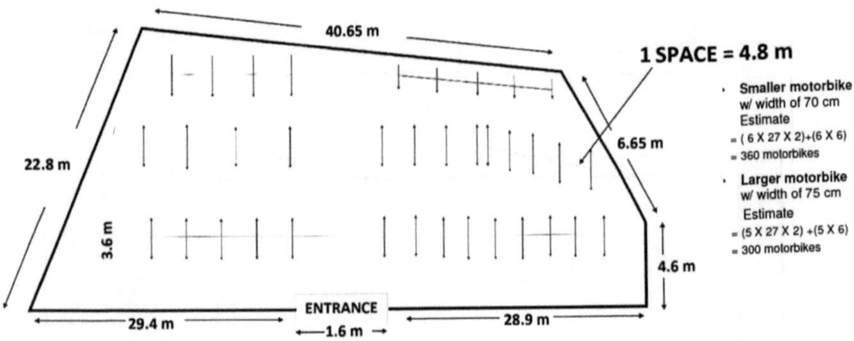

Fig. 50.4 Proposed new design of shopping mall parking lot from Group 4

Group 3 concluded that the highest capacity was reached when motorbikes were parked in 1 stand and reflected this in the proposed design (Fig. 50.3).

Group 4 opted to conduct their investigation in a parking lot next to a shopping mall and proposed to subdivide the parking lots into two divisions to cater for different motorbike sizes, that is, sport motorbikes and standard size motorbikes (scooters). However, the illustration in the proposed design (Fig. 50.4) was not consistent with this idea suggesting difficulties in translating real world considerations into the proposed mathematical model. The report showed no clear understanding of the task. No discussion on the derivation of the new mathematical model was presented. A predicted maximum number of motorbikes was stated based on simple calculation taking into account the area of the parking space and motorbike dimensions. The report suggested the task was interpreted as a calculation task to figure out the maximum capacity of the parking space rather than re-designing it.

5 Discussion

Pre-service teachers activated their knowledge on measurement of area and real-world considerations such as relationships between various dimensions of the motorbikes and the capacity of the parking space whilst engaged in this modelling project.

Parking arrangements were considered as a variable but angles and sine of the angle were overlooked as variables. In identifying the limitations of the proposed model, all groups cited accuracy in measurement as an area for refining the model. Group 1 for instance suggested that increasing the number of motorbikes for their investigations could increase the accuracy of the proposed mathematical model. Increasing the number of data was impractical and it did not add insights into refining a mathematical model. The fact that all groups did not clearly state the simplifications of conditions and assumptions made from the real-word situations to the mathematical models might explain difficulties in identifying factors causing limitations to the proposed models. The validation stage is known to be challenging in the mathematical modelling process (Maaß 2007). In this study, Groups 2 and 3 showed better attempts at incorporating real-world considerations into their mathematical model. This concurs with similar findings from a study by Kaiser et al. (2010) who identified difficulties with "evaluating and validating results, making wrong and oversimplified assumptions, and in re-interpreting the results into the real world" (p. 438). Evidence from an earlier study (Widjaja 2010) involving other Indonesian pre-service teachers on a *Cooling of Coffee Project* also suggested re-interpreting and linking variables in the mathematical model back to the real world data were particularly challenging.

6 Concluding Remarks

This study sought to investigate the building of an awareness of mathematical modelling among pre-service teachers in Indonesia as they undertook a personally meaningful modelling task. The task of *Re-designing a Parking Lot* served as a rich platform to build awareness of mathematical modelling. Pre-service teachers engaged in interactive and engaging discussions about re-designing their self-chosen parking space. During the gallery walk, pre-service teachers took up the role of authoritative evaluators of their own work and that of their peers as suggested by Doerr (2007). Multiple ways of re-designing parking lots for motorbikes were observed with the area of each other parking space as the main strategy in determining the optimum capacity of the parking space. Pertinent variables such as parking positions, distance between motorbikes, and motorbike dimensions were identified. However, only one group took into consideration the key variable parking arrangements (curb angle between two motorbikes) in re-designing the parking lot. In retrospect, more prompts could have been provided to help groups consider angles to expand their strategy.

The validation stage, in particular establishing the link between real-world conditions and mathematical solutions, was challenging for these pre-service teachers. Clarifying assumptions and simplifications of conditions from real-world contexts at the beginning stages of modelling and revisiting these towards the end of the modelling stages might be helpful to overcome this problem. Moreover, it is important to revisit the assumptions to establish links between assumptions, variables chosen and proposed mathematical models. Pre-service teachers might benefit more from explicit discussion of assumptions (see Stillman 2008) and examples of modelling processes from modelling research literature. More time should be spent in discussing these critical aspects with pre-service teachers.

Finally, it must be pointed out that the following are limitations of this study: only one modelling task was used with the pre-service teachers and the data analysed were from a small sample. Increasing the sample size and varied tasks are important for validation of the results of this study.

Acknowledgements The author thanks the pre-service teachers from Universitas Sanata Dharma, Indonesia for their support. Contribution of the National Institute of Education Singapore in supporting the author to present a related short paper with early findings at ICTMA 15 Conference in Melbourne is acknowledged.

References

Ang, K. C. (2009). *Mathematical modelling in the secondary and junior college classroom*. Singapore: Prentice Hall.

Blum, W. (1993). Mathematical modelling in mathematics education and instruction. In T. Breiteig, I. Huntley, & G. Kaiser (Eds.), *Teaching and learning mathematics in context* (pp. 3–14). Chichester: Ellis Horwood.

Burkhardt, H. (2006). Modelling in mathematics classrooms: Reflections on past development and the future. *ZDM, 38*(2), 178–195.

Doerr, H. M. (2007). What knowledge do teachers need for teaching mathematics through applications and modelling? In W. Blum, P. L. Galbraith, H.-W. Henn, & M. Niss (Eds.), *Modelling and applications in mathematics education* (pp. 69–78). New York: Springer.

Dolk, M., Widjaja, W., Zonneveld, E., & Fauzan, A. (2010). Examining teacher's role in relation to their beliefs and expectations about students' thinking in design research. In R. K. Sembiring, K. Hoogland, & M. Dolk (Eds.), *A decade of PMRI in Indonesia* (pp. 175–187). Bandung: APS International.

English, L. (2006). Mathematical modelling in the primary school: Children's construction of a consumer guide. *Educational Studies in Mathematics, 63*(3), 303–323.

English, L. (2010). Modelling with complex data in the primary school. In R. Lesh, P. L. Galbraith, C. R. Haines, & A. Hurford (Eds.), *Modelling students' mathematical modelling competencies* (pp. 287–299). New York: Springer.

Freudenthal, H. (1983). *Didactical phenomenology of mathematical structures*. Dordrecht: Reidel.

Kaiser, G., Schwarz, B., & Tiedemann, S. (2010). Future teachers' professional knowledge on modelling. In R. Lesh, P. L. Galbraith, C. R. Haines, & A. Hurford (Eds.), *Modelling students' mathematical modelling competencies* (pp. 433–444). New York: Springer.

Lamon, S. J., Parker, W. A., & Houston, K. (2003). Preface. In S. J. Lamon, W. A. Parker, & K. Houston (Eds.), *Mathematical modelling: A way of life* (pp. ix–x). Chichester: Horwood.

Maaß, K. (2007). Modelling in class: What do we want the students to learn? In C. Haines, P. Galbraith, W. Blum, & S. Khan (Eds.), *Mathematical modelling ICTMA 12: Education, engineering and education* (pp. 63–78). Chichester: Horwood.

Niss, M., Blum, W., & Galbraith, P. (2007). Introduction. In W. Blum, P. Galbraith, H-W. Henn, & M. Niss (Eds.), *Modelling and applications in mathematics education: The 14th ICMI study* (pp. 3–32). New York: Springer.

Sembiring, R. K., Hadi, S., & Dolk, M. (2008). Reforming mathematics learning in Indonesian classrooms through RME. *ZDM – The International Journal of Mathematics Education, 40*(6), 927–939.

Sembiring, R. K., Hoogland, K., & Dolk, M. (2010). *A decade of PMRI in Indonesia*. Utrecht: APS International.

Stillman, G. (2008). Connected mathematics through mathematical modelling and applications. In J. Vincent, R. Pierce, & J. Dowsey (Eds.), *Connected mathematics* (pp. 325–339). Melbourne: MAV.

Stillman, G. (2010). Implementing applications and modelling in secondary school: Issues for teaching and learning. In B. Kaur & J. Dindyal (Eds.), *Mathematical modelling and applications yearbook 2010* (pp. 300–322). Singapore: World Scientific.

Widjaja, W. (2010). Modelling the cooling of coffee: Insights from a preliminary study in Indonesia. In L. Sparrow, B. Kissane, & C. Hurst (Eds.), *Proceedings of the 33rd annual conference of the Mathematics Education Research Group of Australasia* (pp. 626–633). Fremantle: MERGA.

Widjaja, W., Dolk, M., & Fauzan, A. (2010). The role of contexts and teacher's questioning to enhance students' thinking. *Journal of Science and Mathematics Education in Southeast Asia, 33*(2), 168–186.

Part VII
Modelling and Applications in Business and the Lived Environment

Chapter 51
Mathematics and the Pharmacokinetics of Alcohol

Michael Jennings and Peter Adams

Abstract Alcohol is the most widely consumed recreational drug in the world and is commonly regarded as an essential component of Australian culture. However, alcohol is also a major contributor to mortality and morbidity not only in Australia but also around the world. Whilst alcohol-related education is typically considered from social and medical perspectives, it is also an area that provides a range of mathematically rich investigations. This chapter discusses mathematical modelling of blood alcohol content (BAC) levels. The investigations detailed enable teachers of a range of year levels to develop students' mathematical abilities and also their social understanding of alcohol-related issues. The mathematics of the investigations includes functions, graphing, derivatives, integration, and mathematical modelling and reasoning.

1 Introduction

In the last few decades, mathematical modelling of authentic phenomena has become increasingly prominent in school and university curricula. In the Australian state of Queensland, for example, Modelling and Problem Solving now makes up one-third of a student's senior mathematics final grade (QSA 2008), while many university courses require students to develop, analyse, and apply models.

Researchers have interpreted the term *mathematical modelling* in a range of ways. Julie and Mudaly (2007) categorised modelling in an educational context into two types according to purpose: modelling as vehicle and modelling as content. In addition to these two interpretations, Stillman et al. (2008) identified three

M. Jennings (✉) • P. Adams
School of Mathematics and Physics, The University of Queensland,
St Lucia, QLD 4072, Australia
e-mail: msj@maths.uq.edu.au

G.A. Stillman et al. (eds.), *Teaching Mathematical Modelling: Connecting to Research and Practice*, International Perspectives on the Teaching and Learning of Mathematical Modelling, DOI 10.1007/978-94-007-6540-5_51,

other approaches of mathematical modelling that teachers have taken in a range of settings: using contextualised examples to motivate students, using technology to fit curves to models, and emergent modelling, where the model emerges in the process of structuring a problem. Kaiser and Sriraman (2006) classify modelling into three perspectives: realistic or applied modelling, educational modelling, and the model-eliciting approach.

Teachers rarely have classes in which all students have the same mathematical abilities. Even in classes streamed according to previous performance, students show variation in knowledge, skills, and motivation. As a result, teachers typically require a range of classroom activities. In the most conceptually complete form of mathematical modelling, a problem is posed, data collected, and an appropriate model developed. At the other end of the spectrum, students simply insert numbers into an existing model to confirm or refute the suitability of the model. The most appropriate approach will depend on a number of factors including: year level, students' abilities, mathematical content, syllabus requirements, teachers' own likes and abilities, and of course, time.

Mathematical modelling is often considered as an excellent way to motivate students to engage in mathematics (Brown and Edwards 2011; Stillman and Galbraith 2011; Zbiek and Conner 2006). Brown and Edwards (2011) suggest that mathematical modelling activities support three different types of motivation: real world situations appealing to learners, motivation to study mathematics in general, and motivation to learn new mathematics. Stillman and Galbraith (2011) found that mathematics teachers in the Queensland context were often extremely positive towards modelling as it promotes mathematical understanding and thinking rather than 'regurgitation', makes mathematics applicable and demonstrates utility of mathematics, engenders student interest in mathematics promoting engagement, promotes variety in teaching strategies, and promotes success leading to retention of students in mathematics.

2 Background

There is no doubt that alcohol consumption and its impact is an area of major interest to societies and individuals. In 2009–2010 the total quantity of pure drinking alcohol available for consumption in Australia was equivalent to 2.2 standard drinks per person per day for each person aged 15 or more (Australian Bureau of Statistics 2011), and consuming alcohol is widely regarded as an integral part of Australian culture. Data from the 2007–2008 National Health Survey (Australian Bureau of Statistics 2009) showed that 13 % of Australians aged 15 or more had consumed alcohol at a level associated with long term health risks. Furthermore, 27 % of Australians aged 16–85 first consumed alcohol when aged less than 15, and 40 % when aged between 15 and 17.

Alcohol-related harm is a major cause of mortality and morbidity in Australia, causing around 3,000 deaths and 65,000 hospitalisations every year (Chikritzhs et al. 2003). There are numerous well-known negative health and social impacts associated with excessive alcohol consumption, including increased risks of developing a range of cancers, risk of causing damage to the brain (particularly in younger drinkers), negative health impacts on unborn babies when the mother consumes alcohol during pregnancy, road trauma and accidents, violence and sexual promiscuity or assault. In 2004–2005, the annual cost to the Australian community of alcohol-related social problems was estimated at $15.3 billion (Collins and Lapsley 2008).

In an effort to reduce the frequency and impact of excessive alcohol consumption, governments and health bodies run regular education campaigns highlighting both short and long-term risks. In 2008, the Australian Government announced a national strategy to address the high levels of binge drinking among young Australians. In 2011 the Government classed alcohol as a Class 1 carcinogen, joining tobacco and asbestos.

Alcohol-related education is typically considered from social and medical perspectives. However, it is also an area that provides a range of mathematically rich investigations, particularly in the context of developing mathematical models. For example, Siller (2010) and Greefrath et al. (2011) proposed developing models of blood alcohol content using technology. (Interestingly, some of their statements and the graph they present do not accurately represent the pharmacokinetics of alcohol. For example, the rate of metabolism of alcohol tends to be constant irrespective of amount consumed, absorption is more rapid than their graph indicates, and maximum blood concentration is proportional to amount consumed.)

In this chapter we discuss mathematical modelling of the pharmacokinetics of alcohol after consumption, and in particular, Blood Alcohol Content (BAC) levels. (Strictly speaking, pure drinking alcohol is *ethanol*. The term *alcohol* is used in this chapter, to match common usage.) Starting with only the facts that alcohol is soluble in water and typically metabolised at a constant rate, and assuming that consumption is relatively rapid, six investigations are presented, to both increase students' mathematical skills and also develop their understanding of why behaviours such as binge drinking are harmful.

Mathematical content of the case studies includes functions, graphing, derivatives, integration, mathematical modelling and reasoning, all in an authentic context. By assuming that consumed alcohol is instantly absorbed in the digestive system, students are able to deduce the Widmark formula (Posey and Mozayani 2007) which is used in criminal prosecutions of drink drivers. A variation of this formula that does not assume instant absorption of alcohol can also be deduced. Using these formulae and areas under curves, students investigate the effects of periodic binge drinking compared to other consumption patterns.

These activities have been used in two first-year courses at The University of Queensland, taken by mathematics, science and engineering students. The modelling approaches are consistent with Julie and Mudaly's (2007) modelling as vehicle, Stillman et al.'s (2008) contextualised examples, and Kaiser and Sriraman's (2006) realistic modelling. However, the investigations are easily modifiable to allow teachers to use a different modelling perspective. The investigations can also be used with high school students and can run from 1 hour to a week or longer. Below we detail a range of investigations and approaches that teachers can use in their classes from early secondary school through to university. There are also many possible extensions to these investigations, based on information sourced from newspapers or the internet. For example, there are numerous high-profile court cases involving celebrities facing trial for driving under the influence of alcohol. Students can investigate these cases mathematically, and also analyse the social and legal contexts.

3 Investigation 1: Time Taken for BAC to Return to Zero

Goal: Given the following USA guidelines for time taken for BAC to return to zero for males of various body weights and levels of alcohol consumption, deduce a corresponding mathematical model (Table 51.1).

Table 51.1 Approximate time (in hours) for BAC to return to zero for males (Adams 2011)

Number of drinks (n)	Weight, W (pounds) 120	140	160	180	200	220	240	260
1	2	2	2	1.5	1	1	1	1
2	4	3.5	3	3	2.5	2	2	2
3	6	5	4.5	4	3.5	3.5	3	3
4	8	7	6	5.5	5	4.5	4	3.5
5	10	8.5	7.5	6.5	6	5.5	5	4.5

Students will require different levels of prompting, depending on their mathematical confidence and modelling abilities. A guided approach could be as follows; the actual discussion should be student-driven, with the questions below highlighting key points.

- What parameters are given in the table? What happens to time as the values of the parameters change? What mathematical operations can be used to achieve this? (Answer: The number of drinks consumed n, and weight W. As n increases, time t increases. As W increases, t decreases. Mathematically, the former can be achieved by adding or multiplying by n, the latter by subtracting or dividing by W.)
- Do you notice any important properties of some of the data in the table? (Hint: consider the column corresponding to a weight of 240 pounds. Answer: for this weight, the time in hours equals n. For a weight of 120 pounds, time equals $2n$.)
- Find a mathematical relationship between n, W and t that gives correct values for the column corresponding to a weight of 240 pounds. (Answer: $t = \dfrac{240n}{W}$.)
- How accurate is this formula for other weights? (Note: the model is quite accurate, as can be verified by substituting some values into the model. The largest variation between modelled and tabulated times is less than 20 min.)

A less guided approach would be simply to present the table to students and ask them to derive the model by hand or using technology. Note that weights in the table are given in pounds, so a further discussion could be held on conversions between systems of units.

4 Investigation 2: Predicting Your BAC, Part 1

Goal: From first principles, derive an equation that models a person's BAC over time, assuming that consumed alcohol is instantly absorbed.

Experimental scientific research has shown that alcohol is soluble in water (but not in fat), and is typically metabolised by the body at a constant rate. A mathematical model for BAC was derived in the 1920s by the Swedish physician E. Widmark, who conducted much of the early research into alcohol pharmacokinetics. The model, known as the *Widmark formula*, is $B = \frac{A}{rW} \times 100 - Vt$, and is commonly used in modern legal cases. The parameter r is called the *Widmark factor*, and estimates the proportion of body weight that is water. The precise value of r varies between individuals, but reasonable estimates are $r = 0.7$ for males and $r = 0.6$ for females. (The Widmark model assumes that alcohol is instantly absorbed after consumption; this is relaxed in Investigation 5.) Here are some questions that could be used to prompt discussions on deducing the Widmark formula.

- What factors impact on a person's BAC over time? (Answer: Amount of pure alcohol consumed, weight (more correctly, mass), gender, time since consumption, stomach contents – see Investigation 5.)
- Will each of these result in an increase or decrease in BAC? Why? (Note: more consumption leads to increased BACs because more alcohol must be distributed in the body; higher weights lead to lower BACs because there is a larger volume into which the alcohol is distributed; women typically have a lower proportion of body weight that is water compared to men, so will have a higher BAC when other factors are fixed; as the time since consumption increases, BAC will decrease as more alcohol will already have been metabolized.)
- Consider a person with weight W grams and proportion r of their body weight being water, who consumes an amount A grams of pure alcohol. Find their BAC immediately after consuming the alcohol, assuming that absorption is instantaneous. Express your answer as a percentage. (Answer: $B = \frac{A}{rW} \times 100$.)
- Assuming that alcohol is metabolized at a constant rate of V % per hour, find an expression for BAC at time t after consumption. (Answer: the Widmark formula.)
- Verify that the units in the Widmark formula are consistent.
- Typically, the value of V is around 0.015 % per hour, $r = 0.7$ for males and $r = 0.6$ for females. Determine how long it would take for your BAC to go below 0.05 % (the legal limit for driving in Australia on an Open Driver's Licence) and to zero (the legal limit for Learner's Permits and Provisional Licences.) Do this for your current weight and ± 10 kg, for a range of levels of alcohol consumption.
- What does the graph of BAC look like? (Answer: a decreasing linear function.) Interpret the gradient and y-intercept. (Note: y-intercept is the initial BAC, and the gradient is V.)

An alternative approach to this investigation is to use calculus to derive the Widmark formula. Given that alcohol is metabolised at a constant rate of 0.015 % per hour, students can use integration and knowledge of the initial BAC deduced above to derive the formula.

5 Investigation 3: Predicting Your BAC, Part 2

Goal: From first principles, derive an equation that models a person's BAC over time, without assuming that consumed alcohol is instantly absorbed.

Derivation of the Widmark formula (for example, in Investigation 2) relies on the assumption that after alcohol is consumed, it is instantaneously absorbed by the body. In reality, whilst absorption is very rapid, it is not immediate. Instead, the rate at which alcohol is absorbed by the gut at any time is proportional to the amount of alcohol that is present in the gut at that time. Posey and Mozayani (2007) present the following variant of the Widmark formula for BAC at any time *t* after commencing consuming alcohol:

$$B = \frac{A}{rW}(1 - e^{-kt}) \times 100 - Vt.$$

A range of approaches could be taken in this investigation. For students who have studied differential equations, the following key questions could be used to prompt discussions.

- The rate at which alcohol is absorbed into the bloodstream is proportional to the amount in the gut at any time, *G(t)*. Write a DE for *G(t)*. (Answer: $G' = -kG$, noting that the constant must be negative because the amount is decreasing.)
- If a person consumes a quantity *A* of pure alcohol, solve the preceding DE. (Hint: what function has its derivative equal to some constant multiplied by the original function? Answer: $G(t) = Ae^{-kt}$.)
- Write an expression for the amount of alcohol *absorbed* into the body at any time. (Answer: $A - Ae^{-kt}$.)
- Rewrite the Widmark formula to include the new absorption term. (Note: this is the Posey and Mozayani formula, given above.)
- What does *k* represent, and what might affect its value? (Answer: *k* represents the absorption rate of alcohol in the gut, and is influenced by factors such as whether there is food in the stomach prior to consuming alcohol.)
- Compare and contrast the two formulae. What happens to the new absorption term for larger values of *t*? Repeat for larger values of *k*. (Note: as *t* increases, the

absorption term approaches the amount of alcohol consumed, A. For larger values of k, the absorption term approaches A more quickly.)

- The value of k is approximately 2.3 when there is no food in the stomach, and 6 for a full stomach. What does this tell you about eating before consuming alcohol? (Note: alcohol will be absorbed less rapidly if you eat before drinking.)
- It is commonly believed that eating prior to consuming alcohol results more quickly in a zero BAC, compared to not eating first. Is this supported by the mathematics? (Note: there is an impact on the rate of absorption but not on metabolism.)

6 Investigation 4: Maximum BAC Levels

Goal: Investigate the maximum BAC and the time at which it occurs.

Posey and Mozayani's formula (see Investigation 3) can be used to complete this investigation. Here are some key points that could be used to prompt discussions.

- Graphically investigate the maximum BAC and the time at which it occurs.
- Use calculus to find an expression for the maximum BAC and the corresponding time. (Note: find the derivative B' and solve $B'=0$.)
- Investigate the impact of different levels of consumption, heights, weights and genders, and the presence (or otherwise) of food in the stomach.
- Collect information on some commonly prescribed drugs, and report on the role of mathematics in the pharmacological data. (Note: most fact sheets report such things as the maximum concentration, the corresponding time, and the drug half-life.)

There are many interesting extensions to this investigation. For example, students could be presented with a scenario in which an individual records a certain BAC at a given time after consuming alcohol, and be asked to extrapolate back to determine whether the individual had a peak BAC higher than a certain level, say 0.05 %. Students would need to know the height, weight and gender of the person. This type of calculation is relatively common in court cases relating to driving under the influence of alcohol.

7 Investigation 5: Time for BAC to Reach a Certain Level

Goal: Investigate the time at which BAC reaches certain levels (such as the legal level for driving a vehicle), without assuming that consumed alcohol is instantly absorbed.

Estimating the time at which BAC reaches certain levels (such as 0.05 % or zero) is easy when using the Widmark formula. However, this is substantially more challenging when using Posey and Mozayani's formula as it is not possible to isolate t. Students could use technology or apply root finding algorithms to find approximate values of t. The following key questions could be used to prompt discussions.

- How can BAC be *estimated* if it cannot be calculated exactly? (Note: estimate, check, and improve, use technology)
- What value for t might be chosen as the initial value?
- Does the time taken to return to 0 % depend on whether you drink on an empty or full stomach? (Answer: no; however, stomach contents do affect the rate of absorption, and hence the maximum BAC and the time taken to reach that value.)

As in Investigation 5, students can be presented with details of a particular case and deduce (for example) the times during which it would be illegal to drive a motor vehicle.

8 Investigation 6: Areas Under BAC Curves

> Goal: Investigate areas under the curve for BAC graphs, particularly in the context of binge drinking.

Areas under blood concentration curves (AUCs) are important in pharmacology. By combining the key pharmacological parameters of *concentration* and *time*, the AUC quantifies the "total exposure of the body to the drug". In the context of BAC, the AUC allows some interesting observations to be made. The following questions could be used to prompt discussions; they are suitable even for students who have not studied calculus.

- Collect information on some commonly prescribed drugs, and report on the role of mathematics in some of the pharmacological data. (Note: most fact sheets report such things as the AUC of the concentration curve.)
- Sketch the BAC graph using the Widmark formula and calculate AUC for n drinks. (Note: this calculation can be undertaken using the area of a triangle.)
- Compare the AUC for different levels of consumption. What do you notice about how the AUC changes? Why? (Note: the AUC rises with the square of the number of drinks consumed, as there is an increase in both the maximum BAC (height) and the time to return to zero (width).)
- Compare the AUC for a person consuming two drinks a day for each day in a week, with the AUC of another person who consumes 14 standard drinks in a binge drinking session on 1 day in the week. (Note: the AUC for the binge drinker is 7 times that of the other person.)

- What are the implications of your results for binge drinking? (Note: AUC quantifies the total exposure of the body to alcohol. Because the AUC rises with the square of the amount consumed, binge drinking results in a much larger potential negative impact of alcohol on the body as compared to moderate consumption.)

For students who have studied calculus, the above calculations can be repeated using integration to find the AUC for the Widmark BAC graph, then repeated using Posey and Mozayani's formula, for the case when the stomach contains food and when it does not.

9 Conclusion

The investigations provide opportunities to motivate students to engage in mathematics due to their real world appeal. Whilst the modelling approaches are consistent with Julie and Mudaly's (2007) modelling as vehicle, Stillman et al.'s (2008) contextualised examples, and Kaiser and Sriraman's (2006) realistic modelling, the investigations can be approached from other modelling perspectives. These investigations allow teachers, of both upper secondary and first year university students, to promote mathematical understanding and thinking rather than 'regurgitation', they make mathematics applicable, demonstrate utility of mathematics, and promote variety in teaching strategies. They may also engender student interest in mathematics promoting engagement and success, leading to retention of students in mathematics.

References

Adams, P. (2011). *SCIE1000 lecture notes*. Brisbane: The University of Queensland.

Australian Bureau of Statistics. (2009). *Australian social trends. 4102.0*, 10/12/2009.

Australian Bureau of Statistics. (2011). *Apparent consumption of alcohol, Australia. 2009–10.* 4307.0.55.001, 3/6/2011.

Brown, J., & Edwards, I. (2011). Modelling tasks: Insights into mathematical understanding. In G. Kaiser, W. Blum, R. Borromeo Ferri, & G. Stillman (Eds.), *Trends in teaching and learning of mathematical modelling* (pp. 187–197). New York: Springer.

Chikritzhs, T., Catalano, P., Stockwell, T., Donath, S., Ngo, H., Young, D., et al. (2003). *Australian alcohol indicators, 1990–2001: Patterns of alcohol use and related harms for Australian states and territories*. Perth: National Drug Research Institute, Curtin University of Technology.

Collins, D., & Lapsley, H. (2008). *The cost of tobacco, alcohol and illicit drug abuse to Australian society in 2004/05*. Canberra: Commonwealth of Australia.

Greefrath, G., Siller, H.-S., & Weitendorf, J. (2011). Modelling considering the influence of technology. In G. Kaiser, W. Blum, R. Borromeo Ferri, & G. Stillman (Eds.), *Trends in teaching and learning of mathematical modelling* (pp. 315–329). New York: Springer.

Julie, C., & Mudaly, V. (2007). Mathematical modelling of social issues in school mathematics in South Africa. In W. Blum, P. Galbraith, H.-W. Henn, & M. Niss (Eds.), *Modelling and applications in mathematics education: The 14th ICMI study* (pp. 503–510). New York: Springer.

Kaiser, G., & Sriraman, B. (2006). A global survey of international perspectives on modelling in mathematics education. *ZDM, 38*(3), 302–310.

Posey, D., & Mozayani, A. (2007). The estimation of BAC, Widmark revisited. *Forensic Science, Medicine, and Pathology, 3*, 33–39.

Queensland Studies Authority. (2008). *Mathematics A [B, C] senior syllabus 2008*. Brisbane: Queensland Studies Authority.

Siller, H.-S. (2010). Consuming alcohol – A topic for teaching mathematics? In A. Araújo, A. Fernandes, A. Azevedo, & J. Rodrigues (Eds.), *Proceedings of EIMI 2010 conference* (pp. 481–490). Lisbon/Bedford: Centro Internacional de Matemática/COMAP.

Stillman, G., & Galbraith, P. (2011). Evolution of applications and modelling in a senior secondary curriculum. In G. Kaiser, W. Blum, R. Borromeo Ferri, & G. Stillman (Eds.), *Trends in teaching and learning of mathematical modelling* (pp. 165–180). New York: Springer.

Stillman, G., Brown, J., & Galbraith, P. (2008). Research into the teaching and learning of applications and modelling in Australasia. In H. Forgasz, A. Barkatsas, A. Bishop, B. Clarke, S. Keast, W. T. Seah, et al. (Eds.), *Research in mathematics education in Australasia 2004–2007* (pp. 141–164). Rotterdam: Sense.

Zbiek, R. M., & Conner, A. (2006). Beyond motivation: Exploring mathematical modeling as a context for deepening students' understandings of curricular mathematics. *Educational Studies in Mathematics, 63*(1), 89–112.

Chapter 52
Beyond the Modelling Process: An Example to Study the Logistic Model of Customer Lifetime Value in Business Marketing

Issic K.C. Leung

Abstract Mathematical modelling is a popular tool used to solve many quantitative business problems. It is a challenge to teach mathematical modelling skills to students, in non-mathematics majors, from business schools for example, who will potentially be employed to tackle business problems raised in market competition. In discussing the method to estimate customer lifetime value (CLV), as an example, we focus on the dynamical relationship among variables rather than simply setting up a formula from which the subject can be readily solved. Such a dynamical system approach exhibits the logistic nature of the CLV model. The pedagogical implication of learning this logistic property is that the learnt technique is applicable in various market scenarios. In showing this model, it is asserted that mathematical modelling is not merely a variation of problem solving.

1 Introduction

Mathematical modelling is a popular approach for solving many quantitative problems in business. Taking investment optimization as an example, in a highly competitive market environment the survival of an enterprise depends on its ability to capture recurrent revenue generated by its customers. Senior management frequently questions the worthiness of high spending on promotions and other marketing tactics in order to keep customers loyal. When answering the question of worthiness in general, inevitably the solution represented by the unknown of an equation showing the relation between revenue and cost is sought. In other words, the quantitative answer to a modelling equation which reflects the real market situation must be found.

I.K.C. Leung (✉)
Department of Mathematics and Information Technology,
The Hong Kong Institute of Education, Hong Kong
e-mail: ikcleung@ied.edu.hk

G.A. Stillman et al. (eds.), *Teaching Mathematical Modelling: Connecting to Research and Practice*, International Perspectives on the Teaching and Learning of Mathematical Modelling, DOI 10.1007/978-94-007-6540-5_52,

In business school, it is a challenge for teachers to teach mathematical modelling to non-mathematics major students who will work in various business sectors after graduation. Business professionals feel that it is absurd for mathematicians to be concerned with abstract and niche concepts of Lie groups or locally convex spaces, for example. For them, this is totally disconnected from what they have learnt since university education in the real world. My personal experience verifies this double discontinuity described by Klein (2004a, b) really exists. There is a relational gap between school mathematics and university mathematics as well as when we return to teach school mathematics after earning a university degree. Mathematics professionals will experience similar disconnections and hurdles in how to apply learnt knowledge practically when employed, after university education, in various business sectors like banking, telecommunications, or other service industries. To close this gap a little, university teachers have to teach undergraduate students how to use mathematics properly in this area. Applied mathematicians play an irreplaceable role in filling the gap such as the disconnection between university knowledge of mathematical modelling and real-life application in various industries.

One common way to fill such a gap is to have students practise relevant problem-solving skills. Exposing students to many real-life problem scenarios is useful. However, as Sweller et al. (2010) claim, teaching general problem-solving skills is not a substitute for teaching mathematics. How can students be effectively taught real-life problem-solving skills at the same time as being taught the mathematical concepts and knowledge of modelling? One suggestion is that instead of teaching students particular problem-solving skills, we may consider teaching them genuine mathematical modelling skills. Applying modelling skills in tackling the challenge in real business situations leads to solving a particular problem (e.g., solving the modelling equation of business enquiry). While building the model, solid mathematical knowledge is required. In such a learning process, learners have to grasp deep and wide mathematics knowledge, in order that they can use this when encountering real-life problems both at university and in work. Hence, a more holistic learning sequence for modelling could be: (i) identifying real-life problems; (ii) establishing an appropriate modelling equation; (iii) solving the modelling equation in general hence, solving the particular real-life problem; (iv) using the result to make some estimations or business forecasting; and (v) fine-tuning the modelling skills and modifying the modelling equation. A fundamental question is: what is the difference between modelling and problem solving? Besides, is getting the numerical answer the sole goal of learning the modelling process?

2 Theoretical Frameworks

Doerr and English (2003) defined a model as a system of elements, operations, relationships, and rules that can be used to describe, explain, or predict the behaviour of some other familiar system. In problem solving, we do not necessarily need any system of elements, operations, relationships, or rules to discuss or explain.

Many problematic results (expressed by an equation or a formula) are simply represented by a solved solution. Such a solution when obtained via a problem-solving approach is represented in the form of equality. They are mostly parameter dependent but not variable (e.g., time) dependent. That means the equation consists of some unknown parameters and quantities, and real-life situations and conditions will numerically provide the values of such unknowns, except for one subject unknown needing to be found. In modelling, in contrast, the equality among unknowns is not compulsorily directly established but the interrelationship is set up among dependent and independent variables, represented by a modelling equation. Solving such an equation leads to the answer sought. Taking the well-known business problem of finding the value of a customer within a period in the service industry, we may intuitively set up an immediate quantitative solution to the problem of customer value furnished by the difference between the revenue generated and the marketing cost, represented by the equality: *Customer's value over a period of service = revenue generated by the customer – cost to keep the customer loyal to the service provider,* or simply: *customer value = revenue generated – cost.* This shows that customer value is a direct result of a single subtraction by substituting the known quantities: *revenue* and *cost*. However, it is difficult to estimate future value of such a customer because future magnitudes of the two unknowns on the right-hand side are not easy to predict. In more mathematical terminology, to estimate the future value of such a customer when this value is time dependent is a difficult task. Two different approaches for finding such customer value are discussed in coming sections.

2.1 *Models of Customer Life Time Value*

In their paper, Kotler and Armstrong (1996) defined a profitable customer as a person, household, or company whose revenue over time exceeds, by an acceptable amount, the company costs of attracting, selling, and servicing the customer. This excess is called the customer lifetime value (CLV). A positive CLV is then necessary for a company's survival. Berger and Nasr (1998) discussed a CLV equation originally offered by Dwyer (1989) when the market scenario is relatively simple and stable:

$$CLV = C\sum_{i=0}^{n}\left[\frac{r^{i}}{(1+d)^{i}}\right] - M\sum_{i=1}^{n}\left[\frac{r^{i-1}}{(1+d)^{i-0.5}}\right] \tag{52.1}$$

In this equation they defined the following: C as the yearly gross contribution = Revenue – Cost of sales; M as the promotion cost per customer per year; n as the length in years of the period over which cash flows are to be projected; r as the yearly retention rate; d as the yearly discount rate (appropriate for marketing investment). Explanation of setting up Eq. (52.1) is given by Dwyer (1989) (also see Reichheld 1996).

To mathematicians, it seems an already solved resultant equality rather than an equation showing the relationship among the unknown variables and parameters. It simply shows a solution that has already been solved, which is a formula type of equality rather than the original unsolved equation. When mathematicians talk about modelling, they do not directly establish Eq. (52.1) as a well-made type of solution but the system of equations itself. Solving such a system of equations (a dynamical system if the variables are time dependent) will give Eq. (52.1), as a solution, representing the CLV we want. This is one of the differences between problem-solving and modelling. The next question is: does the equation of dynamical modelling to estimate CLV exist? If so, what will it be, and what is the solution?

By definition, CLV can be summed up as the periodical net revenue generated by a particular customer over the entire duration of service. In the service industry, a household may subscribe to the same service over a very long time. Assuming that the gross revenue is larger than the total cost, the profit P, as a function of time t, per such customer will be periodically generated. We can represent the time dependent, periodic profit by $P(t), P(t+h), P(t+2h), \ldots,$ respectively for such a discrete case, where h is the time interval. We then have the increase in profit over the time period $h: P(t+h) - P(t);$ the gross contribution (GC) is proportional to $P(t)$ over a period of h such that $GC = BP(t)h,$ B is a constant; the marketing cost (M) to retain a customer is proportional to $P(t)$, over a period of h, gives $M = CP(t)h,$ C is a constant. Then the profit over a period of h is given by the expression $P(t+h) - P(t) = GC - M$. Now we consider the case when h is small, to get $P(t+h) - P(t) = (B-C)P(t)h$ and obtain a linear ordinary differential equation (ODE) $\frac{1}{P(t)}\frac{d}{dt}P(t) = B - C,$ where the ratio of the rate of change of profit and the profit at the time t is a constant. For a more realistic market situation, we can further suppose that the right-hand side is non-linear. Instead of $B-C$, we make it a non-linear function, $f(P(t))$ or simply $f(P)$. It is natural to assume that the rate of profit generated per customer will start with a slow manner initially, and then grow faster when the customer adopts the services provided, and is eventually slowed down again when consumption is reaching its maximum. Possessing such an effect of diminishing returns, $P(t)$ will approach its maximum P^* as the equilibrium stage after a long service period. Then $f(P)$ can be assumed to be proportional to $P^* - P$. Hence, the above linear ODE can be modified by putting $f(P) = \alpha\left(1 - \frac{P}{P^*}\right)$, where, α is a constant. We finally arrive at the non-linear, ordinary differential equation

$$\frac{d}{dt}P(t) = \alpha P(t)\left(1 - \frac{P(t)}{P^*}\right). \tag{52.2}$$

This is called the logistics CLV modelling equation. It does not directly lead to the value P by simple substitution, but P is embedded in this parametric equation, where the two parameters, α *and* P^*, are feasibly estimated by the historical data of servicing the customer. Solving the above non-linear logistic differential equation gives the profit, in $P(t)$,

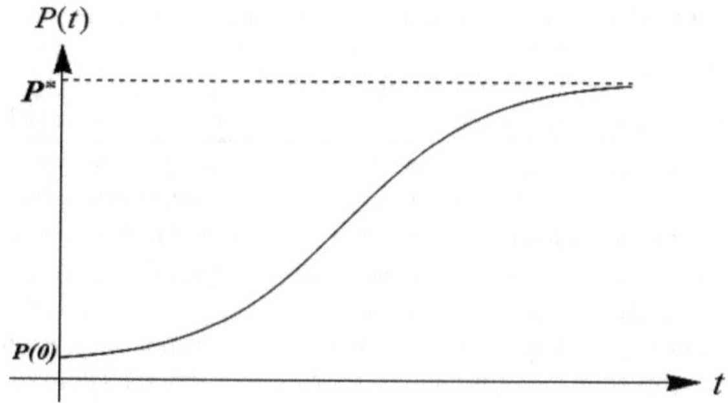

Fig. 52.1 The logistic nature of the profit per customer $P(t)$ over time.

$$P(t) = \frac{P^*}{1+\left(\frac{P^*}{P_0}-1\right)e^{-\alpha t}} = \frac{P^*}{1+Ke^{-\alpha t}}, \tag{52.3}$$

where k is a constant depending on the initial value $P_0 = P(0)$, and the ceiling P^*. The solution curve can be sketched qualitatively (Fig. 52.1).

Integrating the function $P(t)$ over the length of the servicing period T, we shall have the CLV evaluated by using the solution of the logistic differential equation. However, in a realistic case, services are subscribed to in a discrete manner, or for data that can only be collected in discrete time. Putting equal time interval h and letting $t=nh$, and normalizing the time interval, says, monthly, we can set $h=1$, and get from (52.3) that

$$Ke^{-\alpha n} = \frac{P^*}{P_n} - 1 \quad \text{and} \quad Ke^{-\alpha(n+1)} = \frac{P^*}{P_{n+1}} - 1.$$

This gives the discrete modelling equation

$$\frac{P_{n+1}-P_n}{P_n} = \left(\frac{1-e^{\alpha}}{P^*}\right)P_{n+1} + \left(e^{\alpha}-1\right) \tag{52.4}$$

To compare the two approaches to evaluate CLV, we summarise and modify the characteristics of the two different models in the following in Table 52.1, with reference to Zawojewski (2007).

Generalizing Eq. (52.1) into a smaller period of discounted profits, we define the desired effective discount rate d and nominal annual discount rate d^* respectively. The logistic nature can be verified in the light of the characteristic of the exponential property of CLV value function. Then the continuous CLV formula over a period T will be given by:

$$CLV = \pi(0) + \int_0^T r^t \pi(t) e^{-td^*}\,dt = \pi(0) + \int_0^T \pi(t)\left[\frac{r}{1+d}\right]^t dt \tag{52.5}$$

Table 52.1 Comparison of Berger and Nasr's CLV model and the Logistic CLV model

CLV models and the modelling equations	
Berger and Nasr's model	Logistic CLV model
Direct calculation of unknowns from the equation formulated	Setting up the equation, solve it to get the parametric dependent solution
Not necessary to develop a theory	Possible theory developed
Static system establishment: the effect of rate of change is embedded	Dynamical system establishment: incorporate the effect of the rate of change
Static approach focus on "what to do": goes from the "given information" to the "goals"	Dynamic approach focus on "way of thinking" via interpretation and reinterpretation to establish the inter-relationship among variables and parameters

(For detail, see Berger and Nasr 1998). $\pi(t)$ is the profit function, and the term $\frac{r}{1+d}$ also appears in formula (52.1). For validation of the CLV equations (52.1) and (52.5), see illustrations in Berger and Nasr (1998).

2.2 *Pedagogical Implications*

Lesh and Zawojewski (2007) described problem solving as modelling when "a task, or goal-directed activity, becomes a problem (or problematic) when the problem solver needs to develop a more productive mathematical way of thinking about the given situation" (p. 782). The question is of course: how much more is more? Ability to identify the logistic nature of a CLV function by a problem solver is an indicator for his or her development of a more productive way of thinking about the given market scenario. During the modelling procedure, analysis skills will be sharpened. This is an analytic interpretation (logistic nature) rather than pattern recognition (a periodic sum during the service purchased). The model-builder seeks an explanation for phenomena of dependencies and independencies of variables. It is a kind of functional interpretation rather than substituting known values to solve for unknowns from a formula. The most prominent result of modelling, rather than problem solving, is that a modelling equation with a solution has much stronger predictive power. It has a better hypothetical justification in that it predicts potential outcomes and justifies its legitimacy. Such forecasting potential is welcomed by most marketing managers. Solving a problem in a dynamical modelling approach requires an even more productive mathematical way of thinking about the situation.

Applying modelling skills to solve a problem in comparison with applying general problem-solving skill is more appealing to students who are learning mathematical concepts because further steps are taken to analyse the character of the expected solution. The logistic nature of the solution of the CLV equation is a good illustration of how a more holistic way (i.e., more productive mathematical way) of knowing how to solve similar problems in different situations can occur. In a real business environment, such a situation likely exists as a competitive market environment with various promotions offered by competitors.

One of the functional purposes asserted by Doerr and English (2003) about modelling is to use it in some other familiar systems, which indicates a transferral of skills, reinterpretation, and generalisation in the modelling process. A competent modelling skill enables problem solvers to further generalise it, in a relatively straightforward way, to a more sophisticated model to adopt the change of market situation. In particular, when the rate of change in profit made at the current time is the time delay effect of some marketing promotion done before, we can then model Eq. (52.2) by a more realistic time delay ODE for CLV; $\frac{d}{dt}P(t) = \alpha P(t)\left[1-\beta P(t-\tau)\right], P(t) = \phi(t), \quad t \in \left[-\tau, 0\right]$, where τ is the delay and $\phi(t)$ is the initial function. Furthermore, in a keen competitive market situation, it is natural that the system, represented by the modelling equation, will exhibit a sudden stir for a short time during the customers' loyalty period. Such a stir may exist as a competitor promotes a highly discounted subscription fee for potential churners. Thus, a sharp drop in revenue for a particular customer will result for a very short time during the regular service period. We then further generalise the non-linear ODE, where the logistic nature is preserved, to the impulsive ODE for which advanced knowledge of solving is required.

$$\frac{d}{dt}P(t) = \alpha P(t)(1-\beta P(t)), \quad P(t_0^{+}) = P_0 \quad \text{and}$$
$$P(t_k^{+}) - P(t_k^{-}) = c_k\left(P(t_k^{-})\right), \quad |c_k| < 1, \quad k = 1, 2, 3, \ldots.$$

2.3 *From Modelling Process to Theory Development*

As asserted by Newby et al. (2006, p. 25), mathematical theory can be developed in a general looping sequence of three analytical steps:

$$\textit{observation} \rightarrow \textit{explanation} \rightarrow \textit{prediction} \rightarrow \textit{observation} \rightarrow \textit{explanation} \rightarrow \ldots \rightarrow \textit{theory}$$

In reality, there is always discrepancy in the *prediction* stage after we have observed and explained (solved) the corresponding problem. This discrepancy can also be used as a feedback to *observation* for refinement of further explanation then such a loop continues. Observing the spending behaviour from customers' consumption statistics will lead to our conjecture that customers will easily churn without appropriate promotion reaching them. Such promotion is costly and likely to increase over time. When customers learn from the market and shop around for cheaper service, diminishing return on the generated profit of the customers of the service provider or company results. This explains the logistic nature and leads to the establishment of a logistic model equation. The solution acts as a good tool to predict the future value of the customer, hence CLV. Perception wise, problem solvers learn that modelling skills mean looking for a well-made formula and then substituting in some values for unknowns to obtain an answer, though the unknowns are sometimes not easy to obtain. Education professionals may find that, when teaching mathematical modelling to non-mathematics majors, students tend to use a ready-to-evaluate type of formula to find the answer but are reluctant to develop a modelling equation from the very beginning where variables are interrelated. Though in our example of logistic CLV, the solution itself is not easy to be found because it is beyond the standard knowledge of an elementary course of ordinary differential equations, such an approach in solving practical problems in a business course can still be introduced. In introducing the logistic nature, students are not merely told of the characteristic of diminishing return of the solution curve, rather, such a logistic property can be deduced mathematically from the basic CLV equation.

Gravemeijer (1994) defined four levels of the modelling process when a problem solver tackles a real life problem, namely: (1) the situational level, (2) the referential level, (3) the general level, and (4) the formal level. In brief, in the formal level, problem solvers will reason the situation in terms of formal mathematical relations and principles. However, as mathematics learners, we are keen to know if we can further generalise the mathematical relations rigorously to an analytic system from which related theory can possibly be developed. Can we go one more level beyond the formal level in Gravemeijer's hierarchy of the modelling process? What kind of mathematical process will be exhibited at this extended level? From the CLV model we are discussing, we can initially summarize the possible characteristics of such an extended level: it is likely to be an establishment of a generalised, parametric dependent equation (e.g., Eq. (52.2)) where the solution (e.g., Eq. (52.3)) is expressed as an equation, instead of formula (e.g. Eq. (52.1)), and is also parametric dependent. In the extended level, generalisation must go beyond the superficial calculation to develop a related theory that may be used to solve the entire problem of CLV in a holistic way. The circular loop *observation – explanation – prediction* originally defined by Newby et al. (2006) can *lift* up the Gravemeijer's modelling process from the formal level to this extended level where theory can be developed. Because of the intensive analytic nature in such a level, we call it the *analytic level* where mathematics theory is developed for a more holistic sense of solving the real life problem via the modelling process. The following figure schematically shows how the logistic model can be generalised to the time delay model, impulsive model, and discrete model, and reduced to the Berger and Nasr formula. As a whole, these formulate a simple theory of solving the general CLV problem.

All the models shown in Fig. 52.2 serve as a complete explanation to cover various real market situations. It is understood in reality that time measures are discrete, profit generated by a customer exhibits a time delay effect in logistic characteristics, while the competitors' promotion will stir the market in an impulsive manner. With all these generalised versions considered, a theory has been generated, perhaps not the most complete one, to explain and evaluate the CLV problem. Though there might be some more sophisticated model to explain the evaluation

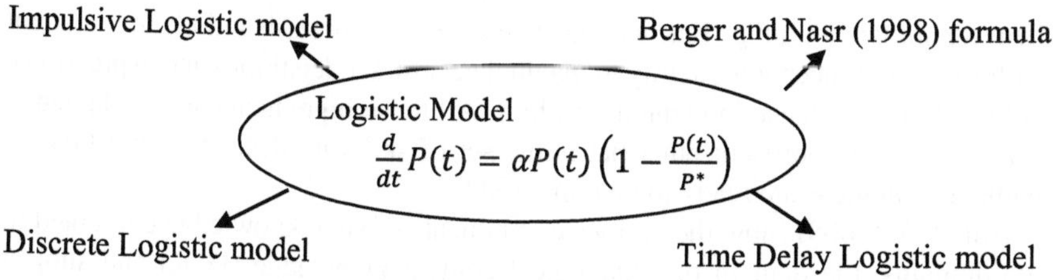

Fig. 52.2 The schematic relationship among the various logistic CLV models

of CLV, such as the stochastic model (e.g., Leung et al. 2008), our analysis has demonstrated how to generate a related theory. We may add an extended level: the analytic level to the modelling process described by Gravemeijer (1994).

3 Conclusion

There are other types of generalisation of the CLV model such as the Stochastic model or the perpetual model. We conclude our discussion for further generalisation but intend to incorporate both problem-solving skills and modelling skills in teacher competency. We note the consensus that the understanding of the logistic nature of a function is essential in deducing the modelling equation of the logistic type. In tackling the problem using Eq. (52.1), students may easily gloss over some very important mathematical knowledge essentially constituting the logical deduction of the non-linear ODE. One may argue that many business students, even in university, may not have the knowledge to solve ODEs. That is true. However, there are computational software tools that provide ready-made solutions of logistic type ODEs. Technically, the answers can be found by supplying the numerical values of those parameters to the equation. We believe that mathematical content knowledge will play a key role in the process of modelling. In teaching modelling courses to business students, we shall not sacrifice the rigorous derivation to arrive at the dynamical system represented in the ODEs, instead of deriving solely the formula solution type of Eq. (52.1). Students will learn that modelling does not directly go for a solution but a precedent stage of dynamical relationship represented by Eq. (52.2). Whether one decides to teach problem-solving skills through the problematic approach as in establishing Eq. (52.1), or the dynamical modelling approach as in establishing Eq. (52.2) and hence the entire theory of tackling the CLV problem depends on, inevitably, the subject content knowledge of the teachers.

Borrowing the claim by Viadero (2004): “teaching mathematics requires a special set of skills” (p. 8), we may analogously say that teaching mathematical modelling requires a special set of skills. Competent teachers must be able to unpack, as defined by Ball et al. (2009), mathematical ideas and be able to scaffold them for student learning. Teachers must be able to distinguish and identify student confusion, misconceptions, ability to manipulate abstract definitions and expressions, and level of confidence in using algorithms. All these ingredients are fundamental to any teaching approach; but what is the set of skills needed, and what kind of mathematical ideas are ready to be unpacked?

Hill et al. (2008) show the various components of such knowledge contained in two domains: the domain of pedagogical content knowledge (PCK) and subject matter knowledge (SMK). SMK consists of the sub-components: (i) *Common content knowledge,* (ii) *Knowledge at the Mathematics horizon*, and (iii) *Specialised content knowledge. Common content knowledge* in the case of developing the theory of mathematical modelling of CLV is knowledge about calculus and differential equations. Knowledge of making use of data to establish unknown parameters in the modelling equation is probably *the knowledge at the mathematics horizon* because teachers must have the ability to connect the real situation such as the sensitivity of retention rates and the reality manifested by recorded data in the process of building the CLV model, and analysis of the solution in the forecasting process. Finally, the most important aspect is knowledge of applying the logistic concept to interpret this whole process of modelling. Knowledge of the consumer market such as the effect of diminishing returns, the concept of the Law of Mass Action in chemical reaction of two substances in justifying the product $P(t)(1\text{-}P(t)/P^*)$ in Eq. (52.2), understanding of the relationship of the derivative of the profit function and the function itself all constitute the third component: the *specialised content knowledge*. Without such knowledge teachers are unlikely to lead students to interpret the CLV in a non-linear, logistic, and realistic manner. Consequently, they will not be able to build up the fundamental logistic ODE (52.2), and in turn, develop the whole theory of CLV.

In light of the many theories of mathematics education, when tackling the problem in a business scenario, how to effectively teach students of non-mathematics majors to use mathematical modelling skills will be a challenge. Higher-level knowledge is essential and a holistic understanding of the subject content is required. Content knowledge is a backup to good pedagogy. Hence, we also believe that content knowledge guides pedagogy (Belfort and Guimaraes 2002; Li et al. 2010), at least in teaching of mathematical modelling under a real business situation.

References

Ball, D. L., Sleep, L., Boerst, T. A., & Bass, H. (2009). Combining the development of practice and the practice of development in teacher education. *The Elementary School Journal, 109*(5), 458–474.

Belfort, E., & Guimaraes, L. C. (2002). The influence of subject matter on teachers' practices: Two case studies. In A. D. Cockburn & E. Nardi (Eds.), *Proceedings of the 26th conference of the International Group for the Psychology of Mathematics Education* (Vol. 1, pp. 73–80). Norwich.

Berger, P. D., & Nasr, N. (1998). Customer lifetime value: Marketing models and application. *Journal of Interactive Marketing, 12*(1), 17–30.

Doerr, H. M., & English, L. D. (2003). A modelling perspective on students' mathematical reasoning about data. *Journal for Research in Mathematics Education, 34*(2), 110–136.

Dwyer, F. R. (1989). Customer lifetime valuation to support marketing decision making. *Journal of Direct Marketing, 8*(2), 73–81.

Gravemeijer, K. P. E. (1994). *Developing realistic mathematics education*. Utrecht: Freudenthal Institute.

Hill, H. C., Ball, D. L., & Schilling, S. G. (2008). Unpacking pedagogical content knowledge: Conceptualizing and measuring teachers' topic-specific knowledge of students. *Journal for Research in Mathematics Education, 39*(4), 372–400.

Klein, F. (2004a). *Elementary mathematics from an advanced standpoint: Vol. 1. Algebra & analysis*. (E. R. Hedrick & C. A. Noble, Trans.). New York: Dover Publications. (Original work published 1908)

Klein, F. (2004b). *Elementary mathematics from an advanced standpoint: Vol. 2: Geometry*. (E. R. Hedrick & C. A. Noble, Trans.). New York: Dover Publications. (Original work published 1908)

Kotler, P., & Armstrong, G. (1996). *Principals of marketing* (7th ed.). Englewood Cliffs: Prentice-Hall.

Lesh, R., & Zawojewski, J. (2007). Problem solving and modelling. In F. Lester (Ed.), *Handbook of research on mathematics teaching and learning* (2nd ed., pp. 763–804). Reston/Charlotte: NCTM/Information Age Publishing.

Leung, H. Y., Ching, W. K., & Leung, I. K. C. (2008). A stochastic optimization model for consecutive promotion. *Quality Technology and Quantitative Management, 5*(4), 403–414.

Li, J., Fan, X., & Zhu, Y. (2010). A framework on mathematical knowledge for teaching. In *Proceedings of the 5th East Asia Regional Conference on Mathematics Education* (*EARCOME 5*), (Vol. 2, pp. 1–8). Tokyo.

Newby, T. J., Stepich, D., Lehman, J., & Russell, J. (2006). *Educational technology for teaching and learning* (3rd ed.). Upper Saddle River: Merrill/Prentice Hall.

Reichheld, F. (1996). *The loyalty effect*. Boston: Harvard Business School Press.

Viadero, D. (2004). Teaching mathematics requires special set of skills. *Education Week, 24*(7), 8.

Sweller, J., Clark, R., & Kirschner, P. (2010). Teaching general problem-solving skills is not a substitute for, or a viable addition to, teaching mathematics. *Notices of the American Mathematical Society, 57*(10), 1303–1304.

Zawojewski, J. (2007). Problem solving versus modeling. In R. Lesh, P. Galbraith, C. R. Haines, & A. Hurford (Eds.), *Modeling students' mathematical modelling competencies* (pp. 237–243). New York: Springer.

Refereeing Process

To ensure the scholastic quality of the chapters in this book, all proposed chapters submitted to the editors for consideration in a timely manner have been subjected to a strict impartial and independent peer review and revision process. Only those chapters that have been accepted after this process as being of appropriate quality (given differences of geographical location) and consistent with the intended aims of the book have been included. The editors wish to thank the following international reviewers:

Ang Keng Cheng, Nanyang Technological University, Singapore
Jussara de Loiola Arraújo, Universidade Federal de Minas Gerais, Brazil
Jonei Barbosa, State University of Feira De Santana, Brazil
Jonas Bergman Ärlebäck, Linköping University, Sweden
Michael Besser, University of Kassel, Germany
Maria Salett Biembengut, Pontifícia Universidade Católica do Rio Grande do Sul, Brazil
Morten Blomhøj, Roskilde University, Denmark
Rita Borromeo Ferri, University of Kassel, Germany
Nils Buchholtz, University of Hamburg, Germany
Phillip Clarkson, Australian Catholic University (Melbourne), Australia
Chan Chun Ming Eric, Nanyang Technological University, Singapore
Dirk De Bock, Katholieke Universiteit Leuven & Hogeschool-Universiteit Brussels, Belgium
Andréia Maria Pereira de Oliveira, State University of Feira de Santana, Brazil
Helen Doerr, Syracuse University, USA
Ann Downton, Australian Catholic University (Melbourne), Australia
Lyn D. English, Queensland University of Technology, Australia
Peter Frejd, Linköping University, Sweden
Peter Galbraith, University of Queensland, Australia
Vince Geiger, Australian Catholic University (Brisbane), Australia
Gilbert Greefrath, University of Münster, Germany

G.A. Stillman et al. (eds.), *Teaching Mathematical Modelling: Connecting to Research and Practice*, International Perspectives on the Teaching and Learning of Mathematical Modelling, DOI 10.1007/978-94-007-6540-5,

Susanne Grünewald, University of Hamburg, Germany
Christopher Haines, City University, London, UK
Jamal Hassain, Mizoram University, India
Toshikazu Ikeda, Yokohama National University, Japan
Michael Jennings, The University of Queensland, Australia
Tinne Kjeldsen, Roskilde University, Denmark
Sebastian Kuntze, Ludwigsburg University of Education, Germany
Janeen Lamb, Australian Catholic University (Brisbane), Australia
Geert Langereis, Eindhoven University of Technology, The Netherlands
Lee Ngan Hoe, Nanyang Technological University, Singapore
Issic K.C. Leung, The Hong Kong Institute of Education, Hong Kong
Matthias Ludwig, Goethe-University Frankfurt, Germany
Katja Maaß, University of Education Freiburg, Germany
Nicolas Mousoulides, Cyprus University of Technology, Cyprus
Kit Ee Dawn Ng, Nanyang Technological University, Singapore
Porras Päivi, Saimaa University of Applied Sciences, Finland
Trevor Redmond, AB Paterson College, Gold Coast, Australia
Milton Rosa, Universidade Federal de Ouro Preto (UFOP), Brazil
Mark Schofield, Edge Hill University, UK
Stanislaw Schukajlow, University of Paderborn, Germany
Hans-Stefan Siller, University of Salzburg, Austria
Maxwell Stephens, University of Melbourne, Australia
Pauline Vos, University of Agder, Norway/Amsterdam University of Applied Sciences The Netherlands
Helena Wessels, Stellenbosch University, South Africia
Wanty Widjaja, Deakin University (Burwood), Australia
Mark Winter, University of the Witwatersrand, Johannesburg, South Africa
Jinxing Xie, Tsinghua University, Beijing, China
Shih-Yi Yu, National Changhua University of Education, Taiwan

Index

G.A. Stillman et al. (eds.), *Teaching Mathematical Modelling: Connecting to Research and Practice*, International Perspectives on the Teaching and Learning of Mathematical Modelling, DOI 10.1007/978-94-007-6540-5,

R

S

MIX
Papier aus verantwortungsvollen Quellen
Paper from responsible sources
FSC® C105338

If you have any concerns about our products,
you can contact us on
ProductSafety@springernature.com

In case Publisher is established outside the EU,
the EU authorized representative is:
Springer Nature Customer Service Center GmbH
Europaplatz 3, 69115 Heidelberg, Germany

Printed by Libri Plureos GmbH
in Hamburg, Germany